Classification Matrix:
Techniques for Organizing and Presenting Empirical Data

D0574338

Level of Measurement of Variable(s) o......

		Categorical: Nominal or Ordered Categories	Rank-Order	Interval Scale (or Ratio-Scale)
Univariate Problems		Frequency Distribution, Relative Frequency Distribution; Bar Graph, Pictograph, Circle Graph	Set of Ranks	Frequency Distribution, Grouped Frequency Distribution, Cumulative Frequency Distribution; *Graphs for Discrete Interval Data:* Discrete Graph, Step Function; *Graphs for Continuous Interval Data:* Histogram, Frequency Polygon, Cumulative Frequency Graph
Bivariate Problems	Correlational Problems	Bivariate Frequency Distribution: The Crosstabulation Table	Two Sets of Ranks	Scatterdiagram
	Experimental and Pseudo-Experimental Problems: Independent Groups	*Independent and Dependent Variable Categorical:* The Crosstabulation Table		*Independent Variable Categorical, Dependent Variable Interval:* Paired Frequency Polygons, Paired Cumulative Frequency Graphs. *Independent and Dependent Variable Interval:* Scatterdiagram
	Dependent Measures: Pretest and Posttest or Matched Groups	Crosstabulation Table		Paired Cumulative Frequency Graphs, Paired Frequency Polygons

Type of Problem

Introductory Statistics

Introductory Statistics

James A. Twaite

Columbia University

Jane A. Monroe

Columbia University

Scott, Foresman and Company

Glenview, Illinois
Dallas, Texas
Oakland, New Jersey
Palo Alto, California
Tucker, Georgia
London, England

Library of Congress Cataloging in Publication Data

Twaite, James A., 1946–
  Introductory statistics.

  Bibliography: p.
  Includes index.
  1. Statistics. I. Monroe, Jane A., 1932–
joint author. II. Title.
QA276.12.T83
ISBN 0-673-15097-6

Acknowledgments for Statistical Tables

(C.1) From Mosteller/Rourke/Thomas, PROBABILITY WITH STATISTICAL APPLICATIONS, © 1970, Addison-Wesley, Reading, Massachusetts. Reprinted with permission. **(C.2)** Adapted from STATISTICAL ANALYSIS, FOURTH EDITION by Allen L. Edwards. Copyright © 1958 © 1969 © 1974 by Allen Edwards. Reprinted by permission of Holt, Rinehart and Winston and the author. **(C.3)** Abridged from Biometrika, Volume 32, 1941 by Catherine M. Thompson and *Statistical Tables for Biological, Agricultural, and Medical Research,* Sixth Edition, 1974 by Ronald A. Fisher and Frank Yates. Reprinted by permission of the Biometrika Trustees and Longman Group Ltd. Reproduced from *Elementary Statistical Methods,* Third Edition, by Helen M. Walker and Joseph Lev, Holt, Rinehart, and Winston, Inc. **(C.4)** Adapted from the table of critical values of D in the Kolmogorov-Smirnov one sample test by F. J. Massey Jr., *Journal of the American Statistical Association,* 1951, 46:70. Copyright © 1951 the American Statistical Association. Reproduced from S. Siegel, NONPARAMETRIC STATISTICS FOR THE BEHAVIORAL SCIENCES, New York: McGraw-Hill Book Company, 1956. Reprinted by permission of the American Statistical Association and McGraw-Hill Book Company. **(C.5)** Table C-5 is taken from Table III of Fisher and Yates: *Statistical Tables for Biological, Agricultural, and Medical Research,* published by Longman Group Ltd., London. (previously published by Oliver & Boyd, Edinburgh), and by permission of the authors and publishers. Reproduced from FUNDAMENTALS OF BEHAVIORAL STATISTICS, Third Edition by Richard P. Runyon and Audrey Haber, 1976, Addison-Wesley Publishing Company. **(C.6)** From FUNDAMENTAL STATISTICS IN PSYCHOLOGY AND EDUCATION, by J. P. Guilford and Benjamin Fruchter. Copyright © 1973 by McGraw-Hill, Inc. Used with permission of McGraw-Hill Book Company. **(C.7)** From INTRODUCTION TO STATISTICAL ANALYSIS by Wilfrid Dixon and F. J. Massey, Jr. Copyright © 1969 by McGraw-Hill, Inc. Used with permission of McGraw-Hill Book Company. **(C.8)** Abridged from *Statistical Tables for Biological, Agricultural, and Medical Research* by Ronald A. Fisher and Frank Yates, Sixth Edition, 1974. Reprinted by permission of Longman Group, Ltd. Reproduced from *Elementary Statistical Methods,* Third Edition, by Helen M. Walker and Joseph Lev, Holt, Rinehart and Winston, Inc. **(C.9)** From CONCEPTS OF STATISTICAL INFERENCE by William Guenther. Copyright © 1965 by McGraw-Hill, Inc. Used with permission of McGraw-Hill Book Company. **(C.10)** From "Tables of the Ordinates and Probability Integral of the Distribution of the Correlation Coefficient in Small Samples" by F. N. David, *Biometrika,* 1938 in CONCEPTS OF STATISTICAL INFERENCE by William Guenther, Copyright © 1965 by McGraw-Hill, Inc. Used with permission of the Biometrika Trustees and the McGraw-Hill Book Company. **(C.11)** Table of Critical Values of D in the Fisher Test, adapted from D. J. Finney, *Biometrika,* Volume 35, pages 149–154, 1948. Reproduced from S. Siegel, NONPARAMETRIC STATISTICS FOR THE BEHAVIORAL SCIENCES, New York: McGraw-Hill Book Company, 1956. Reprinted by permission of the Biometrika Trustees and the McGraw-Hill Book Company. **(C.12)** From *Journal of the American Statistical Association,* September, 1964, pages 927–932. Copyright © 1964 by the American Statistical Association. Reprinted by permission. Reproduced from FUNDAMENTAL RESEARCH STATISTICS FOR THE BEHAVIORAL SCIENCES, by John T. Roscoe, Holt, Rinehart, and Winston, Inc., 1975. **(C.13)** Adapted and abridged from W. H. Kruskal and W. A. Wallis' use of ranks in one criterion variance analysis from the *Journal of the American Statistical Association,* 1952, 47:614–617. Copyright © 1952 by the American Statistical Association. Reproduced from S. Siegel, NONPARAMETRIC STATISTICS FOR THE BEHAVIORAL SCIENCES, New York: McGraw-Hill Book Company, 1956. Reprinted by permission of the American Statistical Association and the McGraw-Hill Book Company. **(C.14)** Reprinted by permission from STATISTICAL METHODS by George W. Snedecor and William Cochran, Sixth Edition, © 1967 by Iowa State University Press, Ames, Iowa 50010. **(C.15)** Adapted from Table 2 of F. Wilcoxen and R. A. Wilcox, SOME RAPID APPROXIMATE STATISTICAL PROCEDURES, rev. ed. Pearl River, New York: American Cyanamid Company, 1964. Reproduced from S. Siegel, NONPARAMETRIC STATISTICS FOR THE BEHAVIORAL SCIENCES, New York: McGraw-Hill Book Company, 1956. Reprinted by permission of the American Cyanamid Company and McGraw-Hill Book Company.

12345678910-VHJ-8584838281807978

Introductory Statistics, by James Twaite and Jane Monroe, is an ideal introduction to the subject, particularly for those students with little or no background in mathematics. Twaite and Monroe both have extensive experience teaching the introductory course. They have carefully designed their text to promote a thorough understanding of both basic statistical concepts and the statistical procedures through which these concepts are applied in specific research situations. The result is a book that combines clarity of presentation, skillful organization, and technical accuracy, with a wide coverage of topics.

A number of special features contribute to the exceptional effectiveness of this book in teaching statistical concepts and procedures:

1 A "Classification Matrix," appearing at the end of each chapter, summarizes the statistical problems covered in the chapter and classifies each problem according to the specific conditions and techniques involved. Particular emphasis is given in each matrix to the relationship between the type of measurement scale represented in the data and the statistical analysis to be carried out.

2 A Student Guide section, also following each chapter, provides students with both a review of important concepts and an opportunity to test their comprehension. In addition to chapter Summaries, Key Terms, and Review Questions, the Student Guide includes a variety of Problem Sets organized according to a systematic breakdown of experimental designs. Students will find this particularly useful in learning the application of statistics to real-life situations.

3 Finally, a Review of Basic Math, Symbols, and Formulas as well as an unusually comprehensive Glossary of Terms are included at the back of the book to facilitate students' computation and overall review.

In short, this is a book designed not merely to describe statistics but to teach it. Twaite and Monroe have made an important contribution to the instruction of statistics: their book successfully encourages enjoyment—through understanding—of this most stimulating field of study.

RICHARD H. LINDEMAN
Columbia University

The authors are grateful to the Literary Executor of the late Sir Ronald A. Fisher, F. R. S., to Dr. Frank Yates, F. R. S., and to Longman Group Ltd., London, for permission to reprint Tables III, IV and VIII, adapted and abridged, from their book *Statistical Tables for Biological, Agricultural and Medical Research.* (6th edition, 1974.)

foreword

Part I

Introduction

Part II

Organizing and Presenting Empirical Data

Table
of
Contents

Part IV
Inferential Statistics

Appendices

Introduction 1

1

Introduction to Statistical Techniques in Empirical Research

A research project begins when an individual asks a question that cannot be answered on the basis of information immediately at hand. The type of question asked will determine where the researcher must seek the necessary information. For subjective questions like "What was the principal cause of the War of 1812?" or "What is the meaning of the swan image in the poetry of W.B. Yeats?," researchers would probably go to a library to find out what other scholars have said on the question. They would combine the subjective impressions of previous workers with their own impressions to reach an answer that they consider satisfactory. This is research, but it is not empirical research. For more objective questions like "Which of these two routes from New York to Boston is the quickest?" or "What is the average reading achievement level in the Newark, N.J. public high schools?," researchers may attempt to find an answer by observing actual events. For example, in determining the quickest route to Boston they might establish two groups of drivers and assign one route to each group. They would then record the time required to make the trip by each driver in each group. Whenever researchers directly or indirectly observe actual events in this manner, they are engaging in empirical research.

The focus of this book is on the elementary statistical techniques that are researchers' basic tools for organizing, presenting, and analyzing empirical data. In this chapter, we have two goals: (1) to introduce you to some of the essential principles and terminology of empirical research; and (2) to provide you with a logical framework that you may use to organize the statistical techniques that will be presented in the chapters that follow.

THE PRINCIPLES AND THE TERMINOLOGY
OF EMPIRICAL RESEARCH

The Scientific Method

In their observations, researchers should be guided by the principles of the *scientific method*. These principles may be broken down into two groups. One group of principles aims at ensuring that researchers obtain a correct answer to their questions. The other group aims at making this answer available to other interested parties. The first set of principles may be summarized in the following way. All observations must be made in a careful and systematic manner and with an effort to control for the effect of irrelevant factors. This set of principles applies to any empirical research question. Consider the researchers attempting to determine which of two routes is the quickest route from New York to Boston. Of course, they would want to make sure that their observations were accurate. Rather than ask drivers how long their trips took, they would actually want to time them. They would want to be certain that their timing device was accurate. They would want to make sure that the drivers using the two routes had comparable weather conditions. They would want to know if there had been any unusual delays on one route or the other. These kinds of considerations are obvious. They apply to any empirical research question.

The other set of principles is particularly relevant to empirical research in psychology, education, or any of the natural or social sciences. If the researchers are scholars, they will be interested in sharing the results of their work with other members of their profession. This requires that all the procedures employed in their research be specified exactly. This is essential for two reasons. First, other members of the scientific community must know exactly what was done if they are to be able to evaluate the results critically and relate the results to their own work and the work of others in the field. Second, other researchers should be able to verify the results of a research project by reproducing, or *replicating,* the original research. In order to replicate a study, they must know exactly what was done. Thus, the exact specification of research procedures provides the basis for the growth of scientific knowledge.

In brief, the specification of research procedures may be broken down into three critical steps: (1) Researchers must clearly specify what it is that they are observing; (2) they must indicate precisely how they go about observing it; and (3) they must describe the group on which these observations are made. In the jargon of empirical research these three steps have special names. When researchers specify what it is that they are observing, they are defining the *variable* (or variables) *of interest.* When they indicate exactly how they will be observing this variable, they are defining the *measurement process.* And when they describe the group on which their observations are made, they are defining the *population of interest.* We now proceed to a more complete explanation of these terms.

3

Variables

In empirical research, we constantly refer to variables. It is essential that you have a clear understanding of this term. A variable is a characteristic that may have different values from individual to individual or from observation to observation. A variable has a name and associated with that name is a *set of values*. It may be the name of a physical characteristic, a psychological trait, an ability, a type of achievement, or any number of other attributes. For example, the name "eye color" is the name of a variable that is a physical characteristic. In a particular study we might establish three values for the variable "eye color": (A) brown, (B) blue, and (C) other. Notice that these values are not numbers. There is no requirement that the values of a variable be numerical scores. In order for a variable to be employed in empirical research, the only requirement is that we be able to *differentiate* between the various values of the variable. Thus, if we were using the variable "eye color" in a study of human subjects, we would need to be able to assign each subject in the study to *one and only one* of the three values of that variable. The value "other" enables us to assign individuals with unusual or hard-to-describe eye colors to one specific value.

There are several different types of variables that may be employed in empirical research. It is important that we be able to distinguish among the different types of variables, for different statistical techniques are appropriate for use with different types of variables. There are four broad classes of variables. These are the categorical variable, the rank-order variable, the interval scale variable, and the ratio scale variable. Moreover, one of these four broad classes has an important subdivision. Categorical variables may be subdivided into one of two types, the pure nominal scale variable and the variable with ordered categories. In the paragraphs that follow, we will describe each of these different types of variables.

Categorical Variables.   A variable is categorical when the values of the variable are not numbers but simply categories into which observations may be classified. As already noted, there are two different types of categorical variables, the *nominal scale variable* and the *ordered categorical variable* with ordered categories. The variable "eye color" just mentioned is an example of a nominal scale variable. The variable is categorical because the values of the variable are the categories (A) brown, (B) blue, and (C) other. The variable is a nominal scale variable because these three categories are not related to each other in any quantitative or qualitative sense. That is, brown is not bigger than blue, and blue is not better than other. The three categories are simply different from each other.

Let us consider a set of nominal scale data. If we observe the 25 children in Mr. White's first-grade class and record their values on the variable "eye color," our results might appear as in Table 1.1. We see from Table 1.1 that John and Helen have the same eye color, brown. They have a different eye color from Bill and Jane, who both have blue eyes. They also have a different eye color

TABLE 1.1
Eye Color of 25 Children in Mr. White's First-Grade Class

Subject	Name	Eye color	Subject	Name	Eye color
1	John	brown	14	Tom	blue
2	Helen	brown	15	Barry	brown
3	Bill	blue	16	Joan	brown
4	Fred	other	17	Shawn	brown
5	Jane	blue	18	Toby	other
6	Willy	brown	19	Larry	brown
7	Dave	blue	20	Dick	blue
8	Joe	other	21	Lisa	brown
9	Nancy	brown	22	Mark	blue
10	Peggy	brown	23	Cathy	brown
11	Janet	blue	24	Jody	brown
12	Chris	blue	25	Brian	brown
13	Owen	other			

from Fred, whose eyes were neither brown nor blue but some other color. Because the values of the variable "eye color" are not related to each other in any quantitative sense, we cannot say that John's eye color is bigger than Bill's eye color or that John's eye color is better than Bill's eye color. Certainly, we cannot say that John's eye color is one unit greater than Bill's eye color. We have no units to work with. We can only say that they have different eye colors. We use the word nominal to refer to this type of variable because nominal is derived from the Latin word for name. The values of such a variable are simply the names of the different categories. Other examples of nominal scale variables include sex, with the values (A) male and (B) female; and marital status, with the values (A) single, (B) married, (C) divorced or separated, or (D) widowed.

The other type of categorical variable is that in which the categories are related to each other in some quantitative sense. For example, we often find items in questionnaires that ask the respondents to indicate the extent to which they agree with a particular statement by circling or checking one of several response categories like the following: strongly agree, agree, neutral, disagree, or strongly disagree. An example of this type of questionnaire follows.

Please indicate your attitude toward each of the following statements by checking the response option that is closest to your own feeling.
1. A strategic arms limitation agreement   _____ strongly agree
   would increase the prestige of the   _____ agree
   United States overseas.   _____ neutral
                                                 _____ disagree
                                                 _____ strongly disagree

5

Introduction to Statistical Techniques

2. A strategic arms limitation agreement would jeopardize the military readiness of the United States.

    _____ strongly agree
    _____ agree
    _____ neutral
    _____ disagree
    _____ strongly disagree

Alternatively, a questionnaire might require respondents to indicate the frequency with which they perform certain behaviors by checking one of these response options: always, usually, sometimes, rarely, never. In each of these cases, the variables are categorical. Each subject will have a score that is simply one of the response option categories. However, these categorical variables are not nominal scale variables because the categories provided are related to each other. The categories are arranged in a clear order. The order may be from strongly agree to strongly disagree, it may be from always to never, or it may be from the smallest category to the largest. When the categories of a categorical variable are ordered in this manner, we refer to the variable as an ordered categorical variable.

Observations on an ordered categorical variable provide us with more information than observations on a nominal scale variable. For example, let us assume that a hypothetical subject 1 responded to question 1 on strategic arms limitation with the response option "agree," and the hypothetical subject 2 responded to the same item with the response option "strongly disagree." Given this information, we know that the two subjects have *different* values on the variable, just as we would know if the variable were a nominal scale variable. However, we know more. We also know that subject 1 indicated a *greater* degree of agreement with the statement than subject 2. It is this additional element of comparison along a dimension that distinguishes an ordered categorical variable from a nominal scale variable. Note that the ordered categorical variable does not enable us to say *how much more* subject 1 agrees with the statement than does subject 2. We only know that subject 1 does agree more than subject 2.

Ordered categorical variables are frequently employed in survey questionnaires. However, they often appear in other research situations as well. For example, suppose a winemaster were interested in determining which of four types of wine was most popular. He might conduct an experiment in which subjects tasted each wine and indicated their preference. If the four wines could be placed in a clear order along some dimension, such as from driest to sweetest, then the categorical variable "wine preferred" would be an ordered categorical variable.

In order to illustrate the properties of ordered categorical variables, let us suppose that our winemaster ascertained the preference of 25 wine drinkers for four wines labeled "very dry," "dry," "sweet," and "very sweet." Let us further suppose that, of the 25 subjects, 14 preferred the very dry wine; 9 preferred the dry wine; 2 preferred the sweet wine, and none preferred the very sweet wine. Because the categories are clearly ordered along a dimension from very dry to

6

very sweet, we know that all 14 of the subjects in the very dry category prefer a wine that is drier than the wine preferred by the 9 subjects in the dry category or the 2 subjects in the sweet category. However, we cannot differentiate between the subjects falling into any one category. There may very well be differences between the 14 subjects choosing the very dry wine in terms of the precise amount of dryness they consider ideal. However, our method of measuring the dryness they prefer is rather crude in the sense that it cannot uncover these differences.

We can think of the data from this study as it is represented in Figure 1.1. There we depict our 25 subjects as they might be located along a dimension or continuum of preference ranging from most dry to most sweet. Each subject's actual preference is represented by an X on this continuum. As we have represented these 25 subjects, no two are exactly the same in terms of their actual preferred level of dryness. Above this continuum of actual preferred level of dryness we have shown the four wines used in the experiment to measure preferred dryness. Each of these four wines has an amount of dryness corresponding to a point on the continuum. This is indicated by the dotted line connecting each glass to the underlying continuum. We do not know whether the four wines are equally distant from each other on the continuum. They may well not be. We only know that they are definitely in order along the dryness dimension.

Given this situation, we assume that each subject will select the wine that is closest to his or her actual preferred level of dryness. We have indicated this by the arrows directed from each X to one of the four possible choices. When a subject makes a selection, the winemaster has measured that subject on preferred dryness, using an ordered categorical scale with four values. In practice, the winemaster would never see the underlying continuum of preferred dryness. Only the ordered categorical measurement of the variable would be seen.

However, it is important that we keep in mind the existence of this underlying dimension. It is important when we consider the statistical techniques appropriate for use with ordered categorical data. In general, the statistical techniques that are appropriate to nominal scale variables are also appropriate for use with ordered categorical variables. However, because of the additional character of order associated with the ordered categorical variable, there are certain techniques that are not applicable to nominal scale data but that are applicable to ordered categorical data. These techniques will be pointed out at the appropriate points in this text.

Rank-Order Variables. A second general class of variables is the *rank-order* or *ordinal scale variable*. With rank-order variables, there are no fixed categories into which observations may fall. Rather, observations are compared to each other and put in order, perhaps from best to worst or from biggest to smallest. An example of a rank-order variable is rank in class. If we were considering the rank in class of the 15 students in Ms. Jones' tenth-grade English class, our data might be presented as in Table 1.2. Note that each student, that is, each observa-

7

tion in the data set, has been assigned a rank that locates that observation relative to the others.

Ideally, a set of rank-order data will have as many different ranks as there are observations in the data set. In the case of Ms. Jones' class, there were 15 possible positions or ranks, one for each student in the class. There are occasions, however, when it may be impossible to distinguish between two observations. That is, there may be ties. When two observations are considered equal, they are listed as having the same rank. Thus, in our example, Billy and Sally, who are tied for the position of second best student in the class, share the rank 2. Notice that in this situation we do not have anyone ranked number 3. The next student down the list, Helen, is not ranked 3, because she is *not* third best. There are 3 people ahead of her, so she is really the fourth best student in the class. For this reason, her rank is 4. This is the same reasoning that we use in races. If there is a tie for second place, there is no third place finisher. Looking at Ms. Jones' class again, we see that the two sets of ties have had no effect on the rank of the last

FIGURE 1.1
Measurement of Preferred Dryness
Using a Scale with Four Ordered Categories

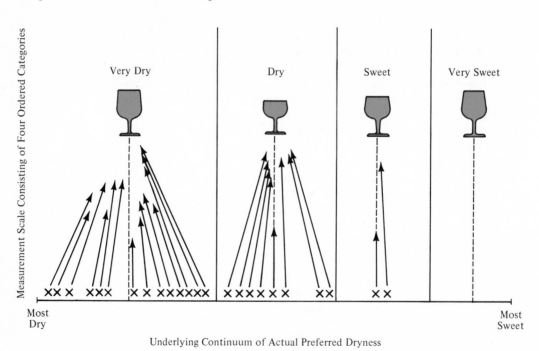

TABLE 1.2
Rank Order of Students in Ms. Jones' English Class

Name	Rank	Name	Rank
John	1	Doris	8
Billy	2	Jane	10
Sally	2	Ruth	10
Helen	4	Dick	12
Sam	5	Joe	13
Betty	6	Paul	14
Leona	7	Ted	15
Tom	8		

person in the class. Ted is still ranked 15 out of 15. The only situation in which the rank of the last person may be different from the number of observations is if a tie occurs for last place. If Paul and Ted were considered equal, they would share the rank 14.*

When we are dealing with rank-order data, we can compare the relative position of different observations. In Table 1.2, for example, we can see that John is ranked higher than Billy, and Billy is ranked higher than Helen. In this respect, rank-order data are similar to ordered categorical data. The idea of order is present in each type of data. The difference is that with ordered categorical data there are a limited number of categories, and no attempt is made to differentiate between observations falling in the same category. With rank-order data, an attempt is made to differentiate between all the observations, so that two observations occupy the same rank only when it is impossible to determine which should have a higher position.

The classic example of a rank order variable is the geologist's technique for measuring the hardness of minerals. A mineral is ranked ahead of all those minerals that it will scratch and behind all those that can scratch it. A diamond is ranked number one because no mineral can scratch it. Steel is ranked below the diamond because the diamond will scratch it. In using this scale, we can always say that a higher ranked mineral is harder than a lower ranked mineral. We cannot, however, say *how much* harder.

---

* Later in this book, you will learn about statistical tests employed in the analysis of rank-order data. There you will see that for computational purposes it is often necessary to handle ties in a slightly different fashion from the common method just shown. Often each of the observations that are tied will be assigned *the average* of the ranks they occupy. Had this procedure been employed here, Billy and Sally would have each been assigned the rank 2.5, because together they occupy both the second and third ranks, and the average of 2 and 3 is 2.5.

Introduction to Statistical Techniques

The statistical techniques employed with rank-order data are different from the techniques employed with ordered categorical data. It is therefore important that you be able to distinguish between these two types of variables.

Interval Scale Variables. A third general category of variables is the *interval scale variable.* One characteristic of an interval variable is that the values of the variable are *numerical scores.* The use of numerical scores means that we can do more than just put scores in order. We can take any two scores and subtract one from the other to find out by *how much* they differ from each other. The classic example of an interval variable is temperature read in degrees centigrade or degrees Farenheit. Consider the two temperatures 40°C and 30°C. We know that these two temperatures are different from each other, as we would if temperature were a categorical variable. We know that the temperature of 40°C is higher (or warmer) than the temperature of 30°C, as we would if temperature were a rank-order variable. But we also know that the temperature of 40°C is exactly 10°C warmer than the temperature of 30°C. Thus, interval data tell us everything that categorical data or ordinal data tell us, plus the added information of the amount by which two scores differ.

Another characteristic of a true interval variable is that the *unit of measure* is the same size throughout the entire range of possible values of the variable. On the centigrade temperature scale the unit of measure is the degree. Because the degree represents the same difference in temperature anywhere on the scale, we know that the difference between temperatures of −10°C and 0°C is exactly the same as the difference between temperatures of 20°C and 30°C. In each case, the difference is 10°C. This is illustrated by the number line in Figure 1.2.

FIGURE 1.2
A True Interval Scale:
Temperature in Degrees Centigrade

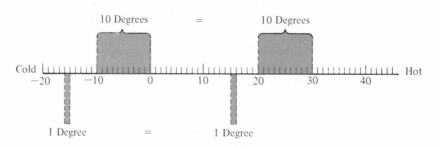

On a true interval scale, a unit of measure is the same size, regardless of the position on the scale in which it occurs.

In psychology, education, and the social sciences, we often encounter variables whose values are numerical scores. We do not often encounter variables that have a truly uniform unit of measure. A case in point is the numerical scores obtained by students on an achievement test. Suppose Dr. Brown gave a 100-item, multiple-choice exam as a final in his statistics course. If each item was scored 0 for a wrong answer and 1 for a right answer, then a given student could score anywhere from 0 to 100 on the test. If there were 50 students in Dr. Brown's class, the test scores might appear as in Table 1.3. These numerical scores appear to have the characteristics of an interval scale. In fact, it would be surprising if they truly did. The one point units of measurement will be equal if and only if each and every one of the 100 exam questions represents the same amount of statistics achievement. In practice, this is unlikely. Some of the questions on the exam would be more difficult than others, requiring greater amounts of knowledge of statistics. Nevertheless, in assigning a value of one point to each question, Dr. Brown has assumed that each question represents the same amount of statistics achievement, just as each degree centigrade represents the same amount of temperature.

On the basis of this type of assumption, we typically treat scores of this nature as if the variable test score were a true interval variable. Thus, referring to Table 1.3, we would say that students 4 and 5, Ann and Michael, have shown the same amount of statistics achievement because they each scored 93. Technically speaking, they may not have exactly the same amount of achievement. They may have missed different questions. Perhaps the questions that Ann missed represented relatively little knowledge of statistics. Perhaps her mistakes were careless errors caused by her boredom with these easy questions. If this were true, her actual statistics achievement might be different from that of Michael. However, when we construct an achievement test, we try to keep the possibility of such errors to a minimum. When we actually use such test scores, we assume these two students have the same achievement. We also assume that a given number of points represents the same amount of statistics achievement on one part of the scale as another.

First, consider two students who have done well on the exam. Jim (student 1) has a score of 98. Ann (student 4) has a score of 93. Their scores are 5 points apart. Now consider two other students who did not do quite so well. Grace (student 6) has a score of 72. Andy (student 7) has a score of 67. Thus, Grace and Andy have scores that are 5 points apart, just as Jim and Ann have scores that are 5 points apart. Because we are assuming that our test is well constructed and our variable is very nearly a true interval variable, we can say that Jim differs from Ann by the same amount of statistics achievement that Grace differs from Andy. In other words, we are locating our students along a number line that is laid off in equal units or intervals, just as we located temperatures on a number line in Figure 1.2. It is this concept of equal units of measurement, equal intervals, that leads to the name interval variable. With many psychological and educational tests, we assume the existence of equal intervals. This assumption is not

11

TABLE 1.3
Final Examination Scores of the 50 Students
in Dr. Brown's Statistics Class

Student	Name	Score	Student	Name	Score
1	Jim	98	26	Dick	70
2	John	86	27	Tom	75
3	Peter	78	28	Bruce	84
4	Ann	93	29	Lee	85
5	Michael	93	30	Carmen	79
6	Grace	72	31	Iris	74
7	Andy	67	32	Consuela	71
8	Karl	54	33	Muriel	77
9	Sandra	64	34	Donna	81
10	George	76	35	Miriam	90
11	Leona	80	36	Fred	57
12	Cathy	79	37	Rose	85
13	Dorothy	69	38	Vincent	85
14	Angela	59	39	Paul	84
15	Hester	75	40	Annette	60
16	Rufus	83	41	Margie	90
17	Myron	82	42	Lynn	91
18	Henry	60	43	Mary	63
19	Sam	77	44	Jane	98
20	Sally	84	45	Daisy	79
21	Don	81	46	Roberto	87
22	Stanley	82	47	Fern	88
23	Pablo	79	48	Gary	74
24	Laurel	68	49	Jerome	80
25	Ruth	73	50	Gloria	70

always absolutely true, but in most cases it is a reasonable approximation of the truth.

Ratio Scale Variables. The fourth general class of variable is the *ratio scale variable.* The ratio scale variable has all the characteristics of the interval scale, plus an additional feature. The ratio scale has a true zero point. Statistics achievement is at best an interval scale. A student scoring zero on Dr. Brown's exam should not be assumed to have no knowledge of statistics whatsoever. On the other hand, the amount of money in the school budget is a ratio scale variable. If the value of this variable is zero, then there is really no money in the budget. Ratio scale variables are employed in statistical analysis in psychology and the social sciences somewhat less frequently than interval scale variables. The techniques employed with ratio scale variables are the same techniques that are employed

Introduction

with interval scale variables. Accordingly, these two types of variables are considered together in this text.

You should have a clear idea of the properties of categorical, rank-order, and interval scale variables. If you can examine a set of data and determine which of these types of variables are involved, then you will have made the first important step toward knowing how to analyze the data from your own research work. As indicated previously, different types of variables call for different statistical techniques. Therefore, before you can decide what to do with your data, you have to be able to decide what kind of data it is. At the conclusion of this chapter, in the student guide section, you will find a set of problem situations designed to give you practice in determining the type of variable involved in various research situations.

Measurement

How do we know that John has brown eyes? How do we know that Helen is ranked fourth among the students in Ms. Jones' English class? How do we know that the high temperature on December 15 was 5°C? Whenever we determine the value of a particular variable in a particular subject or at a particular point in time, we are said to have *measured* that variable. The process of measurement may be simple or complex, depending on what the variable is and upon what group we are measuring it. In any case, unless we can obtain a true measure of the variable of interest, none of our statistical techniques mean anything. In empirical research, it is a truism that "You can't get good results without good data." But what do we mean by "good data"? Typically, we mean that our measurement procedure is (1) reliable and (2) valid.

Reliability. When we speak of the *reliability* of a particular measurement process we are referring to the consistency or repeatability of its results. For example, consider the measurement of eye color in human subjects using the three-category variable described previously. In all probability, the researcher would employ a judge who would observe the subjects in the study and determine whether their eye color should be classified as "brown," "blue," or "other." This measurement procedure would be perfectly reliable if each subject were always placed in the same category, no matter how many times the judge made the decision. With a variable like eye color, most subjects would present no problem, provided that we made sure our judge had good color perception. However, there would probably be some difficulty with certain subjects. For example, one subject might have eyes that are "grey with a distinct blue cast." In this case, the judge's decision might not be an easy one. The judge might place this subject in the "other" category one day and in the "blue" category another day. The decision might be influenced by personal factors, such as the judge's mood or fatigue. The decision might also be influenced by situational factors, such as the type of lighting at the time the determination was made.

13

As another example of what we mean by reliability, let us consider the measurement of statistics achievement by Dr. Brown's 100-item, multiple-choice exam. If this exam were perfectly reliable, subjects taking the exam for a second time or taking an alternative form of the exam would obtain the same score each time, providing they had not looked up answers or done further studying in the interim period. With a written exam, there would be no judge whose fatigue or mood might result in differences in scoring. However, the subjects themselves will undergo changes in energy level and mood. On one occasion a student may be feeling fine and score very high. On the next day, the same student might feel sick and score poorly, even though the achievement level in statistics remains unchanged. Situational factors may also make an achievement test unreliable. If the class were to take the exam in a bright, well-ventilated room, with ample workspace, they would probably achieve higher scores than if they were to take the exam in a dark, stuffy, overcrowded room. Even aspects of the test format may contribute to unreliability. If the exam is multiple-choice, a student could easily determine the correct answer to a question but write the wrong choice code on the answer sheet. A careless error of this kind could cause the score to differ from one administration to another.

Finally, even a measurement process as objective as the measurement of temperature may not be totally reliable. The thermometer employed may have some imperfection in construction that results in different readings at different times for the same true air temperature. The technician employed to read the thermometer may be careless or nearsighted. Even in the physical sciences, 100 percent reliability is an ideal not often achieved. In education, psychology, and the social sciences this ideal is typically rather far from the reality of the research situation. Nevertheless, it is an important ideal toward which the researcher should always strive, for without reasonably reliable measurement procedures, there is no point in conducting empirical research.

Validity. When we speak of the *validity* of a particular measurement process, we are concerned with the extent to which that process is measuring what we really want to measure, rather than some other variable. In the case of a variable like eye color, validity does not present a great problem. But consider the variable "statistics achievement." When Dr. Brown develops his 100-item, multiple-choice test, he must be very careful to construct questions that measure statistics achievement and not something else. For example, suppose Dr. Brown were to include questions that require complicated algebraic manipulations to obtain the answer. On these questions, students with strong mathematics aptitude but relatively little knowledge of statistics might perform *better* than students with modest math aptitude but a great deal of knowledge of statistics. In this case, the test would no longer be measuring statistics achievement alone; it would be measuring math aptitude as well. The test might still be reliable in the sense that a given student would obtain the same or similar scores from one administration to the next, but the validity of the test as a measure of statistics achievement would suffer. In reality, of course, it is very difficult to create an achievement test that measures only achievement and not aptitude. In fact, it is difficult to create a test that is a pure measure of any one psychological variable. Nevertheless, in planning a research study, careful consideration must be given to the issue of validity. Unless the researcher can develop a measurement process that produces a reasonably valid measure of the variable of interest, the research will be of little value.*

Operational Definitions.   Because issues of measurement are so important in empirical research, it is extremely important that researchers define each experimental variable employed in their study precisely, in terms of the method by which the variable is measured. This is known as providing an *operational definition* of the variable of interest. The operational definition gives the reader of a research report the information needed to judge whether the researcher has achieved a reliable and valid measure of the variable of interest. The requirements of a good operational definition will vary, depending on the variable. Let us consider several examples. First, consider a variable that would seem to be straightforward, the variable "age." At first glance, it would seem that this variable would be self-explanatory. Everyone knows how old they are. However, if our researchers simply asked subjects their age, they might run into trouble. One difficulty is that some subjects might simply lie about their age. Another problem is that subjects might not all interpret the question the same way. Most subjects would probably give their age in years *at the time of their last birthday,* even if their next birthday were only a few weeks off. However, a few subjects would look ahead and round off their age *to the nearest year.* If this occurred, our researchers would not be measuring the same variable for all subjects. Such an error could influence their results, especially if they were working with children. Thus, it is important that the experimenters carefully describe how information on age was obtained and exactly what question was asked. If the research article stated that "subjects were asked their age," the reader would be able to see that the difficulties just noted could be present. On the other hand, if the article stated that "age to the nearest whole year was obtained from official class records," then the reader of the article would have more confidence in the consistency of measurement of this variable. In either case, the operational definition serves to provide the information required by the reader to make a reasonable judgment regarding the reliability and validity of the measurement of the variable. Often you will find that in the process of developing a clear operational definition, you will recognize a potential problem area and correct it before wasting a great deal of time and energy collecting data.

   In some laboratory studies in psychology and the social sciences, variables are measured by means of rather elaborate mechanical or electrical apparatus.

---

* The reader should be aware that there is a great deal more to the topics of reliability and validity than we are able to consider in this text on introductory statistics. The interested reader is referred to Lindeman, R. and Merenda, P., *Educational Measurement* (2nd ed.). Glenview, Illinois: Scott, Foresman and Company, 1979.

In cases like this, providing an adequate operational definition of a variable may be quite a demanding task. One will frequently find several pages of detailed specifications of procedures and apparatus in articles describing such studies. Often the specification is so exact that the brand names and model numbers of pieces of equipment are supplied. In these situations, the operational definition is of particular importance with respect to replication.

When the variable of interest is measured by a psychometric test, the operational definition is particularly important. Suppose we are measuring the variable "intelligence." Standardized IQ tests vary widely in terms of the average performance of subjects. They also differ in the way individual scores spread out on either side of this average. It is well known among psychologists and teachers that a bright sixth-grader is likely to get an IQ score on the Stanford-Binet considerably higher than the score he or she would get on the Wechsler Intelligence Scale for Children (WISC). A set of scores based on one of these two tests would not be directly comparable to a set of scores based on the other. Thus, any researchers wishing to employ a standardized test as an instrument to measure intelligence would need to specify this test. If there were several forms of the test, they would specify which form they had used. With most standardized tests, the test authors have provided a manual containing descriptive material. When the exact form of the test used is specified in a research article, the reader may refer to the manual to get the important facts about the test. Often test manuals will present numerical indices of the reliability and validity of the test. These indices are generally referred to as *reliability coefficients* and *validity coefficients*. We will not consider these indices in detail here for they are the proper subject matter of a measurement textbook rather than a statistics textbook. Here we simply note that good researchers will select a psychometric test with proven reliability and validity and will provide their readers with the reliability and validity coefficients to substantiate this fact.

The Population of Interest

In addition to providing a good operational definition of each of the variables to be measured, researchers must always specify exactly with what group they will be observing this variable. Whenever a research study is completed and a conclusion is drawn, there are two questions that may be asked. The first of these questions has to do with the *internal validity* of the study. Suppose experimenter Smith has tested a group of boys and a group of girls on the variable mathematics aptitude. On the basis of her results, she concludes that boys are more able in this area. In this case, we might question whether the test used to determine math aptitude was valid and reliable. If the test was not valid and reliable, we would not be sure that Smith's group of boys really had higher scores on the variable "math aptitude" than her group of girls. When we question whether the differences between two experimental groups are real, we are questioning the internal validity of the study. Here, the question is "Has Smith really shown that the boys *in her study* have higher math aptitude than the girls?"

In addition to internal validity, we may also question the *external validity* of a study. When we speak of external validity, we are asking: "Assuming that Smith has really shown that *her group* of boys had more ability in math than *her group* of girls, does this mean that boys *in general* have more math ability than girls *in general*?" In other words, we are asking whether we can generalize from what Smith observed to make a broader statement about a larger population of subjects.

In order for the reader to be able to judge the external validity of a particular study, it is necessary that the researcher provide a complete description of the population of interest, the group to which the results of the study are to be applied. It is important to note that the population of interest may be people, rats, monkeys, institutions, and so on. The term population does not apply to human subjects alone. In psychology and the social sciences, we often refer to human subjects, but a rat may be considered a subject as well.

In the example of the math aptitude of boys and girls, experimenter Smith would not simply state that she had used a group of boys and a group of girls. Rather, she would write something like this: "Subjects included 25 sixth-grade boys and 25 sixth-grade girls selected at random from the class roster of the public Henry Barnard Elementary School in New Rochelle, New York. New Rochelle is a suburban community. This particular elementary school serves a rather diverse student population, with children of low, moderate, and high socioeconomic status represented about equally." With a description like this at their disposal, readers can judge for themselves the extent to which the results of Smith's work may be generalized beyond the actual children involved in the study. Readers would realize that generalization to second-grade youngsters in an inner-city school situation would probably be unjustified. Readers would also probably not generalize Smith's results to groups of exceptional children, such as low achievers or gifted students. They would probably not do this because Smith specified that the children in her study were *randomly selected* from among all the sixth-graders at the school. Random selection in this case means that any sixth-grade boy in the school would have the same chance of being included as a subject as any other sixth-grade boy. The same would hold for the girls. When a sample of subjects is used, the process of random selection helps to ensure that the sample actually measured in the study is representative of the population of interest. In this case, Smith randomly selected a sample of 25 boys to represent the population of sixth-grade boys at this school and a sample of 25 girls to represent the population of sixth-grade girls at the school.

The importance of random selection to the question of external validity may be demonstrated by reference to a survey conducted at the time of the 1948 presidential election. Pre-election polls seemed to indicate that Dewey would win by a landslide; yet, Truman actually won the election. The reason

17

is not that the voters changed their minds between the time of the polls and the time of the election. Rather, the subjects selected for the polls did not represent a random sample of the electorate as a whole. In one famous poll, the voters surveyed were selected from the subscription lists of a well-known national magazine. Although this poll may well have represented accurately the feelings of the readers of this magazine, it did not represent accurately the feelings of the electorate as a whole, the logical population of interest. Sampling bias of this kind may occur very easily in educational and psychological research. This is why it is essential for researchers to specify the population of interest of their study *and* the method used to select subjects from that population to be measured. Only in this way does the reader of a research report have the information necessary to determine the external validity of the study. We will discuss sampling error in more detail when we examine statistical inference in later chapters.

A CLASSIFICATION SCHEME
FOR ELEMENTARY STATISTICAL PROBLEMS

Now that you have had an introduction to the basic principles and terminology of empirical research, we can begin to deal with the application of statistical techniques to empirical data. In most instances, researchers will want to do two things with their data. First, they will want to organize and present the data in a form that will be convenient and meaningful to their readers. This typically involves the use of various forms of tables and graphs. In addition, they will probably want to make summarizing statements or draw conclusions from their data. This typically involves the calculation of one or more numerical indices known as *statistics*.

There are many different kinds of tables and graphs and many different types of statistics. The most important problem facing researchers is frequently the problem of deciding which techniques to employ with their particular set of data. At this point, we would like to develop a matrix for classifying statistical problems. This matrix will be helpful in determining which techniques are appropriate for a given set of data. It will also serve as a framework for organizing the material in the book. It is based on a series of dimensions according to which the more elementary statistical problems may be classified. You have already seen one of these dimensions, the type of variable with which we are concerned. When we determine whether we have a categorical, rank-order, interval, or ratio scale variable or variables, we determine the "level of measurement" of the variable(s) of interest. The techniques appropriate for categorical data are typically different from the techniques appropriate for rank-order or interval data. Thus, we need to be able to determine the level of measurement of the variable of interest before we can decide how to table the data, or how to graph the data, or what statistics to use with the data.

In this section of the chapter, we consider several additional distinctions that can be made among statistical problems. To choose the appropriate techniques, we must be able to classify our problem on the basis of these new dimensions as well. To illustrate these new dimensions and to give the reader an idea of the diversity of problems that can be dealt with by statistical techniques, we present these five research questions:

1 What is Hank Aaron's lifetime batting average?
2 What proportion of the television viewing audience watches station WXYZ at 10:00 PM on Saturday?
3 Is there a relationship between political party membership and opinion on the Equal Rights Amendment among residents of Tranquility, New York?
4 Is my new method for teaching Spanish really more effective than the old one?
5 Can my remedial reading program raise the reading achievement scores of eighth-graders?

Each of these questions is a statistical question. In each case we can seek out some empirical evidence bearing on the question. In each case we can apply some statistical techniques to these empirical data to help us arrive at an answer. Yet, these five questions differ greatly from one another. They differ in content, of course. But they differ from one another in important other ways as well.

Descriptive Problems vs. Inferential Problems

The first major distinction we will discuss is the distinction between descriptive problems and inferential problems. This distinction can be seen by comparing research question 1 with research question 2. The batting average question represents a *descriptive* problem. When we compute a player's batting average, we are of course seeking to describe an aspect of his play. However, in describing an aspect of his play, we are really describing a specific variable in a specific population. The variable of interest is hitting. It has two values here: (1) hit and (2) not hit. The population of interest is the population of all the official times at bat charged to Hank Aaron. The average we compute is a convenient way to describe the variable "hitting" in this population. If we find that Hank Aaron's average is .333, it tells us that he had a hit 333 times out of every 1,000 times he was officially at bat or roughly once in every three times. To obtain this average, the baseball statistician needs two figures: the total number of hits of the player and the total number of official times he came to bat. Suppose Hank Aaron had 3,210 hits in 9,630 official times at bat. The statistician would divide 3,210 by 9,630 to obtain the average .333. There are two key concepts involved here. One has to do with why the average is computed and the other with how it is computed.

19

With regard to the why question, the statistician could simply have reported Hank Aaron's total hits and at bats, without computing the average. However, if we heard, "Hank Aaron has had 3,210 hits in 9,630 official at bats," it would not be apparent immediately whether this was good or bad. The numbers are just too large. The batting average reduces these numbers to a simple, easily understood index whose meaning is immediately apparent. Most descriptive statistical techniques share this quality of reducing data to simpler terms for ease of understanding.

Now with regard to the how question, the key point is that when the statistician computes the average, the result is based on *every one* of the player's official times at bat. In the language of empirical research, we would say that the population of interest here consists of all Hank Aaron's official times at bat. Each time at bat is said to be a member or element of the population of interest. Thus, when the statistician computes the statistic that we call the batting average, it is based on every member or every element of the population of interest. This means that the batting average is an exact description of the variable of interest within the population of interest. It is not an estimate. It does not require the statistician to make any inferences beyond the data in hand. For this reason, it is a descriptive statistic.

Now consider research question 2, concerning the proportion of the television audience watching station WXYZ at 10:00 PM on Saturday. In this case, the variable of interest might be called "watching WXYZ." The values of this variable would be (1) yes and (2) no. The population of interest would be all those individuals within the WXYZ viewing area who were watching television at 10:00 PM. Think about how you would answer this question with respect to this variable in this population. The information you would need would not be available to you simply by looking it up in a record book, as a ball player's hits and times at bat might be. You would have to find it out for yourself. One possible way to do this would be to visit every home within the range of station WXYZ. You could then ask the residents if they were watching television and if they were watching station WXYZ. If you did this, you could compute the answer to the question in a manner similar to the way the baseball statistician computes a batting average. You would take the number of people watching station WXYZ and divide by the total number of subjects who were watching television. This would give you the exact proportion of the total viewing audience who were watching station WXYZ. If you followed this course of action, the problem would be a descriptive problem, like research question 1, for your answer would have been computed on the basis of the entire population of interest.

But would this course of action be feasible? What if the station in question were in an urban area? The potential viewing audience might consist of several million people. The cost, in time and money, of contacting all of the subjects in the population of interest would be enormous. For this reason, the descriptive approach is used rather infrequently when the question asked concerns a large population of interest.

A more reasonable approach to research question 2 would be the *inferential* approach. Instead of trying to contact all the subjects in the population of interest, we would select a smaller number of these subjects, a sample. As noted previously, we would probably employ a random sampling procedure to ensure that our sample was representative of the greater population. We would ask the subjects in our sample the same questions that we would have asked everyone in the population if we were using the descriptive approach: "Are you watching television?" and "Are you watching WXYZ?" We would also compute the proportion of people in our sample watching WXYZ by dividing the number in the sample watching WXYZ by the total number in the sample watching television at all. This computed proportion would be the exact proportion watching station WXYZ within our sample, but we would use this exact proportion within the sample as an *estimate* of the exact proportion within the entire population of television viewers. Thus, our answer to research question 2 would provide an *estimate* of the exact proportion of the whole population of television viewers watching WXYZ. It would not provide an exact description of a whole population. Rather, it would consist of an exact description of a sample drawn from the population, coupled with an *inference* that what is true of the part is probably *near* to what is true of the whole. The process of random sampling involves an element of chance, so samples are scarcely ever *exactly* like the population from which they are drawn. Our sample may happen to include a proportion of WXYZ viewers that is somewhat larger or somewhat smaller than the proportion of WXYZ viewers in the entire population of television viewers. Any difference between the estimate computed on the basis of the sample observations and the actual proportion of WXYZ viewers in the population is said to be because of sampling error. By careful sampling procedures, this error can be held within reasonable limits. Later in this book, you will learn about techniques that allow you to make precise judgments about how close an estimated answer based on a sample is likely to be to the true answer in an entire population. For now, however, it is only important that you understand the difference between *describing* a whole population and *making an inference* about a whole population based on a sample from that population.

In such an inferential problem, the attribute of the population that we seek to estimate is referred to as the *parameter* of interest. The estimate of the parameter, computed using data in the sample, is referred to as a *statistic*. Thus, in the example just considered, the actual proportion of all the viewers who were watching station WXYZ at 10:00 PM on Saturday is the population parameter of interest. The proportion of the viewers in our sample who were watching is the statistic that we use as an estimate of this parameter. We will review this distinction in our discussion of inferential statistics in Part 4.

Introduction to Statistical Techniques

Univariate Problems vs. Problems with Two or More Variables

In research questions 1 and 2, we are dealing with a single variable. We refer to such problems as *univariate* problems. Most often, however, our statistical problems will involve two or more variables. Research questions 3, 4, and 5 each involve two variables. We refer to these problems as *bivariate* problems.

In research question 3, the two variables are obvious, for they are named in the question itself. If we are seeking to determine whether there is a relationship between political party membership and opinion on the Equal Rights Amendment, it stands to reason that we are going to gather information on two variables. Political party membership is a categorical variable with the values (A) Republican, (B) Democrat, and (C) Independent. Opinion on the ERA is also a categorical variable. It has the values (A) in favor, (B) opposed, and (C) no opinion. Because the population of interest for this question consists of the residents of a rather small town, we might employ a descriptive approach to this problem. We could obtain a list of all registered voters in this town. Then we could contact these voters and ask each one two questions: "What is your political party?" and "What is your opinion on the ERA?" On the basis of the responses of these voters, we could compute the proportion of Republicans in favor and compare this figure to the proportion of the Democrats in favor. This would help us to see whether there is any relationship between these two variables in this population. Had this same question been asked with reference to the population of a larger town, we might have employed an inferential approach. In this case, we would have selected a random sample of the voters from the town and asked each subject in the sample the same two questions. Then we would have used our results to make an inference about the population. Whether a descriptive or an inferential approach was employed, however, we would have a bivariate problem.

In research question 4, there are also two variables of interest. However, they are not quite so obvious in this case. How would you go about determining whether the old teaching method or the new teaching method is more effective? Logically, we would need to try out both methods under comparable circumstances to see how well people do under each. One procedure would be to take two randomly selected groups of students who were just beginning to study the language. We could then use the new method with Group A and the old method with Group B. (Method would also be randomly assigned.) At the end of the program, we could give the members of each group the same test of Spanish achievement. If the performance on this test in Group A was considerably higher than the performance in Group B, it would be an indication that the new method is better. When we plan our experiment in this manner, each subject will have a value on each of two variables. One of these variables is the categorical variable we call "group." This variable has the values (A) new method and (B) old method. The other variable is "score on test of Spanish achievement." We assume that this is an interval variable with numerical scores for values.

In research question 5, there are two variables of interest as well. Here our concern is whether the treatment, a remedial reading program, can *improve* reading achievement. In order to get an idea of whether a student has improved, we must know his or her achievement level before the treatment, and we must compare it to the achievement level after the treatment. In this situation, we would probably have reading achievement scores on record to use as a before measure. If not, we could give an achievement test before the start of our remedial reading program. This type of test is usually referred to as a pretest. Then, at the conclusion of the program, we could administer another form of the achievement test. This would be our posttest. Thus, we would have measured two variables: pretest reading achievement and posttest reading achievement.

You may be thinking at this point that questions 3, 4, and 5 seem quite different from each other even though they all involve two variables. If so, you are right. At this point, we will consider the important differences that distinguish these problems. We begin by comparing research question 3 to research question 4. Question 3 represents a *correlational* problem and question 4 represents an *experimental* problem.

Correlational Problems vs. Experimental Problems

In seeking an answer to research question 3, we selected a sample of subjects and asked each subject two questions, one to determine party and one to determine voting behavior. These two questions were asked at the same point in time, *after* the selection of the sample. We have no control over the answers to these questions. We do not pick a subject and assign that subject to the Republican value of the variable party membership, nor do we pick a subject and assign that subject to the "no opinion" value of the variable opinion. In this situation, we simply observe two variables, we do not have any control over either of them. We refer to such a problem as a bivariate *correlational* problem.

In contrast, in seeking an answer to research question 4 we do have control over one of the two variables. Through our random assignment procedure, we determine which value of the variable "group" each subject will have. This does not mean that we are free to place the bright subjects in Group A and the dull subjects in Group B. On the contrary, the random assignment procedure is a technique aimed at making Groups A and B comparable. However, from the point of view of the subject, we know that he or she is a member of Group A or a member of Group B for one reason only, because the random assignment procedure placed him or her there. In this situation, the experimenter is said to have manipulated the variable "group." Such a problem is referred to as an *experimental* problem. In an experimental problem, the variable that is manipulated by the experimenter is called the *independent variable*. Thus, in the procedure outlined to answer question 4, group is the independent variable. This independent variable is a categorical variable with two values: A and B, i.e., new method and old method. The purpose of the study is to find out whether subjects in one group perform better than those in the other on the Spanish achievement test. The score on the achievement test is the *dependent variable.* If the experiment is successful, the experimenter will have evidence that Spanish achievement depends upon group membership. That is, the experiment will have shown that Spanish achievement scores are higher for Group A subjects, who have the new method, than for Group B subjects, who have the old method.

The Issue of Causation in Correlational and Experimental Problems.   The distinction between correlational and experimental problems is extremely important in empirical research. In a correlational study, the researcher may show that two variables are related, but it is not always clear which of the variables is influencing the other. For example, in question 3 we may show that there is a relationship between party membership and opinion on the ERA. We may find that a higher proportion of Democrats were in favor of the ERA than Republicans. But can we tell for sure that being a Democrat tends to cause one to be in favor of the ERA? Not really. It could be that a person very much in favor of the ERA would become a Democrat because this voter perceived the Democrats as being more attractive on this important issue. In this case, we can legitimately conceive of party membership influencing opinion or opinion influencing party membership. This is known as *reciprocal causation.* A correlational study does not typically allow us to determine *direction of causation.*

On the other hand, an experimental study does deal with the issue of causation. For example, in question 4 if we show that there is some relationship between group membership and Spanish achievement, there is no doubt about which variable does the influencing and which is influenced. It is clear that the independent variable, group membership, is influencing the dependent variable, Spanish achievement. Why? There are two reasons. First, we know that an individual's group membership was determined by the random assignment procedure *before* Spanish achievement. This is important because we know that one thing cannot be the cause of another unless it comes first in time. Second, we know that Spanish achievement could not determine group membership because we really know that group membership was determined by one thing and one thing only: by our random assignment procedure.

It should be noted at this point that there are occasions in which the variables in a correlational study are such that the potential direction of causation may be only one way. For example, suppose research question 3 were modified to become "Is there a relationship between sex and opinion on the equal rights amendment among the residents of Tranquility, New York?" In this case, it is clear that one variable, sex, will be determined for each subject long before the determination of the second variable, opinion. It is therefore impossible that there should be reciprocal causation between these two variables. One's sex may influence one's opinion, but one's opinion can hardly influence one's sex. If there is a causal relationship here, the direction of causation must be from sex to opinion. Nevertheless, it should be clearly understood that the single direction of potential causation in this case is because of the particular variables of interest in this correlational study. Furthermore, even when there is only one possible direction of causation in a correlational study, the researchers will not be able to establish a definite causal relationship as they would in an experimental study. This is because of the possible influence of intervening variables that would be controlled for in an experimental study but not in a correlational study. We will consider the issue of causation in correlational studies in some detail in Chapter 3. For the present, the important point is the reader's ability to distinguish correlational problems from experimental problems.

Pseudo-Experimental Studies.   It should be noted here that in addition to correlational studies and experimental studies there is a middle category that we shall refer to as the *pseudo-experimental* study. The pseudo-experimental study is in some respects similar to the experimental study and in other respects similar to the correlational study. It is similar to the experimental study and different from the correlational study in that the experimenter begins by obtaining two or more distinguishable groups of subjects, who are subsequently measured on some other variable of interest. The pseudo-experimental study is different from the experimental study, however, in that subjects are not placed in a particular group via the random assignment procedure of the experimenter. Rather, they are naturally members of the group.

An example of a pseudo-experimental study would be as follows. Suppose the researchers were interested in determining whether men had different opinions from women on the ERA. To answer this question, they randomly selected a group of 50 men and another group of 50 women. They asked members of each group their opinion on the ERA and compared the responses in the two groups. This study would be neither correlational nor experimental. It is not correlational because the researchers have not measured one group of subjects on two different variables at one time. They did not take a single, randomly selected group of voters and measure the voters on the two variables "sex" and "opinion." They set up their study with an initial group of males and an initial group of females and measured each member of each group on the variable "opinion." The researchers exercised some control over one of the two variables in the study. At least they could be sure that their subjects would not turn out to be all of a single sex. However, they did not exercise the kind of control they would exercise in an experimental study. They did not assign some subjects to be males and some to be females. The value a subject has on the variable "sex" was determined naturally, rather than by the experimenter.

The appropriate statistical techniques for data derived from pseudo-experimental research studies have been the subject of much controversy and some confusion. We will not become involved in this controversy in this text. Instead, we simply make the following observations. Although pseudo-experimental data does not always satisfy the strict mathematical assumptions that underlie certain of the techniques appropriate for true experimental data, it has nevertheless

25

become common practice to analyze the results of pseudo-experimental studies using these techniques. Accordingly, the scheme for the classification of elementary statistical problems presented in this text will consider experimental and pseudo-experimental studies as a single category. Nevertheless, the true experimental study is capable of generating much more powerful statements regarding the question of causation than the pseudo-experimental study. Just as intervening variables may affect the results of correlational studies, so they may affect the results of pseudo-experimental studies. Only in the true experimental study, where the experimenter actually determines the values that subjects will have on one of the variables, can we make definite statements regarding causation.

These points will be considered in more detail later. For now, the reader should be sure to remember that experimental and pseudo-experimental problems are similar in that two or more groups of subjects are selected by the researcher initially and that these two types of problems are typically handled using the same techniques.

Independent Measures vs. Dependent Measures

Research question 4 represents a problem in which we are dealing with two sets of *independent measures*. You will recall that the two variables involved in the study were the experimentally controlled variable "group" and the variable of interest, Spanish achievement. Because the total number of subjects included in the study are divided into two groups, we have one set of measures on Spanish achievement that belong to Group A and a second set of measures on Spanish achievement that belong to Group B. These two sets of measures of Spanish achievement are said to be *independent* because the scores of subjects in one group have no influence on the scores of subjects in the other group. The two groups consist of different subjects selected at random. Note that the word "independent" used in this context has a meaning somewhat different from its meaning in the term "independent variable" considered earlier. Unfortunately, this word is used in several different contexts in the field of statistics. You should try to keep these different contexts clear. When we say that these two sets of scores are independent, we refer to the fact that at no point in the experiment do the subjects in the two groups do anything together or share any common characteristic or experience.

To answer research question 4, we simply compare the average achievement score in Group A to the average achievement score in Group B at the end of the training period. Because both groups began training with no knowledge of Spanish and because the only difference between the groups that we know of is the difference in teaching method, we assume that any marked difference between the average achievement scores in the two groups reflects a difference in the effectiveness of training methods. In short, if the average achievement score in Group A turns out to be substantially higher than the average achievement

26

score in Group B, we have evidence that the new method is superior. Later in this text, you will learn statistical techniques that will allow you to decide just exactly how much of a difference between the average achievement scores in the two groups we would want to see before we concluded that the new method was better.

The Pretest and Posttest Research Situation.   Now let us compare the situation envisioned in research question 4 to that envisioned in research question 5. In question 5, we have a set of measures on the variable reading achievement at the time of the pretest and a set of measures on the variable reading achievement at the time of the posttest. Thus, it could be said that we have two sets of measures, just as we have two sets of measures in research question 4. However, in research question 5, the two sets of measures of the variable of interest are not independent of each other. We do not have two separate groups of randomly selected subjects here. We have a *single group of subjects* in which each subject has been measured *twice.*

We say that these two sets of measures are *dependent* because each subject's posttest achievement score not only reflects the improvement made as a result of the remedial reading program, but also depends upon reading achievement level at the start of the program, as measured by the pretest achievement score. Because our research question asks whether the remedial reading program can improve scores, we would not analyze our data by making a direct comparison of the average pretest score to the average posttest score. Instead, we would look at the improvement of each subject from pretest to posttest. That is, for each subject, we would subtract the pretest score from the posttest score to obtain a measure of the amount of improvement in achievement that occurred as a result of the program. In such a case, the difference between the posttest measure and the pretest measure is typically referred to as a gain score. Later in this text, you will learn techniques for determining just how large the average gain from pretest to posttest would have to be before we would feel justified in concluding that the remedial reading program is generally effective.

Up to this point, we have considered three different classes of bivariate research problems: correlational problems, experimental and pseudo-experimental problems involving the comparison of two or more independent groups, and pretest and posttest problems involving dependent measures. We use different statistical techniques with each of these different types of problems. It is therefore important that you be able to distinguish between them. Before moving on to our classification matrix, we must make special note of one additional group of bivariate problems, the matched group problem.

The Matched Group Research Situation.   The matched group research situation is unique. It is an experimental situation, for the researcher manipulates the values of one of the variables by establishing two separate groups of subjects prior to the experiment. However, it is not like the experimental situation de-

27

scribed earlier, for the two groups are not independent. There are times when researchers will ensure that two groups of subjects will be comparable with respect to some variable by *matching* subjects in the two groups on that variable. The effect of this procedure is to create a dependency between the scores in the two groups, similar to the dependency between the two sets of measures in the pretest and posttest situation. This in turn means that the data from matched groups research situations is analyzed using the same techniques used in the pretest and posttest situation.

Here is an example of the matching procedure. Suppose that in answering our question about the relative effectiveness of the two methods of teaching Spanish, we wished to be absolutely sure that our two groups of subjects were comparable with respect to language ability. In this case, we might give all the subjects in the study a test of their language learning ability before deciding upon the composition of Groups A and B. We could then rank all the subjects according to language ability scores, from best to worst. From this ranking, we would select pairs of subjects, so that the top two students were in one pair, the next two students were in a second pair, and so on down the ranks. We would then place one member of each pair in Group A and the other member of each pair in Group B. We would decide which member of each pair to place in a given group by flipping a coin. Groups formed in this way are matched groups. Which of the two groups would become Group A and which Group B would be decided by the toss of a coin.

In this experiment, the two sets of measures of the dependent variable Spanish achievement are related to each other for the following reason. For each subject in Group A with a particular level of ability, there is a paired subject in Group B with a similar level of ability. To the extent that a subject's Spanish achievement depends on ability, subjects paired on ability will tend to have similar achievement scores. Thus, the two sets of measures of Spanish achievement cannot be considered completely separate or independent of each other. As you will see, the statistical procedures used with matched groups data are the same as the procedures used with pretest and posttest data.

THE CLASSIFICATION MATRIX

On the basis of the distinctions illustrated by these research questions, we have developed a matrix to assist you in determining the statistical techniques appropriate for a given problem. This matrix is presented in Table 1.4. We will present this matrix at the end of each chapter in the text to summarize the techniques presented.

Actually, we will use three matrices like Table 1.4, each corresponding to one of the three remaining parts of this text. One matrix will summarize the techniques for organizing and presenting data that are covered in Part 2. The second matrix will summarize the descriptive statistics presented in Part 3. The

28

TABLE 1.4
A Classification Matrix for Elementary Statistical Techniques

Level of Measurement of Variable(s) of Interest

	Categorical: Nominal or Ordered Categories	Rank-Order	Interval Scale (or Ratio-Scale)
Univariate Problems			
Correlational Problems			
Experimental and Pseudo-Experimental Problems: Independent Groups			
Dependent Measures: Pretest and Posttest or Matched Groups			

Type of Problem

Bivariate Problems

third matrix will summarize the inferential statistics covered in Part 4. Remember that the techniques presented in Parts 2 and 3 are descriptive techniques, but they are often used to describe data in inferential problems as well.

Each matrix will look like Table 1.4; the actual contents of the matrix at the end of any chapter will depend on which part that chapter is in. In order to use the matrix to classify a particular problem, you will need to answer several questions. These questions all relate to the various distinctions among statistical problems discussed in this chapter. The first question will direct you to either the two descriptive matrices or the inferential matrix. The next three questions will direct you to a specific cell in that matrix. We suggest you proceed in the following order:

1 Is the problem descriptive or inferential?
2 Is the problem univariate or bivariate?
3 If the problem is bivariate, into which of these three categories may it be placed?
   a Is it a correlational problem, in which a single group is measured on two variables at one point in time?
   b Is it an experimental or pseudo-experimental problem, in which two or more independent groups are formed and then compared on some variable of interest?
   c Is it a problem involving two sets of dependent measures, such as a pretest and posttest problem or a matched groups experimental problem?
4 What is the level of measurement of the variable(s) of interest?

If you can answer these four questions, you will be able to locate any problem that you may encounter in this text in one of the cells of our matrix. Of course, there are more complicated statistical problems that are beyond the scope of this text and that may not fit conveniently into the matrix. For example, there are correlational problems in which the two variables that have been measured have different levels of measurement. Also, there are problems in which researchers consider three or more variables in a single research design. Where we confront problems that are beyond the scope of this text, appropriate references will be provided. In an introductory course, however, we feel that the range of problems included in the classification matrix is sufficient. In order to give you practice in classifying statistical problems according to this system, we provide exercises in the student guide section following this chapter.

Key Terms

Be sure you can define each of these terms.

the scientific method
variable
measurement process
population of interest
categorical variable
nominal scale variable
ordered categorical variable
rank-order variable
ordinal scale variable
interval scale variable
ratio scale variable
reliability
validity
operational definition
internal validity
external validity
random selection
descriptive problem
inferential problem
parameter
statistic
univariate problem
bivariate problem
correlational problem
experimental problem
pseudo-experimental problem
independent variable
dependent variable
independent measures
dependent measures

1

Student Guide

Summary

In this first chapter, we have introduced basic considerations for doing empirical research. We have outlined some essential procedures, defined terms, and presented a logical framework that will be useful as a guide to the selection of statistical techniques appropriate to a variety of research situations.

We have noted the basic steps that must be taken when a research question is raised.

We must identify the variable or variables involved, choose our measurement process, and define the population on which the variables will be measured. Each variable must be defined operationally in a manner clear enough to allow our study to be replicated. We must take care that our measurements are reliable and valid. We must decide whether we will measure our entire population or use a sample to make inferences about the population from which it was drawn.

We also examined several dimensions of research problems. These dimensions can be used to classify a variety of problems and help us to choose the statistical techniques appropriate to analyzing them. One thing we must determine is whether we have a descriptive problem, in which we are describing a whole population, or an inferential problem, in which we use a sample to make inferences about the population from which it was drawn. We must also classify our problem as univariate or bivariate. If we have a bivariate problem, we must decide if it is correlational, in which we are merely observing the relationship between two variables; experimental or pseudo-experimental, in which we are comparing two or more independent groups on some variable of interest; or a problem with two sets of dependent measures, such as a matched groups or pretest/posttest problem. We must also determine the level of measurement of our variable or variables. We have to decide if we have a categorical variable, whose values are simply categories; a rank-order variable, whose values are ranks; or an interval or ratio scale variable, whose values are numerical scores. When we have classified our problem along all of these dimensions, we can locate it in our classification matrix (Table 1.4).

Review Questions

To review the concepts presented in Chapter 1, choose the best answer to each of the following questions.

1 Which one of the following situations best illustrates the concept of replicating an experiment?
_____a School Psychologist Diaz works at a reading center for illiterate adults. She comes across a recent article that shows that a new method of teaching reading to first-graders is more successful than a conventional method. She is very excited about this result and decides to run the same experiment on illiterate adults.
_____b School Psychologist Diaz reads an article that states that a new method of teaching reading to first-graders is more successful than an old method. She is very excited about this result and wants to see for herself that it is true. She selects the same number of first-grade subjects to study, and runs the same experiment that is mentioned in the article.

2 A student is given an arithmetic test. She is then shown her mistakes. After correcting all her mistakes she is then given the test again. The two measures of arithmetic achievement constitute

32

_____a two independent observations.
_____b two dependent observations.

3 A basketball coach has a team with 12 players. Each player is given a shirt with a number on the back. In this situation the 12 numbers could be considered
_____a an interval scale.
_____b an ordinal scale.
_____c a nominal scale.

4 An investigator is interested in studying the variable "scholastic aptitude" for high-school seniors in the United States. He has available scores from 5,782 seniors. This set of scores could be called
_____a the population of interest.
_____b a sample from the population of interest.

5 In research we are often concerned that our results can be generalized to a broad population. In other words, we are concerned with
_____a external validity.
_____b internal validity.

6 Which one of the following would be an inferential problem?
_____a giving a final examination to see how well your class learned the material taught
_____b trying to determine if your school system was reading at the national average by testing 30 students out of the 2,000 students in your system

7 Which of the following questions would be an example of an experimental problem?
_____a Do students with a special arithmetic program do better on an arithmetic achievement test than students who have the regular program?
_____b Do boys do better than girls on an arithmetic achievement test?

8 A diet clinic is interested in the nutritional state of all incoming clients. This situation would be classified as
_____a a univariate problem.
_____b a bivariate problem.

9 A study is designed to measure the effect of homework on arithmetic achievement. Students will be assigned to one of two groups. One group will get no homework. The other group will get 30 minutes of homework per day. At the end of the experiment, both groups will take the same arithmetic test. The independent variable of this study is
_____a arithmetic test score.
_____b homework group.

33

10 An investigator wants to know if elementary school children watch more television than teenagers. He goes to a school near his home and asks for volunteers. He gets 50 elementary children and 30 teenagers to participate in his study. The 80 children can be called

_____a two random samples.

_____b two samples.

11 A test of 25 equally weighted multiple-choice items is designed to measure statistics achievement. Scores on the instrument will be numbers from 0 to 25. A score of 0 indicates no correct answers; however, it does not indicate an absolute lack of statistics achievement. For this reason, we might reasonably classify this scale as

_____a a ratio scale.

_____b an interval scale.

12 An investigator measures the amount of weight a mother gains during her pregnancy and the weight of her baby. She wants to know if weight gain in pregnancy helps to explain the birth weight of the baby. This situation is an example of

_____a an experimental problem.

_____b a correlation problem.

13 A statistics achievement test is written in English but used with a group of students for whom English is a second language. The instructor might be concerned that the results of this test would measure ability to read English rather than statistics achievement. In this case, the instructor is concerned about

_____a reliability.

_____b validity.

Problems for Chapter 1: Set 1

 1 For each of the following examples, state the population of interest, name the variable, list the value set of the measurement scale, and tell the level of measurement of the scale.

   a classifying blood type in human adults
      Sample answer: The population of interest is human adults. Blood types are classified as A, B, O, and AB. These types are names of categories so we have a nominal scale.

   b dividing a class in statistics into those who score less than 50 on a final exam and those who score at least 50

   c measuring statistics aptitude in graduate students in psychology by using a 50-question, multiple-choice test

   d measuring degree of blondness in 20 blond students by comparing each student to every other student

   e classifying elementary schools as public or private

f measuring short-term memory by recording to the nearest second how long it takes an individual to recall designs that have been viewed a specified time earlier

g measuring handwriting ability in small children by a name-writing task that consists of seeing whether or not a child can write his or her name

h measuring handwriting ability in small children by use of a word list of 10 words that children are asked to copy. One point is given for each correct word

i measuring economic status of a family according to whether or not the father of the family is employed

j measuring economic status of a family according to family income

k measuring economic status of a family by using a list of jobs and professions that are ranked according to economic status

l measuring success of high schools according to the number of students who go on to college

m using caloric intake of American adults as a measure of daily diet

n identifying car color preference in American adults

2 Describe how intelligence could be measured using
a a nominal scale,
b an ordinal scale, and
c an interval scale.

3 Suppose you wish to measure the achievement of a group of students who have been "taught" a section of material. You design 10 questions on the material and weight them equally, giving a scale from 0 to 10. Each question does not measure the same amount of achievement. What are some defects of the scale? Explain.

4 Suppose you wish to investigate the variable "height" of 10 individuals, not all the same size.
a Assume you have no measuring instrument of any kind. Describe how you could measure the individuals. Write down the numbers you might record. What do these numbers tell you?
b Next, assume you have a small stick available. Now how would you measure the individuals? Write down numbers you might record. What do these numbers tell you?

Set 2

For Problems 1 through 10 in this set, answer the classificatory questions presented in this chapter:
A Is the problem descriptive or inferential?
B Is the problem a univariate or a bivariate problem?
C If the problem is bivariate, is it an experimental problem, a pseudo-experimental problem, or a problem with dependent measures?
D What is the level of measurement of each of the variables of interest?

35

1 What is the incidence of smokers in teenage America? A researcher surveys a large sample of teenagers in America and asks them how many cigarettes they smoke each day.

2 What is the favorite car color in America? A researcher asks all the car manufacturers in the United States which color they sell the most.

3 A sociologist is comparing college professors and plumbers on the variable of job satisfaction. He selects a random sample of plumbers and a random sample of college professors and measures each on the variable "job satisfaction," using an instrument that provides a numerical scale from 0 to 20 that reflects the extent to which a person is satisfied with his or her job.

4 To answer the question, "How did the town mayoralty election turn out?," the actual vote count for each of four candidates from all the voting precincts in the town is added up.

5 A well-known pollster samples 1,000 voters and asks them how they feel about the manner in which the president is doing his job. To do this, the pollster uses a rating scale from 1 to 4. The number 4 means doing a good job, the number 1 means doing a poor job. Each interviewee selects the number that is most representative of his or her feeling.

6 To study the effects of a training program that purports to teach leadership skills to politicians, an investigator selects 20 politicians and divides them into two groups in such a way that they are matched according to initial ability. One group is trained according to the program and the other is not. All subjects are measured at the end of the training period on a leadership scale that provides numerical scores.

7 A teacher wants to find out whether or not a reading program that she has developed is better than the program she is currently using, in order to decide whether to use the new program in the future. At the beginning of the year she randomly divides her class into two groups. The children are all at the same level of reading ability. One group uses the current program, and the other uses her new program. At the end of the term, all students will be given a reading test that measures level of reading ability.

8 An investigator is interested in the effect of smoking marijuana on intellectual functioning. He selects a random sample of people who are nonsmokers and who agree to participate in the study. At random, half of the subjects are selected to receive a placebo cigarette to smoke. The other half are given a marijuana cigarette to smoke. Fifteen minutes later all subjects are given a test that involves reaction to a certain stimuli. The time it takes the subject to react is recorded.

9 Are twelfth-graders more vocationally mature than ninth-graders? To study this question, an investigator develops an instrument that measures the ex-

tent to which a person is able to select a career that is compatible with his or her interests and abilities. The higher the score on the instrument, the more vocationally mature is the subject. To answer her question, the investigator selects a sample of ninth-grade students and measures them using her instrument. The same students are contacted in their senior year and remeasured. A comparison of the results is made.

10 To study whether or not ninth-graders are more or less apt to correctly select their career than are twelfth-graders, an investigator selects a sample of ninth-graders and asks them to select the career of their choice. Then, 15 years later, he recontacts the same group and finds out what career they are in. He records the number of subjects in the sample who are in the career they specified in ninth grade. He follows a similar procedure with an independent group of twelfth-graders. The two samples are compared with respect to the proportion of subjects actually involved in the same career they had chosen 15 years earlier.

11 For a study in community living, volunteers are requested from the undergraduate classes of a large university. From the 150 students who wish to participate, a random sample of 20 men and 20 women are chosen. The sample is then randomly divided into two "communities": a group of 10 men and 10 women who are assigned to Living Unit A and a group of 10 men and 10 women who are assigned to Living Unit B. The two plans represent different philosophies about community structure and management. The purpose of the study is to determine which of the two plans leads to more positive attitudes, acceptance, and liking among community members. Each group is to live in a community, governed by its plan, for one month. At the end of the month, all subjects will be interviewed individually and will be asked to answer a 20-item scale about each other member of his or her community. The total number of positive answers the subjects gives constitute the "attitude score." Thus, scores could range from 0 to 380.

  a   i   Classify the research situation on the basis of the dimensions outlined in this chapter.

       ii Name the variable(s) and define the population(s) of subjects.

      iii Which are the independent variables in the research and which are the dependent variables?

      iv For each variable, decide what is the level of measurement.

  b How would your answers change if the researcher had selected 20 men and 20 women randomly from the university population and randomly assigned them to communities?

  c How would your answers change if the researcher had randomly selected his subjects from a pool of all subjects who attend universities?

Introduction to Statistical Techniques

Organizing and Presenting Empirical Data

2

2

Organizing and Presenting Empirical Data: Univariate Problems

Once we have operationally defined our variable of interest and measured this variable within a population of interest, we need to decide how to organize and present the data in a way that will tell us something meaningful. Most often, raw data will consist of a list of observations that may be listed on yellow pads, filled in on survey questionnaires, or recorded in some other way. Such a set of raw data is most often laid out in the order in which the measurements were made. An example of such a set of raw data for a categorical variable is found in Table 1.1, in which we recorded the eye color of the 25 children in Mr. White's class. An example of such a set of raw data for an interval variable is found in Table 1.3, where we recorded the scores of 50 subjects on Dr. Brown's final examination in statistics. It should be clear from these tables that unless we have very few observations raw data of this type are difficult to interpret. In most instances, we begin to organize such raw data by constructing a frequency distribution. In addition, it is often useful to present a set of data graphically. In this chapter, we consider the use of frequency distributions and graphs for univariate problems. First, we consider categorical data, then interval data. We do not consider rank-order data here, for a rank ordering of a set of observations such as that presented in Table 1.2 cannot be presented any more succinctly.

FREQUENCY DISTRIBUTIONS AND GRAPHS FOR CATEGORICAL DATA

Frequency Distributions With Absolute Frequencies

The purpose of a *frequency distribution* is to tell us how many times each of the various values of a variable occurred within a given set of data. Thus, each *value* of the variable listed in the frequency distribution will have associated with it a *frequency,* which is the number of observations from the total set of all the observations that had that specific value.

For example, in Chapter 1 we defined the categorical variable "eye color" as having three values: brown, blue, and other. A frequency distribution on this nominal variable would tell us how many of our subjects had been classified as having brown eyes, how many had been classified as having blue eyes, and how many had been classified as belonging to the "other" category. To construct the frequency distribution of eye color for the set of data contained in Table 1.1, we need only list our three values and count up how many subjects were measured as having each of these values. If the frequency distribution is to be constructed by hand, the counting process is generally done using *tallies.*

First, we would make a list of all the values of the variable that occur. Then, we would go through the raw data subject by subject. For each subject, we would decide the eye color. Then, we would go to the list of values and place a tally next to this color. When we had checked and tallied each subject, we would have a rough work sheet like this:

Brown ~~HHT~~ ~~HHT~~ |||
Blue ~~HHT~~ |||
Other ||||

We would then count the tallies in each row to obtain the frequency corresponding to each value of eye color. The tallying procedure may seem unnecessarily tedious. You may be wondering why we simply do not run through the raw data quickly three times, counting up the browns, blues, and others. There are several reasons. First, not all data sets are as small as this one. We might well have many more subjects (observations) than we have here. Also, we may be dealing with a variable that has many more values than the variable "eye color" as we have defined it here. In such cases, tallying is a good safeguard against error. Of course, few researchers construct frequency distributions by hand today. It is much more convenient to use the computer to do this routine work of data organization.

If we were working by hand, however, after we had counted the tallies we would be ready to lay out the frequency distribution in its final form, as shown in Table 2.1. The advantage of the frequency distribution over the list of raw data can be seen by comparing the frequency distribution in Table 2.1 to the list of observations in Table 1.1. The frequency distribution is more economical. It takes up less space. Second, the frequency distribution conveys the general pattern of the data more quickly, for the reader can take in the three variable values and their associated frequencies at a single glance. It is much more difficult to get an idea of which eye color occurred most frequently by looking at the raw data. If you attempt to do so, you will probably discover yourself either counting or trying to estimate the number of observations of the most frequent

TABLE 2.1
Frequency Distribution of Eye Color
in Mr. White's First-Grade Class

Value (color)	Frequency of value
Brown	13
Blue	8
Other	4
Total	25

value or of all the values. If you are presented with a frequency distribution, this work has already been done for you.

Relative Frequency Distributions

Often, instead of presenting simply the absolute frequency of each value, we show the proportion or percentage of the total number of observations in the data set that fall at each of the possible values of the variable. The proportion of scores occurring at a given value is referred to as the *relative frequency* of that value. The word relative is used to indicate that we are interested in the frequency of scores for a particular value *relative to* the total number of scores in the data set. The idea of relative frequency is also clear in the following formula for obtaining the relative frequency of a given value:

$$\frac{\text{Relative frequency of a}}{\text{given score value}} = \frac{\text{The absolute frequency of that value}}{\text{Total number of observations}}.$$

This, of course, is just the formula for a proportion, written out in the terminology of the frequency distribution. Thus, the relative frequency of the value brown in the above frequency distribution would be 13 divided by 25 or .52. This tells us that 52 hundredths of our total sample was classified as having brown eyes. If we multiply this proportion by 100, we have expressed the same idea as a percentage. That is, 52 percent of our sample had brown eyes.

To construct a *relative frequency distribution,* we find the relative frequency for each value of the variable and table the relative frequencies opposite the values to which they correspond. The relative frequency distribution for our set of observations on eye color would appear as in Table 2.2. The relative frequencies may appear as either proportions adding up to 1.00 or as percentages adding up to 100.00 percent. If we choose percentages, we simply move the decimal point in the proportion two places to the right, which really means that we have multiplied each proportion by 100. When presenting a relative frequency dis-

42

tribution, one should always indicate the total number of observations upon which the relative frequencies are based. Thus, in Table 2.2 we have indicated that $N = 25$, which means that the proportions are based on 25 observations. Often, both the absolute and relative frequencies of the values will be given. We could have combined Tables 2.1 and 2.2 under the title "The Number and Percentage of Subjects of Three Different Eye Color Values in Mr. White's First-Grade Class."

Why is relative frequency used instead of or in addition to absolute frequency? The most basic reason is that we are accustomed to thinking in terms of proportions and percentages. This is especially important when we have large numbers of observations. As we noted with regard to the batting average question in Chapter 1, it makes more sense to us to know that a batter gets 333 hits out of every 1,000 times at bat than it does for us to know that he got exactly 3,210 hits in 9,630 times at bat. Similarly, it would make more sense to us to know that 30.31 percent of a group of 287 subjects were married than to know that 87 out of 287 subjects were married. Another important reason for using relative frequencies has to do with the comparison of several sets of scores. This will be considered in detail in Chapter 4.

Graphic Representations of Categorical Data

Even after we have organized a set of data into a frequency distribution or a relative frequency distribution, we may find it useful to present the data in a graphic form. Often a graphic representation of a distribution will convey to the reader a general notion of how a set of observations is distributed over the range of possible values more quickly than a frequency distribution. Also, a graphic representation generally has more impact than a frequency distribution alone.

There are many forms of graphic representation of data. Two of the most common forms employed with categorical data are *bar graphs* and *circle graphs*. In a bar graph, we draw a bar for each value of the variable. These bars stand

TABLE 2.2
Relative Frequency Distribution of Eye Color
in Mr. White's First-Grade Class ($N = 25$)

Value (color)	Proportion
Brown	.52
Blue	.32
Other	.16
Total	1.00

FIGURE 2.1
The Distribution of Eye Color
for 25 First-Graders in Mr. White's Class

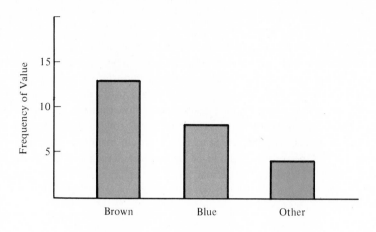

along the horizontal axis of the graph. The height of each bar, measured along the vertical axis, represents the frequency of the value that corresponds to that bar. A bar graph representing the distribution of the nominal scale variable "eye color" might appear as in Figure 2.1. Because the values of nominal scale data are not quantitatively related to each other, the bars need not appear in any special order. If we were to construct a bar graph representing a set of ordered categorical data, we would arrange the bars on the graph in the same order as the categories they represent.

When we construct a bar graph, we will decide on the width of the bars according to our judgment on the most effective way to convey our point. In general, all the bars will have the same width, but we are free to make the bars wide or narrow. Our own sense of aesthetics will come into play. Occasionally, such considerations lead to the use of a special form of bar graph called a *pictograph,* in which the frequencies of the various values of a variable are represented by small pictures.

An extremely useful technique for presenting relative frequency distributions for categorical data is the familiar circle graph. In this form of graphic presentation, all of the observations in our distribution are represented by the entire area contained in a circle. This circle is then divided up into as many segments as there are values or categories of the variable in such a way that each segment represents the proportion of the total number of observations that fall

44

into a given category. The circle graph representing our distribution of eye colors is shown in Figure 2.2. The circle graph is like a bar graph of a relative frequency distribution in several ways. First, the area of one segment in a circle graph and the area of one bar in a bar graph may be viewed similarly. We have already seen that the area of any segment of the circle graph represents the proportion of the total of all the observations falling into a particular category. We can look at the area of a bar in the bar graph in the same way. Because the height of each bar in the bar graph represents the proportion of observations falling into a given category, and because the widths of the bars are equal, we can consider the area of each bar in the bar graph as representing the proportion of the total falling into a particular category.

A second related point concerns the total of all the areas. The total of *all the areas of all the segments* of the circle graph represents the sum of the proportion of observations falling into all of the categories of the variable, or 100 percent of our observations. Similarly, we can think of the total of *all the areas of all the bars* of the bar graph as representing 100 percent of all the observations.

The difference between the circle graph and the bar graph of a relative frequency distribution is that the division of a single circle into a series of segments makes it much easier for the reader to see that what we are doing is taking 100 percent of the frequencies and showing the proportions of this total that fall into each of the different categories.

FIGURE 2.2
Circle Graph of the Proportion of the 25 First-Graders
in Mr. White's Class Having Each of Three Eye Colors

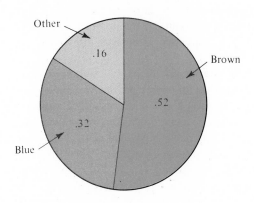

FREQUENCY DISTRIBUTIONS
AND GRAPHS
FOR INTERVAL DATA

Discrete Interval Data and Continuous Interval Data

In Chapter 1 we considered the nature of variables that have numerical scores as values. We pointed out that when we obtain a numerical score for a subject, we are locating that subject along a number line that is, at least in theory, laid off in equal units or intervals. This is why data in the form of numerical scores are often called equal interval data. At this point we need to distinguish between two different kinds of interval variables: *discrete interval variables* and *continuous interval variables.* Discrete interval variables and continuous interval variables both possess all the characteristics of interval scale data presented in Chapter 1. That is, the values of both discrete interval variables and continuous interval variables are numerical scores and these scores form a scale that is laid off in equal units or intervals. However, there are differences between the way we obtain the numerical scores for discrete variables and the way we obtain the numerical scores for continuous variables. Also, there are differences in the techniques used to organize and present the two types of variables.

Discrete Interval Variables. A discrete interval variable is distinguished from a continuous interval variable in that measures on a discrete interval variable may be obtained directly by counting *how many* of something there are. An example of a discrete interval variable is "the number of hearts in a five-card poker hand." If we were taking observations on this variable, we would deal a number of poker hands and for each hand dealt we would *count* and record the number of hearts. Because there are five cards in a poker hand, this number could be any whole number from zero to five. It would have to be a whole number, for each card would either be a heart or not a heart. If it is a heart, we count it, and we add one to the number we record as our observation. If it is not a heart, we do not count it, and we add nothing to the number we record. It is impossible to obtain a fraction of a heart. Scores on true discrete interval variables always increase in whole units.

Another discrete interval variable is "the number of children in the home." If we were obtaining a set of scores on this variable for the homes in a given community, we could proceed in several ways. One way would be to go to each home at a time when children are most likely to be there. Then, we could count the children directly. More likely, we would contact a member of the household and ask how many children were living there. Then, this member of the household would be doing the counting for us. Either way, however, the notion of actual counting is present. Here again, we would never obtain a fraction of a child. There may be one child, two, three, or more, but there can never be a

46

score for a given household of 3.5 children. Scores on discrete interval variables may increase only in discrete steps of whole units.

A third example of a discrete interval variable is "the number of students registered in introductory psychology in the university." If we were recording observations on this variable for a number of different universities, we would probably obtain class rosters and count the names that appeared. In some schools, the number of names might be quite large. However, the number would always be a whole number. It would not be possible to have a fraction of a student registered for a class. Other examples of discrete interval variables are "the number of completed passes in a football game" and "the number of times out of 25 rolls that a six-sided die lands with a single dot facing up."

Continuous Interval Variables.   In contrast to discrete interval variables, measures on a continuous interval variable may not be obtained by counting the number of whole units present. Rather, measures on a continuous interval variable must be obtained by estimating as accurately as possible *how much* of something there is. An example of a continuous interval variable is "body weight." If we wanted to obtain scores on this variable for a number of different subjects, we would have to use a mechanical scale of some sort. In reading the scale, we would find that our set of scores would contain many fractions. How precisely we could measure each subject's weight would depend on the quality of our scale. However, even if we had a very precise scale, it would always be possible for us to imagine the possibility of obtaining more accurate measures. Notice the difference between discrete and continuous variables in this respect. Whereas the numerical scores on discrete variables jump in neat steps, a whole unit at a time, the numerical scores on continuous variables may increase in imperceptibly small intervals. For this reason, when we are considering a continuous interval variable, we will generally specify how precise our measurements are. For example, we might specify "weight to the nearest kilogram," "age to the nearest month," "waist measurement to the nearest centimeter," or "reaction time to the nearest millisecond."

The Case of Discrete Measures on a Continuous Variable.   In differentiating between discrete and continuous interval variables, it is important to realize that some measuring instruments will yield scores that are discrete, even though the variable being measured is continuous. A common example of an instrument that yields a set of discrete measures on a continuous variable is a digital clock. Suppose we are using such a clock to measure the time it takes a subject to finish a work task. Suppose further that the clock gives us times in minutes. Every 60 seconds the clock switches and the score in complete minutes is increased by one. If one subject finishes immediately after the clock has switched and another finishes just before the next switch, both will have the same score. Yet, it is clear that they did not take the same amount of time to finish. It is

47

simply that their actual scores have been given in *complete minutes,* so that all the time up to but not including the next whole minute is not counted. If we needed to have more precise measures, we could obtain a more precise timing device, one that showed elapsed time in complete seconds. This new instrument would differentiate between the two subjects just described. However, even with this new timer, it is still possible to conceive of two subjects having the same score in whole seconds who did not finish at precisely the same instant. Because there are some very accurate timepieces available, we can often obtain very accurate measures on continuous variables involving time. The same is true of many physical attributes, such as weight or length. However, no matter how precise our measuring instrument is, it is always possible to conceptualize the underlying variable as more finely broken down.

We frequently have the same situation when we use paper-and-pencil tests as a measure of some psychological ability or educational achievement. Consider, for example, the case of Dr. Brown's final exam in statistics cited in Chapter 1 as an example of measurement with numerical scores (see Table 1.3). We noted that this examination consisted of 100 multiple-choice questions. The answers to each of these questions would be either right or wrong. Each correct answer would result in the addition of one whole point to the student's score. Each student would obtain a final score that could be 0, 1, 2, or any whole number up to 100. If we were recording student grades, we would count the number of correct answers to obtain a score for each student. Thus, the test scores themselves fit the definition of discrete interval data. Our final set of scores would contain no fractions.

However, we must ask ourselves what it is that Dr. Brown is attempting to find out when he gives this test. Of course, one thing he will most certainly find out is how many questions each of his students got right. But is this really his fundamental concern? What Dr. Brown is really trying to do is to measure his students' achievement in the area of statistics. To accomplish this job of measurement, he has devised a measuring instrument that consists of 100 questions that students may answer correctly or incorrectly. If he has constructed his test carefully, there will be a high degree of correspondence between students' scores on the test and the actual amount of statistics that they have learned. However, students' statistics achievement is conceptually different from their test scores per se. We would not conceptualize the variable statistics achievement as an attribute that is neatly divided into 100 little packages. Underlying the discrete set of measures yielded by Dr. Brown's test is a variable, like time or weight, that we can conceive of as being continuous.

Thus, a particular student can be thought of as having a "true score" on the variable statistics achievement that might well fall between two points on the scale. Moreover, if two people answer the same number of questions correctly, they may or may not have precisely identical true scores. All we know for sure is that their true statistics achievement is close enough so that the measuring instrument that Dr. Brown created cannot distinguish between them. The situation here is like that of the two subjects in the work task example who got the same time in whole minutes. In each case, we are measuring a continuous variable in discrete units. However, if you plan to use paper-and-pencil tests designed to measure psychological traits or educational achievement, you should be aware that these measuring instruments are crude in comparison to scales, clocks, and metersticks.

The differing nature of discrete interval variables and continuous interval variables are reflected in the techniques we use when we organize and present sets of scores on the two types of variables. We now proceed to look at these techniques, beginning with those for discrete interval data.

Frequency Distributions for Discrete Interval Variables

What we said regarding frequency distributions for categorical data applies to frequency distributions for discrete interval data as well. The frequency distribution still tells us how many times each of the various values of a given variable occurred within a given set of data. Moreover, the frequency distribution is constructed in the same way. If we wanted to construct the frequency distribution for the variable "number of hearts in a five-card poker hand," we would note that a given hand could contain anywhere from no hearts to five hearts. We could then list the *six* possible score values from zero to five inclusive. Then, we would proceed exactly as we did in the case of the categorical variable "eye color." We would go down our list of observations of five-card poker hands, one by one, and for each observation we would place a tally next to the appropriate score value. When we had run down the entire list of observations, we would count the tallies to obtain the frequency with which each value occurred. The frequency distribution could be presented as a relative frequency distribution as well, if we chose to do so. A frequency distribution based on a sample of 200 poker hands is presented in Table 2.3. Both absolute and relative frequencies are included in the table.

Combining Score Values.   If we wanted to construct a frequency distribution for the variable "number of children in the household," the procedure would be very similar. However, there would probably be one small difference with respect to the way in which we listed the score values to be included in the distribution. Remember that when we are talking about the number of hearts in a five-card poker hand, the maximum score value for the variable is five. When we talk about the number of children in the household, however, there is no such fixed upper limit on the values that the variable can take on. Relatively few homes would have more than, say, 6 or 7 children, although there would probably be some homes in which there were 8, or 10, or even more. In laying out the score values we wish to include in our frequency distribution, therefore, it would probably not be very useful to list every number from zero to the largest number of children actually observed in any one home. If we did, and if the largest num-

49

TABLE 2.3

Frequency Distribution and Relative Frequency Distribution
for the Number of Hearts in a Five-Card Poker Hand ($N = 200$ Hands)

Values for number of hearts in a hand	Frequency of value	Relative frequency of value
0	44	.22
1	80	.40
2	54	.27
3	16	.08
4	4	.02
5	2	.01
Total	200	1.00

ber were something like 14, then most of the score values from 6 up would have a low frequency. Some of the larger values, such as 12 and 13, might even have a frequency of zero. For this reason, we would probably select a reasonable number, like 6, as a good point at which to locate the uppermost score value in the distribution. We would label this last value "6 or more" and would include in this category all the homes with 6 or more children. An example of a frequency distribution for number of children in the household is presented in Table 2.4.

In some ways, this procedure of combining the higher values is similar to the way in which we used the "other" category in our frequency distribution of the nominal scale variable "eye color." It is similar because we are "lumping" together several different scores under a single score value heading. In the case of both the number of children and the eye color distributions, this lumping together has the effect of eliminating from the list those score values that have a low frequency associated with them. However, there are important differences between the procedures. In the number of children example, we are dealing with an interval variable and we are lumping together scores at one end of the interval scale. In a sense, this destroys some of the properties of the scale. For example, we cannot say that those homes with "6 or more children" have 2 more children than those homes with 4 children. Only those homes with exactly 6 children have 2 more children than homes with 4 children, and when we combine all the homes with 6 or more children in a single category, we lose track of how many of these homes have exactly 6 and how many have more. This effect of combining values is especially important when we consider methods for the graphic representation of interval data.

This concern over the effect of combining data in this manner does not apply to nominal data. Consider the way in which we grouped our observations

Organizing and Presenting Empirical Data

of eye color into brown, blue, and other categories. We cannot say that any one of these categories is lower on a scale than any other category. The scale does not have high and low ends. Therefore, combining several different colors into the category "other" has no effect on the properties of the nominal scale. Those observations in the "other" category are still different from those observations in the blue or brown category, just as they were before the combining took place. Thus, the grouping has no effect on the properties of the scale.

Grouped Frequency Distributions.   Another type of grouping is employed when the variable we are interested in has a large number of different score values. For example, if we wanted to construct a frequency distribution for the discrete interval variable "number of students registered in introductory psychology in the university," we would most likely find that the number of students taking introductory psychology at a given institution at a given time would range from 15 or 20 students at a small college to several hundred or even a thousand at a large university. Because of the wide range of score values associated with this variable, it would not make any sense to include every single value in our list of score values. If we did, the list could be a thousand score values long, and there would be many score values in the list that would contain no scores or only a few scores. To obtain a manageable and meaningful frequency distribution on this variable, it would be necessary to use a *grouped frequency distribution.* In a grouped frequency distribution, we do not count the number of observations falling at each and every discrete score value. Rather, we count the observations falling into score value intervals, each of which contains a number of discrete score values. In this case, we might want to group the score values from 0 to 49 in the first interval, so that all the schools with 0 to 49 students taking introduc-

TABLE 2.4
Frequency Distribution and Relative Frequency Distribution
of the Number of Children in the Household ($N = 90$ Families)

Number of children in the household	Frequency of value	Relative frequency of value
0	9	.100
1	15	.167
2	20	.222
3	25	.278
4	10	.111
5	6	.067
6 or more	5	.055
Totals	90	1.000

TABLE 2.5

Grouped Frequency Distribution of Number of Students
Enrolled in Introductory Psychology
in 900 Institutions of Higher Education

Number of students enrolled	Number of institutions having this many students enrolled	Percentage of institutions out of 900 having this many enrolled
0– 49	63	7.00
50– 99	205	22.78
100–149	213	23.67
150–199	175	19.44
200–249	104	11.56
250–299	72	8.00
300–349	29	3.22
350–399	23	2.56
400–449	10	1.11
450–499	2	0.22
500–549	0	0.00
550–599	2	0.22
600–649	1	0.11
650–699	1	0.11
	900	100.00

tory psychology would be included here. We could then group all the score values from 50 to 99 in the second interval, and proceed upward in equal steps of 50 until we had included the highest observation recorded. These score intervals are often referred to as classes or class intervals. The lowest score value in each interval is often called the lower class limit of that interval and the highest score value in the interval the upper class limit of the interval.

In grouped frequency distributions, it is a common practice to make each of the class intervals contain the same number of discrete score values. This preserves the equal interval property of the scale. If we had recorded the number of students enrolled in introductory psychology in a sample of 900 schools, then the grouped frequency distribution might appear as in Table 2.5.

Grouping data in this fashion results in some loss of information. When we look at the grouped frequency distribution, we no longer know exactly how many observations there were for each individual score value. That is, we can no longer make a statement like "7 schools had exactly 59 students registered." We are willing to sacrifice this specific information to get a better view of the broader picture.

There may be cases in which we wish to present a frequency distribution in which we have not only grouped scores into intervals throughout the range of the distribution, but also included an initial or final class interval that is open-ended. For example, suppose one school among the 900 surveyed had 1,067 students taking introductory psychology, but that no other school had more than 475 students. We certainly would not want to continue to list score value intervals 50 units wide from 450–499 all the way up through the interval 1050–1099. If we did, we would be listing many intervals that contained no scores at all. Instead, we might choose to label our highest category "400 students or more." This would serve the same function as the "6 or more" category in the number of children example. That is, it would avoid having a large number of intervals containing either a low frequency or a frequency of zero. The grouped frequency distribution would then appear as in Table 2.6. If we do decide to include an initial or final interval that is different in length from the other intervals, however, we do sacrifice our equal interval scale to some extent. This is a serious problem when we attempt to represent such distributions graphically.

Although each of the frequency distributions presented in this section is a distribution on a discrete interval variable, the distributions could not be constructed in precisely the same manner. The decision as to whether a frequency distribution should show every single score value individually or whether values should be grouped depends on the particular set of data involved. There are no hard and fast rules for determining when a grouped distribution should be used. Because the purpose of a frequency distribution is to present a set of data in a form that is economical with regard to space and meaningful to the reader, it is desirable to keep the number of intervals in a frequency distribution rather

---

TABLE 2.6
Grouped Frequency Distribution of the Number of Students
Enrolled in Introductory Psychology in 900 Institutions of Higher Education
by Frequency and Percentage

Number of students enrolled	Number of institutions having this many students enrolled	Percentage of institutions out of 900 having his many enrolled
0– 49	63	7.00
50– 99	205	22.78
100–149	213	23.67
150–199	175	19.44
200–249	104	11.56
250–299	72	8.00
300–349	29	3.22
350–399	23	2.56
400 or More	16	1.78
	900	100.00

small. Some statisticians feel that 12 to 15 intervals is a reasonable number. Others feel that this is too large a number for the eye to take in at a glance. In the final analysis, the decision to group or not to group—as well as the decision as to how many intervals to create—depends on the variable involved and the amount of detail the researcher requires. These in turn depend on the nature of the research and the sophistication of the readers for whom the distribution is intended.

Frequency Distributions for Continuous Interval Variables

Frequency distributions for continuous interval variables often appear identical to frequency distributions for discrete interval variables. The questions of when and how to group scores, already considered with reference to discrete interval data, also apply to continuous interval data. Thus, the frequency distribution of the weights of a sample of 200 adult American males might appear as in Table 2.7. Notice that we have made the frequency distribution economical by employing weight categories or classes that contain a range of 10 kilograms. In this respect, the distribution is grouped, like the distribution for the variable "number of students enrolled in introductory psychology."

Real Class Limits.   Although frequency distributions for discrete interval data and frequency distributions for continuous interval data often appear similar, we must remember that there is a difference between the two types of data. This difference must be considered when we construct and interpret frequency distributions. Consider the first score interval in Table 2.5. Because the number

---

TABLE 2.7
Number and Percentage of 200 Adult American Males
Falling into Different Weight Categories

Weight to nearest kilogram	Number of males	Percentage
55– 64	12	6.0
65– 74	46	23.0
75– 84	51	25.5
85– 94	40	20.0
95–104	30	15.0
105–114	16	8.0
115–124	4	2.0
125–134	1	.5
Total	200	100.0

Organizing and Presenting Empirical Data

of students enrolled in introductory psychology is a discrete interval variable, we know that only schools that have between 0 students (the lower class limit) and 49 students (the upper class limit) are included in this interval. Now consider the first score interval in Table 2.7, the frequency distribution for the continuous variable weight of American males. Can we say that the 12 men in this category weigh at least 55 kilos and not more than 64 kilos? A quick glance at the frequency distribution might cause one to answer "yes." But the correct answer is "no." It is true that the first interval is labeled 55 kilos to 64 kilos. However, each observation on this variable has been made *to the nearest kilo.* Thus, it is possible that one or more of the 12 observations falling into the first weight category was actually measured as 54.92 kilos or 54.50 kilos and that it was rounded up to make it 55 kilos to the nearest whole kilo. It is also possible that one or more of the 12 observations was 64.17 kilos or 64.49 kilos and that the measurement was rounded down to make it 64 kilos to the nearest whole kilo.

For this reason, when we deal with continuous interval data, class limits like 55 kilos and 64 kilos are referred to as *apparent class limits.* These apparent class limits are not necessarily the same as the *real class limits,* which are the actual points on the measurement scale that divide one score value from another. In the case of the weight of adult American males, the real class limits are 54.5 kilos, 64.5 kilos, and so on, up to 134.5 kilos.

The key to constructing the frequency distribution correctly is to be aware of the way in which the variable has been measured. When we are dealing with continuous interval variables, we generally have a choice regarding the measurement scale we will use. As we indicated in our discussion of operational definitions in Chapter 1, the variable "age" may be defined in various ways. Given a specific group of subjects, the frequency distribution of the variable "age" will be very different when age is defined as "to the nearest year" than it would be when age is defined as "at the time of last birthday." Consider the set of 20 observations presented in Table 2.8. These observations are accurate to the nearest day.

If we look down this list of ages, we see that the youngest child is Sue, who is 7 years, 10 months, and 3 days old. The oldest child is Les, who is 11 years, 8 months, and 6 days old. You should recognize that if we use the measurement scale of "age to the nearest year," we will begin our frequency distribution with a score interval called "8," because Sue's age, rounded up, is 8 years. However, if we use the measurement scale of "age at last birthday," our frequency distribution will begin with the score interval "7," because Sue was 7 on her last birthday. Before reading on, take a minute to think about Les' age. What would the highest score interval be if we used the "age to the nearest year" scale? What would it be using the "age at last birthday" scale? In Table 2.9, we have presented the two frequency distributions. Notice how different the two distributions look. They begin and end at different points and there are different numbers of children in categories that have the same name. For example, there

TABLE 2.8
A Set of 20 Observations of Age to the Nearest Day

Child	Age			Child	Age		
Sara	9 years,	3 months,	1 day	Sue	7 years,	10 months,	3 days
Bill	10 years,	6 months,	0 days	Roberta	8 years,	3 months,	4 days
Bob	8 years,	9 months,	17 days	Pablo	8 years,	7 months,	1 day
Jane	9 years,	0 months,	9 days	Tom	9 years,	1 month,	19 days
Sally	7 years,	11 months,	16 days	Les	11 years,	8 months,	6 days
Wendy	8 years,	5 months,	2 days	Peter	9 years,	5 months,	26 days
Toby	8 years,	7 months,	1 day	Maria	8 years,	3 months,	12 days
Mark	8 years,	9 months,	29 days	Ann	9 years,	7 months,	12 days
David	9 years,	1 month,	14 days	Sandra	7 years,	10 months,	21 days
Jose	9 years,	6 months,	12 days	Chris	10 years,	5 months,	4 days

seem to be 6 8-year-olds in the "nearest year" frequency distribution, but 7 8-year-olds in the "last birthday" distribution. Yet the children are the same. Only the scales are different. In the first frequency distribution, we reported "age to the nearest year." In this distribution, the category labeled 8 included all children from 7 years, 6 months, and 0 days up to 8 years, 5 months, and 29 days, i.e., from 7 1/2 years up to but not including 8 1/2 years. In the second frequency distribution, we reported "age at last birthday." In this new distribution, the category labeled 8 included all children from 8 years, 0 months, and 0 days up to 8 years, 11 months, and 29 days, i.e., all children who had reached their eighth birthday but had not yet reached their ninth birthday. Both distributions are correct as defined. As indicated in our discussion of operational definitions in Chapter 1, it is up to the researcher to indicate clearly how the variable has been measured and reported. It is up to the reader to take note of this information in interpreting the frequency distribution.

Frequency Distributions Formed from a Set of Discrete Measures on a Continuous Variable. In our age example, we had a set of continuous data measured all the way down to the nearest day. We looked at two ways of constructing frequency distributions. One of these, the "age at last birthday," involved truncating ages so that only complete years were counted. At this point, you should see that through this process of rounding off or truncating scores we are grouping scores that are really different from one another under identical headings. In a sense, we are performing a task that is done automatically for us when our measuring instrument yields discrete observations on a continuous variable. If you will recall the example of the digital clock that presents measurements in complete minutes, you will see that this clock is truncating the variable time in a manner similar to what we did ourselves in constructing the frequency distribu-

Organizing and Presenting Empirical Data

tion of "age at last birthday." Of course, the clock is dealing with smaller units of time, but the principle is the same. The only difference is that in our age example we had measurements more accurate than we needed, so we did the truncating. In the timing example, we are given a set of measures that are discrete. We know in theory that the variable is continuous and that the measures have been truncated, but we never see anything except the discrete reading.

It is important to consider this aspect of discrete measurements on continuous variables when we construct frequency distributions for such variables. Just as we had to define the scale used in presenting our frequency distributions for the ages of 20 children, we have to define the scale used when we construct a frequency distribution of a continuous variable for which we have a set of discrete scores. In the case of a digital timing device, we would need to indicate that time was expressed in complete minutes or complete seconds. In the case of the examination grades for Dr. Brown's statistics class, we could label our scale "number of correct responses." However, in so labeling the scale, we need to be aware that the underlying variable, statistics achievement, is continuous. A grouped frequency distribution for the statistics test data is presented in Table 2.10. For the moment, ignore the last two columns of this table. These will be considered shortly. Consider the first column of the table. What are the apparent lower and upper class limits of the first score interval? What are the real limits? If you can conceptualize the underlying continuous variable "statistics achievement," you will agree that even though it is impossible to get a fraction of a question right, it *is* possible for an individual to have a *true score* on statistics achievement that is not a whole number. In fact, it is likely. We have scores in whole numbers only because our measuring instrument, the examination, is constructed that way. Thus, a person with a score of 78 may be thought of as having a true score of anywhere from 77.5 up to, but not including, 78.5. For this reason, the real limits of the first score interval are 53.5 and 58.5.

TABLE 2.9
A Comparison of Two Frequency Distributions
Based on the Same Set of Data

Age of 20 children to nearest year			Age of 20 children at last birthday		
Age	Number	Percentage	Age	Number	Percentage
8	6	30	7	3	15
9	9	45	8	7	35
10	3	15	9	7	35
11	1	5	10	2	10
12	1	5	11	1	5
Total	20	100	Total	20	100

TABLE 2.10
Grouped Frequency Distribution of Grades
on Dr. Brown's Final Exam in Statistics for 50 Subjects

Exam scores in 5-point intervals	Frequency of score interval	Relative frequency of score interval	Cumulative frequency	Cumulative percentage
54–58	2	.04	2	4
59–63	4	.08	6	12
64–68	3	.06	9	18
69–73	6	.12	15	30
74–78	8	.16	23	46
79–83	11	.22	34	68
84–88	9	.18	43	86
89–93	5	.10	48	96
94–98	2	.04	50	100
Totals	50	1.00		

Cumulative Frequency Distributions for Interval Data

Whether we are dealing with discrete interval data or continuous interval data, it is often useful to know not just the frequency or percentage of observations falling into each of the various score value intervals, but also the *cumulative frequency* or *cumulative percentage,* the number or percentage of observations falling *at or below* a particular interval. For example, in the case of the distribution of Dr. Brown's statistics exam scores, it would be useful for all students in the class to know how many students scored at or below their score. This would give them an idea of where they stood in relation to the rest of the class. For this purpose, we use a *cumulative frequency distribution.* In the cumulative frequency distribution, next to each score interval we table not the number of observations falling into that interval, but the number falling into that interval and all lower intervals. This is illustrated in Table 2.10 in the column labeled "cumulative frequency." The number 6 tabled adjacent to the score value interval 59–63 represents the 4 scores actually falling in this second interval plus the 2 scores that fell in the first score interval. The remainder of the entries in this column may be obtained similarly, by adding the number of observations in each interval to the cumulative frequency for the previous interval. We may also be interested in cumulative percentage, which tell us the percentage of all the scores in a distribution that fall at or below a given score interval. The last column of Table 2.10, labeled "cumulative percentage," illustrates a *cumulative percentage distribution.* Each entry in this column is obtained by dividing the cumulative frequency for a given score interval by the total number of observations and then multiplying by 100 to turn this proportion into a percentage:

Organizing and Presenting Empirical Data

$$\text{Cumulative percentage for an interval} = \frac{\text{Cumulative frequency for the interval}}{\text{Total number of observations}} \times 100$$

In Table 2.10, for example, we see that 46 percent of the subjects taking Dr. Brown's exam scored at or below the score interval 74–78. The cumulative relative frequency distribution is particularly useful when we are interested in determining the percentile rank of a given score. This will be discussed in Chapter 4.

Graphing Discrete Interval Data

When we are dealing with a set of discrete interval data that contains a relatively small range of values, we typically construct ungrouped frequency distributions, as in the frequency distribution for the number of hearts in a five-card poker hand, presented in Table 2.3. For such frequency distributions, the *discrete graph* is the appropriate device for representing the frequency distribution graphically and the *step function* is the appropriate device for representing the cumulative frequency distribution graphically.

The Discrete Graph.   The discrete graph representing our distribution of the number of hearts in a five-card poker hand is presented in Figure 2.3. Take a moment to consider how the discrete graph is constructed. Along the horizontal axis of the discrete graph are listed the various values of the discrete interval variable. The distances between any two points on this horizontal scale are all equal, reflecting the fact that we are dealing with an interval variable. The vertical axis of the discrete graph represents the frequency with which each of these values occurred. The graph is constructed so that above each value of the variable there is a vertical line segment. The length of this line segment is proportional to the frequency with which that particular value occurred. For example, above the score value 0 there is a line segment that rises until it is even with the number 44 on the vertical axis, because the value 0 occurred 44 times in our sample of 200 poker hands. Note that the line segments above each value are entirely separate from each other. This is to emphasize that the number of hearts in a five-card poker hand is a discrete variable. The graph is telling us that 44 observations fell at exactly 0, 80 observations fell at exactly 1, and so on. No observations fell between 0 and 1 or between any two adjacent points on the scale. None could fall in between, for these are discrete data. As indicated before, there cannot be 3 1/2 hearts in a poker hand. Thus, there can be no line segments in the discrete graph except over whole numbers. When we present a data set graphically, we should always attempt to employ a graphic technique that does not distort the nature of the variable under consideration.

The Step Function.   The step function is used to represent discrete interval data when we are concerned with a cumulative frequency distribution of the

FIGURE 2.3
Discrete Graph for the Number of Hearts
in a Five-Card Poker Hand for 200 Different Hands

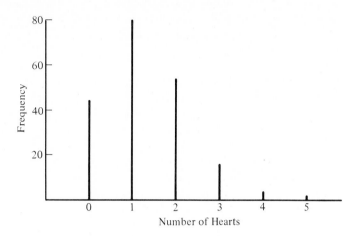

data. The step function corresponding to our distribution of the number of hearts in a five-card poker hand is presented in Figure 2.4. The step function shows us how many observations in the data set fall at or below each score value. Referring to Table 2.3, you will note that out of the 200 poker hands in our sample, 44 contained 0 hearts. Therefore, we draw a line segment 44 units above the horizontal axis, beginning at the score value 0 on the horizontal scale and extending horizontally up to but not including the score value 1. Thus, the step function also reflects the discrete nature of the variable. Each line segment is flat because no osbservations occur anywhere except when we reach whole numbers. Thus, 44 observations have occurred at 0, and the same 44 observations have occurred at 0.99. When we reach exactly 1, however, we add the number of hands in which there was 1 heart. Therefore, the graph jumps up a step, to 124. The graph then remains flat until we reach the value for exactly 2 hearts, when the next jump occurs. The final step in the function always represents the total number of observations in the data set. It extends horizontally from the highest score value in the data set (in this case, 5) to a point just short of where the next score value would have been, had more values been represented in the distribution. (Here the next score value would have been 6.)

The discrete graph and step function are appropriate techniques only for discrete data that have not been grouped in any way. The question of what to do with discrete data that have been grouped is considered in a later section.

Graphing Continuous Interval Data

For continuous interval data, the *histogram,* the *frequency polygon,* and the *cumulative frequency graph* are the most common forms of graphic representation. The histogram and the frequency polygon are analogous to the bar graph for categorical data and to the discrete graph for discrete interval data. It is a graphic representation of the frequency distribution. The cumulative frequency graph is the analogue of the step function for discrete interval data. It is a graphic representation of the cumulative frequency distribution.

The Histogram. The frequency distribution for the continuous interval variable "weight of 200 American males" was presented in Table 2.7. The histogram

---

FIGURE 2.4
Step Function Representing the Cumulative Frequency Distribution
of the Number of Hearts in a Five-Card Poker Hand ($N = 200$ Hands)

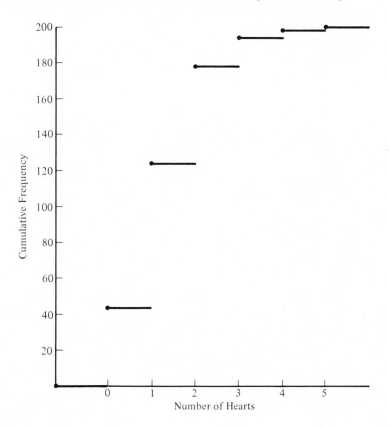

Univariate Problems

used to represent this distribution graphically is presented in Figure 2.5. You can see from the histogram that the various score intervals of the frequency distribution are laid out along the horizontal axis. The frequency of observations falling into each of these intervals can be read from the vertical axis. Thus, we see that 12 men fell in the weight class between 54.5 kilos and 64.5 kilograms, 46 men fell in the weight class between 64.5 kilograms and 74.5 kilograms, and so on.

The histogram differs from the bar graph and the discrete graph in the same way that continuous interval data differ from categorical data or discrete interval data. Recall the categorical variable "eye color." As we defined that variable, there were exactly three values: brown, blue, and other. Every subject placed in the "brown" category has *exactly* the same score on the variable "eye color" as everyone else in the category. It is true that there are different shades of brown, but the variable "eye color" does not take these into account. There are three and only three values for this variable, and all observations in a single category have exactly the same value. Now consider the discrete interval variable number of hearts in a five-card poker hand. It has exactly 6 values, the whole numbers from zero through five inclusive. Each hand with four hearts has exactly the same score as every other hand with this score value. This is *not* the case with the continuous variable "weight." For the variable "weight," there are an infinite number of possible values contained in each weight interval. Even if we had constructed our frequency distribution so that each weight interval covered a range of only one half of a kilogram, there would still be an infinite number of possible weights in each interval. No matter how precisely we measure weight, it is always conceivable to measure it more precisely. Therefore, it is highly unlikely that all of the 12 subjects falling into the 55 to 64 kilogram weight class will have exactly the same weight. In constructing the histogram, we make the assumption that all of the observations in each interval are spread uniformly over the length of the interval. Of course, we realize that this assumption is not precisely true for every score interval. In large distributions with many score values, however, the distortion will be small.

When we draw the histogram, we indicate the continuous nature of the variable by joining all the bars together. Each bar extends horizontally from the lower real limit of the score interval to the upper real limit of the interval. Because the upper real limit of each interval is the lower real limit of the next interval, the bars are connected. The bars may be labeled either at the real limits of the intervals or at the midpoints of the intervals. The midpoint of the first score interval in Figure 2.5 is 59.5 kilograms. The midpoint of an interval is often referred to as the *class mark*.

The height of each bar in the histogram represents the number of observations falling in a particular score interval. It can also represent the proportion of all observations falling in the interval if we choose to use relative frequencies. The height of each bar is constant across the entire interval. This is a reflection of the assumption that all the scores falling in a given interval are spread evenly throughout the interval. If all of the score intervals are equal in length, then the

FIGURE 2.5
Histogram of the Frequency Distribution
for the Weight of 200 Adult American Males

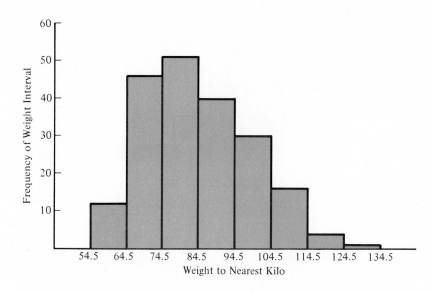

area of each bar is also proportional to the number of observations falling into that score interval. For this reason, it is recommended that histograms be used only to represent distributions in which all the score intervals are of *equal length.*

We indicated previously that the key to constructing frequency distributions properly is being aware of the way in which the variable of interest is defined. In Figure 2.6, we see the two histograms corresponding to our frequency distributions of "age to the nearest year" and "age at the time of last birthday" presented in Table 2.9. You will notice in Figure 2.6 that the two histograms differ with regard to the way the bars have been positioned along the horizontal axis. Can you explain why? What are the real limits of the intervals in the case of the variable age to the nearest year? What are the real limits of the intervals in the case of the variable "age at time of last birthday"? If these questions are troublesome for you, ask yourself what true ages fall into the first interval in each case. You will realize that if an individual's age "to the nearest year" is 8, his or her true age may be anywhere from 7 1/2 up to but not including 8 1/2. The histogram for age to the nearest year is labeled at the class marks. For the first interval, the class mark is 8, the midpoint of the interval. However, the bar extends from the lower real limit of 7.5 to the upper real limit of 8.5, indicating that the three people falling in this first interval can have true ages anywhere in

63

FIGURE 2.6
Two Histograms for Distributions of Age

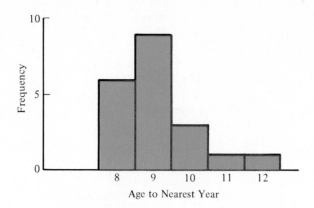

Age to Nearest Year

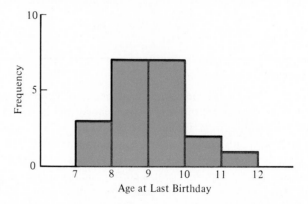

Age at Last Birthday

this range. On the other hand, when we measure "age at last birthday," a person classified as 7 could be anywhere from exactly 7 all the way to just short of 8. Therefore, in the distribution of "age at last birthday," the lower real limit of the first interval is 7, and the upper real limit is 8. The histogram for "age at last birthday" is labeled at the real limits of the intervals. In each case the labels were kept as whole numbers, but they were located at different points on the scale to fit the definition of the variable of interest.

The Frequency Polygon. An alternative method of representing a set of continuous interval data graphically is the frequency polygon. The frequency polygon is similar to the histogram, except that frequencies are graphed as points located over the midpoints of the intervals, and these points are then connected to each other. The frequency polygon of the variable "weight of 200 American males" is presented in Figure 2.7, so that you can compare it to the histogram representing the same data, presented in Figure 2.5. You will note that the ends of the frequency polygon are "tied down" to the horizontal axis at the points 49.5 and 139.5. That is, 49.5 lies the length of an interval below the midpoint of the interval in which the first frequencies occurred, and 139.5 lies the length of an interval above the midpoint of the interval in which the last frequencies occurred. The histogram and frequency polygon are equally good techniques for representing a set of continuous interval data graphically. The histogram is more often used when single distrigutions are represented. The frequency polygon, as we shall see, is more frequently used in experimental and pseudo-experimental situations, when the scores of two separate groups of subjects are to be compared.

The Cumulative Frequency Graph. The cumulative frequency graph is used to represent continuous interval data when we are interested in the cumulative frequency distribution of the data. It allows us to see how many observations in a

---

FIGURE 2.7
Frequency Polygon of the Distribution of Weight
of 200 Adult American Males

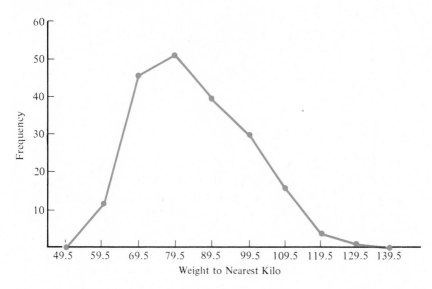

Univariate Problems

data set fall at or below a given point on the scale. This is most useful when we have a distribution of scores and we are interested in finding out how one score compares to the rest of the scores. We have a good example of such a situation in the frequency distribution of scores of Dr. Brown's statistics exam presented in Table 2.10. In Figure 2.8 we have presented the cumulative frequency graph for this set of data. Consider Figure 2.8 carefully. You will note that the scale of exam scores has been laid out along the horizontal axis of the graph. Over the upper real limit of each score value interval we place a dot indicating the number of students who achieved *this score or lower* on the exam. These cumulative

FIGURE 2.8
Cumulative Frequency Graph of 50 Final Exam Scores
for Dr. Brown's Statistics Course

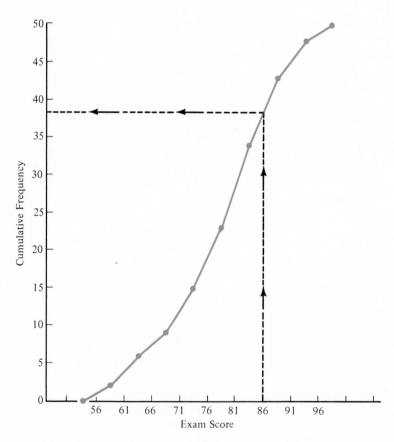

Organizing and Presenting Empirical Data

frequencies are labeled along the vertical axis. As you can see from the graph, we have plotted the cumulative frequency 0 at 53.5. This is the lower real limit of the first interval of the distribution, and none of the students scored any lower than this. We plot a 2 over the score value 58.5. This is the upper real limit of the first interval, and this dot represents the 2 students who scored between 53.5 and 58.5. We plot a 6 at 63.5. It represents all of the 6 students scoring up through 63.5. You will recall that the 6 was obtained by adding the 4 scores that fell in the second interval to the 2 scores that fell in the first interval. We continue this process until we have plotted points over each of the real limits of the distribution. When we come to the upper real limit of the last score interval, 98.5, we will have accounted for all of the observations in the data set. When all the points have been plotted, they can be connected by straight lines.

The useful thing about the cumulative frequency graph is that it enables us to obtain a quick estimate of the number of observations falling at or below *any score* at all, not just the end points of intervals. If we wish to estimate the number of scores falling at or below 86, we can do so as follows. First, locate the score 86 on the horizontal axis. From this point, draw a vertical line straight up until it intersects with the cumulative frequency graph. From this point, draw a horizontal line segment straight across until it intersects with the vertical axis. The point at which this horizontal line intersects with the vertical axis is our estimate of the number of people scoring at or below 86. The two lines that have to be drawn are indicated by dotted lines on Figure 2.8. You will see that the horizontal line intersects the vertical axis at approximately 38. Thus, we would estimate that about 38 of the 50 students in the class had scored 86 or less on Dr. Brown's exam. This, of course, would be useful information to any students who actually did score 86, for it would give them a good idea of where they stood in relation to the rest of the class. Very often, a cumulative frequency graph is presented with cumulative percentages labeled on the vertical axis. If this is done, the graph can be used in the same manner to estimate the percentage of all observations falling below a certain point. This will be considered in more detail in Chapter 4, where we discuss percentile ranks.

Graphing Grouped Interval Data

In constructing frequency distributions for various interval variables, we found several cases in which it was useful to group scores. Grouping scores may result in destroying one or more of the special properties of the interval scale. If this occurs, it renders inappropriate certain of the graphic techniques we have considered in earlier sections of this chapter.

Representing Discrete Interval Data That Have Been Grouped. Data may be grouped in equal intervals or unequal intervals. When we deal with discrete interval data, either method of grouping will cause us to modify the graphic techniques we employ.

67

As an example of discrete interval data that have been grouped into equal intervals, we consider once again the variable "number of students enrolled in introductory psychology" (see Table 2.5). You will recall that this was a discrete interval variable that had a large number of discrete score values spread over a wide range. For this reason, we constructed a frequency distribution in which scores were grouped throughout the range of the variable. This distribution is presented in Table 2.5. As already noted, this form of grouping results in the loss of information. We would not be able to draw a discrete graph because we no longer know exactly how many schools fall at each of the possible discrete score values from 0 to 700. This is just as well, because it would be difficult to draw a graph with 701 different value labels across the horizontal axis. So how can we represent such data?

In situations where a wide range of discrete interval data has been grouped into equal intervals, it is customary to represent the data graphically as if they were continuous interval data. That is, we use a histogram or a frequency polygon. The histogram corresponding to the frequency distribution presented in Table 2.5 is presented in Figure 2.9. Of course, it might be argued that this procedure results in a distortion of the data. Although it is clearly impossible to have 149.5 students registered for introductory psychology, 149.5 does exist as a point on the horizontal axis of the histogram. However, in a set of data of this size, the distortion is small. You will note that in Figure 2.9 we have attempted to avoid confusion by labeling the horizontal axis not at the real end points of the class intervals, but at the whole numbers just above these points.

Now let us consider a discrete interval variable that has been grouped into unequal intervals, the variable "number of children in the household" (see Table 2.4). You will recall that there were relatively few score values here, but that some of the higher values occurred so infrequently that it was convenient for us to include in the final category all those homes with six or more children. Remember also that this process of "collapsing" the highest discrete score values into a single final category has the effect of destroying the interval scale. When discrete interval data are collapsed in such a manner, it is not appropriate to use any graphic technique that assumes interval scale data. Thus, it would not be appropriate to use either a discrete graph, a histogram, or a frequency polygon to represent these data. You may wonder why it would be incorrect to use a discrete graph with the final category clearly labeled "6 or more." The reason is that readers who frequently use these various techniques of graphic representation tend to associate certain types of graphs with the corresponding types of data. Individuals reading the title of the graph as a discrete interval graph might assume we were dealing with a discrete interval variable. Having made this assumption, they might fail to notice the label "6 or more" under the final category. If so, they would come away with an inaccurate picture of the data.

Under these circumstances, a good way to proceed is to represent the data using a bar graph as shown in Figure 2.10. Because the bar graph is suitable for categorical data, using the name bar graph does not imply that the data have

FIGURE 2.9
Histogram for the Number of Students Enrolled in Introductory Psychology
in 900 Institutions of Higher Education

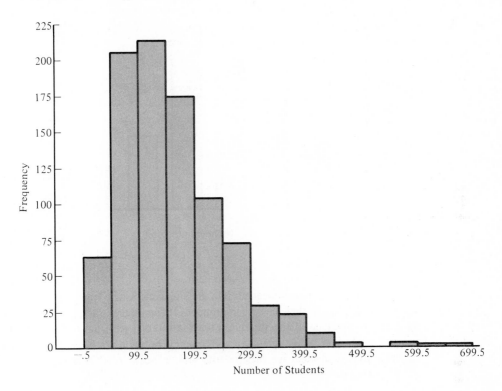

interval scale properties. Therefore, someone reading the title "bar graph" would
not be tempted to make any unwarranted assumptions.

In summary, when scores on a discrete interval variable are grouped into
equal intervals over the entire range of the variable, the interval scale has been
preserved. Then we represent the data graphically using a histogram or frequency
polygon. However, when scores on a discrete interval variable are grouped into
unequal intervals, the interval scale has been destroyed. In this case, we use a bar
graph to represent the data.

Representing Continuous Interval Data That Have Been Grouped.   Because continuous interval data can always be measured more accurately than we have
measured it, we are really always dealing with grouped data when we are dealing
with continuous interval data. The rules that determine the selection of graphing

69

FIGURE 2.10
Bar Graph Showing the Number of Children in the Family
for 90 Families

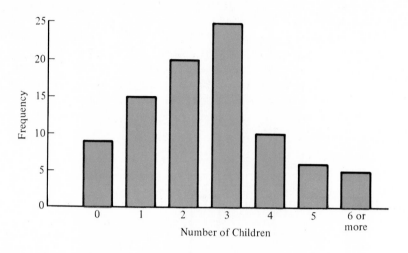

techniques for continuous interval data are the same as the rules that apply to discrete interval data that have been grouped. If the intervals are of equal length over the entire range of the variable, then the interval scale has been preserved and we use either a histogram or a frequency polygon. However, if unequal intervals are established, we sacrifice the interval scale. In this case, we should use a bar graph.

Using a Smooth Curve to Represent Large Data Sets

One final and very important point must be made regarding the graphic representation of interval scale data. When researchers are dealing with very large sets of data, they frequently represent the data with a *smooth curve,* rather than a histogram or frequency polygon. To see why this is a reasonable procedure to use when we have a large number of cases, we will consider the relative frequency distribution of the continuous interval variable "height" for a group of 10,000 American males. In Figure 2.11, we have sketched several histograms representing this data set. In Figure 2.11(a), we have sketched a histogram in which the score interval is 10 cm. long. In Figure 2.11(b), we have sketched the same data, but we have reduced the size of the score interval from 10 cm. to 5 cm. You will notice that with this change in the size of the interval, the histogram becomes "smoother," i.e., the jumps up and down between class intervals are not as great. Similarly, Figure 2.11(c) is the histogram representing the relative frequency

70

distribution when the width of a class interval has been reduced to 1 cm. At this point, you will note that it becomes difficult to distinguish individual class intervals. In fact, if you hold the figure several feet away from your eyes, the histogram may appear more like a single area than like a series of individual bars. In Figure 2.11(d), we have emphasized the appearance of uniformity by sketching in a smooth curve over the top of the histogram. As we make the class interval smaller and smaller, the actual histogram becomes less and less distinguishable from a smooth curve. Because height is a continuous interval variable that in theory may always be measured more accurately, it is possible to envision a situation in which the class interval has become infinitely small. If this were the case, the relative frequency distribution would appear as it does in Figure 2.11(d), a smooth curve. It is quite common to find large sets of interval scale data represented in this manner.

FIGURE 2.11
Relative Frequency Distribution of Height of American Males
Using Different Interval Sizes ($N = 10,000$)

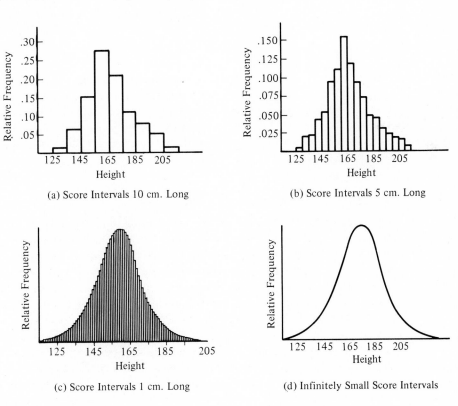

(a) Score Intervals 10 cm. Long

(b) Score Intervals 5 cm. Long

(c) Score Intervals 1 cm. Long

(d) Infinitely Small Score Intervals

THE CLASSIFICATION MATRIX

In this chapter, we have considered a variety of techniques we can use to organize and present empirical data in univariate problems. In Table 2.11, we present our matrix with the techniques covered thus far entered in the appropriate cells.

Organizing and Presenting Empirical Data

TABLE 2.11
Classification Matrix:
Techniques for Organizing and Presenting Empirical Data

Level of Measurement of Variable(s) of Interest

Type of Problem		Categorical: Nominal or Ordered Categories	Rank-Order	Interval Scale (or Ratio-Scale)
Univariate Problems		Frequency Distribution, Relative Frequency Distribution; Bar Graph, Pictograph, Circle Graph	Set of Ranks	Frequency Distribution, Grouped Frequency Distribution, Cumulative Frequency Distribution; *Graphs for Discrete Interval Data:* Discrete Graph, Step Function; *Graphs for Continuous Interval Data:* Histogram, Frequency Polygon, Cumulative Frequency Graph
Bivariate Problems	Correlational Problems			
	Experimental and Pseudo-Experimental Problems: Independent Groups			
	Dependent Measures: Pretest and Posttest or Matched Groups			

Key Terms

Be sure you can define each of these terms.

frequency distribution
frequency
relative frequency
relative frequency distribution
bar graph
circle graph
pictograph
discrete interval variable
continuous interval variable
grouped frequency distribution
cumulative frequency
cumulative percentage
cumulative frequency distribution
cumulative percentage distribution
apparent class limits
real class limits
discrete graph
step function
histogram
frequency polygon
cumulative frequency graph
class mark
smooth curve

Summary

In Chapter 2, we discussed methods for organizing and presenting a set of measurements taken on a single variable. We suggested that these methods are appropriate for both descriptive and inferential problems.

We introduced the frequency distribution, a table that shows the values of a variable and how frequently each value appears in the data set. We saw that this table can reflect either the number of times each value appears or the relative frequency with which each value appears. We pointed out that the frequency distribution is appropriate regardless of the scale of measurement. We also noted that with interval data, we are sometimes interested in the number or percentage of scores occurring at

2

Student
Guide

or below a certain value. In this case, we can make a cumulative frequency or cumulative percentage distribution.

We introduced several graphs that are used to represent frequency distributions. We saw that because we want our graph to reflect the level of measurement as well as the frequency of values, we use different graphs for different levels of measurement. Thus, we choose bar graphs or circle graphs for nominal scale variables or ordered categories. We pointed out that no graph can describe a set of rank-order data more succinctly than a simple listing of ranks.

In our discussion of interval scale variables, we considered whether a variable is a discrete interval variable, that is, one that is measured by counting in discrete, whole-number steps, or a continuous interval variable, that is, one that in principle can be measured as finely as we choose. If our variable is discrete, we use a discrete graph to represent the frequency distribution and the step function to represent the cumulative frequency distribution. For continuous variables, we use the histogram or frequency polygon for frequency distributions and the cumulative frequency graph for cumulative frequency distributions. We also saw that if we have an interval variable with many different score values, we frequently combine score values into larger intervals and make a grouped frequency distribution.

Finally, we saw that if we have a discrete interval variable with many score values and a large data set, our discrete graph will resemble a smooth curve. With a continuous variable and a large data set, our graph will also resemble a smooth curve as our measurements become more precise. Thus, researchers frequently use smooth curves to represent large sets of data.

Review Questions

To review the concepts presented in Chapter 2, choose the best answer to each of the following questions.

1 One reason we use a frequency polygon instead of a histogram is that the frequency polygon
   _____a reflects the continuous nature of the scale.
   _____b makes it easier to compare two distributions.

2 If we are picturing the distribution of measures on a discrete interval variable, we want the picture to show that it is impossible to have scores between recorded values. For this reason, we draw
   _____a a frequency polygon.
   _____b a discrete graph.

3 If we took millions of observations on the variable number of children in a family with values 0, 1, 2, 3, 4, 5, and 6 or more, we would expect the graph of the distribution to look like
   _____a a discrete graph.
   _____b a smooth curve.

75

4 If we take measurements on some variable using a scale that has a unit that can be made as small as we please, we say we are measuring

_____ a a discrete variable.

_____ b a continuous variable.

5 A data set consists of 530 measures on the variable birth weight of newborn babies. The measures are taken to the nearest gram. In this situation, the variable is

_____ a a discrete interval variable.

_____ b a continuous interval variable.

6 To summarize the data in the above situation, we should prepare a

_____ a grouped frequency distribution.

_____ b frequency distribution.

7 To get a good picture of the distribution of birth weights mentioned in 5, we should make

_____ a a histogram.

_____ b a bar graph.

8 To get a picture of the proportion of the sample at or below any weight, we should draw

_____ a a relative frequency polygon.

_____ b a cumulative percentage graph.

9 If we wanted to report to someone the possible true weights of babies recorded as weighing from 3,000 grams through 3,499 grams, we should report

_____ a the real class limits of this interval.

_____ b the apparent class limits of this interval.

10 If we wanted to compare the distribution of our sample of 530 babies to another distribution of 894 babies, we should prepare

_____ a a frequency distribution.

_____ b a relative frequency distribution.

11 We have asked a sample of 400 people to express their feeling about a movie that they previewed. Each person is to select a number from 1 to 5, where 1 means "I disliked it strongly" and 5 means "I liked it a lot," to indicate his or her feeling. To summarize these results, we should construct

_____ a a bar graph.

_____ b a frequency polygon.

12 The bars of a histogram sit over

_____ a the apparent class limits.

_____ b the real class limits.

76

13 The number of occurrences of a value in a data set is called
_____a the frequency of the value.
_____b the relative frequency of the value.

14 In a cumulative frequency graph, the point on the scale where the graph will be at zero is
_____a the lower apparent limit of the lowest interval.
_____b the lower real limit of the lowest interval.

15 In order to use a histogram to picture a grouped frequency distribution, we must be sure that
_____a our variable is continuous.
_____b we have equal size intervals on our scale.

16 If an interval in a grouped frequency distribution has no recorded scores, then in the frequency polygon picturing this distribution, we will see
_____a the polygon tied to zero at the midpoint of the interval.
_____b the polygon tied to zero at the lower limit of the interval.

17 One disadvantage of a grouped frequency distribution is that when we group the scores into larger intervals
_____a we destroy the continuity of the scale.
_____b we lose information about the actual scores that were observed.

18 The cumulative frequency graph tells the number of scores in a distribution at or below any score value. For this reason, the cumulative graph
_____a should be tied down to zero at the upper real limit of the highest interval.
_____b should be tied to zero only at the lower real limit of the lowest interval.

Problems for Chapter 2: Set 1

1 In a study, 50 subjects were classified by sex. There were 35 males and 15 females. What is the relative frequency of each value of this variable?

2 In another study, 42 subjects were classified on the variable "handedness." There were 10 left-handers and 32 right-handers. What is the relative frequency of each value of this variable?

3 Researchers classified 237 subjects on the variable "favorite color." The breakdown by percentages, to three decimal places, is as follows:
    Red—.401, Blue—.249, Yellow—.131, Green—.072, Other—.148
Using absolute frequencies, make a frequency distribution for the data. Use the frequency distribution to make a bar graph.

4 A department in a large university offers the following budget figures.

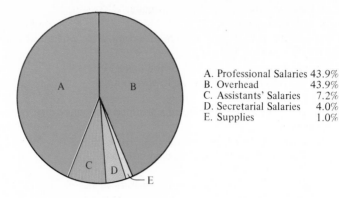

A. Professional Salaries 43.9%
B. Overhead   43.9%
C. Assistants' Salaries 7.2%
D. Secretarial Salaries 4.0%
E. Supplies   1.0%

Approximately how many dollars are allocated for each category if the total budget is $209,500? Show these results in a frequency distribution.

5 A researcher questioned 188 subjects as to whether they favored the passage of the Equal Rights Amendment. Of the 188, 64 say they favor its passage, 76 say they do not favor its passage, 23 say they have no opinion, and the rest of the sample refuses to answer. Summarize these results. Do the data suggest that the public favors the adoption of the Equal Rights Amendment?

Set 2

1 Decide which of the following are discrete interval variables and which are continuous interval variables. Which, if any, of the discrete interval variables might be considered to be measuring traits that have an underlying continuous nature?
  a short-term memory as measured by recording the time it takes a subject to recall a response to a given stimulus, to the nearest tenth of a second
  b the possible number of heads recorded in 20 tosses of a coin
  c family economic status as measured by family income to the nearest $100
  d achievement in statistics as measured by the number of correct responses on a standardized exam
  e effectiveness of a drug as measured by the number of patients cured
  f schooling as measured by the nearest number of whole years in school
  g number of births per day at St. Mark's hospital
  h shooting accuracy as described by number of clay pigeons broken
  i fishing ability as measured by number of trout caught on opening day
  j fishing ability as measured by weight of largest fish caught

2a Use the data contained in Data Set A (at the end of this problem set) to make a frequency distribution and a relative frequency distribution of

Organizing and Presenting Empirical Data

income for 125 families in Tranquility, New York. Group the data in the following manner:

Make the first interval contain all those families whose income is at least $4,000 but less than $6,000, make the second interval contain all those families whose income is at least $6,000 but less than $8,000, and so on.

b Use the relative frequency distribution to answer the following.
  i   What are the real limits of the intervals?
  ii   What percentage of the families have incomes of less than $10,000?
  iii What percentage of the families have incomes of $26,000 or more?
  iv   What percentage of the families have incomes of at least $16,000 but less than $22,000?
  v   Make a histogram of the results and answer questions i–iv.
  vi   How would the presentation change if we had rounded to the nearest $2,000?

3a Use the data in Data Set A to make a frequency distribution, a relative frequency distribution, and a cumulative relative frequency distribution for the variable "years of education of head of household" for 125 families in Tranquility, New York.
b Which type of graph would be most appropriate to represent the distribution?
c If a histogram is drawn, how do we know what the real limits are?
d Suppose years of education is measured to the nearest year. Draw a cumulative relative frequency graph.
e Use the cumulative graph to estimate answers for the following.
  i   No father has fewer than _____ years of education.
  ii   No father has more than _____ years of education.
  iii A father who has 12.6 years of education is average because _____ percent of all fathers have no more than 12.6 years of schooling.
  iv   Approximately what percentage of all fathers have had more than 16 years of schooling?
  v   What percentage of all fathers have had at least 12 years of schooling but no more than 16 years of schooling?

4a For Table 2.5, the grouped frequency distribution of number of students enrolled in 900 institutions of higher education, give the apparent limits and the real limits for each of the first four intervals.
b For Table 2.7, the frequency distribution of number and percentage of 200 adult American males falling into different weight categories, give the apparent limits and the real limits for each interval. What information do these numbers give us?
c For Figure 2.5, the histogram of the frequency distribution for the weight of 200 adult American males, what numbers on the scale are the endpoints of the bars? What information do these numbers give?

5 Using Table 2.7, the frequency distribution of number and percentage of 200 American males falling into different weight categories, make a cumulative relative frequency graph for the distribution of weights.

79

a How are the real limits used in constructing this graph?
b Use your graph to estimate answers for the following.
   i   What percentage of the measures are less than or equal to 74.5 kilo-
       grams?
   ii What percentage of the measures are more than 94.5 kilograms?
   iii Estimate the percentage of the measures that are less than 79 kilo-
       grams.
   iv Estimate the percentage of the measures that are between 69 kilo-
       grams and 99 kilograms.
   v   Estimate the percentage of the measures that are above 119 kilo-
       grams.
   vi What percentage of the men weigh more than 86 kilograms?
   vii Estimate the weight that marks off the lighest 50 percent of the
       population of men.
   viii Estimate the weights that mark off the lightest 25 percent of the
       men and the heaviest 25 percent of the men.

6 Use Figure 2.8, the cumulative frequency graph for 50 final exam scores for
   Dr. Brown's statistics class, to answer the following questions.
   a How many scores are less than or equal to 63.5?
   b How many scores are less than or equal to 83.5?
   c How many scores are greater than 73.5?
   d How many scores are between 68.5 and 88.5?
   e The lowest 15 scores are at or below which score value?
   f The top 10 scores are above which score value?

7 Redraw Figure 2.8 using relative frequencies rather than absolute frequencies. (You may want to refer to the frequency distribution in Table 2.10.)
   Use the graph to estimate answers to the following:
   a If a score is in the lowest 10 percent of all scores, it is no more than
      which score?
   b If a score is in the top 25 percent of all scores, it is at least which score?
   c If a score is higher than 50 percent of all other scores, then that score is
      which number?
   d What percentage of the class is at least as high as a person whose score is
      90?
   e What percentage of the class is no higher than a person with a score of
      75?

8a How could your graph drawn in problem 7 be used to answer the ques-
    tions of problem 6?
 b How could Figure 2.8 be used to answer the questions in problem 7?

9 A test of 10 multiple-choice items (each item has 5 options) is given to a
   total of 200 students at the beginning of a course. The following is a frequency distribution of the results.

80

Organizing and Presenting Empirical Data

Number of correct answers	Frequency of score
0	21
1	54
2	61
3	40
4	18
5	5
6	1
7	0
8	0
9	0
10	0
Total	200

a Make a discrete graph for the distribution of scores.
b Use the graph to answer the following questions.
   i How many scores are at or below 2?
   ii How many scores are below 2?
   iii How many scores are at 2?
   iv How many scores are more than 4?
   v How many scores are 5 or more?
   vi How many scores are at least 1 but no more than 3?
   vii The lowest 37.5 percent of scores are which values?
   viii The highest 3 percent of scores are which values?
   ix The lowest 25 percent of scores are which scores?
   x The highest 25 percent of scores are which scores?
   xi What percentage of scores are at or below 8?

10 Following are some thought experiments that you can do to practice representing a frequency distribution by a smooth curve.
  a Consider the variable "family income" in the United States population. Sketch a smooth curve that suggests how the frequency distribution for this variable might look.
  b Suppose you have 100 pennies and you throw them into the air. When they land, you count the number of heads. You then repeat this experiment many, many times. For the variable "the number of heads in 100 tosses of a coin," sketch a smooth curve that suggests how the frequency distribution might look.
  c Think about the distribution of scholastic aptitude scores for all high-school seniors in a given year. Sketch a smooth curve that suggests what this distribution looks like.
  d Suppose you have a test consisting of 100 questions. Each question has 4 choices. Suppose that 10,000 people take the test and every person uses a random guessing method to choose the answers. Sketch a smooth curve that represents the frequency distribution of the results.

DATA SET A

Years of Education of the Head of Household and Annual Income
for 125 Families in Tranquility, New York

Subject	Head of household years of education	Annual income
1	12	$14,100
2	16	21,900
3	9	13,800
4	5	4,000
5	17	18,100
6	18	19,500
7	13	7,800
8	14	14,000
9	21	31,000
10	20	26,000
11	5	12,000
12	6	6,000
13	20	29,000
14	19	21,000
15	10	4,000
16	10	16,000
17	19	22,000
18	12	26,000
19	17	21,000
20	13	22,000
21	14	24,000
22	19	27,000
23	10	16,000
24	12	24,000
25	7	6,500
26	10	6,300
27	14	14,500
28	13	11,500
29	14	21,900
30	10	8,000
31	10	17,900
32	15	25,800
33	15	27,000
34	7	8,200
35	13	25,000
36	13	13,000
37	19	29,500
38	14	21,400
39	14	16,200
40	10	10,100
41	10	21,900
42	15	24,000
43	7	10,000
44	19	31,000
45	7	11,000

Organizing and Presenting Empirical Data

Subject	Head of household years of education	Annual income
46	8	8,000
47	11	7,000
48	12	23,000
49	18	21,500
50	16	28,500
51	17	22,200
52	14	17,000
53	13	15,000
54	12	17,900
55	11	18,900
56	10	10,200
57	14	20,900
58	16	27,500
59	16	23,500
60	15	16,000
61	18	22,000
62	10	14,600
63	8	8,200
64	9	15,000
65	10	25,800
66	12	16,400
67	12	17,800
68	12	11,500
69	16	23,000
70	16	27,500
71	16	19,000
72	12	12,000
73	15	17,000
74	16	26,500
75	7	14,000
76	8	11,000
77	8	13,000
78	9	8,000
79	9	9,000
80	9	12,000
81	11	8,000
82	11	10,500
83	12	12,000
84	12	18,000
85	12	13,000
86	13	18,000
87	18	24,000
88	17	23,800
89	16	24,100
90	16	19,000
91	15	17,900
92	13	20,000
93	12	18,900
94	12	19,800

Subject	Head of household years of education	Annual income
95	9	10,000
96	17	25,000
97	18	28,000
98	16	24,500
99	16	20,000
100	17	25,500
101	16	26,100
102	13	21,000
103	12	15,200
104	12	15,900
105	15	19,000
106	11	17,000
107	10	11,800
108	17	26,500
109	12	22,000
110	15	20,500
111	12	21,500
112	14	18,000
113	14	20,000
114	11	15,000
115	10	12,800
116	16	20,500
117	11	11,800
118	11	13,000
119	15	22,000
120	12	20,400
121	12	14,900
122	16	25,900
123	12	10,500
124	5	8,000
125	13	16,000

Organizing and Presenting Empirical Data

3

Organizing and Presenting Empirical Data: Bivariate Problems

In Chapter 1, we noted the distinction between univariate problems and bivariate problems. In Chapter 2, we considered techniques for organizing and presenting empirical data in univariate problems. In this chapter, we will be considering techniques for organizing and presenting empirical data in bivariate problems. You should recall from Chapter 1 that our classification matrix for elementary statistical problems contains three separate classes of bivariate problems. One class of bivariate problems we referred to as bivariate correlational problems. In this type of problem, two variables are observed in a given population at a particular time, with no attempt to exercise any control over either of the variables. A second class of bivariate problems includes those experimental and pseudo-experimental problems in which two or more independent groups are compared on some variable of interest. In these problems, one of the variables is a "group" variable, for the experimenter sets up two or more groups of subjects initially. Then the experimenter measures subjects in each of the previously established groups on a second variable in order to compare the groups with respect to this second variable, the variable of interest. Finally, the third class of bivariate problems, those problems involving two sets of dependent measures, includes both the pretest and posttest type of situation and the experimental situation in which groups have been matched. In general, one set of techniques for organization and presentation of data is used for bivariate correlational problems and a second set of techniques is appropriate for bivariate problems involving the comparison of independent groups or two sets of dependent measures. We consider both sets of techniques in this chapter, beginning with the techniques appropriate for bivariate correlational studies.

ORGANIZING AND PRESENTING DATA
IN BIVARIATE CORRELATIONAL STUDIES

Within the heading of bivariate correlational studies, we must consider the several different types of variables with which we might be dealing: categorical variables, including nominal scale and ordered categorical variables; rank-order variables; or interval or ratio scale variables. We will consider each of these situations in turn.

Correlational Studies Involving Two Categorical Variables

In Chapter 1, we considered several examples of correlational studies. We considered the research question, "Is there a relationship between political party membership and opinion on the Equal Rights Amendment among residents of Tranquility, New York?" We also considered the question, "Is there a relationship between sex and opinion on the Equal Rights Amendment among residents of Tranquility, New York?" Let us take some time here to consider the second of these two questions. As noted previously, each of the two variables involved in this question is a categorical variable. We will therefore use this research question to illustrate the procedures used in organizing and presenting bivariate categorical data.

Because the population of interest in this situation is the population of voters in a rather small town, we might consider a descriptive approach. That is, we might attempt to measure the entire population of interest on the two variables. To do so, we would go to Tranquility and obtain a list of registered voters from the board of election. We would attempt to contact all of the people on the list, and we would ask each voter contacted two questions: "What is your sex?" and "What is your opinion on the Equal Rights Amendment?" Had the population of interest been very large, we would probably have opted for the inferential approach, selecting a sample of voters to represent the larger population. From the point of view of organizing and describing the resulting set of data, there would be no difference. In either case, we would employ a *crosstabulation table*. Furthermore, we would use a crosstabulation with ordered categorical variables as well as with nominal scale variables.

The Crosstabulation Table. Suppose we find that there are 188 registered voters in Tranquility. Of these 188, we are unable to contact 21, and another 4 refuse to answer one or both of the questions. Thus, we finally obtain a set of 163 subjects on whom we have measured both variables. We need a device that summarizes the way our subjects score on *both* of the variables upon which they have been measured. Such a device is referred to in general as a *bivariate frequency distribution*. In the present example, the bivariate frequency distribution would tell us such things as the number of males who were in favor, the number of males who were opposed, and so on. In the case of categorical data, such a bivariate distribution is known as a crosstabulation table.

A crosstabulation table is quite easy to construct. We begin by determining the number of values of each of the two variables of interest. Here, we have two values for the variable sex (male and female) and three values for the variable opinion (in favor, opposed, and no opinion). This means that there are a total of two times three, or six, different combinations of values that a given subject might have. That is, a subject can be: 1) male and in favor; 2) male and opposed; 3) male and have no opinion; 4) female and in favor; 5) female and opposed; and 6) female and have no opinion. A convenient way to represent these various combinations of variable values is to form a rectangular table in which the values of one variable are listed along the top and the values of the other variable are listed along the side. Table 3.1 shows a rough work table used in constructing the crosstabulation table for this problem. Table 3.2 shows the completed cross-tabulation table. Both the work table and the completed bivariate distribution are laid out with the values of one variable listed across the top and the values of the other variable listed down the side. The choice of which variable to place along the top and which to place along the side is arbitrary. We have chosen to place sex on top and opinion along the side. We therefore refer to sex as the *column variable* and opinion as the *row variable*.

Referring to Table 3.1, you will note that there are six boxes in the work table, one for each of the six possible combinations of scores on the two variables. These boxes are called *cells*. To determine which of the six possible com-

TABLE 3.1
Work Table Used in Construction of Crosstabulation Table
Relating Sex to Opinion on the Equal Rights Amendment

		Sex	
		Male	Female
Opinion	In favor	卌 卌 卌 卌 卌 卌 III	卌 卌 卌 卌 卌 卌 I
	Opposed	卌 卌 卌 卌 卌 卌 卌 卌 卌 卌 III	卌 卌 卌 卌 III
	No opinion	卌 卌 卌	卌 III

binations is represented by a particular cell, we look at the column and row headings that pertain to that cell. Thus, if we were concerned with the upper left-hand cell in Table 3.1, we would look to see that this cell lies in the column headed "Male" and in the row headed "In Favor." We therefore know that this cell is for those subjects who were both male and in favor. Once we have laid out the six cells, we can go through our data set to determine how many subjects fall in each of the six cells. We check each subject's score on *both* variables to determine which one of the six possible *combinations of scores* the subject had. We then place a tally in the cell of our work table corresponding to that combination. Once we have located all the subjects with tallies, we count the tallies in each cell to determine how many subjects had each of the six combinations. We enter these six numbers in the appropriate cells of our finished crosstabulation, presented in Table 3.2.

Take a moment to consider Table 3.2. The numbers presented in each of the six cells are called *cell frequencies.* As we noted before, these cell frequencies tell us how many subjects fell into each of the different combinations of variable values. Consider the first *column* of cells in the table. This column contains all the subjects who were classified as male: the 33 males in favor, the 53 males opposed, and the 15 males with no opinion. This column constitutes a *conditional distribution.* It is the distribution of opinions among male subjects, or the distribution of opinions among subjects *on the condition* that they are male. Similarly, the second column is also a conditional distribution. It is the distribution of opinions among subjects on the condition that they are female.

TABLE 3.2
Crosstabulation of Opinion on Equal Rights Amendment by Sex
for 163 Voters in Tranquility, New York

		Sex		
		Male	Female	Row totals
Opinion	In favor	33	31	64
	Opposed	53	23	76
	No opinion	15	8	23
	Column totals	101	62	163

Organizing and Presenting Empirical Data

If we add up all the cell frequencies in the first column, we obtain the *marginal frequency* for that column. This *column marginal* is 101. It represents the total number of males in the data set, summed up over all three of the different opinions they might have. Similarly, if we add up all the cell frequencies in the second column, we obtain the marginal frequency for this column. This column marginal is 62. It represents the total number of females in the data set. Taken together these two column marginals represent the *marginal distribution* for the variable "sex." This distribution tells us how many subjects there were of each sex, regardless of how they voted. It is really the univariate distribution of the variable "sex" among subjects in our data set. Because every subject in the data set was either male or female, the sum of the 101 males and the 62 females is equal to the total number of subjects in the data set, 163.

Now consider the first *row* of cells in the table. This row contains all the subjects who were in favor. This row also constitutes a conditional distribution. It is the distribution of sex among subjects who were in favor, or the distribution of sex *on the condition* of being in favor. Similarly, the second row is the distribution of sex among subjects who were opposed and the third row is the distribution of sex among subjects who had no opinion. If we add up the cell frequencies across the rows, we obtain a new set of marginal frequencies. There are three of these *row marginals*. The row marginal for the first row is 64. It represents the total number of subjects who were in favor, regardless of sex. The row marginals for the second and third rows are 76 and 23. Thus, 76 subjects were opposed and 23 subjects had no opinion. Taken together, the three row marginals constitute the marginal distribution for the variable "opinion." This distribution tells us how many subjects had each of the three possible opinions, regardless of sex. It is therefore the univariate distribution of the variable "opinion." Because every subject in the data set had one of these three opinions, the sum of the three row marginals will be 163, just as the sum of the two column marginals was 163.

Interpretation of Crosstabulation Tables. Once we have laid out the data as in Table 3.2, we are left with the problem of how to interpret them. What information is there in the table to help us decide whether there is some relationship between sex and opinion on the Equal Rights Amendment?

If we just look at the absolute frequencies in the cells, we may be misled. For example, we might note that 33 males were in favor, and only 31 females were in favor. Does this mean that the men in our data set were stronger supporters of the Equal Rights Amendment than the women? A moment's thought tells us that this is clearly not the case. It is true that more men were in favor than women. But more men also were opposed and more men had no opinion. This is simply because our data set included more men than women. It might have been possible for us to have selected groups of men and women of equal size. But we could not do that in this case. We attempted to contact all of the registered voters in Tranquility, and in this town there were more men registered

89

than women. Thus, our data set reflects the sex composition of the town's registered voters. Because the number of men is not the same as the number of women, it makes no sense to compare the absolute frequency of 33 to the absolute frequency of 31. Instead, we must use *relative frequencies*. We should ask, "What proportion of the men were in favor?" and "What proportion of the women were in favor?" When we have these proportions, we have two figures that are comparable.

Referring to Table 3.2, we see that 33 men were in favor of the amendment, out of the total of 101 men represented. Thus, the proportion of men who were in favor is 33 divided by 101, or .3267. On the other hand, 31 women were in favor of the amendment, out of the total of 62 women represented. Thus, the proportion of women in favor is 31 divided by 62, or .5000. Comparing the two proportions, we see that less than one third of the men were in favor, but exactly one half of the women were in favor. Thus, within our data set, men were less likely to be in favor than women. We can easily calculate the proportion of subjects of each sex who had each of the three different opinions. We simply divide each of the cell frequencies by the corresponding column total. We can also calculate the proportion of *all voters* who had each of the three different opinions. To do this, we divide each of the row totals by the total number of subjects, 163. Thus, the proportion of all subjects who were in favor is 64 divided by 163, or .3926. Table 3.3 presents the bivariate frequency distribution for this set of data, along with the proportions that reflect the conditional distributions of opinion for each sex. As is typical in research, each of the proportions has been multiplied by 100, so that it is expressed as a percentage.

Notice that in each case, the percentages presented in Table 3.3 are the percentages of men, women, and all voters who had each of the three possible opinions. Thus, each of the three columns of percentages totals 100 percent. These percentages are referred to as *column percentages,* for each percentage is obtained by dividing the frequency in a given cell (or a given row marginal) by the column marginal, and then multiplying the resulting proportions by 100. Note that we could just as easily have computed *row percentages*. That is, we could have divided the number of subjects in a given cell by the row marginal corresponding to that cell. Consider for a moment what this alternative procedure would mean.

If we divided the number of men in favor by the total number of all voters in favor, what question would we be answering? We would really be finding out the proportion of all those in favor who were men. That is, of the 64 individuals who were in favor, 33 of them were men. Dividing 33 by 64, we get .5156, which means that 51.56 percent of all those in favor were men. Similarly, dividing the number of women in favor by the number of all voters in favor, we would obtain the proportion of all those in favor who were women. This proportion is 31 divided by 64, or .4844. Thus 48.44 percent of all those in favor were women. Notice that the two percentages, 51.56 and 48.44, add up to 100 percent. This is because everyone who was in favor was either a man or a woman.

TABLE 3.3
Number and Percentage of 163 Male and Female Voters in Tranquility, N.Y.
With Various Opinions on the Equal Rights Amendment

		Sex					
		Males		Females		Totals	
		N	Percentage	N	Percentage	N	Percentage
Opinion	In favor	33	32.67	31	50.00	64	39.26
	Opposed	53	52.48	23	37.10	76	46.63
	No opinion	15	14.85	8	12.90	23	14.11
	Totals	101	100.00	62	100.00	163	100.00

Thus, row percentages add up to 100 percent when we add across the rows, just as column percentages add up to 100 percent when we add down the columns. Obviously, we could compute row percentages for those opposed and those having no opinion as well. In each case, these percentages would tell us the percentage of those having a certain opinion who were males, or the percentage who were females.

We could also calculate the percentage of the total number of subjects who were males and the percentage who were females, irrespective of their opinion. We would do this by dividing the *column totals* by the *total number of subjects,* 163. If we did this, we would really be determining the relative frequency distribution of the variable sex in this group.

But would we be interested in the row percentages in this analysis? Would these percentages be as useful to us as the column percentages? Is it important for us to know that 51.56 percent of those who were in favor were males or that 34.78 percent of those who had no opinion were females? Not really. As we indicated before, what we are really interested in here are the proportions of voters of each sex who had each of the three different opinions. It is these proportions that make sense. In the next section, we will explain why these are the proportions that make sense.

The Issue of Causation in Correlational Problems. The question is why does it make more sense to us to compute the percentage of all men who were in favor than it does to compute the percentage of all those in favor who were men? What is there about the way we have been conceptualizing this problem that makes the first percentage interesting, and the second irrelevant?

91

The answer lies in an implicit assumption that we have been making all along, ever since we asked whether there was a relationship between sex and opinion on the Equal Rights Amendment. This assumption has to do with the nature of the relationship between two variables. The assumption is that sex may have an influence on opinion, but that opinion cannot influence sex. In this situation, the assumption is obviously valid. Because one's sex is determined long before one forms an opinion, it is impossible that opinion on the Equal Rights Amendment would influence sex. However, it is quite possible that sex would have an effect on opinion on the Amendment. In fact, when we asked the question, "Is there a relationship between sex and opinion on the Equal Rights Amendment?," we were really asking "Does sex influence opinion?" In other words, there was only one possible direction of influence or *direction of causation* in this situation. We knew before looking at any data that opinion could not influence sex. But there was the possibility that sex could influence opinion. The analysis we carried out aimed at helping us to decide if this potential influence really existed.

In correlational problems where there is a clear direction of causation, we sometimes use the terms independent variable and dependent variable, just as if we were dealing with an experimental situation. We refer to the variable that may be influencing the other as the independent variable and to the variable that may be influenced by the other as the dependent variable. In using these terms, the researcher should be careful that there is only one possible direction of causation. Remember that in an experimental situation the values of the independent variable are determined by the experimenter when subjects are selected or assigned to groups. Thus, in an experimental situation, there is no way that the independent variable may be influenced by the dependent variable. However, in correlational problems, where the experimenter manipulates neither variable, it is not always clear that one variable is doing the influencing and the other is being influenced.

What if we had been considering the question of the relationship between political party membership and opinion on the ERA? As we indicated in Chapter 1, with these two variables it is possible to imagine that the direction of influence could be in either direction. We can imagine how being a Republican or Democrat could influence an individual's position on the Amendment, but we can also imagine how an individual's opinion on the Amendment could influence political party preference. When we examined the relationship between sex and opinion on the Amendment, one of the two variables was clearly determined *before* the other. Sex is determined before opinion. However, when we examine the relationship between political party preference and opinion on the Amendment, there is no certain time order. Both variables are subject to change. People sometimes support issues because their political party has adopted a view on the matter, but they sometimes change parties because their view on an issue differs from the party position. In this situation, we say that there is a possibility of *reciprocal causation*. Whenever reciprocal causation is a possibility, we do not use the terms independent and dependent variable.

92

Furthermore, in a situation where the possibility of reciprocal causation exists, it is not always clear which percentages we should compute in the bivariate frequency distribution. Only when we are justified in assuming that one variable is the independent variable and the other is the dependent variable can we be certain which percentages make sense. To see why knowing the independent and dependent variable in a correlational problem determines which percentages to compute, consider once again the example of the relationship between sex and opinion among the 163 voters from Tranquility, New York. The answer is simple. We looked at men's opinions on the issue, we looked at women's opinions on the issue, and we *compared* the pattern of opinions in these two groups. As indicated previously, in order to make the comparison fair, we had to compare the proportion of men with a given opinion to the proportion of women with that opinion. Thus, the direction of causation that we infer determines which groups we will compare, and the group comparisons that we will make determine which percentages will be instructive.

In any bivariate correlational problem in which the potential direction of causation is clear, we will always look at subjects falling into the various categories of the independent variable separately in order to determine how subjects in each of these categories distribute over the various values of the dependent variable. This means that we will consider as many conditional distributions as there are values of the independent variable. In this problem, there were two values of the independent variable, male and female. Therefore, we considered two conditional distributions: the conditional distribution of opinion for males and the conditional distribution of opinion for females. In order to make a fair comparison of the two distributions, we expressed these conditional distributions as relative frequency distributions. Because we had chosen to make sex the column variable, we obtained the relative frequencies we needed by computing column percentages. If we had chosen to make sex the row variable, we would have obtained the relative frequencies we needed by computing row percentages. Thus, it is not important which variable we choose to make the column variable and which we choose to make the row variable. What matters is the decision we make as to which variable is the independent variable. If the column variable is the independent variable, we will compute column percentages. If the row variable is the independent variable, we will compute row percentages. Most often we find that researchers will make the independent variable the column variable. This is simply a matter of tradition. If it is not possible to determine which is the independent variable, we may wish to calculate both row and column percentages.

Correlational Studies Involving Two Rank-Order Variables

In Chapter 1, we illustrated the nature of rank-order data with the example of the rank in class of the 15 students in Ms. Jones' English class. Now suppose that these same 15 students were also taking Mr. Smith's math class and that the students have also been ranked on the basis of their performance in math. These two sets of ranks constitute a set of bivariate data because each subject has a score on each of two variables. The data are rank-order data, for each of the two scores is a rank. The study is correlational rather than experimental because a single group of subjects has been selected and measured on two variables at one point in time. There was no "group" variable at the start of the study.

Bivariate rank-order data are typically displayed by placing the two sets of ranks adjacent to one another so that the scores of each subject lie opposite each other. In Table 3.4, we have presented the set of bivariate rank-order data corresponding to the ranking of the 15 students in English and math.

Consider first the ranks on English. You will note that there is a slight difference between the ranking of English performance presented in Table 1.2 and that presented here in Table 3.4. The difference has to do with the case of ties. In Table 1.2, we treated ties as we do in everyday life. Where two subjects were tied, we awarded each of them the next available rank. Thus, Billy and Sally, who were tied for second behind John, were both assigned the rank 2. This is common practice in races, where two runners would be said to be tied for second place. Although the next person in the ranking would be assigned the rank 4 rather than 3, we would still regard the two individuals who tied for second as having the rank 2. In Table 3.4, a different procedure is used to handle ties. Rather than giving each of the tied subjects the highest available rank, we assign each subject the average of the ranks that they would have occupied, had they not been tied. If two subjects are tied for second, we recognize that they would have taken the second and third rankings, so we assign to each of them the average of 2 and 3, or 2.5. In the case of the rankings in English, three pairs of subjects were tied. Billy and Sally are tied right behind the top-ranked student, John. Therefore, they are both assigned the rank 2.5, the average of ranks 2 and 3. Similarly, Tom and Doris are both assigned the rank 8.5 (the average of the ranks 8 and 9) and Jane and Ruth are both assigned the rank 10.5 (the average of ranks 10 and 11). In the case of the rankings in math, there were two sets of ties, one of which was a three-way tie. John and Sam are both assigned the rank 1.5 (the average of ranks 1 and 2). Betty, Leona, and Doris were all assigned the rank 8 (the average of ranks 7, 8, and 9). This procedure of averaging ranks in the case of ties is employed whenever we compare two sets of ranks. It is used primarily to increase the precision of certain descriptive and inferential statistics that may be computed on bivariate rank-order data. These statistics will be considered later, in Chapters 5 and 9. For now, we are concerned with what we can learn about bivariate rank-order data simply by inspecting two sets of ranks.

If you look carefully at Table 3.4, you can observe several things about these students. Consider John. He is ranked first in English, tied for first in math. That is, he has ranked high on both measures. Now consider Ted. His rank is 15 in English and 14 in math. He has done poorly on both measures. If you consider all the students, comparing their performance in English to their per-

94

TABLE 3.4
Rank Order of 15 Students in English and Math

Student	Rank in English	Rank in math
John	1	1.5
Billy	2.5	6
Sally	2.5	4
Helen	4	5
Sam	5	1.5
Betty	6	8
Leona	7	8
Tom	8.5	12
Doris	8.5	8
Jane	10.5	3
Ruth	10.5	13
Dick	12	11
Joe	13	15
Paul	14	10
Ted	15	14

formance in math, you will see that there is a tendency for those students who rank high in the one subject to rank high in the other as well. Correspondingly, there is a tendency for those students who rank low on one subject to rank low on the other as well. Of course, this tendency does not hold up with every single subject. Jane, whose rank is 10.5 in English, has the rank 3 in math. However, this discrepancy seems a bit unusual in this set of data, for there is generally a close correspondence between a given subject's rank in one subject and his or her rank in the other. When this situation exists in a set of bivariate ordinal data, the two sets of ranks are said to be *positively related*.

Two sets of ranks may also be *negatively related*. As you may have guessed, a negative relationship implies that a high rank on one of the two measures tends to be accompanied by a *low* rank on the other. As an example of such a set of data, consider the following situation. I am a political reporter and I have assembled a list of nine goals of congressional action. Included among these goals are controversial issues like maintaining a balanced budget, attaining full employment, ensuring equal employment opportunity, strengthening the military, and so on. Now suppose that I find two congressmen, one very conservative and the other very liberal. I ask each of these two politicians to rank the issues in the list from most important to least important. Given the differing political philosophies of the two people who are doing the judging, we would expect that their rankings would differ. In fact, we suspect that a goal considered of high priority by the conservative may be considered of low priority by the liberal. In Table 3.5, we present two sets of ranks as they might appear in this situation.

TABLE 3.5
Nine Political Objectives Ranked in Order of Importance
by Two Politicians

Issue	Rank given issue by conservative	Rank given issue by liberal
Maintaining balanced federal budget	1	7
Controlling inflation	2	8
Achieving a surplus in international payments	3	6
Strengthening military forces	4	9
Reducing crime in the streets	5	5
Conserving natural resources	6	4
Ensuring equal employment opportunity	7	1
Providing quality education to the poor	8	3
Ensuring full employment	9	2

In interpreting this set of bivariate rank-order data, first note that our observations are no longer subjects, rather they are issues. They are bivariate data because each issue has had assigned to it two scores, one from the conservative congressman and the other from the liberal congressman. As we expected, we note a tendency for an item ranked high by the conservative to be ranked low by the liberal, and vice versa. In the case of this set of bivariate rank-order data, this tendency is quite clear. When the data are organized in this way, we can see by inspection that the two sets of ranks are negatively related.

Later, in Chapter 5, we will consider statistical procedures that allow us to quantify the relationship between two sets of ranks. For now, however, let us move on to consider techniques for the presentation and interpretation of bivariate data in which the variables have interval (or ratio) measurement scales.

Correlational Studies Involving Two Interval Variables

As an example of a correlational study involving two interval variables, consider an industrial psychologist who is interested in the relationship between Visual Motor Integration (VMI) and employee work ratings among the production workers in a certain optical instruments company. VMI is similar to what we commonly refer to as eye-hand coordination. It is reasonable to believe that workers with better coordination may perform better on the job. Our psychologist administers a test of VMI to each of the 20 employees. She then obtains work ratings for each of these employees from the production supervisor. As in

96

the case of sex vs. opinion on the ERA, we can identify an independent and a dependent variable here. We assume that VMI is fully developed prior to the age at which the worker begins employment. We further assume that VMI could conceivably influence employee performance and, therefore, employee work ratings, but that it is most unlikely that an employee's work rating would affect VMI. For these reasons, we treat VMI as an independent variable, and employee work rating as a dependent variable. Thus, we are really asking if VMI affects performance, as measured by work ratings.

We measure VMI using a test upon which scores may range from 0 to 20. Employee work ratings are scores that range from 0, reflecting unsatisfactory performance, to 10, reflecting excellent performance. Both sets of scores are assumed to be true interval data. Once we have obtained this set of bivariate interval data, there are several procedures that may be used to help us reach a conclusion regarding the possible relationship between VMI and employee work rating. Here we consider the most basic of these procedures, a technique for organizing and presenting the data in such a way that the raw data themselves shed light on the question.

The Scatterdiagram.   What is the best technique for organizing and presenting bivariate interval data? Can we make a crosstabulation table, as we did in presenting the bivariate data for the categorical variables "sex" and "opinion" on the ERA? Before answering, remember that the interval variables "VMI" and "work rating" may take on numerous values. The data for our group of 20 production workers are shown in Table 3.6. You will readily see that it would be impractical to use these data to construct a crosstabulation table similar to the one we constructed for sex and opinion on the ERA. Among the 20 subjects in the study, there are 12 different score values on VMI and 8 different score values on work rating. Thus, the crosstabulation table that we would construct on the basis of these bivariate data would contain $12 \times 8 = 96$ different cells. Moreover, most of these cells would be *empty!* Obviously, the table would be inordinately large and extremely difficult to interpret. One approach to the problem would be to group the scores on each of the two variables, just as we grouped the scores in the case of some univariate frequency distributions. Often this is the best approach to the problem, and we will consider this grouping procedure in a later section. However, when we group scores, we lose information. Another possibility would be to use the numerical scores to create two sets of ranks, which we could compare as indicated earlier in this chapter. But this procedure would also result in the loss of information. There is a technique, however, that allows us to present bivariate interval data without losing our interval scale, the *scatterdiagram.*

The scatterdiagram representing data on VMI and work rating for the 20 workers in our work rating study is presented in Figure 3.1. In constructing the scatterdiagram, we lay out scales representing each of the two variables in such a way that one of the two scales lies along a horizontal line and the other lies

TABLE 3.6
Visual Motor Integration (VMI) and Work Rating Scores
for 20 Production Workers

Subject	VMI score	Work rating
1	11	6
2	12	7
3	9	5
4	18	9
5	15	6
6	6	4
7	1	2
8	8	4
9	18	8
10	12	7
11	10	6
12	10	6
13	9	7
14	15	8
15	19	9
16	14	7
17	18	5
18	15	8
19	19	10
20	20	8

along a vertical line. Those readers who have studied geometry will probably remember that the horizontal line is referred to as the "X-axis" and the vertical line is referred to as the "Y-axis." Together, these intersecting axes form a rectangular *coordinate system.* This coordinate system is useful to us in bivariate problems because we can represent any subject's scores on *both* of the variables of interest by a *single point* on the coordinate system. Thus, there will be one point on the scatterdiagram for each subject.

In our scatterdiagram, we have chosen to place the scale for VMI along the horizontal or X-axis. The lower values of the variable lie on the left side of the scale and the higher values lie on the right side of the scale. Thus, a subject's score on VMI is given by the *horizontal* position of his point. If his or her score on the VMI is 18, the corresponding point will appear in the diagram in a position directly above the value 18 on the horizontal scale. Referring to the scatterdiagram for a moment, you will see that there are 3 points that lie vertically above the value 18 on the X-axis. These 3 points represent 3 different subjects who have one thing in common: They all scored 18 on the VMI test. These points represent subjects 4, 9, and 17 from Table 3.6.

Organizing and Presenting Empirical Data

Similarly, each subject's score on the variable work rating is represented by the position of his or her point with reference to this scale. We have placed the scale for work rating along the vertical or *Y*-axis. The lower values of the variable lie toward the bottom of the scale and the higher values lie toward the top of the scale. Thus, a subject's score on the variable "work rating" is given by the vertical position of the corresponding point. If the score on work rating is 9, then the point will appear in the diagram in a position directly across from the value 9 on the vertical scale. Referring back to Figure 3.1 once again, you will notice that one of the 2 points that lie above 18 on the VMI scale also lies directly across from 9 on the work rating scale. Thus, the individual represented by this point had the scores 18 on the VMI variable and 9 on the work rating variable. Referring back to Table 3.6, you may determine that these are the scores of subject 4.

To make sure you understand the coordinate system, check to see where you would locate the point representing subject 1 (Table 3.6). This subject had 11 on the VMI test, and a work rating of 6. To locate this point in the scatterdiagram, we first locate the value 11 on the VMI scale (horizontal). Then we go straight up from this point until we find the point directly across from the value

FIGURE 3.1
Scatterdiagram of Work Rating by Visual Motor Integration Score for 20 Production Workers

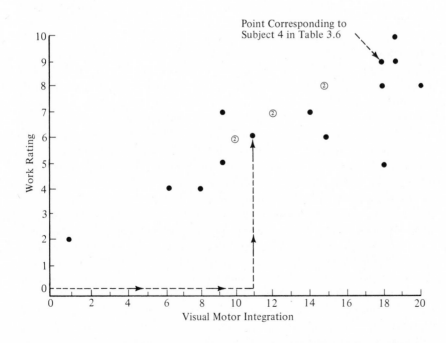

6 on the work rating scale (vertical). The procedure is indicated on the scatterdiagram by two dotted arrows leading to the point representing subject 1. Now perform the same steps to locate the points corresponding to subjects 2 and 3. Do you find points in the scatterdiagram where you believe they should be?

Two more points need to be made in connection with the construction of the scatterdiagram. First, you will notice that there are several points on the scatterdiagram that have small circles drawn around the number 2. This is a way of indicating that two subjects had identical scores on both of the variables. Thus, their respective points would have to be placed in exactly the same location on the scatterdiagram. If we simply drew in the two points one on top of the other, it would be impossible for the reader to tell that there were two subjects whose scores were given by this location. We therefore use the number 2 to indicate that there were two subjects at this particular point. Had there been three, four, or more subjects at any point, we would have used the number 3, 4, and so on.

The second point concerns the way in which we chose to lay out our variables. We placed VMI along the X-axis and work rating along the Y-axis. We could have chosen to reverse the positions of the two scales. The reason we placed them as we did has to do with the question of causation. Remember that we had determined that it made sense to view VMI as the independent variable in this case and work rating as the dependent variable. It has become customary in constructing scatterdiagrams to lay out the independent variable along the horizontal axis. There is no special theoretical reason for this; it is just common practice.

Interpreting Scatterdiagrams.   It is the shape of the scatter of points in the scatterdiagram that helps us to determine whether or not the two variables plotted are related to each other. In our VMI vs. work rating example, we laid out our axes such that low values on the variable VMI would be represented by points on the left side of the scatterdiagram, and high values on this variable would be represented by points on the right. Analogously, low values on the variable work rating would be represented by points near the bottom of the scatterdiagram, and high values on this variable would be represented by points near the top. Thus, an individual with a low score on VMI and a low score on work rating would be represented by a point in the lower left of the scatter of points. Similarly, an individual with a low score on VMI but a high score on work rating would be represented by a point in the upper left of the scatter of points; an individual with a high score on VMI and a low score on work rating would be represented by a point in the lower right of the scatterdiagram; and an individual with a high score on both variables would be represented by a point in the upper right. These relationships have been summarized in Figure 3.2.

Now consider what we mean when we say that two variables are related to one another. We mean that certain scores on one variable tend to be found in conjunction with certain scores on the other. For example, it might be that low

FIGURE 3.2
Verbal Description of Location of Points in Scatterdiagram

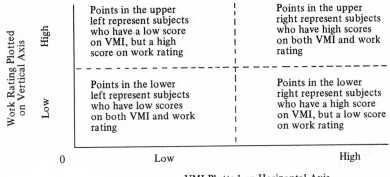

scores on VMI tend to be found in conjunction with low scores on work rating, and high scores on VMI tend to be found with high scores on work rating. That is, if an individual has a low score on VMI, the chances are good that he or she will also have a low work rating. If the individual has a high score on VMI, then the chances are good that he or she will also have a high work rating. This situation is referred to as a *positive linear relationship* between the two variables.

This idea of a positive relationship here is similar to the idea of a positive relationship in a set of bivariate ordinal data, considered earlier in this chapter. There we noted that a positive relationship implied a tendency for an individual who had a high *rank* on one variable to have a high rank on the other. Here, in reference to interval data, we say that a positive relationship implies a tendency for a high *score* on one variable to be associated with a high score on the other. Similarly, low scores will tend to be associated with low scores, and medium scores with medium scores. The word *linear* refers to the fact that bivariate interval data related in this manner will produce a scatterdiagram in which points seem to stretch out lengthwise around an imaginary line.

Referring back to Figure 3.2, it should be clear that when two interval variables have a positive linear relationship, most of the points in the scatterdiagram should fall in the lower left, indicating a low score on both variables, or the upper right, indicating a high score on both variables. These are the sections of the scatterdiagram in which scores on the two variables "match." Now look at Figure 3.1, the actual scatterdiagram from our set of observations. Imagine a pair of lines drawn through the approximate middle of the scatter of points at right angles to each other. If you like, go ahead and sketch in these two lines to form the four sections (quadrants) shown in Figure 3.2. In which of the

101

four quadrants do the majority of the points fall? Do the low scores on VMI tend to be associated with the low scores on work rating? Do the high scores seem to "go together" as well? It is quite clear in this example that the answer to these questions is "yes." The majority of scores fall in the lower left and upper right quadrants. Low scores on one variable are most often associated with low scores on the other. High scores on one variable are most often associated with high scores on the other. Thus, it seems clear that among this group of 20 subjects there is a positive relationship between VMI and work rating. Moreover, because of our initial decision to regard VMI as the independent variable, we would probably conclude that VMI has a positive influence upon job performance. In this group, the higher the VMI score one has, the better the work rating is likely to be.

Of course, the positive relationship between VMI and work rating is not perfect. We can see from Figure 3.1 that there are observations in the upper left and lower right quadrants. There was one subject with a relatively low score of 10 on the VMI test who had a relatively high score of 8 on work rating and there was a subject with a score of 18 on VMI whose work rating was only 5. Thus, our conclusion that there is a positive relationship between the two variables is a subjective one, based on our impression that in a large majority of cases scores on the two variables tend to "go together." Later in this book, we will consider other techniques that provide more objective answers to the question of whether two interval or ratio scale variables are linearly related to each other. We will learn how to compute a numerical index, called a correlation coefficient, that summarizes the degree to which two interval or ratio scale variables are linearly related. For now, however, the primary concern is that you understand what it means to say that two such variables are related. The scatterdiagram is the best way to convey this understanding.

Other Types of Linear Relationships.   Two interval variables may also have a *negative linear relationship.* In this case, high values on one variable tend to occur in conjunction with low values on the other. To illustrate this situation, consider the relationship between waist size in centimeters and score on a test of physical stamina, such as a decathlon. Suppose we go into a boy's high-school physical education class for subjects. We obtain everyone's waist measurement. Then we arrange for all the subjects to participate in the events of the decathlon. Under the Olympic scoring system, contestants receive a certain number of points for each event. The better they do, the more points they receive. We add up the total number of points the subjects receive over the 10 events. We use this point total as a measure of overall performance in the decathlon competition. The better the overall performance, the higher the point total. A scatterdiagram representing a hypothetical group of 30 physical education students is presented in Figure 3.3. How does this scatterdiagram differ from the scatterdiagram presented in Figure 3.1? If an individual has a relatively high score on the variable "waist size," what kind of score is he likely to have on the variable "total points

102

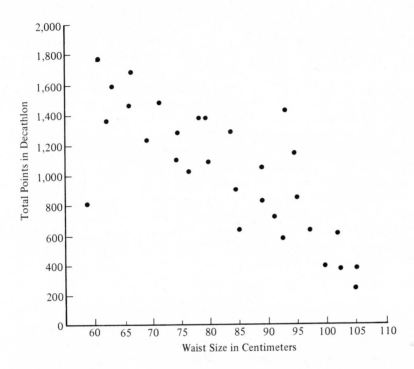

in decathlon?" What if he has a relatively low score on the variable "waist size"?
Referring to Figure 3.3, we find that most of the points in this scatterdiagram
fall in the upper left and the lower right. Thus, the points are falling in a manner
opposite to the way in which they fell in the case of VMI vs. work rating. Points
in the scatterdiagram also seem to fall along a straight line, but this time the
imaginary line would run from the upper left of the scatterdiagram to the lower
right. This means that those subjects who have the lowest scores on the variable
"waist size" tend to have the highest scores on the variable "total points in decathlon." Similarly, those subjects with the largest waist sizes tend to have the
lowest point totals. This makes sense, for we expect relatively thin people to
perform better in such an event than relatively fat people. In this situation, we
say that the two variables are negatively related, because a high score on one of
the variables tends to go with a low score on the other. Such a situation is also
referred to as an *inverse linear relationship.*

It is also possible for two interval variables to have *no linear relationship.* In this situation, a high score on one of the two variables will not tend to be found with any particular type of score on the other variable. In this case, a subject's score on one of the two variables may not give us any hint as to how he or she might have scored on the other variable. As an example of this situation, consider the relationship between height in centimeters and college grade point average. Suppose we go to a university dormitory and find a group of 50 female students. We measure the height of each of the subjects, and we obtain her grade point average from the registrar's office. The scatterdiagram representing a hypothetical group of 50 female university students is presented in Figure 3.4. How does this scatterdiagram compare to those presented in Figures 3.1 and 3.3? If we knew that an individual had a relatively low score on the variable "height," would this give us an idea about where her grade point average would be? What if we knew she had a relatively high score on the variable height? Referring to Figure 3.4, we see that the points do not tend to fall in the lower left and upper right, as they did in the scatterdiagram of VMI vs. work rating. Nor do they tend to fall in the upper left and lower right, as they did in the scatterdiagram of waist size vs. points in decathlon. Instead, the points in Figure 3.4 seem to fall randomly, forming a circular shaped scatter of points. No linear relationship exists between these two variables.

Grouping Data in Bivariate Studies

There are several situations in which it is useful to group data to form a crosstabulation table. If both of the variables in a bivariate situation are categorical, the researchers may find that there are too many categories for one or both of the variables to produce a compact, intelligible table. Generally speaking, crosstabulation tables are easiest to interpret when there are few cells. Thus, it may make sense to combine categories on a particular variable before constructing the table. In this case, the researcher should be careful to combine those categories that are most alike. For example, if we had defined a variable called political party that had 13 values including a broad spectrum of large and small parties, it would probably make sense to place the various conservative parties in one group and the various liberal parties in another.

It may also make sense in some bivariate studies to group ordinal or interval data so that a crosstabulation table can be constructed. The crosstabulation table enables us to summarize a large amount of data in a small space. Moreover, crosstabulation tables with relevant conditional percentages are concrete and, thus, more easily interpreted by most readers than two sets of ranks or a scatterdiagram.

Let us consider how we might go about constructing a crosstabulation table from a set of bivariate interval data. The first step would be to group each of the two sets of scores into a reasonable number of categories. You will recall that the crosstabulation table will have as many cells as the product of the number of

Organizing and Presenting Empirical Data

FIGURE 3.4
Scatterdiagram of Grade Point Average vs. Height
for 50 Female University Students

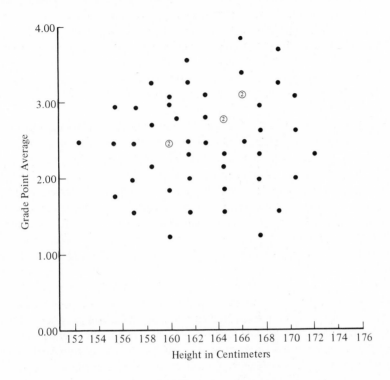

values of one variable and the number of values of the other variable. Therefore, our purpose in grouping is to reduce the number of cells in the table to a reasonable number. In grouping interval data to construct crosstabulation tables, we may follow the procedures outlined in Chapter 2 for constructing grouped frequency distributions. However, in order to keep down the number of cells in the table, we will often use fewer categories for each of the two variables to be crosstabulated than we might use if we were constructing univariate grouped frequency distributions for each of the two variables separately. Generally, the manner in which we choose to group the variables will depend on the particular nature of the variables employed. Consider the way in which we might group our data if we wished to construct a crosstabulation table for the data on the relationship between VMI score and work rating.

With regard to the independent variable, VMI score, we might use these three categories:

1 Low (corresponding to the values 0 to 9),
2 Medium (corresponding to the values 10 to 14), and
3 High (corresponding to the values 15 to 20).

With regard to the dependent variable, work rating, we might use two categories:

1 Low (corresponding to the values 0 to 6) and
2 High (corresponding to the values 7 to 10).

Notice that in neither case did we use equal intervals in grouping these data. Rather, an effort was made to include an equivalent number of cases in each category of each variable and to reduce the total number of categories to a reasonable number.

When we have decided upon the categories for our two variables, we can go down our list of data and assign each subject the appropriate categorical value for each of the two variables. This procedure is illustrated in Table 3.7. Once we have completed the transition from raw interval data to grouped categories for all subjects, we can construct the crosstabulation table, following the procedures outlined previously. The resulting table is presented, along with appropriate percentages, in Table 3.8. Notice that, as before, we computed percentages of the dependent variable (work rating) in each category of the independent variable (VMI). It is clear from Table 3.8 that we can make a number of statements that we could not have made very easily on the basis of the scatterdiagram. For example, we can say that of the 5 workers who did score low on VMI, only 1 (20.00 percent) received a high work rating. In contrast, of the 9 workers who had high VMI scores, 7 (77.78 percent) received high ratings. Comparing percentages in this manner makes the relationship between VMI and work rating concrete. It is clear that a worker has a much greater chance of receiving a high work rating with a high VMI score than with a low VMI score. Of course, we should come to the same conclusion regarding the relationship between these two variables whether we are looking at a scatterdiagram or a crosstabulation table. However, it may be more convenient to present the data in the form of a crosstabulation table. This would depend on such factors as the level of sophistication of the intended readers.

Bivariate rank-order data may also be collapsed into categories and crosstabulated. Furthermore, the collapsing procedure may be used when the researcher has bivariate data in which each variable has a different level of measurement. One variable may be categorical and the other rank-order, one variable may be rank-order and the other interval or ratio scale, and so on. In these situations, we frequently group data to form a crosstabulation table.

TABLE 3.7
Transition from Raw Interval Data to Grouped Data
for VMI Score and Work Rating

Subject number	Raw score VMI	Category of VMI	Raw score work rating	Category of work rating
1	11	Medium	6	Low
2	12	Medium	7	High
3	9	Low	5	Low
4	18	High	9	High
5	15	High	6	Low
6	6	Low	4	Low
7	1	Low	2	Low
8	8	Low	4	Low
9	18	High	8	High
10	12	Medium	7	High
11	10	Medium	6	Low
12	10	Medium	6	Low
13	9	Low	7	High
14	15	High	8	High
15	19	High	9	High
16	14	Medium	7	High
17	18	High	5	Low
18	15	High	8	High
19	19	High	10	High
20	20	High	8	High

ORGANIZING AND PRESENTING DATA IN EXPERIMENTAL AND PSEUDO-EXPERIMENTAL STUDIES

In experimental and pseudo-experimental studies, we always have two or more distinguishable groups of subjects who are being compared with respect to some variable. When we organize or present data graphically in these problems, we are using techniques that make it easy to compare the groups. These techniques are the same, whether the study is a true experiment or a pseudo-experimental problem. However, the conclusions that we may draw based on the comparison may not be the same for an experimental study as for a pseudo-experimental study. This is because of the potential influence of intervening variables. A well-planned experimental study will control for the possible effects of intervening variables and allow us to make strong statements regarding causation. This is not the case

107

TABLE 3.8
Number and Percentage of 20 Production Workers of Three Different VMI Levels
Who Were Rated High or Low by Their Supervisors

		VMI							
		Low		Medium		High		Total	
		N	Percentage	N	Percentage	N	Percentage	N	Percentage
Rating	Low	4	80.00	3	50.00	2	22.22	9	45.00
	High	1	20.00	3	50.00	7	77.78	11	55.00
	Total	5	100.00	6	100.00	9	100.00	20	100.00

in the pseudo-experimental study. Before we consider the techniques used to compare independent groups in experimental and pseudo-experimental studies, let us take a moment to consider the role of intervening variables.

Intervening Variables and the Issue of Causation in Experimental and Pseudo-Experimental Studies

In true experimental studies, groups are formed by random assignment. In Chapter 1, we considered the question "Is my new method for teaching Spanish really more effective than the old one?" And we outlined the procedure that might be employed to answer this question. We indicated that the experimenter would probably use random assignment to form two groups, one of which would receive the new method while the other received the old method. Through the process of random assignment, the experimenter is attempting to make the two groups as similar as possible with respect to every variable except the critical variable, teaching method. In addition, the experimenter will be careful to ensure that the treatments given to the two groups are identical in every respect except teaching method. Given this approach, we can be reasonably confident that if we find any differences in Spanish achievement between the two groups, they must be because of the difference in method. Thus, the experimental nature of the study enables us to make strong statements regarding causation.

Now let us consider a pseudo-experimental situation. Suppose we wished to determine whether males differ from females with respect to Spanish achievement. Suppose we select a sample of males and a sample of females from among the population of students who have never studied Spanish. We give each group

108

the same Spanish course, then we test all subjects in both groups. Let us assume that we find that the females score higher than the males. Are we justified in concluding that being born a female causes one to do better in Spanish? No. We are justified in concluding that the women in our study did better than the men, but we are not justified in stating that sex causes differences in Spanish achievement. The reason is that other factors may be involved here. Perhaps the culture in which we live makes women more interested in foreign languages and men more interested in some other school subject. If this were true, women would tend to perform better in Spanish, but not really because they were women. Rather, they would tend to perform better because an *intervening variable,* culture, affected them differently from the way it affected the male subjects. In a true experiment, intervening variables are controlled. In a pseudo-experimental situation, they are not. Thus, any conclusion regarding causation in a pseudo-experimental study must be made with great caution. With this caution in mind, we proceed to consider the techniques of group comparison that are used in both experimental and pseudo-experimental studies. We illustrate these techniques by referring to the problem of comparing two methods of teaching Spanish. In experimental and pseudo-experimental studies, there will always be a categorical variable referring to group that we regard as the independent variable. Which techniques we employ to organize and present the data will depend on the level of measurement of the second variable, the variable on which the groups are to be compared.

Group Comparisons on Categorical Variables

In our study of Spanish teaching methods, it is possible that we would employ a categorical measure of Spanish achievement. We might, for example, conduct an oral examination for each student in which the student could be judged as "able to communicate" or "unable to communicate" in Spanish. Had we chosen to employ such a measure as our dependent variable, we would present our data in the form of a crosstabulation table, much as we did in the ERA study. Such a table might appear as in Table 3.9.

However, it is often the case in experimental studies that the dependent variable is regarded as an interval variable. Here, it is quite possible that Spanish achievement would be measured using a 25-item, multiple-choice examination, so that each student tested would have a numerical score ranging from 0 to 25. Of course, we would have the option of collapsing these data into a limited number of categories to form a crosstabulation table. We might establish the categories "15 or more questions right" vs. "less than 15 questions right." On the other hand, recognizing that we lose some information in collapsing categories in this manner, we might wish to employ a technique in which numerical scores could be retained. In this case, the key question would be how to present the distribution of Spanish achievement scores in group A and in group B in such a way that they could be easily compared.

TABLE 3.9
Number and Percentage of Group A and Group B Students
Judged "Able to Communicate" or "Unable to Communicate" in Spanish

Spanish Achievement Rating	Group				
	A (New teaching method)		B (Old teaching method)		
	N	Percentage	N	Percentage	
Able to communicate	40	80.00	24	48.00	
Unable to communicate	10	20.00	26	52.00	
Total	50	100.00	50	100.00	

One possibility is to construct the frequency distributions for each group and to place these distributions side by side. This method is illustrated in Table 3.10. By looking over these two frequency distributions, the reader can see that there are relatively more students in group A than in group B who obtained high scores on the achievement test. The process of comparison can be made easier for the reader if the data are presented in graphic form. The technique most frequently employed in the graphic comparison of two frequency distributions involves the use of frequency polygons. The frequency polygons representing these two distributions of Spanish achievement are presented in Figure 3.5. You will recall that frequency polygons serve the same function as histograms, but they are constructed somewhat differently, giving them a jagged appearance rather than the step-like appearance of the histogram. We use frequency polygons rather than histograms when we are comparing two distributions because their jagged appearance makes it easier for the reader to distinguish between two figures drawn on the same set of axes. If we drew two histograms on the same set of axes, the rectangular bars would make it difficult to distinguish between the two distributions where they overlapped. The two distributions graphed in Figure 3.5, however, are easily distinguishable from one another. It is quite clear to us in looking at Figure 3.5 that group A performed better on the Spanish achievement test than did group B.

Of course, there is no established rule that prescribes exactly which techniques should be used to present data when group comparisons are called for. In the Spanish teaching methods study, we might have chosen to present the data for groups A and B in the form of two cumulative frequency distributions

Organizing and Presenting Empirical Data

tabled side by side. Similarly, we might have constructed the cumulative frequency graphs for the distributions of Spanish achievement test scores on the two groups. By drawing these two cumulative frequency graphs on the same set of axes, we could further facilitate the comparison of the two groups.

When "Group" is an Interval Scale Variable

It is possible for an experimental study to have interval scale variables for both the independent variable and the dependent variable. An example of such a problem would be a situation in which a psychologist is concerned with the relationship between intensity of stimulus and reaction time. Suppose the psychologist has an apparatus that enables him to control precisely the decibel level

TABLE 3.10
Frequency Distributions of Spanish Achievement
for 50 Subjects in Group A and 50 Subjects in Group B

Score on 25-item Spanish test	Frequency of score in group A	Frequency of score in group B
1	0	0
2	0	1
3	0	2
4	0	3
5	0	3
6	1	4
7	1	5
8	2	6
9	1	7
10	2	6
11	3	4
12	3	2
13	4	2
14	4	0
15	5	2
16	6	1
17	4	2
18	4	0
19	2	0
20	3	0
21	2	0
22	1	0
23	2	0
24	0	0
25	0	0
	50	50

FIGURE 3.5
Frequency Polygons of the Variable Spanish Achievement for 50 Subjects
in Group A and 50 Subjects in Group B

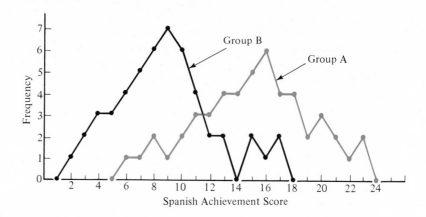

of an auditory stimulus. He may decide to run an experiment in which subjects
will be exposed to differing levels of the stimulus to see if subjects exposed to
louder sounds respond more quickly. The experimenter randomly assigns 10
subjects to each of 10 different treatment groups. The subjects in the first group
receive a stimulus of 10 decibels, and subjects in successive groups receive stimuli
of 20, 30, 40, 50, and so on, up to 100 decibels. In all cases, the average reaction
time of each subject in each of these different groups is recorded.

This is clearly an experimental study, for the researcher has assigned subjects to different treatment groups. He has established the independent variable,
auditory stimulus intensity, as having 10 different values. Moreover, the values
of the independent variable are not simply categories, like the sex variable of
the ERA example or the group variable of the teaching method example. Instead, the values of the independent variable are numerical scores that may be
located along an interval scale. In this study the dependent variable, reaction
time, is also an interval variable. Thus, we have a bivariate experimental problem
with two interval scale variables. In our previous discussion of correlational
problems, we showed how a scatterdiagram could be used to present the data
from a bivariate correlational problem with two interval scale variables. We can
use the same technique in the bivariate experimental problem. Figure 3.6 presents a hypothetical scatterdiagram for the data from the reaction time study.
You will notice that this scatterdiagram differs slightly in appearance from those

Organizing and Presenting Empirical Data

in Figures 3.1, 3.3, and 3.4. This is because the independent variable (stimulus intensity) has been manipulated by the experimenter. There are 10 subjects in each of the groups on the independent variable, and all 10 subjects in each group have exactly the same score. This would not be the case in a correlational study, where neither variable was controlled. Later on in this book, we will consider methods used in drawing inferences from data such as those presented in Figure 3.6. For now, it is sufficient that you understand that there is a difference between correlational problems involving two interval variables and experimental problems involving two interval variables, and that the scatterdiagram may be used to present the data in either case.

FIGURE 3.6
Scatterdiagram of Reaction Time vs. Stimulus Intensity
for 10 Groups of Subjects ($N = 100$)

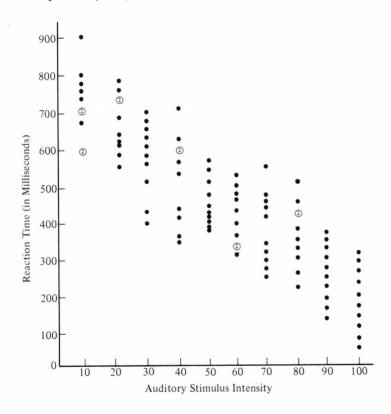

113

ORGANIZING AND PRESENTING
DATA IN STUDIES
INVOLVING TWO SETS
OF DEPENDENT MEASURES

As already noted, the techniques of organization and presentation of data that are used in studies involving the comparison of independent groups are generally applicable to studies involving two sets of dependent measures. For example, in the pretest and posttest research situation, we might employ paired cumulative frequency graphs to compare the performance of subjects at the time of the pretest to the performance of the same subjects at the time of the posttest. This type of comparison is illustrated in Figure 3.7, which shows the hypothetical pretest and posttest distributions of reading scores for a group of 50 remedial reading students. These paired cumulative graphs clearly show the overall superiority of subjects at the time of the posttreatment testing. If this is unclear to you, consider the following question: "What number of students scored less than 40 on the reading test at each testing?" Using the cumulative graphs to estimate

FIGURE 3.7
Cumulative Frequency Graphs of Pretreatment and Posttreatment
Reading Scores for 50 Remedial Reading Students

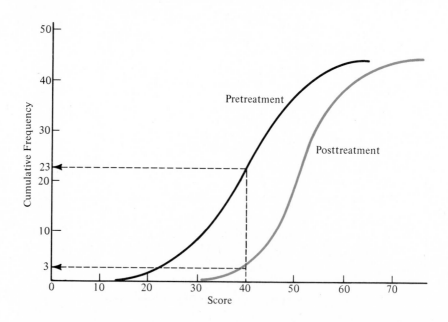

Organizing and Presenting Empirical Data

these numbers, we find that only 3 subjects scored less than 40 at the time of the posttreatment testing, compared to 23 subjects at the time of pretreatment testing. The same pattern holds throughout the distributions of pretreatment and posttreatment scores. For any given reading score, a greater number of subjects fell *below* it at the time of the pretreatment than at the time of the posttreatment.

In the experimental problem in which there is a dependency between the two sets of measures because of matching, the groups may still be compared using paired frequency polygons. The issue of dependency in group comparisons becomes more important when we move from techniques of graphic presentation to inferential statistics.

THE CLASSIFICATION MATRIX

In this chapter, we completed our discussion of techniques for organizing and presenting empirical data. We use these techniques to describe data in both descriptive and inferential problems. Table 3.11 summarizes the techniques discussed in Part 2.

TABLE 3.11
Classification Matrix:
Techniques for Organizing and Presenting Empirical Data

		Level of Measurement of Variable(s) of Interest		
		Categorical: Nominal or Ordered Categories	Rank-Order	Interval Scale (or Ratio-Scale)
Univariate Problems		Frequency Distribution, Relative Frequency Distribution; Bar Graph, Pictograph, Circle Graph	Set of Ranks	Frequency Distribution, Grouped Frequency Distribution, Cumulative Frequency Distribution; *Graphs for Discrete Interval Data:* Discrete Graph, Step Function; *Graphs for Continuous Interval Data:* Histogram, Frequency Polygon, Cumulative Frequency Graph
Bivariate Problems	Correlational Problems	Bivariate Frequency Distribution: The Crosstabulation Table	Two Sets of Ranks	Scatterdiagram
Bivariate Problems	Experimental and Pseudo-Experimental Problems: Independent Groups	*Independent and Dependent Variable Categorical:* The Crosstabulation Table		*Independent Variable Categorical, Dependent Variable Interval:* Paired Frequency Polygons, Paired Cumulative Frequency Graphs. *Independent and Dependent Variable Interval:* Scatterdiagram
Bivariate Problems	Dependent Measures: Pretest and Posttest or Matched Groups	Crosstabulation Table		Paired Cumulative Frequency Graphs, Paired Frequency Polygons

Type of Problem

Key Terms

Be sure you can define each of these terms.

crosstabulation table
bivariate frequency distribution
conditional distribution
marginal frequency
marginal distribution
direction of causation
reciprocal causation
positive relationship
negative relationship
scatterdiagram
coordinate system
linear relationship
unrelated variables
intervening variables

Summary

In Chapter 3, we considered the organization and presentation of bivariate data. We looked at correlational problems, in which two variables are observed with no attempt to control either one; experimental and pseudo-experimental problems, in which two or more groups of independent measures are compared; and problems in which two groups of dependent measures are compared.

We noted that in a bivariate problem it is sometimes possible to identify one of the two variables as the independent variable and the other as the dependent variable. We recalled from Chapter 1 that in an experimental problem, one variable is clearly the independent variable because it is manipulated by the experimenter. In other bivariate problems, however, this distinction is not always clear. In some correlational or pseudo-experimental problems, it is possible to set up a plausible direction of causation, but only in the true experimental situation can we infer causation.

When the variables are both nominal, both ordered categories, or both interval (or ratio) variables, we summarize the data from

3

Student
Guide

a bivariate correlational problem with a bivariate frequency distribution in order to examine the relationship between the two variables. When the variables are both nominal or both ordered categories, we make a crosstabulation table. In the case of interval data, we make a scatterdiagram, a graph of the bivariate frequency distribution. For rank-order data, we examine the relationship between the two variables by listing the pairs of ranks together.

In any case, the purpose of these methods is to see which values of one variable occur with the various values of the other; that is, to see if a relationship exists between the two variables. With crosstabulation tables, we look at the pattern of cell frequencies to determine if there is a relationship. With rank-order and interval data, we examine the pattern of paired values. If high values of one variable go with high values of the other (and low with low), we say there is a positive relationship between the variables. If high values on one variable go with low values on the other and vice versa, we have a negative relationship. In the case of interval data, we noted that when the points in the scatterdiagram are clustered around an imaginary line, we have a positive or negative linear relationship. We also examined the situation in which there is no relationship between two variables.

Finally, we looked at experimental and pseudo-experimental problems in which independent groups are compared on some variable of interest. Because our purpose in these situations is to compare the distributions of our variable of interest, we can use the methods for describing univariate distributions, such as the frequency polygon and the cumulative frequency graph. In such cases, we repeat the process for each group and then compare the groups. We can also use these methods to compare two sets of dependent measures.

Review Questions

To review the concepts presented in Chapter 3, choose the best answer to each of the following questions.

1 If we measure 400 people on sex and handedness, we should display the results in
   _____a a crosstabulation table.
   _____b a scatterdiagram.

2 A developmental psychologist studies the relationship between motor coordination and age. For these two variables, there is the possibility for
   _____a a direction of causation.
   _____b reciprocal causation.

3 If we cross-classified 1,000 people on the variables "sex" and "handedness" and observed that the proportion of female left-handers was the same as the proportion of male left-handers, we could say that for the group studied
   _____a the two variables are dependent.
   _____b the two variables are unrelated.

Organizing and Presenting Empirical Data

4 If we are interested in the relationship between the variables "sex" and "reading achievement" in a sample of 500 sixth-graders, we should picture the results in

_____a a scatterdiagram.

_____b two relative frequency polygons.

5 If we are looking at a scatterdiagram picturing the relationship between arithmetic achievement and reading achievement for a sample of sixth-graders, and we are interested in the arithmetic performance for those who read at grade level, we should look at

_____a a conditional distribution of arithmetic scores.

_____b a conditional distribution of reading scores.

6 If we are looking at a bivariate frequency distribution showing the relationship between arithmetic achievement and reading achievement, and we want to see the performance of all the children on reading regardless of their arithmetic achievement, we should look at the

_____a marginal distribution of reading achievement.

_____b conditional distribution of reading achievement at each level of arithmetic.

7 For the term positive relationship to have meaning, both variables

_____a must be ordinal, interval, or ratio scales.

_____b can be nominal scales.

8 A parabola is an example of

_____a a linear relationship.

_____b a nonlinear relationship.

9 If visual acuity is correlated with age for ages ranging from birth to death, we would expect to see

_____a a positive relationship.

_____b a nonlinear relationship.

10 Suppose we are looking at a scatterdiagram showing a negative relationship between test anxiety and performance on a statistics exam. We could interpret this to mean that

_____a the more anxious you are about tests, the better is your performance.

_____b those people with little text anxiety tend to get high scores.

11 In the above example, we should say

_____a being test anxious causes low scores on the test.

_____b nothing about causation.

12 For a group of 1,000 mothers, we have recorded the average number of cigarettes smoked per day during her pregnancy and the weight of her baby at birth. A scatterdiagram shows that as the average number of cigarettes smoked increases, the weight of the baby decreases. We could say that the two variables are

_____a negatively related.

_____b positively related.

13 In a two variable problem, if one variable is a treatment variable and the other is a performance variable and if subjects are randomly assigned to levels of treatment, we have

_____a an experimental problem.

_____b a pseudo-experimental problem.

14 In a two variable problem, an uncontrolled variable that has an effect on the subjects' performance is called

_____a an independent variable.

_____b an intervening variable.

15 One factor that differentiates an experimental problem from a pseudo-experimental problem is that in an experimental problem

_____a subjects are randomly assigned to treatments.

_____b intervening variables are uncontrolled.

16 In a crosstabulation table, if for each value of the column variable, the distribution of the row variable is proportionately the same, we say the two variables are

_____a positively related.

_____b unrelated.

Problems for Chapter 3: Set 1

1 Use the data in Table 3.2 to help you answer the following questions.
   a What percentage of the total population are males who are in favor?
   b What percentage of the population is male?
   c What percentage of the population are in favor?
   d What percentage of the male population are in favor?
   e Compare your answers to c and d. Are they the same? What does this mean?
   f Make a relative frequency distribution of the sex variable for those voters who were in favor. That is, for the 64 who were in favor, tell what proportion are male and what proportion are female. We call this the conditional distribution for the sex variable for those in the population who were in favor.

120

g Write the conditional distribution for the sex variable for those who were opposed and the distribution for those who had no opinion.

h Compare the three conditional distributions of the sex variable prepared in f and g. Are they the same? What does this illustrate?

2 The following information on the variables "sex of degree recipients" and "university school" was collected from a large university:

	Sex of Recipient	
	Male	Female
Agriculture	4%	1%
Psychology	9%	6%
Engineering	25%	11%
Science	23%	8%
Nursing	4%	9%

(row label: School in University)

Answer the following questions about the crosstabulation table.

a What percentage of the graduates are male agriculture graduates?

b What percentage of the males are in the school of agriculture?

c What percentage of the graduates are male psychologists?

d What percentage of the male graduates are in psychology?

e What percentage of the psychology graduates are male?

f Make a conditional distribution of school in university for males and for females.

g Write the marginal distribution of school in university.

h Compare the distributions of f and g.

i Is there a relationship between sex of degree recipient and school in university?

3 In looking at the statistics for his football team, a manager notices the following table, which shows the breakdown of the number of passes thrown by his quarterback by the outcome of the game.

	Outcome of Game		
	Win	Lose	Tie
Less than 30 passes	47	15	1
At least 30 passes	12	33	3

(row label: Number of Passes)

a What percentage of games won show the quarterback throwing at least 30 passes?

b Of the total number of games played, in what percentage did the quarter-back throw at least 30 passes?

c Write down the two conditional distributions for the variable outcome of game: one for less than 30 passes and one for at least 30 passes.

d Write down the marginal distribution for the variable outcome of game.

e Is there a relationship between the number of passes thrown by the quarterback and the outcome of the game?

f Would you say that the number of passes thrown by the quarterback *causes* the outcome of the game to be win, lose, or tie?

4 A large city school system administers a qualifying test to candidates who wish to become supervisors. An affirmative action committee raises an objection to this test. They say the test discriminates against women. To support their claim, they present the results of the supervisors' test for a typical year. They are as follows:

Of 270 women applicants, 129 passed the exam.

Of 1101 men applicants, 718 passed the exam.

a Organize a crosstabulation table for these results.

b Do the data suggest that performance on the exam is dependent on sex of the applicant? Give evidence.

c Do these results support the affirmative action committee's complaint?

Set 2

1 For the following pairs of variables, decide whether it is possible to establish a possible direction of causation. If it is, identify the independent variable.

a high-school rank	vs.	success in college
b outcome of game	vs.	number of passes
c age	vs.	sex
d pretest	vs.	posttest
e intelligence	vs.	leadership ability
f level of education	vs.	degree of racial prejudice
g mental alertness	vs.	sleep deprivation
h level of education	vs.	sex
i height	vs.	weight
j drinking coffee	vs.	smoking cigarettes
k smoking cigarettes	vs.	lung cancer
l age	vs.	visual acuity
m age	vs.	speed at running 100-yard dash
n hair color	vs.	eye color

Set 3

1 a What makes a relationship linear? Nonlinear?

b What makes a relationship between two variables positive?

c What makes a relationship between two variables negative?

d When would we say there is no relationship between the variables?

Organizing and Presenting Empirical Data

2 For each of the scatterdiagrams presented below, decide whether the two variables are related; if related, whether the relationship is linear or not linear; and if linear, whether it is positive or negative.

a)

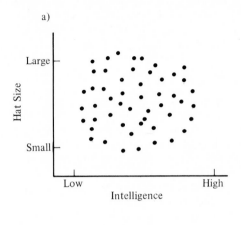

b)

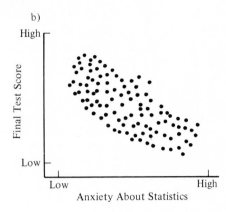

c)

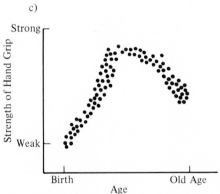

d)

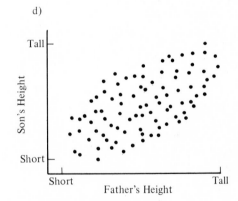

3 For the following pairs of variables, indicate whether or not the two would be related; if related, whether or not the relationship would be linear or not linear; and if linear, whether it would be positive or negative.
a age vs. time to run a mile
b age vs. height
c age vs. intelligence
d intelligence vs. shoe size
e improvement scores from pretest to posttest on a 50-item arithmetic test vs. intelligence
f temperature vs. amount of clothing worn
g length of a side of a square vs. area of a square
h length of a side of a square vs. perimeter of a square

4 a Using the data in Data Set A (see Chapter 2 problems), make a scatterdiagram for years of education vs. annual income.

123

b Divide the scatter of points between 12 and 13 years of education and between $16,000 and $18,000 on annual income. How can you tell that there is a relationship between years of education and annual income?

c Look at the distribution of annual income for those with 10 years of education, 12 years of education, and 16 years of education. How do these conditional distributions compare? How does this help to show that annual income and years of education are related?

Set 4

1 An experimenter believes that eye-hand coordination is an important factor in children's ability to print legibly. To gain support for the possible truth of her hunch, she devises a training program in eye-hand coordination. She needs evidence that her training program really does improve eye-hand co-ordination. She takes 40 young children and administers a test of eye-hand coordination, called the Visual Motor Integration test. She then trains the children, and at the end of the training period, she retests them, using an alternate form of the Visual Motor Integration test. Following are the results:

| | Frequency of Score Values | |
VMI score values	Pretest	Posttest
2	1	
3	4	
4	3	
5	1	
6	6	2
7	8	2
8	4	9
9	8	9
10	2	7
11	3	3
12		3
13		3
14		1
15		0
16		1
Totals	40	40

a What do these results suggest about the two distributions of VMI scores?

b Draw a frequency polygon for each distribution. Place both graphs on the same set of axes and compare the shapes of the graphs. What does this say about performance on the posttest vs. performance on the pretest?

2 A nutritionist is interested in studying newborn babies. She has two hospitals from different environments available for her study. One variable of interest is the weights of the newborns. She decides that each hospital could

Organizing and Presenting Empirical Data

be used to define a separate population of babies. One question of interest is whether the two populations of babies are different on the variable "birthweight." To help her answer the question, she would like to look at the two distributions. They are presented below:

Frequency Distributions for Weights of Newborn Babies
From Hospital I ($N = 200$) and Hospital II ($N = 250$)

Weight intervals	Hospital I	Hospital II
(measured to nearest tenth of a kilogram)	Frequency	Frequency
2.3 - 2.4	1	0
2.5 - 2.6	4	2
2.7 - 2.8	13	10
2.9 - 3.0	50	35
3.1 - 3.2	65	50
3.3 - 3.4	37	80
3.5 - 3.6	12	35
3.7 - 3.8	8	18
3.9 - 4.0	6	1
4.1 - 4.2	4	19
Total	200	250

a Draw two frequency polygons, one for each distribution, on the same axis. What difference do you observe in the two distributions? Why might this difference be misleading?

b Next draw two relative frequency polygons on the same axis. What difference do you observe in the two distributions?

c Do the data suggest any differences in the two populations of babies on the variable "birthweight"?

3 A sample of 67 subjects consists of 30 males and 37 females. All subjects are given an anxiety measure that classifies them as high, medium, or low anxiety. The results show that of the males, 10 are high, 10 are medium, and 10 are low. For the females, 5 are high, 20 are medium, and 12 are low. Use bar graphs to show how the two groups compare on the variable "anxiety." Do men appear to be more anxious than women?

4 An experimenter has 87 people divided into an experimental group ($N = 42$) and a control group ($N = 45$). She is interested in seeing how the two groups compare on a problem-solving task. She administers the task to each subject and scores a pass if the problem is solved in five minutes and a fail otherwise. Next, she trains the experimental group in problem-solving techniques. She spends an equivalent amount of time with the control group but does not train them in problem-solving techniques. At the end of the experiment, both groups are given another problem-solving task. The results are:

Bivariate Problems

		Pass	Fail	Totals
Experimental	Pre	11	31	42
	Post	25	17	42
Control	Pre	15	30	45
	Post	18	27	45

a Make a bar graph that compares experimental to control on pretest.
b Make a bar graph that compares experimental to control on posttest.
c Make a bar graph that compares experimental pre to experimental post and control pre to control post.
d Which group is better on pretest?
e Which group is better on posttest?
f Which group seems to improve the most?
g To what do you attribute this improvement?

Set 5

In Set 2 of the problems for Chapter 1, a set of situations is presented. We classi-fied the problems as to type of problem presented and level of measurement of the variables. In Chapters 2 and 3, we have presented techniques for organizing and graphing collected data sets. To see how some of these techniques fit into our organizational scheme, reread problems 1 through 10 of Set 2, Chapter 1 and briefly describe how the collected data for each situation might be tabled and summarized.

Organizing and Presenting Empirical Data

Descriptive Statistics 13

4

Descriptive
Statistics
for Univariate
Problems

DESCRIPTIVE STATISTICS FOR UNIVARIATE PROBLEMS: AN OVERVIEW

In Part 2 of this book, we considered techniques that we can use to organize and present data. These techniques are designed for efficiency. We want our tables and graphs to convey as much information as possible to a reader in the smallest amount of time and space. In Part 3, we will consider techniques that we can use to describe data with numbers. These numbers are all referred to as *descriptive statistics.* You should remember from Chapter 1 the distinction between descriptive problems and inferential problems. In descriptive problems, we have access to an entire population of observations which we seek to describe. In inferential problems, we have only a sample of observations from a population and we seek to infer something about the larger population from the sample. In the two chapters of Part 3, we are concerned only with descriptive statistics. Like the techniques for organizing and presenting data considered previously, descriptive statistics are designed for efficiency. Often we can convey a great deal of information about a set of data with just a few numerical indices. But what kind of information do we wish to convey with these numbers? As you might expect, this will depend upon the type of problem and the type of data we have.

In this chapter, we are concerned with univariate problems. When we have measured a group of subjects on a single variable, there are typically two types of information that might be useful: information regarding the entire group and information regarding the position of a single subject with respect to the other subjects in the group. To illustrate these two types of information, consider the distribution of final exam scores for the 50 students in Dr. Brown's statistics class (see Table 1.3).

If we were concerned with the effectiveness of Dr. Brown's course, we would be interested in describing the entire set of scores as efficiently as possible. We would want to know how well the typical student mastered the course material. We might also want to know whether all the students performed about the same, or whether some did very well and others did very poorly. On the other hand, if we were concerned with the performance of a single student, we would probably be less concerned with the performance of the class as a whole and more concerned with where that one individual stood relative to the other members of the class.

When we are interested in describing the entire distribution of scores, there are two types of numerical indices that are especially useful. These are measures of central tendency and measures of variability. *Measures of central tendency* are numerical indices that attempt to answer the question "What is the typical score in this distribution of scores?" *Measures of variability* are numbers that attempt to answer questions like: "How different from each other are the scores in the distribution?" or "How do the scores in the distribution spread out around the typical score?" Measures of central tendency and measures of variability are both descriptive statistics because they are used to describe distributions of scores. There are other types of descriptive statistics that describe whole distributions, but here we will concentrate on measures of central tendency and measures of variability.

When we are interested in describing the position of a particular score relative to the other scores in the distribution, we use numerical indices known as *measures of location.* Measures of location are also descriptive statistics, but they describe the position of one score relative to the others rather than describing the whole set of scores.

Measures of Central Tendency

There are several different measures of central tendency. You are probably already acquainted with the terms "mode," "median," and "mean." These are all measures of central tendency. Each is an indicator of what a typical score is, but each employs a different definition of "typical."

One way of thinking about the typical score in a set of scores is to think of it as the score that occurs most often in the set. The *mode* is defined as the most frequently occurring score in a distribution of scores. However, there are other ways to describe a typical score.

Another way of thinking about a typical score in a distribution is to think of it as the score that is exactly in the middle of the distribution, that is, the middle-ranked score. When we think in terms of the middle-ranked score, the measure of central tendency we use is the *median.* The formal definition of median depends upon the type of data we are considering. The median is typically defined in terms of interval scale data. In the case of discrete interval data, we define the median as a score value. It is the score value that has at least 50 percent of the scores in the distribution at or above it and at least 50 percent of the scores at or below it. In the case of continuous interval data, we typically define the median as a point on the measurement scale. The median of a set of scores on a continuous interval variable is the point on the scale above which half the scores occur and below which half the scores occur. We shall see that it is also possible to adapt the concept of the median to data that are in the form of ordered categories as well. We consider the definition of the "median category" later in this chapter.

Perhaps the most common way of thinking about the typical score in a distribution is to think about the average score. The *mean* of a distribution of scores is defined as the numerical average of all the scores in the distribution.

Why do we have three different measures of central tendency? Wouldn't it be easier to get everyone to agree to use just one? The answer to this last question is no, for several reasons. First, not all of the measures of central tendency just noted can be used with all types of data. For example, because the mean is defined as a numerical average, it is clearly impossible to calculate the mean of a set of measures unless they are numerical scores. In the case of categorical data, we do not have numerical scores. Of course, we could assign each value of a categorical variable a numerical value, but this procedure would be arbitrary. Even in the case of ordered categorical variables, we are not justified in assigning the lowest category the value 1, the next lowest the value 2, and so on. We do not know the relative positions of the ordered categories along an equal unit scale. If we did, we would have an interval scale variable, not a categorical variable. In the case of rank-order data, we could regard the ranks as numbers to be averaged. But this would make no sense either, for ranks only relate observations to other observations in the data set. They do not have any external frame of reference. Therefore, the average of a set of ranks provides no measure of the typical performance of the group. For these reasons, it makes no sense to employ the mean as a measure of central tendency if we are considering categorical or rank-order data. We should calculate the mean only if we are considering interval scale data. In addition, even if we have data in the form of numerical scores, there are circumstances in which the mean is not the best measure of central tendency. One purpose of this chapter will be to point out which measures of central tendency may be used with each of the three forms of data we have discussed. You will find that for some kinds of data there is only one appropriate measure of central tendency. For other kinds of data, several different measures are appropriate. When this is the case, we have a choice about which measure to use. Thus, a second purpose of this chapter is to give you some guidance as to which measure is best when several are possible.

Measures of Variability

Just as there are several different measures of central tendency, so there are several different measures of variability. You may already be familiar with the terms "range" and "standard deviation." These are both measures of variability, numerical indices that indicate how the scores in a distribution differ from one another. One way of thinking about the spread of scores in a distribution is to think about how far apart the lowest score and the highest score are. The *range* of a distribution of scores is defined as the distance from the lowest score in the distribution to the highest score. The range depends only on the values of the two most extreme scores. A modification of the range statistic is the *interquartile range,* to be considered later in this chapter.

Another common measure of variability is the *standard deviation.* A precise definition of the standard deviation will also be presented in this chapter. For the moment, you may think about the standard deviation as an approximation of the average amount by which the scores in a distribution differ in either direction from the mean of that distribution. Given these notions regarding the range and the standard deviation, consider the type of data you would need to have in order to employ these measures of variability. Because the range and the standard deviation are both defined in terms of numerical scores, it should be clear that both of these measures are applicable only to interval (or ratio) scale data.

Measures of Location

There are also a number of different indices that may be used to describe the position of a single score with respect to the distribution as a whole. You are already familiar with the concept of rank in class. This is a way of locating an individual in relation to a group. As such, it is a measure of location. A measure of location closely related to rank in class is *percentile rank.* Roughly speaking, the percentile rank of a particular score within a distribution of scores tells us what percentage of all the scores in the distribution fall *below* that score. If you have ever taken a standardized achievement or aptitude test, such as the College Entrance Examination Board Scholastic Aptitude Test (SAT), you have probably used a table to obtain the percentile rank of your own score. It is the percentile rank that tells you where your score falls in relation to the scores of others who have taken the exam. A precise definition of percentile rank will be provided later in this chapter. Another measure of location that we will be considering in this chapter is the *standard score.*

We want to emphasize how important it is to choose measures of central tendency, measures of variability, and measures of location that are appropriate to the particular type of data you have. For this reason, the chapter has been divided into three sections: one for categorical data, one for rank-order data, and one for interval (and ratio scale) data. These sections are of unequal length, because descriptive statistics that are applicable to categorical data and rank-order data are rather limited in comparison to those that are applicable to interval data. As you read the sections that follow, keep in mind that any descriptive statistic that may be used with categorical data and rank-order data may also be used with interval and ratio scale data.

131

DESCRIPTIVE STATISTICS FOR
CATEGORICAL DATA

You will recall that we have defined two types of categorical data: nominal scale data and data in the form of ordered categories. In many cases, the statistical procedures that may be applied to these two types of categorical data are identical. However, there are certain descriptive statistics that may not be used with nominal scale data, but may be used with ordered categorical data. For this reason, we consider the two types of categorical data separately here. We begin with nominal scale data.

Descriptive Statistics for Nominal Scale Data

As an example of nominal scale data, let us reconsider the distribution of eye color of the 25 members of Mr. Smith's first-grade class. The raw data for this distribution are presented in Table 1.1 and the frequency distribution is presented in Table 2.1. Looking at Table 2.1, we see that 13 subjects had brown eyes, 8 subjects had blue eyes, and 4 subjects had eyes of another color.

Measures of Central Tendency for Nominal Scale Data. How can we describe the "typical" eye color of these subjects? You should see very quickly that our choice here is quite limited. Of the three measures of central tendency just mentioned, which can we use?

Clearly, we can use the mode. We defined the mode of a distribution as the score value occurring most frequently in the distribution. In this distribution of eye color, we have three different score values: brown, blue, and other. Of these three score values, we know that the value brown occurred the most often. It occurred 13 times compared to the 8 times that the score blue occurred and the 4 times that the score other occurred. The score value brown is, therefore, the modal score of this distribution of eye color scores. The typical subject in Mr. Smith's class has brown eyes, because more subjects have the score brown than any other score.

Can we use the median as a measure of central tendency with these nominal scale data? Remember that the median is the *middle* score in a set of scores. Considering that the score values in this distribution are the categories brown, blue, and other, is there any way of determining the middle score? No. In order to determine the middle score, we need to be able to line the scores up in order. But the values of a nominal scale variable such as eye color cannot be ordered. As we pointed out in Chapter 1, categories like brown, blue, and other are simply different from one another. They are not quantitatively related at all. Therefore, the median makes no sense as a measure of central tendency for nominal scale data.

What about the mean? Because the mean is the numerical average of all the scores in the distribution, it is clear that the mean can be computed only when we are dealing with a set of numerical scores. With the variable "eye color," we do not have numerical scores. Therefore, we cannot employ the mean as a measure of central tendency with nominal scale data.

In summary, then, when we are concerned with a nominal scale variable, we have no choice regarding measures of central tendency. The mode is the only appropriate descriptive statistic.

Measures of Variability for Nominal Scale Data.   How can we describe the way scores spread out in this distribution of eye color? Can we use either of the two common measures of variability mentioned earlier? No. Both the range and the standard deviation are computed from numerical scores. Thus, they can be used only with interval scale data. Is there a measure that we can use to describe the variability of the distribution of a categorical variable? Actually, there is no one statistic that is generally accepted as the measure of variability for categorical data. However, there is a reasonable index that we might employ. Because the mode is an appropriate measure of central tendency for nominal scale data, we can think of the extent to which scores are clustered together by computing the proportion of scores that lie at the modal score value. If we subtract this proportion from 1.00, we will have a measure of variability. In our eye color example, brown was the modal category. Out of 25 subjects, 13, or 52 percent, had this value. Thus, our measure of variability in this case would be $1.00 - .52 = .48$.

This approach to describing the variability of nominal scale data is simple yet valuable. Once we have indicated that the typical score in the distribution is brown, it is useful to know just how "typical" this value really is. If 95 percent of the subjects had brown eyes, then the score brown would be *more* typical than if only 35 percent of the subjects had brown eyes. Of course, this method of describing variability would not be used in the case of distributions that have no mode.

Measures of Location with Nominal Scale Data.   As you may have guessed by this point, the most common measures of location, such as the percentile rank and the standard score, are not appropriate for use with nominal scale data. Because the score values for such data are unordered categories, we cannot say that John, with the eye color brown, is higher or lower than Bill, with the eye color blue. We can say only that John has the modal score but Bill does not. This enables us to make only one comparison between the two students: John's score is *more typical* than Bill's.

Descriptive Statistics for Ordered Categorical Data

When we move from nominal scale data to data in which there are ordered categories, we have a bit more information to work with. The ordering of the categories makes it possible for us to employ several descriptive statistics with ordered categorical data that we cannot use with nominal scale data. As an example of an

133

ordered categorical variable, let us reconsider the winemaster's experiment described in Chapter 1. In that research project, the winemaster sought to determine the preference of wine drinkers for dry, as opposed to sweet, wines. He did so by asking a group of wine drinkers to indicate which of four wines they preferred when these four wines were clearly ordered in terms of sweetness. Of 25 subjects measured, 14 preferred the "very dry" wine, 9 preferred the "dry" wine, 2 preferred the "sweet" wine, and none preferred the "very sweet" wine.

Measures of Central Tendency for Ordered Categorical Data.   In describing the "typical" wine preference of subjects in this study, we can certainly employ the mode. The largest number of subjects, 14, preferred the "very dry" wine. We can therefore state that "very dry" is the modal score value or category. The use of the mode here is exactly the same as it would be if the data were nominal scale data.

What about other measures of central tendency? We have seen that no other measures of central tendency may be used with nominal scale data. Is this also the case with ordered categorical data? The answer to this question is "it depends." It is sometimes possible to determine a median category with this type of data as well. You will recall that the median is generally thought of as the value of the middle score in a set of scores. When we consider our observations on wine preference, is it possible to find the "middle" observation? Is it possible to find the value of the middle observation? The answers to these questions may surprise you.

In the example we have considered, it is not really possible to determine which of the 25 observations is the middle observation. Yet, it is possible to state that the value of the middle observation is "very dry." That is, it is possible to report that the "median category" for this set of data, the category into which the middle score falls, is the "very dry" category. How can we know the value of the middle score when we do not know which score is the middle score? To answer this question, we must remember that our four ordered categories represent an effort to measure an underlying continuum of preferred dryness in wine. Referring to Figure 1.1, imagine that we could see the wine preference scores laid out as they are in the lower half of this figure. Can you find the "middle" score now? Of course, we find the middle score by counting in 13 scores from either end of the continuum. Which category does this thirteenth score fall in? The "very dry" category.

Of course, we cannot actually tell which of the 14 scores in the "very dry" category is this middle score, because we cannot actually see the underlying continuum. Nevertheless, because the middle score of 25 scores lined up in order will always be the thirteenth score, we know that the middle score would be thirteenth from the "most dry" end of the continuum. Moreover, because our categories are ordered, we know that the 14 scores in the "very dry" category are the first through the fourteenth scores from this end of the continuum. Thus, one of these 14 scores must be the thirteenth, or middle, observation.

Descriptive Statistics

It is not always possible to determine the median category in a set of ordered categorical data. If there is an even number of scores in the set, for example, there will not be a single "middle" score, but rather a pair of two middle scores. If these two scores should happen to fall in different categories, we could not determine a median category. For example, suppose the wine-master had surveyed 20 subjects, 10 of whom preferred the very dry wine and 10 of whom preferred the dry wine. In this case we could not find a median category. In such cases, it is best to report the frequency distribution, noting that subjects preferred the very dry and dry wines in equal numbers.

The mean cannot be used with ordered categorical data any more than it can be used with nominal scale data. Because we do not have numerical scores, we cannot compute an average.

Measures of Variability for Ordered Categorical Data.   For the same reason that we cannot employ the mean with ordered categorical data, we cannot employ the range or standard deviation as measures of variability. These statistics require numerical scores, that is, interval or ratio scale data. We recommend the same measure of variability for ordered categorical data as for nominal scale data, i.e., 1.00 minus the proportion of observations falling in the modal category.

Measures of Location for Ordered Categorical Data.   With ordered categorical data, we can describe the position of a single observation relative to the other observations in the data set using an approximate percentile rank. As indicated before, the concept of percentile rank has to do with the percentage of all the scores in a data set that fall "below" a particular score. If we have a set of ordered categories representing an underlying continuum, we can decide to consider one end of the continuum the upper end and the other the lower end. In our wine tasting example, let us suppose that we are interested in a subject's percentile rank on preferred sweetness. In making this statement, we establish the "most dry" end of the continuum as the low end and the "most sweet" end as the high end. Having made this decision, we might ask, "What is the percentile rank of a subject falling into the 'sweet' category?" Because our categories are ordered, we know that all 14 subjects in the "very dry" category and all 9 subjects in the "dry" category fall further toward the "most dry" end of the continuum than do either of the 2 subjects in the "sweet" category. We can therefore say that at least 23 out of the 25 subjects, or 92 percent, fall below a subject in the "sweet" category on preferred sweetness. However, we cannot distinguish between the 2 subjects in the "sweet" category. We do not know whether the subject we are interested in is above or below the other subject in the same category on preferred sweetness. This situation is illustrated in Figure 4.1. The procedure we employ in this situation is to divide the number of scores in the category of interest in half and add this number to the number of scores that we know to be below the score we are interested in. Thus, in this instance, we say that the percentile rank on preferred sweetness of a subject in the "sweet" category would be 96.

135

FIGURE 4.1

Graphic Representation of Approximate Percentile Rank on Preferred Sweetness of a Subject Who Selected the Sweet Wine

Most Dry   Most Sweet

Very Dry	Dry	Sweet
14	9	2

Low   High

There are 14 subjects who prefer the very dry wine. They all lie below the subject of interest on preferred sweetness.

There are 9 subjects who prefer the dry wine. They, too, are below the subject of interest on preferred sweetness.

There are two subjects who prefer the sweet wine. We do not know which is really further toward the "sweet" end.

There are 23 subjects who are definitely below the subject of interest on preferred sweetness.

We divide the number in this category in half and add it to the 23 we know are below our subject.

So, percentile rank $= \dfrac{23 + 1}{25}(100) = 96.$

In general, the procedure for finding an approximate percentile rank for a score in a set of ordered categorical data is as follows:

Step 1   Find the total number of observations in the set that occur in categories below the category you are interested in.

Step 2   Find the number of observations that occur in the same category.

Step 3   Divide the result of Step 2 in half.

Step 4   Add together the results of Step 1 and Step 3.

Step 5   Divide the result of Step 4 by the total number of observations and multiply by 100.

Using this procedure, how would we determine the approximate percentile rank on preferred sweetness of a subject in the "dry" category? We take the 14 subjects in categories that are lower (i.e., very dry), plus half of the 9 subjects in the "dry" category itself, divide by the total number, 25, and multiply the result by 100:

136

$$\text{Approximate percentile rank of subject in ``dry'' category} = \frac{14 + \frac{1}{2}(9)}{25}(100) = 74.$$

Of course, the more subjects there are in the category of interest, the more approximate our estimate of percentile rank will be. We are really assuming that the specific observation with which we are concerned lies in the middle of all the observations in the same category. We will say more about percentile rank in connection with rank-order and interval scale data.

The concept of the standard score as a measure of location is not applicable to ordered categorical data because standard scores require that the data be in the form of actual numerical scores.

DESCRIPTIVE STATISTICS FOR RANK-ORDER DATA

As our example of rank-order data, we will reconsider the rank in class of the 15 students in Ms. Jones' tenth-grade English class. These data are presented in Table 1.2.

Measures of Central Tendency for Rank-Order Data

Is it possible to describe the "typical" performance of Ms. Jones' students on the basis of the ranking presented in Table 1.2? What kind of information would we need to describe typical performance? We would need scores. And with rank-order data, we really do not have scores. We can see from Table 1.2 that John is rated *better* than Billy and Sally and the other students in the class, but we do not really know *how well* any of the students did. In fact, we do not even know the criterion upon which the ranking is based. The ranks could have been constructed on the basis of the students' grades on a series of examinations, or they could be just Ms. Jones' subjective impression of which students are better than the others. Let's assume for the moment that the rankings are based on Ms. Jones' impressions of the quality of the students' writing. In this case, we know that she feels that John's writing is best in the class. But how good is John's writing or Tom's writing? This may be a special class for gifted students, or it may be a remedial English class. Thus, we have no way of telling how good the "typical" writer is. For this reason, the concept of central tendency as a description of the group does not apply to ordinal measurement. We know where the students stand *in relation to each other*, but we have no way of telling how good or bad they are as a *group*.

137

Measures of Variability for Rank-Order Data

In Chapter 1, we noted that unless two or more observations were tied for last place, a set of rank-order data would always have as many ranks as observations. The concept of rank-order data is based on our ability to differentiate one observation from another. In this example, we must be able to decide that John is better than Billy, Sam is better than Betty, and so forth. However, rank ordering does not require us to decide by *how much* John exceeds Billy or Sam exceeds Betty. Thus, we do not know by how much observations differ from one another. Accordingly, we cannot employ common measures of variability like the range and the standard deviation with rank-order data.

Measures of Location for Rank-Order Data

By definition, rank order locates an individual observation with respect to a group. If we know an individual's rank, we need only specify how many observations there were in order to obtain an idea of how well that individual was performing relative to the whole group. For example, we know from Table 1.2 that Billy's rank was 2. If we say that Billy was ranked second out of 15 students, we can conclude that he has done rather well relative to the whole group. On the opposite end of the scale, we see that Paul was ranked number 14 out of 15 students, so we can conclude that he hasn't done very well relative to the whole group.

Percentile Rank. The concept of percentile rank, discussed in connection with ordered categorical data, may also be applied to rank-order data. The percentile rank is generally a more useful measure of location than simple rank order when we are concerned with large sets of data. As long as there are only a few observations to be ranked, raw numbers alone have meaning. For example, it has considerable meaning for us to know that Sam was ranked fifth out of the 15 students in Ms. Jones' English class. But what if we had been told that Sam was ranked fortieth out of 141 students who are taking English in the entire tenth grade at Sam's school? This statement is more difficult for us to evaluate. It is in situations such as this that we find the percentile rank a helpful measure of location.

The procedure used to determine the percentile rank of an observation in a set of rank-order data is similar to the procedure involved with ordered categorical data:

Step 1	Find the number of all the observations in the set that fall *below* the observation you are interested in.
Step 2	Find the number of all the observations in the set that *equal* the observation you are interested in.
Step 3	Divide the result of Step 2 in half.
Step 4	Add together the results of Step 1 and Step 3.
Step 5	Divide the result of Step 4 by the total number of observations and multiply the result by 100.

Let us consider several examples of this procedure. First, let us calculate the percentile ranks of several members of Ms. Jones' class. Let us begin with Sam, whom we have seen to be ranked fifth in the class of 15 students. We can conceptualize Sam's position with relation to the whole class as indicated in Figure 4.2. Looking at this figure, we find that there are 10 students ranked lower than Sam. Therefore, the result of Step 1 of the procedure is 10. There is only one student, Sam himself, having the rank 5. Therefore, the result of Step 2 is 1. We divide this 1 in half (Step 3) and add the results of Steps 1 and 3: $10 + .5 = 10.5$ (Step 4). Then we divide the result of Step 4 by the total number of observations in the distribution, 15, and multiply by 100:

$$\frac{10.5}{15}(100) = 70 \text{ (Step 5)}.$$

So, Sam's percentile rank is 70.

Now let us consider a slightly different example from the same rank ordering. Let us calculate the percentile rank of Doris, who is tied with Tom for the eighth rank in the class of 15. We can conceptualize Doris' position with relation to the whole class as indicated in Figure 4.3. Because Doris is tied with Tom for the rank 8, there are 2 observations located at the position in the ranking with which we are concerned. Therefore, we divide these 2 in half and add 1 to the 6

FIGURE 4.2
Graphic Representation of Sam's Percentile Rank
in Ms. Jones' English Class

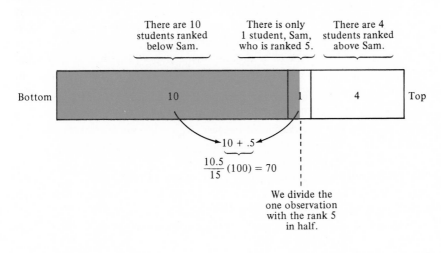

Univariate Problems

FIGURE 4.3
Graphic Representation of Doris' Percentile Rank
in Ms. Jones' English Class

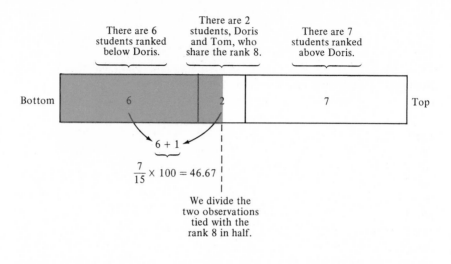

subjects who are ranked below Doris. The result, as indicated in Figure 4.3, is a percentile rank of 46.67.

This procedure for determining percentile rank may be summarized conveniently through the use of a mathematical formula. The following equation is a shorthand way of instructing you to perform the five steps just presented.

$$\text{Percentile Rank} = P = \left[\frac{N_{below} + \frac{1}{2}N_{at}}{N_{total}} \right] \times 100$$

The formula tells us that the percentile rank (denoted P) of a particular observation in a set of observations may be found by adding the number of observations below the observation of interest (N_{below}) to one half of the observations that are equal to the observation of interest (N_{at}), dividing the sum by the total number of observations (N_{total}), and finally multiplying this quotient by 100. As an example of the use of this formula, let us find the percentile equivalent of Sam's rank of 40 among the 141 tenth-grade English students in his entire school. Let us assume that Sam does not share his rank of 40 with anyone. This means that there are $141 - 40 = 101$ students ranked below Sam, and only 1 student, Sam, at his rank. Then:

140

$$P = \left[\frac{N_{\text{below}} + \frac{1}{2}(N_{\text{at}})}{N_{\text{total}}} \right] \times 100$$

$$P = \left[\frac{101 + \frac{1}{2}(1)}{141} \right] \times 100$$

$$P = 71.99$$

Notice that the formula really requires us to perform all the steps indicated previously.

Use of Percentile Ranks in Comparisons. As we noted, one of the reasons for using percentile ranks is simply to enable us to get a better idea of where an individual really does stand in relation to large groups of observations. We have been brought up thinking in terms of percentages, so it means more to us to know that Sam's percentile rank is about 72 than it does for us to know that he is ranked 40 out of 141 students of English. This use of percentile ranks is for ease of understanding. It is exactly analogous to the use of relative frequencies in the construction of frequency distributions. However, there is another reason why percentile ranks are useful. This is in the comparison of where an individual stands with respect to one set of ranks to where the same individual stands with respect to another set of ranks. Let us suppose that we also know that Sam ranks thirty-fourth among the 57 tenth-grade students who are taking solid geometry. Is Sam doing better in English or in geometry? Looking at the ranks alone, it's not easy to decide. Sam is further down the ranks when we look at his rank in English, but there are many more students taking English. It would not be reasonable to compare his rank of 40 in English to his rank of 34 in geometry. We have seen that Sam's percentile rank in English is about 72. Using the formula presented previously, we find that his percentile rank in solid geometry is about 41. Thus, Sam is actually performing somewhat more poorly in relation to his classmates in geometry than he is relative to his classmates in English.

A word of caution is needed here. We *cannot* infer from the comparison that Sam is better in English than in geometry. The reason is that the reference groups used in calculating the two percentile ranks are different. Perhaps only the best 57 students in Sam's class were allowed to take solid geometry in tenth grade. In this case, a percentile rank of 41 in geometry might be something to be really proud of, whereas a rank of 72 in English might not show such an accomplishment. Whenever percentile ranks are used to measure location, we must be careful to specify the reference group. An individual might be ranked number 1 among all the amateur tennis players in the state of Maine, but the same individual would almost certainly not attain the number 1 ranking on the professional tournament circuit.

141

DESCRIPTIVE STATISTICS
FOR INTERVAL DATA

We will be looking at several different sets of interval data in this section. We will be emphasizing the distribution of final examination scores of the 50 students enrolled in Dr. Brown's statistics class, presented in Table 1.3. We have already looked at this set of data. We used it to construct the grouped frequency distribution and the cumulative frequency distribution presented in Table 2.10. You will recall from Chapter 2 that the distribution of exam scores is a set of discrete measures (number of questions correct) on a continuous variable (statistics achievement).

Measures of Central Tendency for Interval Data

When we are dealing with interval data, we are able to use any of the three common measures of central tendency: the mode, the median, or the mean.

The Mode.   We have already defined the mode as the score value that occurs most frequently. In the section on categorical data, we applied this definition to our distribution of eye colors and determined that the most frequent score value was the value brown. But the same definition applies equally well whether our score values are categories like "brown" or numbers. Consider this set of numerical scores:

$$2, 3, 4, 4, 6, 8, 8, 8, 9.$$

It contains 9 scores, but only 6 different score values. There are 2 scores that have the value 4, and 3 scores that have the value 8. Thus, the mode of this set of scores is 8 because there are more scores occurring at 8 than at any other score value. Another way of saying this is to say that the modal score is 8. Now consider this set of scores:

$$2, 3, 4, 5, 6, 8, 9, 11, 13.$$

It contains 9 scores as well, but none of the 9 scores has the same value as any other. No one score occurs more often than the others, for they all occur only once. For this reason, this set of scores has no mode. Let's consider one more of these sets of scores:

$$2, 4, 4, 4, 6, 6, 8, 8, 8.$$

Here we have 9 scores occurring at 4 different values. There are 3 scores that fall at the value 4, and 3 scores that fall at the value 8. The values of 4 and 8 occur with the same frequency. They both occur more often than the other score values. In this case, we consider both 4 and 8 to be the modes. We say that this set of scores has a *bimodal distribution.*

Now let's look at a more realistic distribution of scores, the final exam scores in Dr. Brown's statistics class presented in Table 1.3. Can you find the mode? It's not easy because these are raw data. There are 50 scores and they have not been laid out in the form of a frequency distribution. We did construct a frequency distribution using these data in Chapter 2, but this was a *grouped* frequency distribution. In order to determine the modal score value, we need to know how often each individual score value occurred, so the grouped frequency distribution is not too helpful. With a bit of work, we arrive at the ungrouped frequency distribution presented in Table 4.1. This ungrouped frequency distribution is not as good as the grouped frequency distribution for summarizing the whole set of data quickly, but it is what we need to determine the modal score. From Table 4.1, we see that the 50 scores in the distribution fall at 32 different score values. We see that 4 scores occur at the value 79. No other score value has a frequency as high, so 79 is the mode of this distribution. You can see that it requires a fair bit of work to obtain the mode of a relatively large set of scores by hand. It is common practice today to perform such routine tasks by comput-

TABLE 4.1
Ungrouped Frequency Distribution of Final Examination Scores
of 50 Students in Dr. Brown's Statistics Class

Score value	Frequency	Score value	Frequency
98	2	76	1
93	2	75	2
91	1	74	2
90	2	73	1
88	1	72	1
87	1	71	1
86	1	70	2
85	3	69	1
84	3	68	1
83	1	67	1
82	2	64	1
81	2	63	1
80	2	60	2
79	4	59	1
78	1	57	1
77	2	54	1
			50

er. If no computer is available, it is possible to estimate the mode of a grouped frequency distribution by taking the midpoint of the class interval with the highest frequency.

The Median. We indicated earlier that the median score in a distribution of scores is the score that falls in the middle of the distribution. Obviously, we cannot tell which score is the middle score unless we can line the scores up in ascending or descending order. Here is a set of scores written out in no particular order:

$$8, 27, 3, 15, 4, 9, 6, 11, 31.$$

To find the median, we first put the scores in order:

$$3, 4, 6, 8, 9, 11, 15, 27, 31.$$

The middle score is 9. There are 4 scores above and 4 scores below. Thus, 9 is the median score. It is one way of thinking about the typical score in this set.
   Now consider this set of scores:

$$3, 5, 6, 7, 9, 10, 10, 12.$$

They are arranged in order. What is the median? What is the middle score? We do not really have a single middle score because there happens to be an even number of scores in the set. The 7 and the 9 are the middle 2 scores, but the exact middle of the set falls *between* the 7 and the 9. In the case of an even number of scores, the median score value is defined as the average of the values of the 2 middle scores. In this case, the average of 7 and 9 is 8. Thus, even though there is no actual score in this distribution with the value 8, the median score of the distribution is 8.
   In the examples just presented, we had very few scores in each set. It was easy to locate the middle score or the middle two scores just by inspecting the distributions. We could mentally cross off one score from each end in turn until we were left with the middle score or the middle two scores. However, when we have a large number of scores in a distribution, inspection will not work. We could lay the scores out in order and physically cross off one from each end in turn, but this would be unduly time-consuming. There are two formulas that we can use to find the position of the median score without crossing off scores from the end. One is to be used when there is an odd number of scores and the other when there is an even number of scores. Once we have arranged a set of scores in order, then we begin the process of finding the median by counting to see how many scores there are. Let's consider a small set of scores first:

144

$$3, 7, 9, 9, 11, 13, 16, 17, 17, 17, 19, 20, 22.$$

We count and find that there are 13 scores in this set, an odd number. The position of the median score when there is an odd number of scores can be found using this formula:

$$\text{Position of median} = \frac{\text{Number of scores in set} + 1}{2}.$$

In this example, the formula works out as follows:

$$\text{Position of median} = \frac{13 + 1}{2} = 7.$$

The 7 means that the *seventh score* in the set is the median score. To find the score, we count in from either side of the set exactly 7 scores. The seventh will be the median. In this example, we find that the seventh score from the left (or right) is the score 16. This is the median.

Now consider this set:

$$2, 3, 9, 9, 11, 13, 16, 17, 17, 17, 19, 20, 22, 23.$$

There are 14 scores in this set, an even number. The position of the *two middle scores* when there is an even number of scores can be found as follows:

$$\frac{\text{Position of the lower}}{\text{of the two middle scores}} = \frac{\text{Number of scores in the set}}{2}$$

and

$$\frac{\text{Position of the higher}}{\text{of the two middle scores}} = \frac{\text{Number of scores in the set}}{2} + 1.$$

Applying these rules to our example, we find:

$$\frac{\text{Position of the lower}}{\text{of the two middle scores}} = \frac{14}{2} = 7.$$

That is, the lower of the two middle scores is the seventh score. This means that the upper of the two middle scores is the 7 + 1, or eighth, score. Counting from the left in the set of scores above, we see that the seventh score is 16 and the eighth score is 17. The median score value is then the average of 16 and 17, or 16.5.

These rules for finding the position of the median score are especially useful with large sets of data. Now let's find the median score of the 50 final examination scores in Dr. Brown's statistics class. We can use Table 4.1 as our list of

145

scores laid out in order. We need only remember that any score value that has a frequency greater than one represents more than one actual score at that value. Any counting we do must count frequencies rather than score values. Remember that there are 50 scores in this set. Using the rules for distributions with an even number of scores, we find that the twenty-fifth (50/2) and the twenty-sixth (50/2 + 1) scores are the 2 middle scores. We count frequencies starting with the 2 scores that have the value 98. We find that the twenty-fifth and twenty-sixth scores both fall at the score value 79 (you should count, too, to verify this). Because the 2 middle scores both have the same value, there is no need to average. The median score in this distribution is 79. You may remember that the modal score of this distribution was also 79. This is a coincidence; it is not true of every distribution.

The Mean.   The third measure of central tendency that we can use with interval data is the mean. We indicated previously that the mean of a set of scores is simply the arithmetic average of all the scores in the set, that is, the sum of all the scores divided by the total number of scores in the set.

For example, consider this set of scores:

$$5, 3, 4, 1, 5, 6.$$

The mean may be calculated as follows:

$$\text{Mean} = \frac{5 + 3 + 4 + 1 + 5 + 6}{6}$$

$$\text{Mean} = \frac{24}{6}$$

$$\text{Mean} = 4.0.$$

The 6 in the denominator of this expression represents the number of scores or observations in the set. As you can see, the mean is easy to compute. It is not even necessary for us to arrange the scores in order. The procedure for calculating the mean may be summed up in a shorthand formula that looks like this:

$$\mu = \frac{\overset{\text{all}}{\overset{\text{scores}}{\sum}} X}{N}.$$

The Greek letters are simply instructions that tell us to perform the simple computation just illustrated. The symbols stand for the following:

146

μ: This is "mu." It is the Greek equivalent of our letter "m." It stands for mean.

Σ: This is a capital sigma, the Greek equivalent of our letter "S." This symbol tells us to sum up something. In this case, it tells us to sum up all the scores in the set. For clarity, we have written "all scores" as a superscript above sigma. Whenever we use sigma as a summation sign in this book, it will indicate the direction to sum up all the scores referred to. This kind of summation notation can also be used to direct us to add only selected scores. You need not worry about this now. However, we have included a detailed explanation of the various uses of summation notation in Appendix A.

X: This is the letter X. We use it here to give a name to our variable. Referring to the numerical example just presented, suppose each score represents the number of runs scored in one game of a World Series that lasted 6 games. The X would be a short name for the variable "number of runs scored." In the first game, 5 runs were scored; in the second, 3; and so on. Having found a mean of 4.0, we would know that the average number of runs scored during this World Series was 4.0 runs per game.

N: This stands for the total number of scores in the distribution whose mean we are calculating.

If any of these symbols are unclear to you, just remember that the mean is simply an arithmetic average. Think of what you do when you average numbers and relate the symbols to this process. What about a mean for the distribution of final exam scores in Dr. Brown's class? There are more scores to add up, but the process is the same. Calculate the mean of the exam grades yourself before reading on (a calculator will be helpful). You should find the following:

$$\mu = \frac{\overset{\text{all scores}}{\Sigma} X}{N}$$

$$\mu = \frac{3,889}{50} \quad \begin{array}{l}\text{(The sum of all 50 scores added up is 3,889.)}\\ \text{(There were 50 scores.)}\end{array}$$

$\mu = 77.78.$ (The average score is 77.78.)

Choosing A Measure of Central Tendency. You now know how to determine the mode, the median, and the mean for interval data. The only question that remains is "Which of these measures of central tendency is best?" As you might

147

expect, the answer is "It depends." In the case of the exam scores, all three measures of central tendency were similar. The mode and the median turned out to be 79, and the mean was 77.78. In this instance, it would not make much difference which one we used, for all would give us roughly the same idea about what a "typical" exam score is.

On the other hand, there are situations in which the three measures will be rather different from one another. For example, consider this set of 12 exam scores:

$$10, 13, 84, 87, 88, 88, 89, 89, 89, 91, 92, 92.$$

The mode of this distribution is 89. The median is 88.5 (verify these). But the mean we obtain is 76. Which of these measures of central tendency best describes the "typical" score in the distribution? Does 76 seem like a typical score? No. Not only did no one get the score 76, but the actual score in the set that is closest to 76 is 8 points higher than 76. In fact, most of the scores in this distribution are considerably higher than 76. The mode or median would better describe the position of the bulk of these scores. What happened is that the two extremely low scores "pulled down" the average. They pulled it down to the point that the average or mean score was no longer a good description of the typical score. A distribution in which the bulk of the scores falls in one area, but a few scores lie far away in one direction, is known as a *skewed distribution*. If the few extreme scores lie *below* the bulk of the scores, the distribution is said to be *negatively skewed*. In this case, the mean we calculate will fall below the median and the mean may not be a good measure of central tendency. That is the situation we have with the distribution just described. Alternatively, the few extreme scores may lie *above* the bulk of the scores, in which case the distribution is said to be *positively skewed*. Then the calculated mean will fall above the median.

In Figure 4.4, we have illustrated the relative position of the mean and the median in skewed distributions. Figure 4.4(a) is a negatively skewed distribution and Figure 4.4(b) is a positively skewed distribution. Figures 4.4(c) and 4.4(d) represent distributions that are not skewed, but rather are *symmetrical*. By symmetrical, we refer to the "mirror image" property of these distributions. Imagine drawing a vertical line through the middle of either of these distributions. Then, imagine folding the page along this line, so that one half of the distribution is folded over on top of the other half. The two halves would match exactly. In a symmetrical distribution, every score on one side of the mean is exactly matched by a score on the other side of the mean, the same distance away. In a symmetrical distribution, the mean and the median are identical. When distributions are symmetrical or approximately symmetrical, the mean is a good measure of central tendency. However, when a distribution is seriously skewed, it is preferable to use the median as a measure of central tendency. The mode may be

FIGURE 4.4
The Shape of Several Distributions of Scores

a Negatively skewed

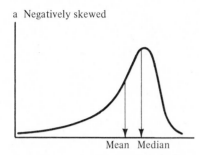

Mean Median

b Positively skewed

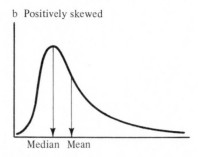

Median Mean

c Symmetrical (rectangular)

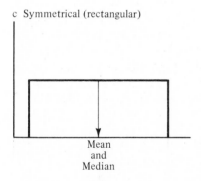

Mean
and
Median

d Symmetrical (normal)

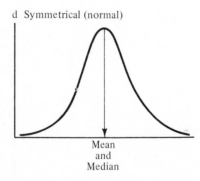

Mean
and
Median

appropriate as well, though it is possible to encounter distributions in which the modal score is itself an extreme score that does not describe the typical score well.

Measures of Variability for Interval Data

The measures of variability most often used with interval data are the range, the interquartile range, and the standard deviation.

The Range.   As noted earlier, the range of a distribution is the distance from the lowest score in the distribution to the highest score in the distribution. In the case of the final examination scores in Dr. Brown's class, the range is from 54 to 98, or 44 points. When using the range as a measure of dispersion, it is

149

more informative to give the lowest and highest scores, rather than just the difference between these scores. To see why this is so, consider these two sets of test scores:

$$0, 3, 5, 7, 9, 9, 12, 15$$

and

$$80, 83, 85, 87, 89, 89, 92, 95.$$

Both of these two sets of scores have the same range, 15 points. However, the variability of the two sets seems to be quite different. Suppose the first set of scores represents the number of correct answers on a 15-item test. One student in the group got nothing right and one student got everything right. These two students are 15 points apart, but these 15 points seem to indicate a great difference in achievement. Now suppose the second set of scores represents the number of correct answers on a 100-item test. The worst score was 80 and the best was 95. These two scores are also 15 points apart, but here 15 points would seem to indicate a smaller difference in achievement. For this reason, one should always specify the lowest and highest scores when giving the range of a distribution.

Even if you are careful to specify the lowest and highest scores in a distribution, there is a major disadvantage in using the range as a measure of variability. The range of a distribution is found by referring to just two scores, the bottom score and the top score. The range tells us nothing about how the scores in between vary. Consider these two sets of test scores:

$$10, 15, 17, 27, 30, 44, 51, 70, 79, 86, 89, 97$$

and

$$10, 62, 69, 70, 70, 73, 74, 75, 75, 75, 75, 97.$$

These two distributions have precisely the same range, from 10 to 97. But would you say that the variability of the two distributions is the same? In the first distribution, the scores are spread out rather evenly over the entire range. A distribution like this implies considerable variability in achievement. In the second distribution, however, the great majority of the scores are grouped tightly together. There is one extremely low score (10) and one very high score (97). A distribution like this implies less variability in achievement over the entire set of scores. Most of the students performed similarly to each other. There just happened to be one student who did very badly and one who did very well. Thus, it could be misleading to report just the range of these two distributions. To do so could create the false impression that the variabilities of the two sets of scores were similar when in fact they are very different. The two measures of variability discussed in the following sections help avoid the problem of extreme scores.

The Interquartile Range. One method of handling extreme scores is to exclude a portion of the most extreme scores from consideration, as we do with the interquartile range. When we use the interquartile range as a measure of variability, we ignore the bottom quarter and the top quarter of the scores, and report the range of what is left. In the case of the two distributions just discussed, this method works quite well. Because there are 12 scores in each set, we exclude the bottom 3 scores and the top 3 scores from consideration. For the first set of scores, we find the interquartile range indicated in Figure 4.5. The range of scores in the middle 50 percent is from 27 to 79, or 52 points. This is the interquartile range of the entire first set. Using the same technique, we find that the interquartile range for the second set is from 70 to 75 points, or just 5 points! In this example then, the interquartile range is successful in differentiating between the variabilities of the two distributions. The interquartile range of the first set of scores is much greater than that of the second set of scores, reflecting the greater variability that we intuitively sensed from looking at the whole distributions.

There are situations, however, in which the interquartile range cannot differentiate between two distributions whose variabilities really do differ. Remember that the interquartile range simply eliminates the most extreme 25 percent of the scores at each end of the distribution from our consideration. If a distribution is such that more than half of the scores are "extreme," even the interquartile range can be misleading. Look at these two sets of scores:

$$8, 10, 15, 16, 17, 17, 84, 85, 87, 89, 93, 97$$

and

$$8, 10, 15, 16, 29, 37, 55, 69, 87, 89, 93, 97.$$

FIGURE 4.5
The Interquartile Range of a Distribution

	Interquartile Range	
10, 15, 17,	(27,) 30, 44, 51, 70, (79,)	86, 89, 97
Bottom 25 percent	Middle 50 percent	Top 25 percent

These two sets of scores have the same range, 8 to 97, and the same interquartile range, 16 to 87. Yet, the variabilities of the two sets of scores do seem to differ. In order to describe the differing variabilities of these two distributions, we need to do more than exclude the most extreme scores and then look at a range. Whenever we use a range statistic, either the whole range or the interquartile range, that statistic is based on only two scores. To discriminate between the variabilities of the two distributions just presented, we need a measure that takes into consideration all the scores. The most frequently used measure of this type is the standard deviation.

Standard Deviation.   At the beginning of this chapter, we suggested that the standard deviation could be thought of as an approximation of the average amount by which the scores in a distribution differ in either direction from the mean of that distribution. The word "standard" suggests the idea of average or typical. The word "deviation" refers to the difference between any given score and the mean. We use the word "approximation" to make it clear to you that the standard deviation is not, in fact, a straight numerical average of all the deviations of all the scores from the mean. If it were, it would probably be called an average deviation. Nevertheless, when you are told that a set of scores has a mean of 50 and a standard deviation of 10, it is helpful to think of that 10 as roughly equivalent to the average amount by which a typical score differs or deviates in either direction from the mean.

One reason why we do not use the numerical average of all the deviations from the mean as a measure of variability is that some deviations are positive and some are negative. All scores that lie above the mean produce positive deviations, and all scores that lie below the mean produce negative deviations. In fact, because the mean is a central point in the distribution, the positive deviations will exactly balance the negative ones. That is, the sum of all the deviations for any distribution of scores will always equal zero. This is illustrated in Figure 4.6 in the column headed Step 3. This figure illustrates the calculation of the standard deviation for a set of scores, but we have also used it to show that the sum of all the deviations from the mean is zero. This is true for any distribution of scores. When we consider variability, we must eliminate the negative signs that cause the sum of the deviations to be zero. This could be done by considering the absolute value of the deviation. This is what we had in mind when we referred to the average amount by which a score differs or deviates from the mean *in either direction*. If we used this method with the set of 10 scores presented in Figure 4.6, we would find that the average of the absolute values of the deviations was $18/10 = 1.8$. However, the standard deviation is computed in a slightly different manner. Rather than eliminating negative signs by considering the absolute value of the deviations, we eliminate negative signs by squaring the deviations. We explain why in the following section.

Descriptive Statistics

The Definitional Formula for the Standard Deviation. There are several different procedures that may be employed to calculate the standard deviation of a set of scores. We will look at two. The first method we will look at is the most instructive in the sense that it gives you an idea of what a standard deviation really is. It is generally not the easiest method to use, although it may be the easiest in the case of a distribution with relatively few scores and a mean that is a whole number. The procedure may be outlined step-by-step, as follows:

Step 1   If you do not already know the mean of the set of scores, calculate it.

Step 2   List all the scores in the distribution in a column.

Step 3   Subtract the mean of the distribution from each score. Each time you perform one of these subtractions, you will have calculated the deviation of that score from the mean. Some deviations will be positive and some negative. List all these deviations in a second column adjacent to the first.

Step 4   Square each of the deviations. This will eliminate all the negative signs. List all the *squared deviations* in a third column.

Step 5   Add up all the squared deviations by totaling the third column. The result is called, logically, the *sum of the squared deviations* from the mean.

Step 6   Divide this sum of the squared deviations by the number of scores in the distribution.

Step 7   Take the square root of the result of Step 6.

An example of this procedure is presented in Figure 4.6, where we calculate the standard deviation of a set of 10 test scores. Each step is labeled with a number corresponding to the numbers in the list of steps just presented. The example was specifically selected because it has a small number of scores and because the mean is a whole number. These features make the standard deviation easy to calculate using this procedure. You will notice that the title of Figure 4.6 contains the words "definitional formula." As in the case of the procedure used to find the mean, this procedure for finding the standard deviation may be summarized in a shorthand formula. We have inserted the elements of the formula into Figure 4.6, but now we would like to define each element. The definitional formula for the standard deviation is:

$$\sigma = \sqrt{\frac{\overset{\text{all}}{\overset{\text{deviations}}{\sum}} (X - \mu)^2}{N}}$$

In this formula, the symbols have these meanings:

X: This has the same meaning as it did when we used it in the formula for the mean. It is the name we give to the variable we are working with, in this case "test score." As you see from the column headed Step 2 in Figure 4.6, this variable name refers to each of the 10 scores in turn.

μ: This is mu, which is the mean of the distribution. For any given distribution, there is only one mean. Therefore, it is not a variable in this formula. You see from the column headed Step 3 that we subtract the same value, $\mu = 7$, from each score. Thus, X changes, but μ does not.

$(X - \mu)^2$: The exponent 2 tells us that each time we subtract μ from one of the X scores, we will square the result. This is what we did in the column headed Step 4. There are as many deviations as there are scores, and as many squared deviations as there are scores as well.

Σ: This is the capital sigma again, the symbol that tells us to add up something. Σ always tells us to add up whatever appears to its right. In the formula for the mean, we had ΣX, which told us to add up the scores. Here we have $\Sigma (X - \mu)^2$, which tells us to add up the squared deviations. This we did in Step 5.

N: This N is still the symbol for the number of scores in the set. There were 10 scores. We divided by 10 in Step 6.

σ: This is the small Greek letter sigma. It stands for standard deviation.

Once again, remember that this somewhat formidable formula is simply a quick way of directing you to perform the seven steps just outlined. We call this formula a definitional formula because it demonstrates symbolically what a standard deviation really is. It shows us clearly that when the scores in a distribution tend to lie far from the mean, the standard deviation is large.

There are other formulas that may be used for obtaining the standard deviation. Here we would like to present one of the "computational" formulas for the standard deviation. We use the name computational because, in most cases, it is easier to find the standard deviation of a distribution using this formula than using the definitional formula. You may well be thinking that it was really quite easy to find the standard deviation of the test scores using the procedure in Figure 4.6. But remember, this set of 10 scores was especially selected so that computation by the definitional formula would be easy. If the mean of the distribution is not a whole number, then the deviations will be fractions. All these fractions would have to be squared. This is a tedious process, especially if there

154

FIGURE 4.6
Example of Procedure Used to Calculate Standard Deviation
by Definitional Formula

Scores made by 10 students on a statistics quiz:

$$4, 4, 5, 6, 7, 8, 9, 9, 9, 9$$

Step 1

$$\mu = \frac{\Sigma X}{N} = \frac{70}{10} = 7$$

Step 2	Step 3	Step 4
X	$(X - \mu) =$ Deviation	$(X - \mu)^2 =$ Squared deviation
4	$4 - 7 = \quad -3$	$(-3)^2 = \quad 9$
4	$4 - 7 = \quad -3$	$(-3)^2 = \quad 9$
5	$5 - 7 = \quad -2$	$(-2)^2 = \quad 4$
6	$6 - 7 = \quad -1$	$(-1)^2 = \quad 1$
7	$7 - 7 = \quad 0$	$(0)^2 = \quad 0$
8	$8 - 7 = \quad 1$	$(1)^2 = \quad 1$
9	$9 - 7 = \quad 2$	$(2)^2 = \quad 4$
9	$9 - 7 = \quad 2$	$(2)^2 = \quad 4$
9	$9 - 7 = \quad 2$	$(2)^2 = \quad 4$
9	$9 - 7 = \quad 2$	$(2)^2 = \quad 4$
	$\Sigma(X - \mu) = \quad 0$	$\Sigma(X - \mu)^2 = \quad 40$

Step 5

Total of third column is
sum of squared deviations.

Step 6

$$\frac{\Sigma(X - \mu)^2}{N} = \frac{40}{10} = 4$$

Sum of squared deviations divided by number
of scores

Step 7

$$\sigma = \sqrt{\frac{\Sigma(X - \mu)^2}{N}} = \sqrt{4} = 2$$

Square root of result of Step 6 is standard
deviation.

155

are many scores. Also, we would have to carry out all these calculations to many decimal places, if we were to be accurate. The computational formula is both easier and more accurate with distributions where the mean is not a whole number or where there are many scores.

A Computational Formula for the Standard Deviation. This time, let's look at the formula first, and then outline the step-by-step procedure. The computational formula for the standard deviation may be written as follows:

$$\sigma = \sqrt{\frac{\overset{\text{all scores}}{\sum} X^2}{N} - \mu^2}$$

Before reading on, try to think of the steps that would be involved in carrying out the directions given by this formula. Now check your thoughts against this outline:

Step 1 If you have not already done so, calculate the mean.
Step 2 List all the scores in a column.
Step 3 Square *each score* and list these squared scores in a second column.
Step 4 Add up the column of *squared scores* to obtain the *sum of the squared scores,* ΣX^2.
Step 5 Divide the result of Step 4 by N, the number of scores.
Step 6 Square the mean of the set of scores.
Step 7 Subtract the squared mean from the result of Step 5.
Step 8 Take the square root of the result of Step 7.

To illustrate this procedure and to demonstrate to you that the definitional and computational formulas for the standard deviation are equivalent, we have recalculated the standard deviation of the distribution of scores made by 10 students on a statistics quiz. This is presented in Figure 4.7. Once again, each step in the calculation is labeled with a number corresponding to the numbers in the list of steps just given. You will see that the standard deviation calculated via this procedure is identical to that obtained using the definitional formula.

As an exercise, you can work out the standard deviation for the set of 50 final exam scores for Dr. Brown's statistics class. You should use a calculator when you do this. Even using the computational formula, it would take some time to obtain the standard deviation for such a large set of scores by hand. We have already found that the mean of this set of scores is 77.78. You should find that ΣX^2 is 307,821. These figures may be plugged into the computational formula, yielding:

156

$$\sigma = \sqrt{\frac{\Sigma X^2}{N} - (\mu)^2}$$

$$\sigma = \sqrt{\frac{307{,}821}{50} - (77.78)^2}$$

$$\sigma = \sqrt{6156.42 - 6049.7284}$$

$$\sigma = \sqrt{106.6916}$$

$$\sigma = 10.33.$$

This means that in Dr. Brown's class, the mean score was 77.78, but it was not unusual for a score to differ from this mean by as much as 10 points on one side or the other.

Some Additional Comments on the Standard Deviation. We would like to note that the computational formula for the standard deviation may be written in several forms other than the form just presented. Although it is not essential for you to know all these equivalent forms, there is one that is sometimes easier to use than the one just given, depending on the information you have on the distribution of interest. This form is:

$$\sigma = \sqrt{\frac{\Sigma X^2 - \frac{(\Sigma X)^2}{N}}{N}}.$$

In this formula, the $(\Sigma X)^2$ directs you to add up all the scores in the set first, and then square this sum. This is different from ΣX^2, which directs you to square *each score* first and then add up the squared scores. You can use this formula as an exercise to find the standard deviation of Dr. Brown's exam scores.

A second point that we would like to emphasize here is that these formulas all yield *descriptive* or *population standard deviations*. You should remember from Chapter 1 that some statistics are descriptive in that they are based on every observation in the entire population. When we compute the mean and the standard deviation of the scores on Dr. Brown's exam, we are computing *descriptive statistics* because we are using every one of the scores in the class to obtain our statistics. Also, we are concerned only with this class. We are not attempting to generalize to any larger group. We should point out, however, that means and standard deviations may also be used inferentially to generalize from a sample to a larger population. In this case, they are called *sample means* and *sample standard deviations*. We make this distinction here because the for-

157

FIGURE 4.7
Example of Procedure Used to Calculate Standard Deviation
by Computational Formula

Scores made by 10 students on a statistics quiz:

$$4, 4, 5, 6, 7, 8, 9, 9, 9, 9$$

Step 1

$$\mu = \frac{\Sigma X}{N} = \frac{70}{10} = 7$$

Step 2	Step 3	Step 5
X	X^2	$\dfrac{\Sigma X^2}{N} = \dfrac{530}{10} = 53$
4	16	
4	16	
5	25	
6	36	Step 6
7	49	
8	64	$(\mu)^2 = (7)^2 = 49$
9	81	
9	81	
9	81	Step 7
9	81	

Step 7

$$\frac{\Sigma X^2}{N} - (\mu)^2 = 53 - 49 = 4$$

Step 4

$$\Sigma X^2 = \qquad 530 = \begin{array}{l}\text{Sum of all the}\\\text{squared scores}\end{array}$$

Step 8

$$\sigma = \sqrt{\frac{\Sigma X^2}{N} - (\mu)^2} = \sqrt{4} = 2$$

mulas for the sample standard deviation are slightly different from the formulas for the population standard deviation. The formulas for the sample standard deviation will be considered in Part 4 of this book.

Measures of Location for Interval Data

The two common measures of location employed with interval data are the percentile ranks and the standard score. The percentile rank, which may be obtained in the same manner for interval data as for ordinal data, locates an individual score

158

in a distribution by indicating the percentage of scores in the distribution that lie below the given score, plus half of the percentage of scores that lie at that score. The standard score is a measure of location unique to interval data. It locates an individual score in a distribution with reference to the mean of the distribution. It tells us how many standard deviation units above or below the mean a given score lies. Let us look at these two measures of location with reference to the distribution of final exam scores for Dr. Brown's class.

Percentile Rank.   Referring to Table 1.3, we see that student 15, Hester, has a score of 75 on Dr. Brown's examination. What is her percentile rank? In order to compute this statistic, we need to know how many students had scores below hers, how many had scores equal to hers, and how many students there were in all. To obtain this information, we can use the ungrouped frequency distribution for this set of data, which is presented in Table 4.1. Counting up the scores from the lowest score, we find that 17 scores lie below Hester's score of 75, and that there are 2 scores (including Hester's) that lie right at the value 75. Applying the formula presented earlier in this chapter, we find

$$P = \left[\frac{N_{below} + \frac{1}{2}(N_{at})}{N_{total}} \right] \times 100$$

$$P = \left[\frac{17 + (\frac{1}{2} \times 2)}{50} \right] \times 100$$

$$P = \frac{18}{50} \times 100 = 36.$$

Hester's percentile rank is, therefore, 36. This formula is easy enough to use when we have an ungrouped frequency distribution to work with. However, we do not always have an ungrouped frequency distribution. You will remember from Chapter 2 that when we are dealing with an interval variable with many different score values, we generally group scores to make the distribution more meaningful to the reader. What can you do if you need to obtain the percentile rank of a score when you have only a grouped frequency distribution? You must estimate. There are several techniques for estimating percentile ranks from grouped data. We will look at two.

Arithmetic Estimation.   This is a good estimation procedure to employ when you need to estimate the percentile rank of only one or two score values in a grouped frequency distribution. The steps involved in this process are as follows:

Step 1   Locate the score interval in which the score value of interest happens to fall.

159

Step 2 Count how many scores fall *below* that interval, and compute what percentage of the total number of scores in the distribution these scores represent.

Step 3 Now find out how many scores lie *within* the interval in which your score falls.

Step 4 Find the *width* of this interval by subtracting the lower real limit of the interval from the upper real limit.

Step 5 Find out how far above the lower limit of this interval your score falls by subtracting the lower real limit of the interval from the score value of interest.

Step 6 Find the ratio of the number obtained in Step 5 to the number obtained in Step 4.

Step 7 Multiply the number of scores in the interval (Step 3) by the result of Step 6.

Step 8 Take the result of Step 7 and find what percentage it is of the total number of scores in the distribution.

Step 9 Add the results of Steps 2 and 8 to obtain your estimate of the percentile rank.

There are many steps here, but each one is simple. As an example, we have used the grouped frequency distribution presented in Table 2.10 to compute an estimate of Hester's percentile rank. Remember that we have already computed Hester's percentile rank from the ungrouped data and found it to be 36. The estimation from grouped data is contained in Figure 4.8. Each step in the actual example is numbered to correspond to the steps just listed. As you will see from this example, the method of arithmetic estimation produces an estimate of 34.8 as Hester's percentile rank. We know her actual rank was 36. The method has estimated rather well.

Graphic Estimation.   When you need to estimate the percentile rank of several score values in a grouped frequency distribution, the method of arithmetic estimation becomes tedious. It may then prove to be easier to estimate percentile ranks graphically using a cumulative percentage graph of the distribution. You will recall that in Chapter 2, we presented a cumulative frequency graph based on the grouped frequency distribution of final exam scores in Dr. Brown's class. There we explained a graphic procedure that may be used with such a cumulative frequency graph to estimate the number of scores falling at or below a particular score value (see Figure 2.8). We also mentioned that it is possible to construct a cumulative percentage graph for a grouped frequency distribution. Such a graph will appear identical to the regular cumulative frequency graph. However, the vertical axis will indicate not the *number* of scores falling at or below a given score value, but rather, the *percentage* of all the scores in the distribution that fall at or below the score value. If we construct a cumulative percentage graph for a grouped frequency distribution, we can then employ the

FIGURE 4.8
Method of Arithmetic Estimation Used to Estimate the Percentile Rank
of Score in a Grouped Frequency

Step 1

From Table 2.10, we see that the score of interest, 75, lies in the interval 74–78.

Step 2

There are 15 scores that fall below this interval. These 15 scores are 30 percent of the total.

Step 3

There are 8 scores in the interval 74–78.

Step 4

The lower real limit is 73.5, the upper real limit 78.5. The width of the interval is 5.

Step 5

$75 - 73.5 = 1.5$. This is how many units above the lower real limit the score value of interest lies.

Step 6

$$\frac{1.5}{5} = .3.$$

Step 7

$.3(8) = 2.4.$

Step 8

$$\frac{2.4}{50} \times 100 = 4.8 \text{ percent.}$$

Step 9

30 percent + 4.8 percent = 34.8 percent. We estimate Hester's percentile rank to be 34.8.

technique presented in Chapter 2 to obtain the percentile rank of a given score value. Generally, these estimates obtained from graphic estimation will be as accurate as the estimates obtained from arithmetic estimation. In Figure 4.9, we present the cumulative percentage graph of the grouped distribution of the exam scores in Dr. Brown's class.

Let us use this graph to estimate once again the percentile rank of Hester, whose score on the exam was 75. To find her percentile rank, we draw a vertical line up from the horizontal axis until it intersects with the curve. Then we draw a horizontal line from this point to the left until it intersects with the vertical axis. Here we read the percentile rank of that score value. We have drawn in the lines used to estimate Hester's percentile rank in Figure 4.9. The arrowheads indicate the direction in which the lines are drawn. We read approximately 35 as Hester's percentile rank from the vertical axis. Thus, our graphic estimation procedure also yields quite a good estimate of Hester's percentile rank. The ad-

FIGURE 4.9
Cumulative Percentage Graph of 50 Final Exam Scores
for Dr. Brown's Statistics Class

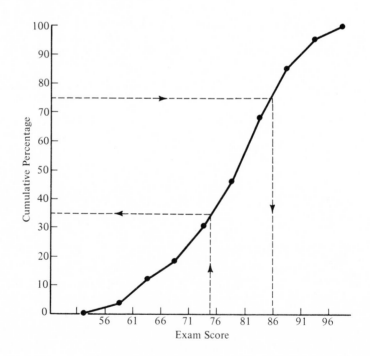

Descriptive Statistics

vantage of this graphic procedure is that once the cumulative graph has been constructed, the percentile ranks of many score values can be estimated quickly. Of course, it takes a little time to construct the cumulative percentage graph, but if you need to find more than half a dozen percentile ranks, it is a worthwhile effort.

Moreover, the cumulative percentage graph can be used in a similar manner to answer other questions. Suppose Hester wanted to know how high she would have had to score in order to reach the seventy-fifth percentile in her class. We can estimate the answer to this question by using the estimation procedure in reverse. We can locate the seventy-fifth percentile on the vertical axis and draw a horizontal line to the right until it intersects the graph. From this point, we draw a vertical line down to meet the horizontal axis. Where the vertical line intersects the horizontal axis, we read our estimate of the 75th percentile score value. We have also drawn in the lines used to estimate the seventy-fifth percentile on Figure 4.9. Once again, the arrows indicate direction. From the horizontal axis, we estimate the seventy-fifth percentile score to be between 85 and 86. Checking the ungrouped frequency distribution in Table 4.1, we find that the seventy-fifth percentile score actually occurs at the value 85. Once again, graphic estimation has done rather well.

Once we have learned to perform these two operations—estimating the percentile rank of a given score and estimating what score has a given percentile rank—we can use these techniques to answer a whole set of questions. For example, we might want to know what percentage of the students in Dr. Brown's class scored *higher* than 80. We can estimate this percentage as follows:

Step 1   Find the percentile rank of the score 80. This will tell us what percentage scored *below* 80. Reading from Figure 4.9, we estimate the percentile rank of the score 80 to be about 53.

Step 2   Subtract the answer to Step 1 from 100. This will give us the percentage scoring *above* 80. Because $100 - 53 = 47$, we estimate that roughly 47 percent of Dr. Brown's class scored above 80.

As another example, suppose we wanted to estimate the percentage of students who scored *between* 56 and 90. We can estimate this percentage as follows:

Step 1   Estimate the percentile ranks of the two scores, 56 and 90. This will tell us what percentage of the scores fall below *each* of these two values. Reading from Figure 4.9, we estimate the percentile rank of the score 90 to be 89 and the percentile rank of the score 56 to be 3.

Step 2   Subtract the percentage of scores that falls below 56 from the percentage that falls below 90. Because all the scores that fall below 56 also fall below 90, the result of this subtraction will give us the

163

percentage of the total number of scores that fall between 56 and 90. In this example, $89 - 3 = 86$, so we estimate that roughly 86 percent of the exam scores fall between the score 56 and the score 90.

You will have a chance to practice using cumulative percentage graphs in the problems at the end of this chapter. We now look at another measure of location for interval data, the standard score.

The Standard Score.   The standard score is a measure that locates a particular score within a set of scores by reference to the mean and the standard deviation of the set of scores. The standard score tells us how many standard deviations above or below the mean of a distribution a particular score lies. If John scored 130 on a particular IQ test that has a mean of 100 and a standard deviation of 15, we could say either that he scored 30 points higher than the mean, or that his score was two standard deviations above the mean. If we chose to say that he was two standard deviations above the mean, we would be using the idea of the standard score. But why would we choose to express the location of an individual's score in terms of means and standard deviations? Why not simply use percentile rank?

There are certain circumstances in which percentile rank does not provide as much information as we would like about the location of a score within a distribution. For example, consider the percentile rank of the score 17 in each of these two sets of quiz scores:

$$1, 1, 2, 3, 3, 3, 4, 4, 5, 6, 6, 6, 6, 6, 7, 7, 7, 17, 17, 20$$

and

$$9, 10, 11, 11, 12, 13, 13, 14, 14, 15, 15, 15, 15, 16, 16, 16, 16, 17, 17, 20.$$

In both distributions, the application of our formula for percentile rank yields the same result for the score value 17. The percentile rank is 90. However, the position of the score 17 relative to the other scores in the first distribution seems quite different from the position of the score 17 in the second distribution. In the first set of scores, 10 points separate the 3 highest scores from the rest. The score 17 seems quite outstanding. In the second distribution, the score 17 still falls near the top of the distribution, but there are many scores that are close to it. The score 17 is still a good score, but it does not seem so outstanding as it seemed in the first distribution. How can we express the location of scores in a way that will differentiate between the position of the score 17 in the first distribution and the position of the score 17 in the second distribution?

In deciding that the score 17 was "outstanding" in the case of the first distribution, we were implicitly comparing the location of the score 17 on the interval scale to the location of a typical score. A more formal way of making the same judgment would be to use the mean of the distribution as a measure of the typical score. This is what we do when we compute a standard score. We determine how far above or below the mean of a distribution a given score lies, and then we express this distance in terms of the number of standard deviation units that it represents. The procedure may be summarized as follows:

$$\text{standard score} = \frac{\text{actual score} - \text{mean of distribution}}{\text{standard deviation of distribution}}.$$

Standard scores are most often referred to as *z-scores*. In differentiating standard scores from the actual scores they represent, we generally refer to the actual score as a "raw score." We denote the raw score as we have all along, with an X. Thus, we can rewrite the procedure just presented in the following symbolic form:

$$z = \frac{X - \mu}{\sigma},$$

where X represents a given raw score in the distribution, μ represents the mean of the distribution, σ represents the standard deviation of the distribution, and z is the name of the standard score equivalent of the raw score. This formula is sometimes known as the standard score transformation equation, because it is used to transform a given raw score into its corresponding standard score.

The standard score tells us how many standard deviation units above or below the mean of the distribution a given raw score lies. You will notice from the formula that standard scores may be positive numbers or negative numbers. If the particular raw score we are dealing with lies *above* the mean of the distribution, then the raw score will be larger than the mean, and $(X - \mu)$ will be a positive number. In this case, the standard score will be a positive number, for a standard deviation can never be negative. On the other hand, if the particular raw score we are dealing with lies *below* the mean of the distribution, then the raw score will be smaller than the mean, and $(X - \mu)$ will be a negative number. If the raw score is exactly equal to the mean of the distribution, the corresponding standard score will be zero. Standard scores typically range from a low of about -3, indicating that the raw score was 3 standard deviation units below the mean, to a high of about $+3$, indicating that the raw score was 3 standard deviation units above the mean. Standard scores outside this range occur less often.

Now let's look back to the two distributions just presented to see what the use of standard scores does for us. First, we use the formulas presented earlier to find the mean and the standard deviation of each of these distributions. Then, we apply the standard score transformation equation to find the standard score equivalent to the score of 17 in each distribution. This has been done in Figure 4.10. We find that the raw score of 17 lies 2.02 standard deviation units above

165

FIGURE 4.10
Standard Score Equivalent of a Raw Score of 17
in Each of Two Distributions

Distribution 1

1, 1, 2, 3, 3, 3, 4, 4, 5, 6, 6, 6, 6, 6, 7, 7, 7, 17, 17, 20

$$\mu_1 = 6.55$$
$$\sigma_1 = 5.18$$

Standard score equivalent
of raw score 17:

$$z = \frac{X - \mu}{\sigma} = \frac{17 - 6.55}{5.18}$$

$$z = +2.02$$

Distribution 2

9, 10, 11, 11, 12, 13, 13, 14, 14, 15, 15, 15, 15, 16, 16, 16, 16, 17, 17, 20

$$\mu_2 = 14.25$$
$$\sigma_2 = 2.62$$

Standard score equivalent
of raw score 17:

$$z = \frac{X - \mu}{\sigma} = \frac{17 - 14.25}{2.62}$$

$$z = +1.05$$

the mean in the first distribution, but only 1.05 standard deviation units above the mean in the second distribution. The larger z-score indicates numerically what we suspected from looking at the distributions, i.e., that a score of 17 in the first distribution was a more outstanding score than was a score of 17 in the second distribution. We could not tell this by comparing raw scores, for both were 17. We could not tell this by comparing percentile ranks, for both were 90. However, by transforming the raw scores into standard scores, we could differentiate between the position of the scores in the two distributions. The score of 17 in the first distribution was more than 2 standard deviations above the mean, but the score of 17 in the second distribution was just more than 1 standard deviation above the mean.

Standard scores are also useful in comparing the position of 2 scores within a single distribution. Consider this distribution of scores:

2, 4, 4, 5, 6, 8, 10, 11, 11, 17, 17, 18, 18, 18, 18, 19, 19, 19, 20, 20.

We note that the raw score of 11 has a percentile rank of 40, and the raw score of 17 has a percentile rank of 50. These percentile ranks are not very far apart. Yet, there is a spread of 6 raw score points between the 2 scores. If we compute the standard score equivalents of these 2 scores, we get a better idea of where the two scores are in relation to each other. This work is shown in Figure 4.11. What we find is that the raw score of 11 lies more than one third of a standard deviation *below* the mean (−0.35), and the raw score of 17 lies close to two thirds of a standard deviation *above* the mean (+0.61). Considering how close the percentile ranks of these two scores were, it seems clear that the use of standard scores has once again improved our ability to discriminate between scores.

Modified Standard Scores.   Given these characteristics of standard scores, you might wonder why these scores are not used to locate individuals with regard to their positions on nationwide examinations like the Scholastic Aptitude Test or the Graduate Record Examinations. In fact, they are, but in a modified form. Those of you who have taken one or more of these examinations know that scores range from a low of 200 to a high of 800 or 900, depending on the particular examination. But we noted that z-scores typically fall within a range from -3 to $+3$. How then can an SAT score of 600 be a standard score? The answer is that SAT scores are z-scores that have been multiplied by 100 and then had 500 added to them. In other words, when these nationwide exam scores are computed, a two-step process is involved. First, the raw scores are transformed to z-scores, using the z-score transformation equation:

$$z = \frac{X - \mu}{\sigma},$$

and then, each z-score thus computed is transformed to an SAT score using a different transformation equation:

$$SAT = z(100) + 500.$$

FIGURE 4.11
Standard Score Equivalent of Two Raw Scores
in a Single Distribution

Distribution
2, 4, 4, 5, 6, 8, 10, 11, 11, 17, 17, 18, 18, 18, 18, 19, 19, 19, 20, 20

$$\mu = 13.20$$
$$\sigma = 6.23$$

Standard score equivalent of raw score 11	Standard score equivalent of raw score 17
$z = \dfrac{X - \mu}{\sigma}$	$z = \dfrac{X - \mu}{\sigma}$
$z = \dfrac{11 - 13.2}{6.23}$	$z = \dfrac{17 - 13.2}{6.23}$
$z = -0.35$	$z = +0.61$

The reason this is done is that individuals who have never taken statistics (and few high-school students have) would not understand standard scores. It is more meaningful for them to read that they have a score of 600 on a test than a score of +1.00. It is much more meaningful to read that they have a score of 400 than a score of −1.00.

Mean and Standard Deviation of a Set of z-Scores.   It is an interesting fact that if we take every raw score in a distribution and transform it into a z-score, the resulting set of z-scores will *always* have a mean of zero and a standard deviation of one. To show you an example of this, we have taken a small set of scores, transformed them all to z-scores, and then computed the mean and the standard deviation of the resulting set of transformed scores. This work is presented in Figure 4.12. This will probably seem like little more than a curiosity to you now. However, keep this idea in the back of your mind, for it will be very important when we come to consider the standard normal curve in Chapter 7. For now, however, we need to consider more descriptive statistics. In Chapter 5, we move on to consider numbers that describe bivariate data.

THE CLASSIFICATION MATRIX

In this chapter, we have considered the descriptive statistics that are used with univariate problems. We looked at measures of central tendency, measures of variability, and measures of location for data with three different levels of measurement. In Table 4.2, we present our matrix once again. This time we have filled in the cells with the descriptive statistics covered in this chapter. Remember that these statistics are often used in conjunction with the techniques covered in Part 2 for descriptive problems and may also be used along with the inferential techniques to be covered in Part 4.

Descriptive Statistics

FIGURE 4.12

The Mean and Standard Deviation of a Set of Standard Scores.
Given the distribution of raw scores: 4, 4, 5, 6, 7, 8, 9, 9, 9
we find: $\mu = 7$, $\sigma = 2$.
We transform the scores to z-scores; then we use the definitional formula
to find the mean and standard deviation of the transformed scores:

X	$z = \dfrac{X - \mu}{\sigma}$	$\mu_z = \dfrac{\Sigma z}{N}$	$(z - \mu_z)$	$(z - \mu_z)^2$	$\sigma_z = \sqrt{\dfrac{(z - \mu_z)^2}{N}}$
4	$\dfrac{4 - 7}{2} = -1.50$	$\mu_z = \dfrac{0}{10}$	-1.50	$+2.25$	$\sigma_z = \sqrt{\dfrac{10}{10}}$
4	$\dfrac{4 - 7}{2} = -1.50$	$\boxed{\mu_z = 0}$	-1.50	$+2.25$	$\sigma_z = \sqrt{1}$
5	$\dfrac{5 - 7}{2} = -1.00$		-1.00	$+1.00$	$\boxed{\sigma_z = 1}$
6	$\dfrac{6 - 7}{2} = -.50$		$-.50$	$+.25$	
7	$\dfrac{7 - 7}{2} = 0$		0	0	
8	$\dfrac{8 - 7}{2} = +.50$		$+.50$	$+.25$	
9	$\dfrac{9 - 7}{2} = +1.00$		$+1.00$	$+1.00$	
9	$\dfrac{9 - 7}{2} = +1.00$		$+1.00$	$+1.00$	
9	$\dfrac{9 - 7}{2} = +1.00$		$+1.00$	$+1.00$	
9	$\dfrac{9 - 7}{2} = +1.00$		$+1.00$	$+1.00$	
	$\Sigma z = 0$		$\Sigma(z - \mu_z) = 0$	$\Sigma(z - \mu_z)^2 = 10$	

TABLE 4.2
Classification Matrix: Descriptive Statistics

Type of Problem		Level of Measurement of Variable(s) of Interest		
		Categorical: Nominal or Ordered Categories	Rank-Order	Interval Scale (or Ratio-Scale)
Univariate Problems		Modal Category Proportion of Observations Falling Outside Modal Category *With Ordered Categories:* Median Category	Percentile Rank	Mode, Median, Mean Range, Interquartile Range, Standard Deviation Percentile Rank, Standard Score
Bivariate Problems	Correlational Problems			
	Experimental and Pseudo-Experimental Problems: Independent Groups			
	Dependent Measures: Pretest and Posttest or Matched Groups			

Be sure you can define each of these terms.

descriptive statistic
measures of central tendency
measures of variability
measures of location
mode
median
mean
range
standard deviation
interquartile range
percentile rank
standard score
bimodal distribution
positively skewed distribution
negatively skewed distribution
symmetrical
definitional formula
computational formula
z-score

Summary

In Chapter 4, we discussed descriptive statistics for univariate problems. Descriptive statistics are numbers that convey information about a set of data. We have looked at three types of descriptive statistics: measures of central tendency, measures of variability, and measures of location. We pointed out that the choice of descriptive statistics depends on the level of measurement of the variable and, in the case of interval variables, on the shape of the frequency distribution.

As measures of central tendency, we examined the mode, the median, and the mean. The mode is the value of the variable that occurs most frequently in the set of data. Thus, it can be appropriately used with any distribution, regardless of the level of measurement. The median is the middle score of a distribution. It is found by ordering the measures from smallest to largest. Thus, we can find a median as long as the scores can be ranked.

4
Student
Guide

The mean is the arithmetic average of the scores in a distribution. Thus, it is an appropriate measure of central tendency only for interval (or ratio) scale data. If a distribution is highly skewed, that is, unsymmetrical, the median may be preferable to the mean as a measure of central tendency.

As measures of variability, the range, the interquartile range, and the standard deviation were examined. The range gives us the smallest and largest values in a distribution. Thus, it can be used whenever the values are ordered from smallest to largest. If we break the range into four parts, with 25 percent of the scores in each part, then the range of the two middle parts is the interquartile range. Thus, the interquartile range is appropriate when the values of a variable can be ordered. The standard deviation is a measure that estimates the average amount that the scores in a distribution deviate from the mean. Thus, it is appropriately used only with interval (or ratio) scale data.

Measures of central tendency and measures of variability describe the entire frequency distribution. To discuss the location of a single score within a frequency distribution, we examined two measures of location. The percentile rank of a score tells what percentage of the total distribution is at or below that score. We can calculate a percentile rank whenever we can order the scores in a distribution from smallest to largest. The standard score, called the z-score, tells us how many standard deviation units a score lies above or below the mean of the distribution of scores. A z-score is appropriately used as a measure of location whenever the mean and standard deviation can be used.

Review Questions

To review the concepts presented in Chapter 4, choose the best answer to each of the following questions.

1 A teacher gives a geometry test to all her students. She wishes to report to the students a score that she feels is typical or representative of the entire group performance. She should report
_____a a measure of central tendency.
_____b a measure of variability.

2 Which one of the following is a score value in a distribution?
_____a The range
_____b The mode

3 Which one of the following is appropriate as a measure of central tendency for any distribution?
_____a The mode
_____b The mean

4 Which one of the following can be represented as a length on an interval scale?
_____a The twenty-fifth percentile
_____b The interquartile range

5 If 51 percent of all the scores in a distribution have the value 10, then we can say that 10 is

_____a the median of the distribution.

_____b the mode of the distribution.

6 A spelling test is given to a class of 25 students. All 25 students spelled 9 out of 10 words correctly. We could say that this distribution of spelling scores has

_____a a standard deviation of zero.

_____b a mean of zero.

7 If the standard deviation for a distribution of scores is equal to zero, we know that for this distribution

_____a every score is equal to zero.

_____b every score has the same value.

8 A descriptive statistic is defined by a formula. This formula tells the meaning of the statistic. We call such a formula

_____a a computational formula.

_____b a definitional formula.

9 A nominal scale has categories coded with the numbers 1 to 5. We classify 68 people into the 5 categories. The statistic that would describe the central tendency of this distribution would be

_____a the mode.

_____b the mean.

10 Suppose you take a test and are told where your score stands relative to all the other scores in the distribution, in terms of cumulative percentage. You then know

_____a your standard score.

_____b your percentile rank.

11 If your percentile rank in a statistics class is 55, then

_____a you know you are above the mean.

_____b you know you are above the median.

12 If your standard score on a statistics test is —.5, you know that you are

_____a below the mean.

_____b below the median.

13 A distribution of scores ranges from 40 to 100. The interquartile range is from 52 to 60. Without further information, we can be sure that

_____a the median is between 52 and 60.

_____b the mean is between 52 and 60.

173

14 Mary is told that her final math test score is as good or better than 95 percent of her classmates. This tells Mary her

_____a percentile rank.

_____b standard score.

15 A teacher posts a cumulative percentage graph of the distribution of scores on her final examination. The students could use this graph to estimate their

_____a percentile rank.

_____b standard score.

16 Jane asks her guidance counselor if she should major in art. Her counselor advises Jane that on a test of artistic aptitude, she scored at the ninetieth percentile. This tells Jane that she

_____a got 90 percent as a score on the test.

_____b got a score as good or better than 90 percent of the tested group.

17 A large class takes a statistics examination. Paul is anxious about his performance on the test. The teacher can best relieve his anxiety by telling him that

_____a his score is 72.

_____b his percentile rank is 85.

18 If for a distribution of scores, $\displaystyle\sum_{\substack{\text{all} \\ \text{scores}}} (X-5) = 0$, then we know for certain that

_____a 5 is the mean score.

_____b the standard deviation is 0.

19 A distribution of scores ranges from 40 to 95. If you add 5 points to the top score and leave all other scores unchanged, you will change the value of

_____a the mean score.

_____b the median score.

20 A distribution of scores for an interval scale variable is changed to a distribution of z-scores. For the distribution of z-scores, we can say

_____a the mean is 0 and the standard deviation is 1.

_____b nothing about the mean and standard deviation without more information.

21 The distribution of family income in the United States is a positively skewed distribution. This would indicate that

_____a the mean family income is smaller than the median.

_____b the median family income is smaller than the mean.

22 A group of brain-damaged children is measured on audiovisual integration (AVI) with a measuring instrument consisting of 30 items. Each item con-

174

sists of several visual patterns and the subjects select one to match an audio pattern that they hear. If the mean AVI score is 8.1 and the standard deviation is 9, we might suspect that the distribution of AVI scores is

_____a positively skewed.

_____b negatively skewed.

23 A teacher gives an examination on material that she has taught. The exam consists of items covering concepts the students must understand in order to move on to the next unit of material. She would want her distribution of test scores to be

_____a positively skewed.

_____b negatively skewed.

Problems for Chapter 4: Set 1

1 Refer to Problem 1 in Set 1 of Chapter 2. What is the mode of the distribution of sex? Why?

2 Refer to Problem 2 in Set 1 of Chapter 2. How would you describe the typical subject on the variable "handedness"?

3 Refer to Problem 3 in Set 1 of Chapter 2. What is the mode of the distribution of favorite colors?

4 Refer to Problem 4 in Set 1 of Chapter 2. How would you describe the distribution of dollars over the budget categories?

Set 2

1 Joe is graduating seventh in a class of 40 students. There are 6 students ahead of him and 33 below him. What is his percentile rank?

2 Sarah's position in the class of 40 students is such that there are 10 students ranked ahead of her and 27 ranked below her. What is her percentile rank in the class?

3 Jim is ranked fortieth in the class of 40. What is his percentile rank in the class?

4 A physical fitness class has 70 members. The students are given a set of tasks to perform and are ranked according to their physical fitness. Joe is tied with 5 other students for the twelfth rank in the class. There are 53 class members less fit than he. Sam is tied with 3 others for the forty-first rank in the class and there are 26 class members less fit than Sam. Based on their percentile ranks, would you conclude that Joe is twice as fit as Sam?

175

1   A distribution of scores is as follows:

$$0, 0, 0, 0, 1, 3, 5, 7, 7, 17.$$

Find the mean, median, and mode of the distribution. Which measure best describes the typical performance?

2   Square each score in Problem 1. Find the average of the squared scores.

3   Subtract 4 from each score in Problem 1. Find the average of the deviation scores.

4   Refer to the data in Data Set A at the end of Chapter 2. You might also refer to the frequency distribution tabulated in Problem 3 in Set 2 of Chapter 2.
a   Calculate the mean number of years of education.
b   Calculate the median number of years of education.
c   What is the mode of the distribution?
d   Which measure best describes the distribution in terms of average number of years of education? Hint: Look at the frequency distribution as well as your answers to a, b, and c.

5   Refer to Problem 9 in Set 2 of Chapter 2.
a   What is the average number of correct answers for the 200 students?
b   What is the median number of correct answers?

6   Consider the following distribution of arithmetic scores:

Scores	Frequency
50–48	15
47–45	20
44–42	11
41–39	9
38–36	7
35–33	3
32–30	1
Total	66

a   Estimate the median arithmetic score.
b   How would you describe the distribution of scores?

7   A statistics teacher finds that on her 50-item final exam, the mean score of the 200 students was 25. The median, however, was only 15. What do these measures say about the difficulty of the exam?

176

Set 4

1 Calculate the standard deviation for the distribution mentioned in Problem 1 of Set 3 in this chapter. First, use the definitional formula, and then, use the computational formula.

2 Calculate the range, the interquartile range, and the standard deviation for the distribution of "the number of years of education" given in Data Set A (see Problem 4 of Set 3 in this chapter). What do these statistics mean?

3 A statistics teacher finds that on his final exam, the mean score for 200 students was 8.4 correct answers out of 30 and the standard deviation was 8.1. What does this suggest about the distributions of test scores?

4 In a high-school athletic league, 2 schools compare the statistics of their baseball teams using the number of runs scored as the variable of comparison. There are 9 teams in the league and each team plays the other once. The results are:

Opponent	School A	School B
1	2	1
2	3	2
3	3	11
4	5	0
5	0	8
6	1	0
7	4	1
Each other	3	1

a Find the mean number of runs scored for each team.
b Find the standard deviation for each distribution.
c Compare the performance of the 2 teams on the variable "number of runs scored per game."

5 a A statistics teacher in a large school randomly assigns her students to a control group and an experimental group. The control group takes the final exam under ordinary conditions and the experimental group takes its exam under conditions that are designed to relieve test anxiety and thus help students concentrate on answering the questions. The control group has a mean score of 35 with a standard deviation of 5; the experimental group has a mean score of 36 with a standard deviation of 15. What does this information tell you about the two distributions of scores?
b Suppose the results had been that the controls had a mean of 35 with a standard deviation of 10 while the experimentals had a mean of 40 with a standard deviation of 10. What could you then say about the two distributions of scores?

1 Refer to Problem 5 in Set 2 of Chapter 2. Use the cumulative relative frequency graph to answer the following questions concerning the weights of
200 males:
a Estimate the percentile rank for a weight of 74.5 kilograms.
b Estimate the percentile rank for a weight of 94.5 kilograms.
c A weight of 79 kilograms has what percentile rank?
d A weight of 86 kilograms has what percentile rank?
e Estimate the median weight for the distribution.
f Estimate the interquartile range for the distribution.

2 The following frequency distribution is a summary of 237 IQ scores on the
Lorge-Thorndike Intelligence Test.

IQ Score	Frequency
80–89	8
90–99	34
100–109	52
110–119	59
120–129	53
130–139	21
140–149	7
150–159	3
Totals	237

Use arithmetic estimation to estimate the following:
a The fiftieth percentile
b The ninetieth percentile
c The percentile rank for 105
d The percentile rank for 146
e The twenty-fifth percentile

3 To compare the answers given from the graph to those given by arithmetic
estimation, estimate the answers for Problem 1 using arithmetic estimation
(see Table 2.7).

4 Suppose the variable X has a mean of 36 and a standard deviation of 6.
a Find the z-score for each of the following values of X.
i $X = 44$ iii $X = 36$ v $X = 31$
ii $X = 27$ iv $X = 39$ vi $X = 57$
b For each standard score given below, find the corresponding value of X.
i $z = 1.00$ iii $z = 0$ v $z = -.67$
ii $z = -1.5$ iv $z = 3.2$ vi $z = -2.1$

5 Below is a scale for the variable X. Directly below it is a scale showing corresponding points on the standard score scale.

178

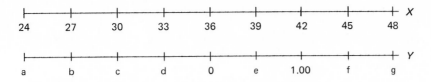

a What is μ_X?
b What is σ_X?
c Label each point a, b, c, d, e, f, g with the appropriate z-score.
d What is the distance from a to d on the z-scale?
   What is the corresponding distance on the X-scale?
e How far is it from b to e on the z-scale?
f Make a rule that relates a length on the z-scale to the corresponding length
   on the X-scale.

6 The formula $C = 5/9(F - 32)$ is the formula relating measurements of temperature on the centigrade scale with corresponding measurements of temperature on the Farenheit scale. A weatherman has been recording temperature for the month of December and reports the following statistics:

$$
\begin{aligned}
\text{Maximum reading} &= 44°F \\
\text{Minimum reading} &= -2°F \\
\text{Mean} &= 27°F \\
\text{Median} &= 25°F \\
\text{Mode} &= 22°F \\
\text{Standard Deviation} &= 9°F
\end{aligned}
$$

a What would these measurements be in centigrade units?
b For the month of December, which one of the following represents the
   more unusual temperature: $14°F$ or $0°C$?

179

5

Descriptive Statistics for Bivariate Data

DESCRIPTIVE STATISTICS FOR BIVARIATE DATA: AN OVERVIEW

In Chapter 3, we considered the techniques used for organizing and presenting data in bivariate problems. You will recall that our classification scheme for elementary statistical problems accommodates several different classes of bivariate problems. There are bivariate correlational problems; experimental and pseudo-experimental problems, involving the comparison of two or more independent groups; and problems involving two sets of dependent measures. Most of this chapter is devoted to the descriptive statistics that apply to bivariate correlational problems. This is because the same descriptive statistics that apply to univariate problems may be applied conveniently to both problems involving the comparison of independent groups and problems involving two sets of dependent measures. For example, in an experimental study in which we sought to compare an experimental group to a control group on reading achievement, we might compute the mean reading achievement score for the experimental group and compare it to the mean score for the control group. These means would be identical to the means computed in a univariate problem. The only difference is that the mean is computed twice, once for each group. The descriptive statistics applicable to univariate problems may also be employed in problems involving two sets of dependent measures. If the experimental and control groups just mentioned had been matched, then we would have two sets of dependent measures rather than measures on two independent groups. However, we would still compare reading achievement in the two groups by comparing the means. Similarly, in a pretest and posttest situation, we might compare the mean pretest score to the mean posttest score, assuming we have interval scale data. Thus, you are already familiar with the descriptive statistics that we employ in studies involving the comparison of independent groups and in studies involving two sets of dependent measures.

With respect to bivariate correlational problems, however, we must examine a new set of descriptive statistics. These statistics, taken as a group, are referred to as *measures of relationship.*

You will recall that in bivariate correlational problems, we are dealing with a single group of subjects that we have measured on two variables. You should remember from Chapter 3 the techniques used in organizing and presenting data from bivariate correlational problems. There we considered the use of crosstabulation tables for categorical data, two sets of ranks for ordinal data, and scatterdiagrams for interval data. We also considered situations in which it was instructive to collapse ordinal or interval variables into variables with a limited number of categorical values and then use a crosstabulation table to organize the data. In all cases, the methods we use to organize bivariate data are designed to shed some light on the relationship that exists between the two variables measured.

In this chapter, we consider numerical indices that summarize the extent to which two variables are related in a given set of bivariate data. There are descriptive statistics for all three types of bivariate data. To describe the relationship between two categorical variables, we typically use Cramér's statistic, more commonly referred to as the *index of contingency.* To express the relationship between two ordinal variables, we generally use the *Spearman rank-order correlation coefficient.* To express the relationship between two interval variables, we generally use the *Pearson product-moment correlation coefficient.* Each of these descriptive statistics yields a single number that describes the relationship between two variables. In the case of the index of contingency, this number may range between 0 and 1.00, with 0 indicating no relationship, and 1.00 indicating a perfect relationship. In the case of the Spearman rank-order correlation and the Pearson product-moment correlation, the number may range from -1.00 through 0 to $+1.00$, covering a continuum from strong negative to strong positive relationships. Here we consider each type of data and each measure of relationship in turn.

DESCRIPTIVE STATISTICS FOR BIVARIATE CORRELATIONAL PROBLEMS

Bivariate Correlational Problems Involving Categorical Data: Cramer's Index of Contingency

In Chapter 3, we considered the crosstabulation of opinion on the Equal Rights Amendment by sex for 163 voters from Tranquility, New York. We learned the names of the different parts of the crosstabulation table. We considered the question of how to determine whether one of the two variables should be regarded as the independent variable. We also showed how we could compare the conditional distributions of the dependent variable for each value of the independent variable to help us determine whether or not a relationship exists between the two variables. In that particular example, we decided it was logical to regard sex as the independent variable and opinion as the dependent variable. In comparing the conditional distributions of the variable "opinion" for the two values of sex, we asked questions like: "Is the proportion of men in favor greater than the proportion of women in favor?" or "Is the proportion of men opposed greater than the proportion of women opposed?" or "Is the proportion of men having no opinion greater than the proportion of women having no opinion?" Once we have answered these questions, we can judge whether sex and opinion are related in this group of subjects. We can make a statement like "52.5 percent of the men in this group were opposed to the ERA compared to only 37.1 percent of the women. This difference between the proportions suggests the variables are related." But what if we were asked the question, "*How strong* is the relationship between sex and opinion?" In this group of subjects, a simple comparison of the proportion of men opposed to the proportion of women opposed would not constitute a sufficient answer, because it ignores the proportions of men and women who were in favor and the proportions who had no opinion. To attempt to answer the question of strength of the relationship, we would really need to compare the proportion of men falling into each category of the variable "opinion" to the proportion of women in the same category. We would need to do this for all categories. This would require us to present several different sets of proportions. It is often preferable to employ some statistical procedure that summarizes *all* of the appropriate comparisons in a single number that describes the extent of the relationship between the two variables. This is what Cramér's index of contingency does. We will explain the procedure for obtaining the index of contingency shortly. First, it is necessary to define one of the elements we use in computing the index. This element is the *expected cell frequency*.

Expected Cell Frequency. To put it simply, the expected frequency of a given cell in a crosstabulation table is the number of observations we would *expect to find* in that cell, *assuming there was no relationship* at all between the two variables. For a moment, pretend that we have conducted our survey of voters in Tranquility, but that we have recorded only the *marginal* frequencies for sex and opinion. These marginal frequencies are shown in boldface type in Table 5.1. As described in Chapter 3, the row marginals tell us how many people out of the total group of 163 had each of the three possible opinions. We see that 64 were in favor, 76 were opposed, and 23 had no opinion. Also, as described in Chapter 3, we can use these row marginal frequencies to compute the proportion or percentage of the *total group* that had each of the three opinions. These percentages are shown in parentheses beneath the absolute frequencies. We see that 39.26

182

TABLE 5.1

Marginal, Expected, and Observed Frequencies for Crosstabulation
of Opinion on ERA by Sex for 163 Residents of Tranquility, New York

Sex

Opinion		Males	Females	Marginal Total
In favor		$Ex = 39.66$ $Ob = 33$	$Ex = 24.34$ $Ob = 31$	**64** (39.26%)
Opposed		$Ex = 47.09$ $Ob = 53$	$Ex = 28.91$ $Ob = 23$	**76** (46.63%)
No opinion		$Ex = 14.25$ $Ob = 15$	$Ex = 8.75$ $Ob = 8$	**23** (14.11%)
Marginal total		**101**	**62**	**163** (100.00%)

Numbers in parentheses are column percentages, i.e., the number in each row is a percentage of the total number in the column.

percent of the total group were in favor, 46.63 percent were opposed, and 14.11 percent had no opinion.

Now assume for the moment that there is in fact no relationship between sex and opinion. We have seen that if this is true, then the conditional distributions of opinion for the males and females would be similar. For example, the proportion of men in favor would have to be the same as the proportion of women in favor. But then each of these proportions would have to be the same as the proportion of the entire group in favor. In this case, that would be 39.26 percent. To restate the argument in slightly different terms, if 39.26 percent of everyone in the group was in favor, and if there is no difference between the men's opinions and the women's opinions, then we would expect that 39.26 percent of the men would be in favor, and 39.26 percent of the women would also be in favor. The same logic would apply to the proportion of men and women opposed, and to the proportion of men and women with no opinion.

But if we expect a certain *proportion* of men to be in favor, and if we know how many men there are in the group we are interested in, then we can compute the *number* of men that we would expect to be in favor. This number would be the expected frequency of the cell containing all the men who were in favor.

183

Here, assuming that there is no relationship between sex and opinion, we would expect 39.26 percent of the 101 men included in our set of data to be in favor. This figure is 39.66. Similarly, we would expect 39.26 percent of the 62 women to be in favor. Then, there would be 24.34 women who were in favor. By the same process of reasoning, we expect 46.63 percent of the 101 men to be opposed, and the same percentage of the 62 women to be opposed. We would also expect that 14.11 percent of both the men and the women would have no opinion.

There is a simple rule that may be used to obtain the expected frequency for any cell of a crosstabulation table. Multiply the marginal total for the *row* in which the cell is located by the marginal total for the *column* in which the cell is located, and divide the result by the total number of observations. In the case of the cell corresponding to males who are in favor, this rule would appear as

$$\text{Expected frequency of males in favor} = \frac{\text{Row marginal} \times \text{Column marginal}}{\text{Total}} = \frac{(64)(101)}{163}.$$

The reason that this rule works is that 64/163 or 39.26 percent represents the percentage of all subjects who were in favor, and 101 represents the total number of males included in the study. Therefore, the product of these two numbers represents the number of males we would expect to be in favor, if there were no difference between the opinions of males and females.

To find the expected frequency of females in favor,

$$\text{Expected frequency of females in favor} = \frac{(64)(62)}{163} = 24.34.$$

In Table 5.1, we have presented the expected frequencies for each of the six cells of the crosstabulation table in the upper left section of the cell. These expected frequencies are labeled *Ex* (denoting "expected"). You will notice that the expected frequencies contain fractions. Of course, there is no way that a fraction of a subject can have an opinion. It is generally the case that it is not possible for the expected frequencies to be duplicated exactly by the actual frequencies that we observe within a given set of data. However, the *less* the two variables in question are related to each other, the *more* closely the actual cell frequencies will resemble these expected frequencies.

In the lower right section of each cell in Table 5.1, we present the *actual* number of subjects whom we observe to fall into the cell. These numbers are most often referred to as *observed cell frequencies*. In Table 5.1, these observed cell frequencies are denoted *Ob*. How do the observed cell frequencies in Table 5.1 compare to the cell frequencies we would expect to find, if the two variables were unrelated? Consider the cell representing males who were in favor. There, we expected to find 39.66 subjects, but we actually observed that 33 subjects fell into the cell. Thus, there were fewer males in favor than we would have expected, if males and females had the same opinions. Likewise, there were more females in favor than we would have expected. If you stop to think about it, you will realize that reporting the differences between the observed frequency in these top two cells is really just another way of saying that the proportion of men in favor is different from the proportion of women in favor. However, these differences between observed frequencies and expected frequencies are easier to use than proportions when we calculate a single descriptive statistic that will summarize the relationship between the two variables over *all* the cells in the table.

The Computation of the Index of Contingency. The calculation of the index of contingency is quite simple, but there are a number of steps involved. An outline of these steps follows:

Step 1   For each cell in the table, subtract the expected frequency from the observed frequency.

Step 2   For each cell, square the result of Step 1. This will eliminate all the negative signs.

Step 3   For each cell, divide the result of Step 2 by the expected frequency *for that cell.*

Step 4   Add up the results of Step 3 for all the cells in the table.*

Step 5   Compare the number of rows in the table to the number of columns. Subtract 1 from the smaller of these 2 numbers.

Step 6   Multiply the result of Step 5 by N, the total number of observations in the data set.

Step 7   Divide the result of Step 4 by the result of Step 6.

Step 8   Take the square root of Step 7.

The shorthand formula that summarizes these step-by-step directions looks like this:

$$\phi' = \sqrt{\frac{\displaystyle\sum_{\text{rows}}^{\text{all}} \sum_{\text{columns}}^{\text{all}} \frac{(Ob - Ex)^2}{Ex}}{N(L-1)}}$$

The symbols have the following meanings:

ϕ':   This is the Greek letter phi with a prime. It is the symbol we use to represent Cramér's index of contingency.

---

* Up to this point, you have calculated a statistic known as the χ^2 (chi-square) statistic. This is a frequently used statistic, and many readers will be familiar with it already. However, it is an inferential statistic rather than a descriptive statistic; thus, we will consider it in Part 4 of this book. For the purpose of the index of contingency, it is an intermediate step.

$$\frac{(Ob - Ex)^2}{Ex}:$$ This tells us to subtract the expected frequency of a cell from the observed frequency, square the result, and divide by the expected frequency of the cell.

$\Sigma\Sigma$: The double summation sign tells us to perform the above operation for every cell in the table and add together the results for all these cells. There are always as many cells as the product of the number of values of one variable and the number of values of the other variable.

N: The total number of observations in the data set.

$(L - 1)$: One less than the number of rows (R) *or* the number of columns (C), whichever is the lesser (L) of these two numbers.

In Figure 5.1, we illustrate the calculation of the index of contingency for the crosstabulation of vote by opinion for the 163 residents of Tranquility, New York. Each step is labeled and the mathematical notation corresponding to each step is presented alongside. We have identified each cell in terms of its row and column position, as described in Chapter 3. As you can see from Figure 5.1, the index of contingency for the table in this set of data is approximately .17. Of course, the .17 is just a number. At this point, we would like to give you a feeling for what an index of contingency of .17 means, in terms of the strength of relationship it represents.

The Interpretation of the Index of Contingency. You should get some idea of the meaning of an index of contingency of .17 from the statement made at the beginning of this chapter, that the index ranges from 0 to 1.00. If $\phi' = 0$, then the two variables that have been crosstabulated are said to be unrelated. If $\phi' = 1.00$, they are said to be perfectly related. To give you a better idea of the meaning of "unrelated" and "perfectly related" and to give you a sense of the meaning of indices of contingency that lie between 0 and 1.00, we will present several different crosstabulation tables along with the calculated value of ϕ' for each table.

Table 5.2 is the crosstabulation of political party preference by level of education for a hypothetical group of 500 high-school teachers. The computed value of ϕ' for this crosstabulation is 0. We have said than an index of contingency of zero indicates that there is no relationship between party preference and education among these 500 teachers. But what does "no relationship" mean? First, think about the procedure employed in calculating the index of contingency. If you think about the formula, you will see that the index of contingency can be zero *only* if the observed frequency is exactly equal to the expected frequency for *every cell* in the table. Calculate the expected frequencies for several cells. Do they equal the observed frequencies? Now, think about what this means in terms of the conditional distributions of party preference for each of the three levels of education. What proportion of the 250 teachers who completed college

FIGURE 5.1
Calculation of Index of Contingency for Crosstabulation of Opinion on ERA
by Sex for 163 Residents of Tranquility, New York

	Step 1	Step 2	Step 3
			$\dfrac{(Ob - Ex)^2}{Ex}$
Cell	$(Ob - Ex)$	$(Ob - Ex)^2$	
1, 1	$33 - 39.66 = -6.66$	$(-6.66)^2 = 44.36$	$\dfrac{44.36}{39.66} = 1.12$
1, 2	$31 - 24.34 = +6.66$	$(+6.66)^2 = 44.36$	$\dfrac{44.36}{24.34} = 1.82$
2, 1	$53 - 47.09 = +5.91$	$(+5.91)^2 = 34.93$	$\dfrac{34.93}{47.09} = 0.74$
2, 2	$23 - 28.91 = -5.91$	$(-5.91)^2 = 34.93$	$\dfrac{34.93}{28.91} = 1.21$
3, 1	$15 - 14.25 = +0.75$	$(-0.75)^2 = 0.56$	$\dfrac{0.56}{14.25} = 0.04$

Step 4

3, 2	$8 - 8.75 = -0.75$	$(-0.75)^2 = 0.56$	$\dfrac{0.56}{8.75} = 0.06$

$$4.99 = \Sigma\Sigma \frac{(Ob - Ex)^2}{Ex}$$

Step 5

This table has 3 rows and 2 columns. The smaller of these numbers is the 2, so $L = 2$ and $L - 1 = 1$

Step 6

N here is 163. Thus, $N(L - 1) = 163 (2 - 1) = 163 (1) = 163$.

Step 7

$$\frac{\Sigma\Sigma \dfrac{(Ob - Ex)^2}{Ex}}{N(L - 1)} = \frac{4.99}{163} = .03$$

Step 8

$$\phi' = \sqrt{\frac{\dfrac{(Ob - Ex)^2}{Ex}}{N(L - 1)}} = \sqrt{.03} = .17$$

TABLE 5.2
Crosstabulation of Political Party Preference by Level of Education
Among 500 High-School Teachers

Level of Education

		Completed college	Have master's degree	Have doctorate	Total
Political Party Preference	Republican	75 (30%)	60 (30%)	15 (30%)	150 (30%)
	Democrat	125 (50%)	100 (50%)	25 (50%)	250 (50%)
	Other	10 (4%)	8 (4%)	2 (4%)	20 (4%)
	Independent	40 (16%)	32 (16%)	8 (16%)	80 (16%)
	Total	250	200	50	500

$$\phi' = 0$$

Numbers in parentheses are column percentages.

but have no higher degree are Republicans? What proportion of the 200 teachers with master's degrees are Republicans? What proportion of the 50 teachers holding doctorates are Republicans? Exactly 30 percent of the teachers in *each* of the three categories of education are Republicans. Similarly, exactly 50 percent of the teachers in each of the three categories of education are Democrats. The same holds true for the "Other" and "Independent" classifications. Thus, in this example, the index of contingency of 0 means that the proportion of teachers falling into any category on the variable party preference will be the same, regardless of which level of education we are considering. There is no relationship between level of education and political party in this example.

Another way of thinking about this is to consider a single teacher selected at random from the group of 500. If we were to pick a teacher at random from the whole group, what would be the chance that this teacher would be a Republican? The chance of being a Republican would be the same as the proportion of Republicans in the entire group of 500. We have seen that this is 30 percent. Now think about a single teacher selected at random from among those 250 teachers who completed college. What is the chance that one of these 250 will be a Republican? This chance would also be 30 percent. And the chance that a teacher selected at random from among the 200 teachers with master's degrees or from among the 50 teachers with doctorates would also be 30 percent. Similarly, the chance of a given teacher being a Democrat is 50 percent, regardless of which category of education he or she falls into; the chance of a teacher supporting an "other" party is 4 percent, regardless of education; and the chance of a teacher being an Independent is 16 percent, regardless of education.

It is often instructive to think of the question of relatedness in the following way. What do we learn about a subject's party preference by knowing level of education? In this problem, the chance that a given subject will fall into a particular category on the variable "party preference" is the same, regardless of what the level of education is. Therefore, knowing a particular subject's education would do nothing to help us guess party preference. If we had to guess a subject's party preference, we would *always* guess Democrat, for 50 percent of the teachers of every level of education are Democrats. This proportion is always the highest of the four categories of party preference. When two variables are unrelated, they are sometimes said to be independent of one another. The term "independent" in this context has nothing whatever to do with the idea of independent variable and dependent variable. We may have good theoretical reasons for regarding one of two variables as the independent (potentially influencing) variable and the other as the dependent variable, yet find that, in fact, the two are unrelated.

Now let's look at the crosstabulation of two categorical variables that are *perfectly* related to one another. Table 5.3 is the crosstabulation of the mood of 200 returning Caribbean vacationers by the predominant weather condition during their respective holidays. The computed value of the index of contingency for this crosstabulation table is 1.00, which indicates a perfect relationship between weather condition and mood. What does this mean? Looking at the row marginals, we see that half of the returning vacationers are in a good mood, a quarter are in a bad mood, and another quarter are in an unspeakably bad mood. If there were no relationship between weather and mood, we would expect that these proportions would be the same when we looked at those who had sunny weather, those who had cloudy weather, and those who had rainy weather. But this is not the case. When we look at the 100 vacationers who had sunny weather, we find that all 100 are in good moods. If there were no relationship between weather and mood, we would have expected to find 50 in a good mood, 25 in a bad mood, and 25 in an unspeakably bad mood. Instead, we find 100 in a good mood and none in bad or unspeakably bad moods. Thus, in the three cells of the first column of the crosstabulation table, there are considerable differences between the actual observed frequencies and the expected frequencies. Differences like these cause the index of contingency to become large.

But we can say much more about this table. Notice that if we specify what weather condition a given vacationer had, we can state positively his or her mood. Whenever the crosstabulation table is such that knowing the category of

TABLE 5.3
Crosstabulation of Mood of 200 Returning Vacationers
by Weather Condition on their Vacations

		Weather Condition			
		Sunny	Cloudy	Rain	Total
Mood	Good	100 (100%)	0 (0%)	0 (0%)	100 (50%)
	Bad	0 (0%)	50 (100%)	0 (0%)	50 (25%)
	Unspeakably bad	0 (0%)	0 (0%)	50 (100%)	50 (25%)
	Total	100	50	50	200

$$\phi' = 1.00$$

a subject on one variable allows you to be certain of the category on the other variable, the two variables are said to be perfectly related. In this case the index of contingency will reach its maximum value, 1.00. In the case of Table 5.3, it is also true that if we know what mood a vacationer is in, we can state positively the type of weather he or she had. That is, if we know a subject's category on *either* of the two variables, we automatically know the category on the other. In this case, we say that the two variables are *mutually determined.*

However, it is not necessary for the two variables to be mutually determined for the index of contingency to reach the value of 1.00. It is only necessary that *one* of the two variables is determined by the other. For example, consider Table 5.4. This is a hypothetical crosstabulation of vote on a referendum to eliminate tax loopholes by annual income for 300 suburban homeowners. The calculated value of ϕ' for this crosstabulation table is 1.00, indicating a perfect relationship between the two variables. If you consider the table, you will realize that if we know a homeowner's income bracket, we automatically know how this homeowner voted. However, if we know the vote, we cannot always be sure of the income bracket. A homeowner voting "no" might fall in *either* the medium or the high income bracket. Thus only one of the two variables needs to be determined by the other to achieve a ϕ' of 1.00.

In Tables 5.2 to 5.4, we have seen the extremes of unrelatedness and relatedness. Obviously, crosstabulation tables as extreme as these occur rarely in practical research situations. To give you a feel for values of the index of contingency lying between these extremes, we present three more tables with values of ϕ' lying between 0 and 1.00. Table 5.5 shows the crosstabulation of two variables that are strongly related but not perfectly related. A group of 200 television watchers from American families were asked whether they would prefer to watch cartoons or a talk show. Their preference was crosstabulated with their age (either under 15 years, or 15 and above). The value of ϕ' for the resulting table is .80, close to the maximum value of 1.00. We can get an idea of what this value of the index of contingency means by considering the marginal and conditional distributions. If we look at the entire group of 200 television viewers, we find that there is an even split between preference for cartoons and preference for the talk show. If a viewer were picked at random from the 200, there would be as great a chance that the viewer would prefer cartoons as the talk show. But if we look at the conditional distributions of viewing preference by age, we find that 90 percent of those under 15 preferred the cartoons, and 90 percent of those 15 or over preferred the talk show. Thus, if we know the age of a given viewer, we are in an excellent position to guess that viewer's preference. Note that we cannot guess a viewer's preference with absolute certainty. If we could, the two variables would be perfectly related, and the index of contingency would be 1.00 rather than .80. When the index of contingency is .80, however,

TABLE 5.4
Crosstabulation of Vote on Eliminating Tax Loopholes
by Income Level for 300 Suburban Homeowners

		Income Level			
		Low (0 – $15,000)	Medium ($15,000 – $30,000)	High (over $30,000)	Total
Vote	Yes (favors bill)	150 (100%)	0 (0%)	0 (0%)	150 (50%)
	No (opposes bill)	0 (0%)	75 (100%)	75 (100%)	150 (50%)
	Total	150	75	75	300

$$\phi' = 1.00$$

Bivariate Problems

TABLE 5.5
Crosstabulation of Television Viewing Preference
by Age for 200 Viewers

		Age		
		Under 15 years	15 or above	Total
Preference	Cartoons	90 (90%)	10 (10%)	100 (50%)
	Talk show	10 (10%)	90 (90%)	100 (50%)
	Total	100	100	200

$$\phi' = .80$$

our ability to determine a subject's category on one variable is likely to be considerably improved if we know the category on the other.

Table 5.6 shows the crosstabulation of two variables that have a moderately strong relationship. A group of 250 governors and big city mayors at a convention in Aspen, Colorado, were asked their opinion on state aid to urban mass transit. Their view of this issue was crosstabulated with their job title. The value of ϕ' for the resulting table is .60. Looking at the marginal and conditional distributions, we find that the mayors are a good deal more likely to favor state aid to urban mass transit than we would expect if there were no relationship between job title and view on this issue. Also, the governors are much more likely to oppose state aid than we would expect if there were no relationship. However, the differences between the conditional distributions of view on state aid and the marginal distribution of view on state aid are not as great as were the differences between the conditional and marginal distributions in the example of Table 5.5. If we were told that a given politician was a mayor, we could refer to the conditional distribution of viewpoint on state aid for the 200 mayors, which would lead us to predict that the politician would favor state aid to urban mass transit. But we would not be as confident in this prediction as we would be in predicting that one of our 200 television viewers would prefer cartoons, if he or she was under 15. Similarly, if we were told that one of our 250 politicians were a governor, we could consult the conditional distribution of viewpoint for the 50 governors, which would lead us to predict that the governor would oppose state aid. There is certainly a relationship here, for the prediction for a governor based on the conditional distribution of governors is different from the prediction we

192

would make for a politician whose job title was not known. However, in predicting that a governor opposes state aid, there would be a considerable (36 percent) chance that we would be wrong.

Table 5.7 shows the crosstabulation of two variables that are just moderately related. A group of 500 students was classified according to hair color and eye color. The value of ϕ' for the resulting table is .41. Comparing the marginal and conditional distributions of hair color, we find that for *some cells* there is a considerable discrepancy between the proportion of students having a given hair color and the proportion we would expect, if eye color and hair color were unrelated. For example, if the two variables were unrelated, we would expect that 20 percent of the subjects of any eye color would have blond hair. In fact, we find that among the 100 subjects with blue eyes, 65 percent have blond hair. On the other hand, among subjects with brown eyes, only 2.5 percent have blond hair. For these cells then, there are large differences between expected and observed frequencies. However, for some *other cells,* the proportion of students actually having a given hair color and the proportion we would expect are quite close. For example, we would expect that 60 percent of those in the "other" category of eye color would have brown hair, and this is exactly what we find. Thus, there is no difference between the observed and expected frequency in this cell. There are several other cells in the table where the observed and expected frequencies are very close, though not exactly equal. In a table like this one, our

TABLE 5.6
Crosstabulation of View on State Aid to Urban Mass Transit
by Job Title for 250 Politicians

		Job Title		
		Big city mayor	State governor	Total
View on State Aid to Transit	Favors	150 (75%)	10 (20%)	160 (64%)
	Opposes	18 (9%)	32 (64%)	50 (20%)
	Uncertain	32 (16%)	8 (16%)	40 (16%)
	Total	200	50	250

$$\phi' = .60$$

Bivariate Problems

TABLE 5.7
Crosstabulation of Hair Color by Eye Color for 500 Students

Eye Color

		Blue	Brown	Other	Total
Hair Color	Blond	65 (65%)	5 (2.5%)	30 (15%)	100 (20%)
	Brown	30 (30%)	150 (75%)	120 (60%)	300 (60%)
	Black	5 (5%)	45 (22.5%)	50 (25%)	100 (20%)
	Total	100	200	200	500

$$\phi' = .41$$

ability to predict a subject's category on one variable from the category on the other variable would vary considerably, depending on which variable we know and which category on this variable the subject falls into. The index of contingency, however, provides us with a summary description of the extent to which the two variables are related for this group of subjects over *all* the cells of the table. In this table, the index of contingency tells us that, on balance, these two variables are just moderately related.

As an example of a crosstabulation of two variables that have only a slight relationship, we return to the crosstabulation of opinion on the ERA by sex for the 163 residents of Tranquility. You will remember that the value of ϕ' for this table was calculated to be .17. We have already looked at the conditional distributions of opinion for each of the two sexes. Turn back to Table 3.2 to find the proportion of males and females with each of the three different opinions. Keeping in mind what you have seen in Tables 5.3 through 5.7, would you now say that the marginal and conditional distributions presented in Table 3.2 are very different from one another? Not really. We can see from Table 3.3 that the conditional distribution of opinion for males does differ from the conditional distribution of opinion for females, so we know that the two variables *are related.* But how much does the chance that a given resident was opposed actually change when we specify the sex of the resident? If we pick a resident at random, the chance that this resident was opposed is 46.63 percent. If we specify that the resident is male, then the chance that he was opposed changes to 52.48 percent. This is not a tremendous difference. Moreover, there is no cell in the table where

194

the proportion of subjects of a particular sex who had a particular opinion differs by more than 11 percentage points from the proportion of subjects in the entire group who had that opinion. Compare this to the change that takes place in the chance that a television viewer will prefer cartoons when we specify that this viewer is under 15 (Table 5.5). Compare it to the change that takes place in the chance that a student will have blond hair when we specify that the student has blue eyes (Table 5.7). These changes are much greater. Thus, we can say that there is a relationship between opinion on the ERA and sex in this group of residents, but this relationship is not very strong. Specifying a particular resident's sex does not result in a great change in the chance that the subject had a particular opinion. This is why ϕ' in this table is only .17.

Bivariate Correlational Problems Involving Rank-Order Data: The Spearman Rank-Order Correlation

In Chapter 3, we considered several sets of bivariate rank-order data. One of these was the ranking of 15 students in their English and math classes. These data illustrated the meaning of a positive relationship between two sets of ranks, for we noticed that students who ranked high in English had a tendency to rank high in math as well. The other example was the ranking of nine political objectives in order of priority by a conservative congressman and a liberal congressman. This last set of data illustrated the meaning of a negative relationship between two sets of ranks, for we noticed that the issues given high priority by the conservative congressman tended to be given much lower priority by the liberal congressman. Here, we move beyond this general conception of the meaning of positive and negative relationships to consider the question of the strength of such relationships. The Spearman rank-order correlation provides us with a numerical index of the strength of relationship in a set of bivariate rank-order data.

The Computation of the Spearman Rank-Order Correlation. The steps involved in the hand calculation of the rank-order correlation may be summarized verbally as follows:

Step 1 Lay out the bivariate rank-order data in such a way that paired ranks are adjacent to each other. This procedure was described in Chapter 3 and illustrated in Tables 3.4 and 3.5. In Table 3.4, each subject was ranked in two classes. Thus, rank in English was written opposite rank in math to represent the pairing of ranks. In Table 3.5, each issue was ranked by two judges. Thus the rank assigned to an issue by one judge was written opposite the rank assigned that issue by the second judge.

Step 2 For each pair of ranks in the data, find the difference between them. In the case of the students ranked in each of two classes,

195

this means that for each student we would subtract the rank achieved in one class from the rank achieved in the other. The smaller these differences are, the stronger is the relationship. This step is most conveniently accomplished by setting up a column headed D, for difference. This is illustrated in Figure 5.2.

Step 3   Square each of these differences. This is also easily done by setting up a column headed D^2.

Step 4   Add up these squared differences and multiply the result by 6.

Step 5   Count the number of pairs of ranks in the data. Call it N.

Step 6   Square the number N and subtract 1 from it.

Step 7   Multiply the result of Step 6 by N once again.

Step 8   Divide the result of Step 4 by the result of Step 7.

Step 9   Subtract the result of Step 8 from 1. This gives you the Spearman rank-order correlation.

The mathematical formula that summarizes these steps looks like this:

$$\rho_s = 1 - \frac{6\Sigma D^2}{N(N^2 - 1)}.$$

The symbols have the following meanings:

ρ_s: The Greek letter rho, often used to refer to correlation, with the subscript s, to indicate that this is the Spearman rank-order correlation.

ΣD^2:   This tells us to find the difference between the paired ranks, square each difference, and add up all these squared differences to get the sum of the squared differences. It corresponds to Steps 1 through 3 just described.

N: This capital N stands for the number of *pairs* of ranks used in the calculation. Here, we have 15 pairs of ranks, one pair for each student. Therefore, $N = 15$.

In Figure 5.2, we have carried out the calculation of ρ_s for the ranks of the 15 students in English and math. Each step is identified and the mathematical notation representing that step is shown next to it.

Interpretation of the Spearman Rank-Order Correlation.   You will note that the calculated value of the Spearman rank-order correlation for the rankings of students in English and math is +.77. As noted before, the rank-order correlation ranges from −1.00 to +1.00, with correlations close to +1.00 indicating strong positive relationships. Thus, we would say that among this group of students, rank in English is rather strongly related to rank in math. This agrees with the impression we had formed when we examined the two sets of ranks for the

FIGURE 5.2
Calculation of Spearman Rank-Order Correlation
for Rank of 15 Students in English and Math

| | Step 1 | Step 2 | Step 3 |
Student	Rank in English	Rank in math	D	D^2
John	1	1.5	+ .5	.25
Billy	2.5	6	+3.5	12.25
Sally	2.5	4	+1.5	2.25
Helen	4	5	+1	1.00
Sam	5	1.5	−3.5	12.25
Betty	6	8	+2	4.00
Leona	7	8	+1	1.00
Tim	8.5	12	+3.5	12.25
Doris	8.5	8	− .5	.25
Jane	10.5	3	−7.5	56.25
Ruth	10.5	13	+2.5	6.25
Dick	12	11	−1	1.00
Joe	13	15	+2	4.00
Paul	14	10	−4	16.00
Ted	15	14	−1	1.00

Step 4

$$\Sigma D^2 = 130.00$$

$$6\Sigma D^2 = 780.00$$

Step 5
There are 15 students, 15 pairs of scores. Therefore, $N = 15$.

Step 6

$$(N^2 - 1) = (15^2 - 1) = (225 - 1) = 224$$

Step 7

$$N(N^2 - 1) = 15(224) = 3360$$

Step 8

$$\frac{6\Sigma D^2}{N(N^2 - 1)} = \frac{780.00}{3360} = .2321$$

Step 9

$$\rho_s = 1 - \frac{6\Sigma D^2}{N(N^2 - 1)} = 1 - .2321 = .7679 = .77$$

first time in Chapter 3. There, we noticed a strong tendency for a student who ranked high in English to rank high in math as well. On the other hand, the fact that the correlation is +.77, and not +1.00, tells us that the relationship between the two rankings is not perfect. This also agrees with what we noticed in Chapter 3. There were students who ranked high in one class but not in the other. Intuitively, we feel that a perfect relationship would exist only if every student had the same rank in English as in math. In fact, this is the case. If you stop and consider the formula used to compute ρ_s, you will see that the only way ρ_s can turn out to be 1.00 is if the whole term ΣD^2 is zero. And the only way that this term can be zero is if *all* the differences between the paired ranks are zero. In other words, the ranks would have to be the same for every student from one class to the other.

What about negative relationships? In the example of the rating of political priorities by the conservative and the liberal congressmen, we obtained a rank-order correlation of $-.83$ (you may verify this calculation for practice). This suggests a strong, but not perfect, negative relationship. Once again, the calculated correlation agrees with the impression we formed by examining the data. We had noticed that the issues given high priority by the conservative congressman tended to be given low priority by the liberal congressman. As you might suspect, in order to obtain a perfect negative correlation, the issue ranked highest by the conservative would have to be ranked lowest by the liberal, the issue ranked second highest by the conservative would have to be ranked second lowest by the liberal, and so on down through all the issues. In other words, the two sets of ranks would have to be exactly opposite from each other. If the rankings by the conservative ran from 1 down through 9, in order, then the corresponding rankings by the liberal would have to run from 9 down through 1, in order. In our example, the two sets of ranks were not quite so diametrically opposed.

Bivariate Correlational Problems Involving Interval Data: The Pearson Product-Moment Correlation

In Chapter 3, we considered several sets of bivariate interval data. The scatter-diagram for work rating vs. VMI score for 20 production workers (Figure 3.1) illustrated the meaning of a positive relationship between two interval variables, the scatterdiagram for total points in the decathlon vs. waist size for 30 physical education students (Figure 3.3) illustrated a negative relationship, and the scatterdiagram for grade point average vs. height for 50 university students (Figure 3.4) illustrated the lack of a strong relationship between two variables. Here, we will consider a descriptive statistic that provides us with a numerical index of the strength of the relationship between two interval variables. This statistic is the Pearson product-moment correlation. Like the Spearman rank-order correlation, the Pearson product-moment correlation may range from -1.00 to $+1.00$, with

—1.00 representing a perfect negative relationship, 0 representing no relationship, and +1.00 representing a perfect positive relationship.

The Computation of the Pearson Product-Moment Correlation. The steps involved in the hand computation of the Pearson product-moment correlation may be summarized verbally as follows:

Step 1   Lay out the data in such a way that the paired scores are adjacent to each other. This is typically done by placing the scores on one variable in a column headed X, and the scores on the other in a column headed Y. If there is any reason to regard one of the two variables as the independent variable, this variable is generally assigned to the column headed X. In Figure 5.3, the bivariate data for work rating vs. VMI score has been laid out in this manner.

Step 2   Establish columns headed X^2 and Y^2. In these columns, list the value of each X-score squared and each Y-score squared, respectively.

Step 3   Establish a fifth column, headed XY. This is known as the cross product column. List here the value of each X-score multiplied by the corresponding Y-score. For example, in Figure 5.3, you will note that for subject 1, the entry in the XY column is 66, the product of the subject's score on X, 11, and his score on Y, 6.

Step 4   Add up each of the 5 columns. This will result in 5 quantities: the sum of the X-scores, the sum of the Y-scores, the sum of the squared X-scores, the sum of the squared Y-scores, and the sum of the cross products. These quantities are designated ΣX, ΣY, ΣX^2, ΣY^2 and ΣXY, respectively.

Step 5   Multiply the sum of the X-scores (ΣX) by the sum of the Y-scores (ΣY).

Step 6   Multiply the sum of the cross products (ΣXY) by N, the number of pairs of scores in the data set.

Step 7   Subtract the result of Step 5 from the result of Step 6.

Step 8   Square the sum of the Y-scores (ΣY) to obtain $(\Sigma Y)^2$, and square the sum of the X-scores (ΣX) to obtain $(\Sigma X)^2$. Note carefully that we are squaring the X and Y totals *after* the original scores have been added up. It is *not* the same as squaring each X-score and Y-score *before* adding. This procedure was already done, in Steps 2 and 4.

Step 9   Multiply the sum of the squared X-scores (ΣX^2) by N and from the result subtract the sum of the X-scores squared $(\Sigma X)^2$.

Step 10 Multiply the sum of the squared Y-scores (ΣY^2) by N and from the result subtract the sum of the Y-scores squared $(\Sigma Y)^2$.

Step 11 Multiply the result of Step 9 by the result of Step 10.

199

Step 12 Take the square root of the result of Step 11.

Step 13 Divide the result of Step 7 by the result of Step 12.

All these steps may be conveniently summarized in the formula:

$$\rho = \frac{N(\Sigma XY) - (\Sigma X)(\Sigma Y)}{\sqrt{[N\Sigma X^2 - (\Sigma X)^2] \cdot [N\Sigma Y^2 - (\Sigma Y)^2]}}$$

where ρ is rho, symbolizing the Pearson product-moment correlation for a population of scores, and the other symbols are as defined earlier. Note again the difference between ΣX^2, the sum of the squared X-scores, obtained in Step 4, and $(\Sigma X)^2$, the sum of the X-scores squared, found in Step 8. In Figure 5.3, we have carried out the entire calculation of the Pearson product-moment correlation for the bivariate data of work rating vs. VMI score for the population of 20 production workers. As usual, we have labeled each step and indicated the mathematical notation corresponding to that step.

As is clear from the number of steps, the calculation of the Pearson product-moment correlation by hand is a fair amount of work. Fortunately, many hand calculators now enable you to do such calculations practically automatically, with relatively little work.

The Interpretation of the Pearson Product-Moment Correlation. In a rough sense, the Pearson correlation may be interpreted as the Spearman. Both correlations have the same range, from -1.00 to $+1.00$. In each case, a correlation close to -1.00 signifies a strong negative relationship between the variables, a correlation close to 0 describes a weak relationship, and a correlation close to $+1.00$ describes a strong positive relationship. Thus, the Pearson correlation of .83 that we computed to describe the relationship between work rating and VMI score for 20 production workers indicates a strong positive relationship between these two variables within this population. That is, workers with high VMI scores tend to receive high scores on work rating as well.

The Spearman Correlation and the Pearson Correlation Compared. The Spearman correlation is a special case of the Pearson correlation. If we were to take a set of bivariate rank-order data and imagine that the ranks in the data were really numerical scores on an interval scale variable, we could apply the Pearson product-moment correlation formula to the data. And if we did compute ρ in this manner, we would find that the value of ρ would be identical to the value of ρ_s computed for the same data set using the Spearman method. This has led some researchers to regard the two methods as interchangeable. Some texts even suggest that the Spearman method may be applied to a set of bivariate interval data as a more convenient computational substitute when the number of pairs of

Descriptive Statistics

FIGURE 5.3
Computation of Pearson Product Moment Correlation
for Work Rating vs. VMI Score for 20 Production Workers

	Step 1		Step 2		Step 3
Subject	X (VMI)	Y (Work Rating)	X^2	Y^2	XY
1	11	6	121	36	66
2	12	7	144	49	84
3	9	5	81	25	45
4	18	9	324	81	162
5	15	6	225	36	90
6	6	4	36	16	24
7	1	2	1	4	2
8	8	4	64	16	32
9	18	8	324	64	144
10	12	7	144	49	84
11	10	6	100	36	60
12	10	6	100	36	60
13	9	7	81	49	63
14	15	8	225	64	120
15	19	9	361	81	171
16	14	7	196	49	98
17	18	5	324	25	90
18	15	8	225	64	120
19	19	10	361	100	*Step 4* 190
20	20	8	400	64	160
	$\Sigma X = 259$	$\Sigma Y = 132$	$\Sigma X^2 = 3{,}837$	$\Sigma Y^2 = 944$	$\Sigma XY = 1{,}865$

Step 5 $\quad (\Sigma X)(\Sigma Y) = 259(132) = 34{,}188$

Step 6 $\quad N = 20 \quad N(\Sigma XY) = 20(1865) = 37{,}300$

Step 7 $\quad N(\Sigma XY) - (\Sigma X)(\Sigma Y) = 37{,}300 - 34{,}188 = 3{,}112$

Step 8 $\quad (\Sigma X)^2 = (259)^2 = 67{,}081 \quad (\Sigma Y)^2 = (132)^2 = 17{,}424$

Step 9 $\quad N\Sigma X^2 - (\Sigma X)^2 = 20(3{,}837) - 67{,}081 = 76{,}740 - 67{,}081 = 9{,}659$

Step 10 $\quad N\Sigma Y^2 - (\Sigma Y)^2 = 20(944) - 17{,}424 = 18{,}880 - 17{,}424 = 1{,}456$

Step 11 $\quad [N\Sigma X^2 - (\Sigma X)^2] \cdot [N\Sigma Y^2 - (\Sigma Y)^2] = [9{,}659] \cdot [1{,}456] = 14{,}063{,}504$

Step 12 $\quad \sqrt{[N\Sigma X^2 - (\Sigma X)^2][N\Sigma Y^2 - (\Sigma Y)^2]} = \sqrt{14{,}063{,}504} = 3{,}750.1339$

Step 13 $\quad \rho = \dfrac{N(\Sigma XY) - (\Sigma X)(\Sigma Y)}{\sqrt{[N\Sigma X^2 - (\Sigma X)^2][N\Sigma Y^2 - (\Sigma Y)^2]}} = \dfrac{3{,}112}{3{,}750.1339} = .8298$

observations is small.* This technique requires us to transform the interval scale scores on the variables X and Y into ranks, and then to compute ρ_s based on the ranks. In Table 5.8, we have performed the transformation on a set of bivariate interval data in which a strong relationship exists between the two variables. The data represent the scores of a group of 10 Air Force trainees on a test of gross motion coordination and a test of fine motor coordination. After deriving the two sets of ranks from the two sets of scores, we computed the Pearson ρ for the data in their original form as well as the Spearman ρ_s for the resulting sets of ranks. What we find is that the two correlation coefficients are similar, but not identical.

When the interval scale scores are transformed into ranks, the two sets of ranks agree perfectly. That is, the subject with the highest score on X also has the highest score on Y, the subject with the second highest score on X also has the second highest score on Y, and so on. The perfect agreement of the two sets of ranks results in a Spearman correlation of +1.00 computed on the rank-ordered scores. However, the Pearson correlation that we obtain using the interval scores is *not* +1.00. It is .96. The Spearman correlation indicates a *perfect* correlation between the ranks, but the Pearson correlation indicates something less than a perfect correlation among the numerical scores. The reason for this is that we have lost information in transforming our interval scale scores to ranks. We feel in research situations in which the variables of interest are truly of interval scale, the Pearson ρ should be used. On the other hand, if the researcher has numerical scores that may not constitute a true interval scale, the researcher will be justified in transforming the scores to ranks and using the Spearman ρ_s. In this case, however, the researchers should make it very clear that the reported correlation is on ranks rather than scores, in order to keep the reader from over-estimating the relationship.

Pearson Correlation With Standard Scores. It is clear from the example presented in Table 5.8 that a perfect Pearson correlation on a set of scores implies *more* than the fact that the highest score on one measure was paired with the highest score on the other measure. It implies a more precise relationship between the sets of scores. When given this information, many students will guess that a perfect Pearson correlation must imply that scores will be equal within each pair. This is not true. It is possible to have a perfect Pearson correlation between two variables that have completely different ranges. For example, let us consider the case of a group of vacuum cleaner salespeople who work for a salary of $20.00 per week plus a commission of $10.00 for each vacuum they sell. In this case, there is a perfect Pearson correlation between the variable "number of vacuums sold" and the variable "weekly income," even though these two variables would have different ranges. However, the student who guesses

* Guilford, J. P., and Fruchter, B. *Fundamental Statistics In Psychology and Education.* New York: McGraw-Hill, 1973.

TABLE 5.8
Gross and Fine Motor Coordination Scores for 10 Air Force Trainees

Subject	Scores		Ranks	
	X (Gross)	Y (Fine)	X (Gross)	Y (Fine)
1	20	24	1	1
2	18	23	2	2
3	14	22	3	3
4	13	16	4	4
5	11	14	5	5
6	10	9	6	6
7	6	8	7	7
8	4	7	8	8
9	2	6	9	9
10	0	1	10	10
	$\rho = .96$		$\rho_s = 1.00$	

that a perfect Pearson correlation implies equal scores within each pair is very close to the truth. You will recall that when we are comparing scores on two variables that have different ranges, it is often useful to standardize the scores. Table 5.9 contains a hypothetical set of data relating weekly salary to number of vacuums sold for 10 salespeople in a particular week. Along with the raw scores for each of the variables, we have presented the standard score equivalents of these raw scores. As you can see, the data are such that when both the X and Y scores have been standardized, the *standard* scores in each pair are all equal.

Applying the formula presented earlier to the raw scores, we calculate the Pearson product-moment correlation for these data to be +1.00. Whenever two sets of scores are such that the z-score equivalents of each pair are equal, the Pearson correlation will be +1.00. Thus, a perfect positive relationship between two interval variables implies that each X-score lies the same number of standard deviation units from the mean of the X-scores as its paired Y-score does from the mean of the Y-scores. In this situation, the scatterdiagram representing the bivariate distribution will lie in a straight line, as indicated in Figure 5.4. Actually, when two variables are perfectly related, all the points in the scatterdiagram will lie in a straight line, whether the scores have been standardized or not. However, if the scores are graphed as z-scores, then the value of each score's X-coordinate will equal its Y-coordinate. If there is a perfect negative relationship between two sets of scores that have been standardized, then the paired X and Y scores will have z-score equivalents that are equal in terms of absolute value, but that

203

TABLE 5.9
Number of Vacuums Sold (X) vs. Weekly Income (Y)
for 10 Vacuum Cleaner Salespeople

Salesperson	Number of vacuums sold (X)	Weekly income (Y)	z_X	z_Y
1	18	200	1.441	1.441
2	16	180	1.081	1.081
3	15	170	0.901	0.901
4	15	170	0.901	0.901
5	12	140	0.360	0.360
6	8	100	−0.360	−0.360
7	5	70	−0.901	−0.901
8	5	70	−0.901	−0.901
9	4	60	−1.081	−1.081
10	2	40	−1.441	−1.441

have opposite signs in each case. In this case, the scatterdiagram will also form a straight line, but the line will slope in the opposite direction.

Because a perfect Pearson correlation of +1.00 or −1.00 occurs when the scatterdiagram for a set of bivariate interval data forms a perfect straight line, we typically say that the Pearson correlation is a measure of the extent to which two interval variables are related *linearly*. It is important to realize that the Pearson correlation reflects the linear relationship between two variables, because it is possible for interval variables to be related to one another in a nonlinear way. The discussion of nonlinear relationships is beyond the scope of this elementary text and, in most research studies in psychology, education, and the social sciences, we are concerned with linear relationships. However, you should be aware that other types of relationships may exist. If we compute a Pearson correlation for a set of bivariate interval data and we find that it is zero, we can say with certainty that the two variables have no linear relationship, but we cannot be sure that they are not related in some other manner.

In connection with the use of z-scores, we note that if z-scores are available for a given set of bivariate interval data, it is possible to compute the Pearson product-moment correlation using a very simple formula:

$$\rho = \frac{\Sigma z_X z_Y}{N}.$$

This formula tells us to multiply each standardized score on the X-variable by the paired standardized score on the Y-variable, add up these products, and divide by the number of pairs of scores. Verify this formula by computing ρ, using the standardized scores in Table 5.9. You should obtain the value +1.00, just as we obtained the value +1.00 when we computed ρ from the raw scores using the computational formula presented originally.

Some Factors Affecting ρ.   You should be aware that it is not correct to make a statement like: "The correlation between intelligence and grade point average is .60." The correct statement would be: "The correlation between intelligence and grade point average for the population of freshmen at Ohio State University is .60." The correlation between two variables will vary from one population to another. In general, the correlation between two variables will be larger in populations that are relatively heterogeneous with respect to these variables. For example, we would expect the correlation between intelligence and grade point average to be larger in a population of high-school students than in a population of college students. We would also expect the correlation between the variables to be larger in a population of college students than in a population of graduate students in psychology. This is because the process of selection works to restrict the range of intelligence represented in populations of students in successively higher levels of study. As the population becomes more homogeneous with respect to intelligence, that factor becomes relatively less important in predicting grades. Other factors, such as motivation, become relatively more important.

The way in which restriction of the range of a variable may affect a correlation coefficient is illustrated in Figure 5.5. There we have presented a hypo-

FIGURE 5.4
Scatterdiagram of Standardized Scores on
Weekly Income and Number of Vacuums Sold

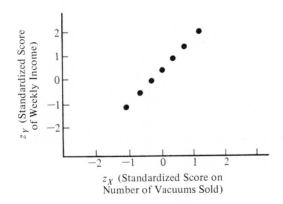

FIGURE 5.5

FIGURE 5.5
The Effect of Restriction of Range on the Correlation Coefficient

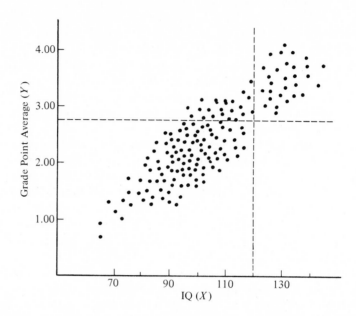

thetical scatterdiagram representing the relationship between intelligence and grade point average for a large heterogeneous population. We can see from the shape of the scatter of points that there seems to be a substantial linear relationship between the two variables. Now look at just that portion of the scatter belonging to students having IQs over 120. (It may be helpful to you to take two sheets of paper to block off most of the scatterdiagram, leaving just the portion in the upper right.) When we consider only these subjects, does the scatter of points still suggest a strong linear relationship? No. It appears to suggest very little relationship at all. For this reason, it is always important to specify carefully the population on which we have calculated ρ.

Estimating One Variable from Another: Linear Regression

Let us take a moment to reconsider the data representing the relationship between VMI and work rating (Figure 5.3). In this data set, the mean VMI score among the 20 workers was 12.95, and the mean supervisor rating was 6.6 (you can verify these means). As noted previously, the Pearson correlation describing the relationship between the two variables was +.83. Now, suppose that a prospective employee walks into the plant seeking work. Before hiring her, we would like to have some idea of how she might perform on the job. Because we know that there is a relationship between VMI score and work rating, we use our test of Visual Motor Integration to measure the candidate on this variable. She scores 18 on VMI. On the basis of this VMI score, what kind of work rating would we expect this candidate to achieve, if she was hired?

We know that VMI and work rating have a positive linear relationship. We also know that this positive relationship implies a tendency for relatively high scores on one of the two variables to be found in association with relatively high scores on the second variable. Furthermore, we know that our job candidate has a relatively high score on VMI, because her score is 18 on a test that ranges from 1 to 20 with a mean of 12.95. Thus, we would expect that this candidate would probably achieve a high work rating if she should be hired. We would predict that her work rating would be above the mean work rating of 6.6. Of course, in making this prediction, we would be aware that there is some chance that the prediction is incorrect. The correlation of +.83 between VMI and work rating is strong, but not perfect. If we check the actual work ratings of those subjects in the original group of 20 workers who had VMI scores of 18, we find that 2 out of 3 had work ratings above the mean rating of 6.6, but that 1 of the 3 had an actual rating below the mean. Thus, our prediction is a statement of what we consider most likely to happen, rather than a statement of something we know for sure. Nevertheless, the ability to make such a prediction would be quite useful for a company seeking to hire good workers, for it would help them to select applicants in such a way as to improve the chance that a particular worker hired will be successful.

In this section, we are going to consider a method of prediction that goes beyond the simple statement: "If this candidate is above average on VMI, he or she will most likely be above average on work rating as well." The method involves the use of a mathematical formula known as a *regression equation.* It enables us to calculate a specific estimate for a subject's score on one variable, the *criterion variable,* based upon that subject's score on the other variable, the *predictor variable.* The method derives from the model of two variables that have a perfect linear relationship, as indicated by a Pearson correlation of ±1.00. You will recall that when a set of bivariate data yields a correlation of ±1.00, then all the points in the scatterdiagram lie along a straight line. When this is the case, we can determine the mathematical equation that describes this straight line. This equation is a regression equation. Having determined the equation, we may use it to calculate exactly the value of the criterion variable that we find paired with the predictor variable of a particular size.

Of course, it is exceedingly rare in research to find a set of bivariate interval data having a perfect linear relationship, so generally the points in a scatterdiagram will not all fall exactly along a straight line. However, when two variables are linearly related in a less than perfect manner, it is possible to determine the equation of the straight line that best "fits" the actual scatter of points. This line is referred to as the "least squares" line, a term that will be described shortly.

Once we have determined the equation of this line, we can use the equation to obtain an estimate or predicted value for a subject's score on the criterion variable from that subject's score on the predictor variable. Thus, we sometimes call this line a *prediction line.* In this case, where the two variables are not perfectly related, we cannot be certain that our predicted value will be exactly correct. However, as you will see, we can generally predict scores on the criterion variable more accurately when we use the regression equation than we can when we do not use the equation.

In considering the nature of the regression equation, we will begin by reviewing some of the basic concepts of graphing a straight line in a coordinate axis system. These will be illustrated with reference to a situation in which two variables have a perfect linear relationship. Then, we will consider the use of the least squares line in making predictions when two variables are less than perfectly related.

The Equation Describing a Perfect Linear Relationship. To demonstrate the use of a mathematical equation to represent a perfect linear relationship, we reconsider the example of the relationship between number of vacuums sold and weekly earnings for our group of 10 vacuum cleaner salespeople. We know that our salespeople start out with an income of $20.00 per week, whether or not they sell any vacuums. In addition to this $20.00, they receive an additional $10.00 in income for each vacuum sold. Thus, we can calculate the exact weekly income of any salesperson using the formula:

$$Y_{PRE} = 20 + 10(X),$$

where Y is weekly income in dollars and X is the number of vacuums sold. But, in the coordinate axis system we use to plot a scatterdiagram, such an equation represents a straight line. In Figure 5.6, we have drawn the scatterdiagram representing the 10 salespeople, using the raw score values for each variable. In the graph, we have drawn the straight line represented by the equation. You will note that the graph provides another way of representing the fact that income increases by $10 for each vacuum sold. Thus, a salesperson who sells 12 vacuums earns $40 more than a salesperson who sells 8 vacuums. In the language of coordinate geometry, 10 is the *slope* of the straight line. The slope represents the change in the Y-variable that is associated with a change of 1 unit in the X-variable. You will also note that even if a salesperson were to sell no vacuums at all, he or she would still receive the $20.00 salary. This is reflected on the graph by the fact that the straight line crosses the Y-axis at the point 20. This point is known as the *Y-intercept.* It is defined as the value of Y when the value of X is zero.

In general, a straight line is represented by an equation of the form:

$$Y = a + bX,$$

FIGURE 5.6

FIGURE 5.6
Scatterdiagram and Regression Equation for Determining Weekly Income
from Number of Vacuums Sold (Raw Scores)

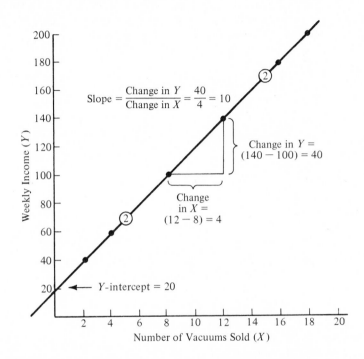

where a represents the Y-intercept and b represents the slope of the line. In any
bivariate problem where X and Y are perfectly related interval scale variables, it
is possible to determine a particular subject's score on Y exactly from the score
on X, provided we know the values of a and b. For example, a salesperson might
wish to know what her weekly income would be if she sold 19 vacuums. She
could substitute the value 19 for X in the equation, yielding

$$Y = 20 + 10(19) = 210.$$

Because we know that the relationship between X and Y is perfect, we know
that the salesperson selling 19 vacuums would have an income of exactly $210.

Prediction When the Linear Relationship is Not Perfect.   When we are dealing
with a set of bivariate interval scale data in which there is a linear relationship
that is less than perfect, we compute the equation of the straight line that

209

"comes the closest" to describing the scatter of points. We will explain exactly what we mean by "comes the closest" below. First, we consider the method used to compute the equation of the line. For the purpose of illustration, let us reconsider the example of work rating (X) versus VMI score (Y) for 20 production workers. You will recall that the correlation between these two variables in this group of subjects was $+.83$. Given the correlation between the variables, we can compute the slope of the best fitting line using the formula:

$$b = \rho \frac{\sigma_Y}{\sigma_X},$$

where ρ is the correlation between the two variables, σ_Y the standard deviation of the Y-scores, and σ_X the standard deviation of the X-scores. In the data set under consideration, σ_Y was 1.90 and σ_X was 4.91 (you can verify these). Thus, the slope of the regression line will be

$$b = \rho \left(\frac{\sigma_Y}{\sigma_X} \right)$$

$$b = .83 \left(\frac{1.90}{4.91} \right)$$

$$b = .32.$$

This tells us that for each additional point on VMI that a worker might score, we would expect the work rating to improve by .32 points. Once we have found the slope of the regression line, we can find the Y-intercept of the line by using the formula

$$a = \mu_Y - b\mu_X,$$

where μ_Y is the mean of the Y-scores and μ_X is the mean of the X-scores. In this case μ_Y was 6.6 and μ_X was 12.95; thus the Y-intercept may be found as follows:

$$a = \mu_Y - b\mu_X$$

$$a = 6.6 - .32\,(12.95)$$

$$a = 2.46.$$

In Figure 5.7, we have reproduced the scatterdiagram for these data (presented originally in Figure 3.1), and we have drawn in the regression line for predicting work rating from VMI score:

210

$$Y_{PRE} = 2.46 + .32(X).$$

You will note that in this case of a less than perfect linear relationship between the two variables, not every point in the scatterdiagram falls exactly on the regression line. Thus, the value of Y, predicted on the basis of a subject's score on X, will not always be exactly equal to that subject's actual score on Y. For example, if we substitute the value 18 for X, we arrive at the following predicted value for Y:

$$Y_{PRE} = 2.46 + .32(18)$$

$$Y_{PRE} = 8.22.$$

This value is the Y-coordinate of the regression line when the X-coordinate is 18. This has been indicated in Figure 5.7. Checking the scatterdiagram itself, we find that there were 3 workers among our group of 20 who had scores of 18 on VMI.

FIGURE 5.7
Scatterdiagram of Work Rating (Y) vs. VMI Score (X) for 20 Production Workers With Least Squares Regression Line for Predicting Y from X

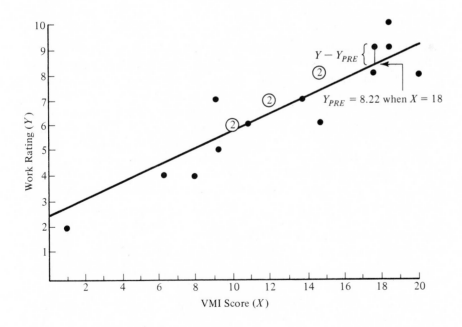

Bivariate Problems

These 3 workers had the following actual scores on work rating: 9, 8, and 5. In the case of workers having a VMI score of 18, then, we can say that the regression estimate was within a point of the actual value for two workers and somewhat further off in the case of the third worker.

The difference between the actual value of the criterion variable and the value of the variable predicted by the regression equation for any subject is known as the *deviation from regression* of that subject. That is, a deviation from regression is equal to $(Y - Y_{PRE})$. Thus, we were just considering the deviations from the regression of the actual Y-scores of the 3 workers who happened to have scores of 18 on the predictor variable. These deviations were $(9 - 8.22) = +.78$, $(8 - 8.22) = -.22$, and $(5 - 8.22) = -3.22$, respectively. The first of these deviations, the $(9 - 8.22) = +.78$, has been sketched in Figure 5.7. There is one such deviation for each of the 20 workers represented in the data set.

The Standard Error of Estimate. Obviously, the smaller these deviations from regression are, the better the regression equation is as a predictor of the criterion variable. For this reason, we use these deviations in computing a numerical index of the effectiveness of the regression equation as a predictor of Y. This index is the *standard error of estimate*, $\sigma_{Y \cdot X}$. It is defined by the formula:

$$\sigma_{Y \cdot X} = \sqrt{\frac{\Sigma(Y - Y_{PRE})^2}{N}}^{*}.$$

You will note that the formula for the standard error of estimate appears rather similar to the formula for the standard deviation. In fact, the difference is that the standard error of estimate considers the deviations of points from the regression estimates, whereas the standard deviation considers the deviation of points from the mean. Just as you may have found it useful to think of the standard deviation in rough terms as the average amount by which observations differ from the mean, so it may be useful to think of the standard error of estimate in rough terms as the average amount by which observations differ from the regression estimates.

At this point, it is possible for us to explain what is meant by the line that "best fits" the scatter of points. It should be clear from the formula for the standard error of estimate that $\sigma_{Y \cdot X}$ will be small when the sum of the squared deviations from the regression line is small. For any given set of bivariate interval data, the regression line obtained using the formulas for the slope and the Y-intercept presented earlier will be the one straight line for which the sum of the

---

* Remember, here we are discussing descriptive statistics. In some books, you will see the standard error of estimate formula with $N-2$ in the denominator. In such cases, the situation is inferential and the standard error is an estimate based on sample data.

Descriptive Statistics

squared deviations is at a minimum. Of course, it would be possible to draw another straight line through the scatter, but no other straight line will produce as
small a sum for the squared deviations. Thus, no other straight line would result
in as small a standard error of estimate as that obtained using these formulas.
That is why the straight line obtained for a set of bivariate interval data by using
these formulas is often referred to as the "least squares" line.

It is possible to calculate the standard error of estimate for the "least squares"
line for a set of bivariate data using the computational formula:

$$\sigma_{Y \cdot X} = \sigma_Y \sqrt{1 - \rho^2}.$$

In this formula, σ_Y represents the standard deviation of the distribution of Y-
scores, and ρ represents the Pearson correlation between the variables X and Y.
From this formula, we can see that the standard error of estimate will be larger
when the variability of the Y-scores about their mean is larger. We can also see
that the standard error of estimate will be smaller when the linear relationship
between X and Y, measured by ρ, is stronger.

The meaning of ρ^2. The computational formula for the standard error of estimate may be rewritten in a way that provides us with a new concept of the
meaning of ρ^2. Because

$$\sigma_Y = \sqrt{\frac{\Sigma(Y - \mu_Y)^2}{N}} \quad \text{and} \quad \sigma_{Y \cdot X} = \sqrt{\frac{\Sigma(Y - Y_{PRE})^2}{N}},$$

we can rewrite the computational formula as follows:

$$\sqrt{\frac{\Sigma(Y - Y_{PRE})^2}{N}} = \sqrt{\frac{(Y - \mu_Y)^2}{N}} \cdot \sqrt{1 - \rho^2}.$$

Squaring and rearranging terms, we find

$$\frac{\dfrac{\Sigma(Y - Y_{PRE})^2}{N}}{\dfrac{\Sigma(Y - \mu_Y)^2}{N}} = 1 - \rho^2 \quad \text{and} \quad \rho^2 = 1 - \frac{\Sigma(Y - Y_{PRE})^2}{\Sigma(Y - \mu_Y)^2}.$$

This is instructive for the following reason. We typically measure error in predic-
tion by a sum of squared deviations of actual from predicted values. Thus, $\Sigma(Y - \mu_Y)^2$ is a measure of the error we make in using μ_Y as the predicted value of Y
for each observation in the data set, and $\Sigma(Y - Y_{PRE})^2$ is a measure of the error
we make in using the regression estimate as the predicted value of Y for each

213

observation in the data set. But when would we use μ_Y as an estimate for Y for every observation? Only when we do not have knowledge of the paired X-scores and the relationship between X and Y. In the absence of knowledge of a predictor, guessing μ_Y as the value of each Y-score will produce a smaller sum of squared errors than the use of any other value. Thus, we can think of $\Sigma(Y - \mu_Y)^2$ as the measure of the maximum total error associated with predicting Y in the absence of any knowledge of the predictor, X. However, when we have the paired X-scores as well for a set of data, we will find the relationship between X and Y and use the regression equation to obtain estimates for Y. Then the error of prediction is represented by $\Sigma(Y - Y_{PRE})^2$, a term that will be smaller than $\Sigma(Y - \mu_Y)^2$ for any bivariate data set in which $\rho \neq 0$.

If we consider $\Sigma(Y - \mu_Y)^2$ as a measure of total error in predicting Y from μ_Y and $\Sigma(Y - Y_{PRE})^2$ as a measure of the error remaining after the use of the regression equation, then the ratio $\Sigma(Y - Y_{PRE})^2 / \Sigma(Y - \mu_Y)^2$ is the proportion of the total error that *remains* after the use of the regression equation. But then

$$ 1 - \frac{\Sigma(Y - Y_{PRE})^2}{\Sigma(Y - \mu_Y)^2} $$

must equal the proportion of the total error of predicting Y from μ_Y that is explained as a result of the use of the regression equation. This is often a helpful manner in which to conceptualize ρ^2. Often, we use the term "variability" instead of error in this context. Then ρ^2 becomes the proportion of the total variability of Y that is explained as a result of the use of the regression equation as a predictor.

DESCRIPTIVE STATISTICS
FOR EXPERIMENTAL
AND PSEUDO-EXPERIMENTAL PROBLEMS

As noted at the beginning of this chapter, the descriptive statistics that would be employed in experimental and pseudo-experimental studies involving the comparison of two or more groups are really the same descriptive statistics that we would use in univariate problems. The only difference is that in these experimental and pseudo-experimental problems we have two or more groups of subjects, so whatever descriptive statistics are employed must be computed for each group under investigation. The type of descriptive statistic to be used remains a function of the type of data we have. If we were to have an experimental study in which treatment and control groups were compared on a categorical variable, we would probably compare the modal categories. This procedure would lead to a statement such as: "Although the largest number of subjects in the experimental group passed the high-school equivalency exam, the largest number of control

214

group subjects failed." If we were to have an experimental study in which treatment and control groups were compared on an interval scale variable, on the other hand, we would compare the means of the two groups. This procedure would lead to a statement such as: "The mean score among 25 experimental subjects on the high-school equivalency exam was 86.4, compared to a mean score of 58.3 among the 20 control group subjects."

When we deal with rank-order data in experimental and pseudo-experimental studies, a technique of comparison that is often used is to rank order all the scores in all the groups combined. If this can be done, we can then separate out the groups once again and compare the overall ranking of the middle score in one group to the overall ranking of the middle score in the other groups. This procedure may be illustrated by the results of a women's cross-country track meet in which 3 teams entered 15 runners each. In Table 5.10, we have listed the overall order of finish for the 45 runners, along with the team (A, B, or C) to which each runner belonged. You will note that the runner finishing in the middle with respect to the members of her own team (i.e., the number 8 runner out of 15) has been designated by an asterisk in the overall ranking. For team A, this runner had the overall rank of 13. For team B, the eighth finisher had an overall rank of 25. And, for team C, the eighth finisher had an overall rank of 36. This comparison gives us an indication that team A performed best. Of course, this technique may be used only in the case where we are able to rank subjects over

TABLE 5.10
A Comparison of Three Cross-Country Teams
Based on Overall Order of Finish

Rank	Team	Rank	Team	Rank	Team
1	A	16	B	31	A
2	A	17	A	32	B
3	B	18	A	33	B
4	C	19	A	34	C
5	A	20	B	35	C
6	A	21	C	36	C*
7	B	22	A	37	C
8	B	23	A	38	B
9	A	24	B	39	C
10	A	25	B*	40	C
11	A	26	B	41	B
12	C	27	C	42	C
13	A*	28	C	43	C
14	A	29	B	44	C
15	B	30	B	45	C

the entire group, as well as within their respective groups. If subjects can be ranked within their respective groups only, no such comparison can be made with rank-order data.

DESCRIPTIVE STATISTICS
FOR PROBLEMS WITH
TWO SETS OF DEPENDENT MEASURES

The suggestions just provided relating to the use of descriptive statistics with experimental and pseudo-experimental problems are also applicable to problems involving dependent measures. Suppose we were dealing with a pretest and post-test design in which subjects were being measured on some categorical variable. We could compare the modal category at the time of the pretest to the modal category at the time of the posttest to describe what change had occurred in the group. If we were dealing instead with an interval scale variable, we could compare the pretest mean to the posttest mean as a method of describing what change had occurred among the subjects studied.

In a pretest and posttest design with rank-order data, it would probably not be possible to employ the method of ranking both pretest and posttest scores as one group and comparing the ranks of the middle-ranked pretest score to the middle-ranked posttest score. This is because the measures would take place at different points in time, so it might not be possible to compare a pretest measure to a posttest measure. For example, if we had a group of runners complete a race before and after a training program and if we recorded only the order of finish each time and not the actual times for the race, we could not determine any one set of rankings that included both pretest and posttest efforts. On the other hand, in a matched groups design using rank-order scale data, subjects could be measured at one time, in which case an overall ranking could be established.

THE CLASSIFICATION MATRIX

In this chapter, we completed our discussion of descriptive statistics by examining bivariate problems. Table 5.11 summarizes the descriptive statistics covered in Part 3. These statistics, along with the techniques for organizing and presenting data discussed in Part 2, provide us with ways to describe sets of data in a variety of problems. Note that when you locate your type of problem in the matrix, you will find the descriptive statistics that should be *considered* for that problem. This does not mean that all the techniques listed in the matrix are automatically appropriate: you must be guided by the additional considerations suggested in the chapters of this book.

216

TABLE 5.11
Classification Matrix: Descriptive Statistics

Level of Measurement of Variable(s) of Interest

Type of Problem		Categorical: Nominal or Ordered Categories	Rank-Order	Interval Scale (or Ratio-Scale)
Univariate Problems		Modal Category Proportion of Observations Falling Outside Modal Category *With Ordered Categories:* Median Category	Percentile Rank	Mode, Median, Mean Range, Interquartile Range, Standard Deviation Percentile Rank, Standard Score
Bivariate Problems	Correlational Problems	Cramer's Index of Contingency	Spearman Rank-Order Correlation	Pearson Product-Moment Correlation
	Experimental and Pseudo-Experimental Problems: Independent Groups	Compare Modes in Different Groups	Rank All Scores in All Groups Taken Together, Then Compare the Overall Rankings of the Middle Score in Each Group	Compare Mode, Median, Mean Compare Range, Inter-quartile Range, Standard Deviation
	Dependent Measures: Pretest and Posttest or Matched Groups	Compare Modes (Either Pretest and Posttest Modes or Modes in the Two Groups)	Use Ranking Method Described in the Cell Above This One	Use Methods Described in the Cell Above This One

Key Terms

Be sure you can define each of these terms.

measures of relationship
Cramér's index of contingency
Spearman rank-order correlation
Pearson product-moment correlation
expected cell frequency
observed cell frequency
regression equation
criterion variable
predictor variable
prediction line
deviation from regression
standard error of estimate

Summary

In Chapter 5, we discussed descriptive statistics for bivariate problems. We recalled that bivariate problems can be classified as correlational, experimental or pseudo-experimental, or a comparison of two groups of dependent measures.

Most of the chapter dealt with descriptive statistics for correlational problems. We recalled that in correlational problems, we have a single set of subjects measured on two variables. In Chapter 3, we studied methods of organizing and presenting this type of data and we pointed out that the purpose of the organization is to see if the two variables are related. In Chapter 5, we presented several indices that tell us the extent to which the two variables are related. We saw that the choice of index depends on the level of measurement of the two variables.

In the case that both variables are either nominal or ordered categories, we calculate Cramér's index of contingency. This number, called ϕ', tells us the extent to which the observed cell frequencies deviate from what we would expect them to be if there was no relationship. This statistic varies from 0 to 1, and the closer it is to 1, the stronger is the relationship between the two variables.

5

Student
Guide

In the case that both variables are rank-order, we calculate the Spearman rank-order correlation, ρ_s; if the variables are interval, we calculate the Pearson product-moment correlation, ρ. Both of these indices vary from -1 to $+1$. If either is positive, then we have a direct relationship between the variables; if negative, the relationship is inverse; and if close to 0, then there is little or no linear relationship between the variables. In the case of ρ, we are measuring the degree and direction of linear relationship.

We also discussed a method for predicting one interval variable from another on the basis of the linear relationship between them. We discussed how to compute the equation of the prediction line that "best fits" our scatter of points. We also saw how to compute the standard error of estimate, based on the deviations of our predicted values from observed values. We pointed out that the square of the correlation coefficient, ρ^2, can be interpreted as the proportion of variation in one variable that can be accounted for by its linear relationship with the other variable.

We looked briefly at experimental and pseudo-experimental problems as well as those with dependent measures. The statistics we use to describe these problems are the same as those we use in univariate problems. We simply compute our univariate statistics for each group and then use these statistics to compare the groups.

Once again, it is important to remember that all of these descriptive techniques can be used in purely descriptive problems as well as in conjunction with the techniques for inferential problems to be discussed in Part 4.

Review Questions

To review the concepts presented in Chapter 5, choose the best answer to each of the following questions.

1 A correlation coefficient is a measure that indicates
  _____a the extent to which one variable causes change in another variable.
  _____b the degree of association between two variables.

2 A correlation coefficient is useful because it helps us decide
  _____a the direction of causation between two variables.
  _____b whether information on one variable improves the accuracy of our prediction of performance on the other variable.

3 A correlation coefficient is a useful statistic to report in some
  _____a univariate problems.
  _____b bivariate problems.

4 For which of the following bivariate problems would the correlation coefficient be an appropriate statistic to report?
  _____a A group of subjects are measured on reading achievement before and after receiving a special reading program.

219

_____b A group of subjects are randomly assigned to an experimental group and a control group. The experimental group gets a special reading program, and the control group does not. Both groups are measured on reading achievement at the end of the experiment.

5   In a recent election, 45 percent of the voters favored a school bond issue and 55 percent were against it. The community consists of families who have children in school and families who have no children in school. To study the relationship between vote on bond issue and type of family, we would calculate
_____a the Pearson product-moment correlation.
_____b the index of contingency.

6   If a population is cross classified on two nominal variables into a 2 by 2 table and the conditional proportions are equal to the marginal proportions, then
_____a $\phi' = 0$.
_____b $\phi' = 1$.

7   If on a certain athletic strength test almost all boys passed and almost all girls failed, then we would expect that the index of contingency relating athletic strength to sex would be close to
_____a 0.
_____b 1.

8   In a bivariate problem, one variable is rank in class and the other is SAT score. If we want to measure the relationship between these two variables, we should
_____a calculate the Pearson product-moment correlation.
_____b rank the SAT scores and calculate the Spearman rank-order correlation.

9   An English teacher ranks a set of compositions from best to worst. To check her system, she has another teacher rank the compositions. As a measure of agreement between the two sets of ranks, she should calculate
_____a the Pearson product-moment correlation.
_____b the Spearman rank-order correlation.

10 If the correlation coefficient between hat size and waist size was found to be zero, we could say that
_____a hat size and waist size are unrelated.
_____b there is no linear relationship between hat size and waist size.

11 The sign of the correlation coefficient tells us
_____a the strength of linear relationship.
_____b the direction of the linear relationship.

220

12 For a group of fifth-graders, reading and arithmetic scores are correlated and the Pearson product-moment correlation is .5. Both sets of scores are changed to standard scores and the correlation is computed again. The new correlation will be

_____a the same as before.
_____b different than before.

13 A group of students are tested and retested on the same exam. The Pearson product-moment correlation is calculated as .89. If both sets of scores are ranked and the correlation of the ranks is calculated, we should not be surprised to see

_____a a higher value.
_____b a negative value.

14 A class of 50 students is given a test consisting of 100 items. The students are shown their mistakes and the correct answers. They are retested on an equivalent test. A "gain" score is computed for each student. The correlation between gain scores and pretest scores will probably be

_____a very strong and positive.
_____b somewhat negative.

15 If a correlation coefficient is equal to −1, it means that the scatterdiagram is

_____a a rising straight line.
_____b a falling straight line.

16 If a scatterdiagram has a rather circular shape, we expect that a computed correlation would be

_____a close to zero.
_____b somewhat positive.

17 A teacher finds that the correlation is .4 between scores on an arithmetic test given at the beginning of the semester and scores on the midterm test in statistics. This result tells the teacher that

_____a 40 percent of the variation of the midterm scores is explained by a linear relationship between the two variables.
_____b 16 percent of the variation of the midterm scores is explained by a linear relationship between the two variables.

18 Suppose the correlation between arithmetic achievement and reading achievement is .7. If we divide the students into high, medium, and low on the arithmetic variable, and recalculate the correlation between arithmetic and reading for those students in the medium category in arithmetic, the new correlation would probably be

_____a less than .7.
_____b the same as before.

19 If we are interested in predicting arithmetic achievement using strength of hand grip and we are told that the standard error of estimate is zero, we know that

_____a there is a perfect linear relationship between these two variables.

_____b it would be useless to try to predict arithmetic achievement using strength of hand grip as a predictor variable.

20 If the prediction line between two variables is $Y = 1 - 2X$ and the standard error of estimate is 0, then the correlation between X and Y is

_____a equal to -1.

_____b unknown without further information.

Problems for Chapter 5: Set 1

1 Refer to Problem 2 in Set 1 of Chapter 3. Calculate the index of contingency for the variables "sex of degree recipient" vs. "university school," if the total number of degree recipients is 2,000. Interpret the result.

2 Refer to Problem 3 in Set 1 of Chapter 3. Calculate the index of contingency for the variables "outcome of game" and "number of passes thrown by the quarterback."

3 Refer to Problem 4 in Set 1 of Chapter 3. Calculate the index of contingency for the variables "performance on the exam" and "sex of the applicant."

4 A medical center has access to the records of 500 patients, 250 of whom are lifelong smokers of cigarettes and 250 of whom are nonsmokers. Of the 250 nonsmokers, the center observed 38 cases of lung cancer; of the 250 smokers, the center observed 150 cases of lung cancer. Is there a relationship between the variables "use of cigarettes" and "lung cancer"?

Set 2

1 A weight watchers' group decided to keep statistics to see if the amount of weight loss is related to the length of time a person is in attendance with the group. There are 9 members in the group. They are ranked according to the amount of weight loss and also according to the length of time in attendance. On weight loss, rank 1 means most weight lost, and on attendance, rank 1 means longest attendance.

Descriptive Statistics

Group member	Rank on weight loss	Rank on attendance
1	4	3
2	5	9
3	9	5
4	6	8
5	3	2
6	7	7
7	8	6
8	2	4
9	1	1

Calculate the Spearman rank-order correlation and describe the relationship between these two variables for these 9 subjects.

2 A recent survey reveals the following information on crimes of violence, ratio of unreported to reported crimes, and robberies of persons in 13 large cities. The cities are ranked from 1 to 13 on each variable. The rank of 1 means the largest number of violent crimes per 1,000, the largest ratio of unreported to reported crimes, or the most robberies of persons per 1,000.

City	Rank on crimes of violence	Rank on ratio of unreported to reported crimes	Rank on robberies of persons
A	1	5	1
B	2	2.5	8
C	3	1	3
D	4.5	7	9
E	4.5	10	4
F	6	4	5
G	7	8	6
H	8	2.5	10
I	9	9	11
J	10	6	13
K	11	13	2
L	12	12	12
M	13	11	7

Are any of these variables related?

Set 3

1 A group of 10 students are given a pretest in arithmetic skills. They are then given instruction and at the end of the instruction period they are given a second arithmetic test. The amount of gain is recorded for each student.
  a Calculate the correlation coefficient between the two variables, pretest and points of gain.

223

b Compute standard scores for each measure in each distribution.
c Calculate the correlation coefficient between the standard score distributions.
The data follow:

Pretest	Points of gain
0	7
0	6
2	5
2	6
2	5
4	3
4	3
5	1
5	3
6	1

2 A group of 16 pupils takes two spelling tests. The scores on the test are the number of words correctly spelled. The results are presented below. For this data set
a Calculate ρ.
b Calculate σ_Y^2.
c What percentage of σ_Y^2 is explained by the linear relationship between the two sets of test scores?
d What percentage of σ_Y^2 is unexplained by the linear relationship between the two sets of test scores?
e Calculate the prediction line to predict the second test score from the first.
f What is the standard error of estimate in part e?
The results:

Pupil	1	2	3	4	5	6	7	8	9	10	11	12	13	14	15	16
Test I (X)	1	2	2	3	3	3	4	4	4	4	5	5	5	6	6	7
Test II (Y)	1	2	3	2	3	4	3	4	5	4	4	5	6	5	6	7

3 A teacher of statistics wishes to determine the extent to which arithmetic ability and performance in statistics are related. She gives a 50-question test on arithmetic skills at the beginning of the semester. At mid-semester, she gives a 35-question test on statistics topics taught in this period. The results are presented below.
a Calculate ρ.
b Calculate the regression line to predict scores on the midterm from arithmetic scores.

c Calculate the standard error of estimate in part b.
d Is score on the arithmetic test a good predictor of performance on the
   midterm test?
e How strong is the linear relationship between these two variables?
The results:

Student	1	2	3	4	5	6	7	8	9	10	11
Arithmetic score	49	42	49	43	49	33	38	40	43	40	49
Midterm score	25	19	30	32	31	25	29	24	26	26	25

Student	12	13	14	15	16	17	18	19	20	21	22
Arithmetic score	47	45	45	42	43	33	34	34	41	47	48
Midterm score	29	29	27	34	32	26	15	22	29	24	30

4 a Refer to Problem 4 in Set 3 of Chapter 3. Calculate the correlation coef-
      ficient for the data in Data Set A (at the end of Chapter 2). In case you
      do not have a calculator, we present some useful summary numbers to
      aid you in this calculation.

$\Sigma X = 1{,}622$ $\Sigma X^2 = 22{,}594$ $\Sigma XY = 31{,}126{,}200$
$\Sigma Y = 2{,}226{,}600$ $\Sigma Y^2 = 44{,}930{,}340{,}000$ $N = 125$

   b To show the effect on the correlation coefficient of a restriction on the
      range, restrict the range of number of years of education to those fami-
      lies where the number of years of education is between 11 years and 15
      years inclusive. Calculate the correlation coefficient between family in-
      come and number of years of education for this group.

$N = 58$ $\Sigma Y = 1{,}012{,}400$ $\Sigma Y^2 = 18{,}980{,}200{,}000$
$\Sigma X = 745$ $\Sigma X^2 = 9{,}667$ $\Sigma XY = 13{,}160{,}300$

   c Next, restrict the range of number of years of education to an extreme
      group: those families where education is 9 years or less and those families
      where education is 17 years or more. Calculate ρ for annual income and
      years of education for this group.

$N = 39$ $Y = 681{,}300$ $Y^2 = 14{,}467{,}310{,}000$
$\Sigma X = 507$ $\Sigma X^2 = 7{,}787$ $\Sigma XY = 10{,}491{,}900$

   d If your interest was to predict annual income using years of education
      as a linear predictor, you could calculate the prediction line. Use the
      data in part a to calculate the slope and intercept for the prediction
      line. Draw this line on the scatterdiagram you prepared for these data
      in Chapter 3.
   e Calculate the standard error of estimate. Interpret this number.
   f Use the prediction line to predict annual income for families with 16
      years of education. Do the same thing for families with 10 years of edu-
      cation.

Set 4

1-10   In the Chapter 1 problems, Set 2, a set of situations was presented. We classified each problem as to type of problem and level of measurement of the variables. At the end of Chapter 3, we reconsidered the situations and described organizational techniques that might be employed to summarize and display collected data. In Chapters 4 and 5, we have presented some descriptive statistics to use for each problem type at each level of measurement. To see how these descriptive statistics fit into our organizational scheme, read again Problems 1 through 10 in Chapter 1, Set 2, and decide what descriptive statistics might be used to summarize the information.

Following is a list of additional research situations.
    Classify the problem according to the dimensions of Chapter 1.
    Identify the variables and classify them as independent or dependent.
    What is the level of measurement for each variable?
    How would you organize the data for presentation?
    What descriptive statistics would be helpful as summaries of the data set?

11   A researcher with the Federal Drug Administration wants to know if there is any relationship between use of saccharine and bladder cancer. He decides to take 100 baby rats and randomly assign them to two groups. One group is fed a standard diet of nutrients and sugar for sweetener and the other is fed the standard diet of nutrients with an equivalent amount of saccharine used in place of sugar. At the end of the experimental period, the researcher will tabulate the number of rats with bladder tumors in each group.

12   A teacher of statistics wants to know whether or not arithmetic ability has anything to do with the level of performance of students that she teaches. She decides that in the next school year she will give her students an arithmetic test at the beginning of the semester and see whether or not these scores have anything to do with the final grades at the end of the year.

13   A school district wants to know how its high school seniors compare to high school seniors nationally. Because all seniors take the SAT, they decide to look at the SAT scores for the last 10 years.

14   At a state fair, 10 lemon pies are entered for the pie baking contest. The fair committee wants to know if the judges are at all consistent in their judgments of the pies. They randomly select two judges to take part in an experiment. Each judge is asked to rank the pies from 1 to 10—1 being the best and 10 being the worst.

15   An investigator wishes to study the relationship between family income and political attitudes. She selects a large sample of families and asks each head-of-household to classify family income as low, low average, high average, or high and to classify political attitude as liberal, conservative, or middle-of-the-road.

Descriptive Statistics

Inferential Statistics

4

6

Introduction to Hypothesis Testing: The Binomial Model

Part 4 of this text deal with inferential statistics. You should recall the distinction between descriptive problems and inferential problems suggested in Chapter 1. We have a descriptive problem when we have measured an entire population and wish to describe that population in terms of some relevant parameter, such as its mean or standard deviation. We have an inferential problem when we have measured only a subset of a particular population. Such a subset is called a sample, and the role of inferential statistics is to use information derived from the sample to help us make a statement about the population as a whole. One way in which a sample may be used to help us make a statement about the population is to calculate a sample statistic as an estimate of the corresponding population parameter.

In Chapter 1, we illustrated the use of a sample statistic as an estimator of a population parameter by reference to the question, "What proportion of the television viewing audience watches WXYZ at 10:00 P.M. on Saturday?" We noted that in most viewing areas, it would be impractical to attempt to contact the entire population of individuals who might have been watching television at that time. There would simply be too many people. Instead, it would make more sense for the researcher seeking to find the answer to this question to select a random sample of individuals residing in the viewing area of interest. Such a sample will generally closely resemble the population from which it is drawn, particularly if the sample is sufficiently large. The researcher would, therefore, calculate the proportion of television viewers in the sample who were watching station WXYZ and then use this proportion as an estimate for the proportion in the population as a whole. In many cases, estimates like this will be close to the true value of the population parameter, but other times they will not be so close. For this reason, the most frequently employed formal approaches to statistical inference involve more than just a single "best estimate" of the parameter of interest. They also attempt to take into account the problem of sampling error. These approaches are known as the *hypothesis testing* approach and the *interval estimation* approach. Of the two approaches, the hypothesis testing approach is the more traditional. However, there are many occasions when the interval estimation approach is more efficient. We will consider both approaches to statistical inference in this book. In recognition of tradition, we begin with hypothesis testing in this chapter and consider interval estimation later, in Chapter 7.

There are a great number of specific techniques that are used to test hypotheses. Which technique to use in a given situation depends on a number of factors, including, of course, the type of statistical problem and the level of measurement of the variable(s) involved. However, all hypothesis testing techniques share a common sequence of logical steps. It is extremely important that you understand the logic of hypothesis testing in general before you begin to learn about all of the various types of statistical tests that are available for use in different research situations. In this chapter, we focus on the logic of hypothesis testing. In illustrating this logic, we consider several examples. First, we will consider a nonquantitative example. Then we consider a simple quantitative example, a binomial problem.

THE LOGIC OF HYPOTHESIS TESTING: THE STRAW MAN

The logic of hypothesis testing is very similar to the logic employed by politicians or debaters when they first set up and then tear down a "straw man." For those readers who are not familiar with this term, we define a straw man as an idea presented for the purpose of *disproving* that idea. For example, one might hear a new candidate for mayor say something like this: "If the incumbent has done such an effective job fighting crime in our town, why did we have a 17 percent increase in the number of armed robberies last year?" This is a rhetorical question designed to make a point about the leadership of the incumbent. If we analyze this rhetorical question, we see that the candidate has really set up an initial hypothesis and then presented evidence that suggests that this hypothesis is probably not correct. The reasoning involves a specific sequence of logical steps. Had our political contender outlined these steps formally, we would find that there are really four separate logical statements involved in his argument. The candidate might have outlined the argument like this:

1 *The initial assumption of the hypothesis:*
"The incumbent claims to have done an effective job of fighting crime in our town. For the moment, let us assume that this claim is the truth."

2 *Development of an expectation based on the initial hypothesis:*
"If the incumbent has really done an effective job fighting crime in our town, then we would expect to see a decrease in the number of crimes committed. If we check the figures and find a decrease in crime, we would believe this claim. But if we see an increase, we would not believe it."

3 *Comparison of expectation to factual evidence observed:*
"When we look at the records, we observe that armed robbery has not decreased. It has increased dramatically."

4 *Conclusion:*
"The discrepancy between what we would expect to find, under the initial assumption, and what we actually find, by observation, causes me to reject that initial assumption. I conclude that the incumbent has actually *not* done an effective job fighting crime."

What the politician has done is this: He has started with a statement that applies to a large area of the incumbent's political behavior, the area having to do with crime. He has taken this statement and *temporarily* assumed that it is true (statement 1). Of course, it is clear that the candidate never really believes that the hypothesis is true. He is actually attempting to show that it is false. For this reason, such a hypothesis is often referred to as a *null hypothesis.* Having established the null hypothesis, the candidate uses the hypothesis to generate an idea of what we might expect to observe, *if* the null hypothesis were really true. *If* the incumbent fought crime effectively, we should see a decrease in crimes (statement 2). Next, he reports an actual observation and compares the observed data to the expectation generated from the initial assumption. Armed robbery has not decreased, but has increased (statement 3). Finally, he points out that the observed data is incompatible with the initial assumption, which causes him to reject this initial or null hypothesis (statement 4).

This four-step procedure corresponds exactly to the logical sequence used in statistical tests. In statistical inference, however, there are two basic differences. First, when we are dealing with a problem involving statistical inference, the null hypothesis will relate to a population parameter, and the evidence that leads us to reject the null hypothesis will be derived from a sample of observations drawn from that population. Second, when we deal with statistical inference, we do not reject the null hypothesis absolutely, but rather with the knowledge that there is always a chance that the null hypothesis is true. When the politician rejects the hypothesis that the incumbent mayor has fought crime, he rejects this hypothesis absolutely. He does not consider the possibility that the incumbent might really have done an effective job fighting crime, but that armed robberies might still have increased. Perhaps murders and rapes and burglaries all decreased while armed robbery increased. That is, it is possible that the observation our politician used to refute the hypothesis was not really representative of the truth. In the language of statistics, we would say that his sample was not representative.

230

In the area of statistical inference, the possibility of such sampling error is taken into account. The researcher knows that there is no statistical test in existence that allows us to reject a hypothesis with 100 percent certainty that the hypothesis rejected is really false. In statistical inference, we reject a hypothesis with a knowledge of *how big a chance* there is that the hypothesis really is true. For this reason, the statistician is concerned with *probability*.

PROBABILITY

Probability may be thought of in two different ways. Most commonly, we think of it as the *chance* that something will happen. If I toss a fair coin, there is a 50 percent chance that it will turn up heads. If I roll a 6-sided die that is not loaded, there is a 1/6 or 16.67 percent chance that the outcome of the roll will be a 4. You can also think of probability in another way. You can think of tossing the fair coin over and over again. Then you would find that about half of the time the outcome of the toss would be a head, and half the time the outcome of the toss would be a tail. Similarly, with the die, one sixth of the time the outcome of a roll would be a 1, one sixth of the time a 2, and so on through 6. It is important for you to make the connection between how often a given outcome will occur over time and the chance that it will occur one particular time. If we toss a coin an infinite number of times and we find that half of the time the result is a head, then the probability that the result of a *single toss* will be a head is one half.

With this idea of probability in mind, let's go back to the politician's argument. The politician simply says, "Armed robbery has gone up, therefore I reject the hypothesis that the incumbent has fought crime effectively." The statistician would not say this. If it were within his or her power to do so, the statistician would make a statement like: "On the basis of my calculations, I can see that *if* the null hypothesis were true, *if* the incumbent had fought crime effectively, there would be only a two percent chance that armed robbery would have increased by an amount as large as the increase we have observed. Because a two percent chance is such a small chance, I reject the hypothesis that the incumbent has fought crime effectively. However, I keep in mind that there is still a two percent chance that I am wrong if I reject the hypothesis." In fact, with a question like the political argument here, it would probably be impossible for the statistician to make such a precise statement. The issue is too complex; there are too many factors to be considered. However, there are many real problems for which the statistician can make such probability statements with at least a fair degree of confidence in the accuracy of the figures. In these problems, the researcher will typically employ a mathematical model based on the assumptions of the null hypothesis to calculate the probability of observing a particular result or set of results. As we consider the field of inferential statistics in Chapters 6 through 11, we will consider many different mathematical models. Some research situations call for one kind of model; other situations call for a different model. In the section that follows, we will describe one simple mathematical model that is frequently used in research situations. This is the binomial model. Then, we will illustrate how this model may be applied to the hypothesis testing technique by reference to a hypothetical research situation.

THE BINOMIAL MODEL

Definition of a Binomial Variable

A *binomial variable* is a variable that has only two values. Examples of binomial variables include the outcome of the toss of a coin (heads or tails), the sex of a newborn child (male or female), and a student's grade in a pass-fail course (pass or fail). A *binomial experiment* may be defined as follows. There is a procedure that may produce one of two possible outcomes, such as the procedure of tossing a coin. There is a probability associated with each of the two possible outcomes. If the coin were fair, the probability of a head would be .50 and the probability of a tail would be .50. There is no requirement that the probabilities of each of the two outcomes be equal, however. The only requirement is that whatever the probability of one outcome is, the probability of the other must equal one minus this probability. This is so simply because only two outcomes are possible. One of them *must* occur, so the probability that one will occur, added together with the probability that the other will occur, must equal 100 percent or 1.00.

In referring to these probabilities, we typically define one of the two possible outcomes as a "success," and we denote the probability of obtaining this outcome as P. Then the other outcome is defined as a "failure," and we denote its probability as Q. Thus, in the binomial problem, $Q = 1 - P$. In the course grade example, if we had defined passing as a success and if we knew that the probability of passing, P, was .8, then the probability of not passing would have to be $Q = 1 - P = 1 - .8 = .2$. Finally, in a binomial experiment, if the experimental procedure that can yield two outcomes is repeated, the probabilities of each of the two possible outcomes must remain the same each time the procedure is repeated. We typically refer to each repetition of the experimental procedure as a *trial*. Thus, in a binomial experiment, P and Q remain constant from trial to trial, regardless of the outcome(s) of previous trials. When the outcome of any given trial does not affect the probability of a success on a succeeding trial, the trials are said to be independent of each other. In order for an experimental procedure to constitute a binomial situation, the trials must be independent.

232

An Example of a Binomial Experiment

Now let's look at an example of a binomial experiment. Suppose that you are concerned with an experimental procedure that may result in either success, S, or failure, F. For example, you may imagine the procedure of tossing a coin and observing whether the result is heads or tails. In this case, we might arbitrarily define the result "heads" as a success and the result "tails" as a failure. Alternatively, you might imagine an individual attempting to predict the winner of a tennis match. In this case, we might designate the result "correct prediction" as a success and the result "incorrect prediction" as a failure. Whatever experimental procedure you may choose to envision, suppose that you repeat this procedure 5 times, and that the probability of a success, P, does not change from one repetition to the other. Then we are talking about a binomial experiment in which there are 5 independent binomial trials. How can such an experiment turn out? How many different outcomes are there?

The Outcomes of a Binomial Experiment. To answer these questions, it is helpful to trace through all the possible outcomes of the experiment trial by trial. A useful device for this purpose is a *tree diagram*. The tree diagram depicting *all* the possible outcomes of the 5 predictions is presented in Figure 6.1. As an aid to understanding this diagram, let us first concentrate on just the first 2 trials and the possible results of these trials. Obviously, the first trial may be either success or failure. These 2 possible outcomes of the first trial are symbolized by the S and F in Figure 6.1 under the heading "1." Regardless of the result of trial 1, the second prediction may be either right or wrong. Thus, on the tree diagram, there are branches from the S possible outcome of trial 1 to both S and F possible outcomes of trial 2, and there are branches from the F possible outcome of trial 1 to both S and F possible outcomes of trial 2. If we now summarize the possible outcomes of the first 2 trials, we see that there are 4 possible patterns.

Success on trial 1, success on trial 2
Success on trial 1, failure on trial 2
Failure on trial 1, success on trial 2
Failure on trial 1, failure on trial 2

These 4 different possible outcomes of the first 2 trials of the experiment can be noted briefly as follows: SS, SF, FS, FF. In this notation, the first letter tells the outcome of the first trial, and the second letter tells the outcome of the second trial. Notice that 2 of these outcomes, SF and FS, differ only in the order in which the success and the failure occur. They are similar in that each of these outcomes contains 1 success and 1 failure. In developing the theory of the binomial model, it is important to differentiate between all of the ordered out-

233

FIGURE 6.1
Tree Diagram Illustrating the 32 Different Ordered Outcomes
of a Binomial Experiment with 5 Trials

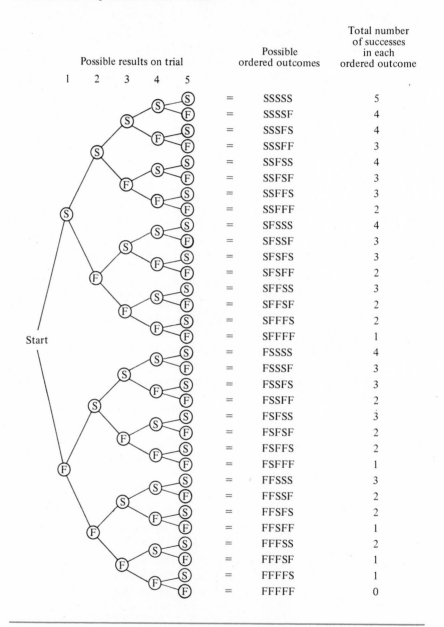

	Possible ordered outcomes	Total number of successes in each ordered outcome
=	SSSSS	5
=	SSSSF	4
=	SSSFS	4
=	SSSFF	3
=	SSFSS	4
=	SSFSF	3
=	SSFFS	3
=	SSFFF	2
=	SFSSS	4
=	SFSSF	3
=	SFSFS	3
=	SFSFF	2
=	SFFSS	3
=	SFFSF	2
=	SFFFS	2
=	SFFFF	1
=	FSSSS	4
=	FSSSF	3
=	FSSFS	3
=	FSSFF	2
=	FSFSS	3
=	FSFSF	2
=	FSFFS	2
=	FSFFF	1
=	FFSSS	3
=	FFSSF	2
=	FFSFS	2
=	FFSFF	1
=	FFFSS	2
=	FFFSF	1
=	FFFFS	1
=	FFFFF	0

Inferential Statistics

comes. However, as you will see later, we are often concerned with the idea of how many successes and failures there are out of the total number of trials, rather than the exact order in which these successes and failures occur.

Now, consider the tree diagram extended fully through the 5 trials. Each additional trial doubles the number of different ordered outcomes that the experiment may produce. With the 5 trials, there are $(2)(2)(2)(2)(2) = 32$ different possible ordered outcomes that may result. In any binomial experiment, the number of different ordered outcomes possible may always be found by raising the number 2 to the nth power, where n equals the number of trials.* Thus, here $(2)(2)(2)(2)(2) = 2^5 = 32$. Of course, you should once again remember that some of these outcomes differ from one another only in the order in which successes and failures occur. If 5 trials occur, the number of successes may be any of the 6 values: $0, 1, 2, 3, 4, 5$. In Figure 6.1, the final column lists the number of successes contained in each of the 32 different ordered outcomes. By counting in this column, we can see how many of the 32 different ordered outcomes contain each of the 6 different possible numbers of successes. This information has been summarized in Table 6.1.

Conveniently, it is possible to determine the number of different ordered outcomes that contain a given number of successes in a binomial problem without producing the tree diagram and then counting up the number of successes in each of the resulting outcomes. Instead, we can use a simple mathematical formula. Call the number of trials in the experiment n, and the number of successes r. Then the number of failures is $n - r$. We can find the number of ordered outcomes that contain r successes by using this formula:

$$\begin{matrix}\text{Number of ordered outcomes} \\ \text{which contain } r \text{ successes} \\ \text{in } n \text{ trials}\end{matrix} = \binom{n}{r} = \frac{n!}{r!(n-r)!}.$$

Some readers will recognize this formula as a formula from high-school algebra. It is for calculating the number of combinations from n objects selected r at a time. The ! is not an exclamation mark but a factorial symbol. This symbol tells you to multiply the number it follows by all smaller integers down to the integer 1. For example, 3! is $3 \times 2 \times 1$; 4! is $4 \times 3 \times 2 \times 1$. The only special rule to remember is that 0! equals 1. Thus, if we wanted to use this method to compute the number of ordered outcomes that contain 1 success in 5 trials we would say

$$\binom{5}{1} = \frac{5!}{(1)!(5-1)!} = \frac{(5 \cdot 4 \cdot 3 \cdot 2 \cdot 1)}{(1)(4 \cdot 3 \cdot 2 \cdot 1)} = 5.$$

* The reason we use a lowercase n here has to do with sampling and will be discussed later.

235

TABLE 6.1
Number of Different Ordered Outcomes Containing
from 0 to 5 Correct Predictions in 5 Trials

Number of successes in 5 trials	Number of ordered outcomes containing this many successes
0	1
1	5
2	10
3	10
4	5
5	1
	32

As you can see, our result agrees with the result obtained by counting specific ordered outcomes from the tree diagram. Similarly, if we wanted to find the number of ordered outcomes containing 5 successes in 5 trials,

$$\binom{5}{5} = \frac{5!}{5!(5-5)!} = \frac{5!}{5!0!} = \frac{(5 \cdot 4 \cdot 3 \cdot 2 \cdot 1)}{(5 \cdot 4 \cdot 3 \cdot 2 \cdot 1)(1)} = 1.$$

As a check on your understanding, compute the number of ordered outcomes containing 0, 2, 3, and 4 successes in 5 trials. Do your results agree with Table 6.1?

In using the binomial model, however, we can do more than just catalogue the different ordered outcomes that are possible and the numbers of successes and failures contained in these different outcomes. We can also determine the *probability* of occurrence of each of these different ordered outcomes.

The Probability of a Particular Ordered Outcome. We can determine the probability of any given ordered outcome using a mathematical theorem called the *multiplication rule.* In its simplest form, this rule tells us that if we know the probability of a particular outcome of some experimental procedure, and if we also know the probability of a given outcome of a second, independent experimental procedure, then we can calculate the probability that *both* of these outcomes will occur on the 2 successive procedures by multiplying together their 2 respective individual probabilities. For the purpose of illustrating this idea, let us once again consider just the first 2 trials of our binomial experiment.

First, assume that the probability of a success is .5 on each trial. This means that, on the average, half the trials will result in successes. If we were to repeat the entire experiment an infinite number of times, half of the outcomes of trial 1

236

would be S. Now just think about the half of the experiments that did produce an S on trial 1. Of this half, one half would also produce an S on trial 2. But half of a half is a quarter. In other words, if we were to repeat the experiment an infinite number of times and consider only the ordered outcomes of the first 2 trials, one fourth of the experiments would have produced the outcome SS at the end of trial 2. This reasoning provides the basis for the multiplication rule. The chance of a success on trial 1 was .5. The chance of a success on the independent trial 2 was .5. Therefore, the chance of a success on trial 1 *and* a success on trial 2 is $(.5) \cdot (.5) = .25$.

Now assume a 70 percent chance of a success on each trial. What is the chance of a success on trial 1 *and* a success on trial 2 under this new assumption? It is $(.7) \cdot (.7) = .49$. If we were to repeat the experiment an infinite number of times with $P = .7$, then 70 percent of the first trials would be successes. Of the 70 percent of the experiments with successes on the first trial, 70 percent of them would also produce successes on the second trial. Therefore, if $P = .7$, the chance of obtaining the outcome SS after 2 trials would be 70 percent of 70 percent, or 49 percent. What would be the chance of obtaining the outcome SF after 2 trials? This question can also be answered using the multiplication rule. If the probability of a success is $P = .7$, then the probability of a failure is $Q = 1 - P = 1 - .7 = .3$. Therefore, according to the multiplication rule, the chance of a success followed by a failure is given by $pr(\text{SF}) = (P) \cdot (Q) = (.7)(.3) = .21$, where *pr* stands for "probability of." Using the same rule, you can describe the probability of the ordered outcomes FS and FF.

The multiplication rule may be extended to include situations involving more than 2 trials. If the probability of a success on a given trial is .5, then the probability of 3 successes on 3 successive trials would be $(.5)(.5)(.5) = .125$. The probability of 5 successes on 5 trials would be $(.5)(.5)(.5)(.5)(.5) = .03125$. In the case where we have 5 trials and $P = .5$, each of the 32 different ordered outcomes has exactly the same probability. This probably makes sense to you intuitively, but it can also be shown using the multiplication rule. If $P = .5$, then $Q = 1 - P = 1 - .5 = .5$ as well, so the probability of both a success and a failure are the same. Thus, whether we consider the ordered outcome SSSSS, or SSSSF, or any of the other 32 different ordered outcomes, the probability of each ordered outcome will always be $(.5)(.5)(.5)(.5)(.5) = .03125$.

If the probability of a success on a given trial is not .5, then the probabilities of the 32 different possible ordered outcomes will not all be the same. If $P = .7$, then $Q = 1 - .7 = .3$. Now, if we want to know the probability of the ordered outcome SSSSS, we find it by $pr(\text{SSSSS}) = (.7)(.7)(.7)(.7)(.7) = .16807$. In Table 6.2, we have listed all of the 32 different possible ordered outcomes of the 5-trial binomial experiment where $P = .7$, and we have used the multiplication rule to calculate the probability of each of these outcomes.

Referring to Table 6.2, you will notice that several of the different ordered outcomes have the same probability. For example, check the probability of each of the 5 ordered outcomes that contain 1 success. You will note that all 5 of these possible outcomes have the same probability, .00567. Why is this? Consider how we use the multiplication rule to find each probability:

$$pr(\text{SFFFF}) = P \cdot Q \cdot Q \cdot Q \cdot Q = (.7)(.3)(.3)(.3)(.3) = .00567$$
$$pr(\text{FSFFF}) = Q \cdot P \cdot Q \cdot Q \cdot Q = (.3)(.7)(.3)(.3)(.3) = .00567$$
$$pr(\text{FFSFF}) = Q \cdot Q \cdot P \cdot Q \cdot Q = (.3)(.3)(.7)(.3)(.3) = .00567$$
$$pr(\text{FFFSF}) = Q \cdot Q \cdot Q \cdot P \cdot Q = (.3)(.3)(.3)(.7)(.3) = .00567$$
$$pr(\text{FFFFS}) = Q \cdot Q \cdot Q \cdot Q \cdot P = (.3)(.3)(.3)(.3)(.7) = .00567$$

In each case, the probability of an ordered outcome containing 1 success is found by multiplying together 1 P and 4 Qs, i.e., 1 (.7) and 4 (.3)s. The only difference is the order in which the P occurs, and in multiplication, this makes no difference in the final answer. This is true of all the sets of ordered outcomes that have a certain number of successes, provided P is constant from trial to trial.

In any binomial experiment, all of the different ordered outcomes containing the same number of successes will have exactly the same probability. If we wish to know the probability of *each* of the outcomes that contains r successes in n trials, we can use the formula:

$$pr \left[\begin{array}{c} \text{Each ordered} \\ \text{outcome containing} \\ r \text{ successes} \end{array} \right] = P^r \cdot Q^{n-r}.$$

Don't let the exponential notation alarm you. All the formula does is group together all the successes and all the failures and use exponents to tell you how many times to include their probabilities in the multiplication. In the case where we want to know the probability of each ordered outcome containing $r = 2$ successes in $n = 5$ trials, where $P = .7$,

$$pr \left[\begin{array}{c} \text{Each ordered} \\ \text{outcome with} \\ 2 \text{ successes} \\ \text{out of 5 trials} \end{array} \right] = (.7)^2 (.3)^{5-2} = (.7)(.7)(.3)(.3)(.3) = .01323.$$

If we had wanted the probability of each ordered outcome containing $r = 3$ successes in $n = 5$ trials, where $P = .7$,

$$pr \left[\begin{array}{c} \text{Each ordered} \\ \text{outcome with} \\ 3 \text{ successes} \\ \text{out of 5 trials} \end{array} \right] = (.7)^3 (.3)^{5-3} = (.7)(.7)(.7)(.3)(.3) = .03087.$$

238

TABLE 6.2
Probabilities of Each of the 32 Different Ordered Outcomes
of the Binomial Experiment, with $n = 5$ and $P = .7$

Ordered outcome	Multiplication	Probability of outcome	Total number of successes in outcome
SSSSS	(.7) (.7) (.7) (.7) (.7)	.16807	5
SSSSF	(.7) (.7) (.7) (.7) (.3)	.07203	4
SSSFS	(.7) (.7) (.7) (.3) (.7)	.07203	4
SSSFF	(.7) (.7) (.7) (.3) (.3)	.03087	3
SSFSS	(.7) (.7) (.3) (.7) (.7)	.07203	4
SSFSF	(.7) (.7) (.3) (.7) (.3)	.03087	3
SSFFS	(.7) (.7) (.3) (.3) (.7)	.03087	3
SSFFF	(.7) (.7) (.3) (.3) (.3)	.01323	2
SFSSS	(.7) (.3) (.7) (.7) (.7)	.07203	4
SFSSF	(.7) (.3) (.7) (.7) (.3)	.03087	3
SFSFS	(.7) (.3) (.7) (.3) (.7)	.03087	3
SFSFF	(.7) (.3) (.7) (.3) (.3)	.01323	2
SFFSS	(.7) (.3) (.3) (.7) (.7)	.03087	3
SFFSF	(.7) (.3) (.3) (.7) (.3)	.01323	2
SFFFS	(.7) (.3) (.3) (.3) (.7)	.01323	2
SFFFF	(.7) (.3) (.3) (.3) (.3)	.00567	1
FSSSS	(.3) (.7) (.7) (.7) (.7)	.07203	4
FSSSF	(.3) (.7) (.7) (.7) (.3)	.03087	3
FSSFS	(.3) (.7) (.7) (.3) (.7)	.03087	3
FSSFF	(.3) (.7) (.7) (.3) (.3)	.01323	2
FSFSS	(.3) (.7) (.3) (.7) (.7)	.03087	3
FSFSF	(.3) (.7) (.3) (.7) (.3)	.01323	2
FSFFS	(.3) (.7) (.3) (.3) (.7)	.01323	2
FSFFF	(.3) (.7) (.3) (.3) (.3)	.00567	1
FFSSS	(.3) (.3) (.7) (.7) (.7)	.03087	3
FFSSF	(.3) (.3) (.7) (.7) (.3)	.01323	2
FFSFS	(.3) (.3) (.7) (.3) (.7)	.01323	2
FFSFF	(.3) (.3) (.7) (.3) (.3)	.00567	1
FFFSS	(.3) (.3) (.3) (.7) (.7)	.01323	2
FFFSF	(.3) (.3) (.3) (.7) (.3)	.00567	1
FFFFS	(.3) (.3) (.3) (.3) (.7)	.00567	1
FFFFF	(.3) (.3) (.3) (.3) (.3)	.00243	0
		1.00000	

The Probability of a Given Number of Successes in a Given Number of Trials.
Up to this point, we have shown how to use a tree diagram or the combination
formula to determine how many different ordered outcomes of a binomial experiment will contain r successes in n trials. We have also shown how to use the
multiplication rule to determine the probability of *each* of the ordered outcomes

239

TABLE 6.3
Calculation of Probability of Obtaining r Successes out of $n = 5$ Trials in a Binomial Experiment Where $P = .5$

Number of successes (r)	Number of ordered outcomes containing this number of successes $\binom{n}{r}$	$\times$	Probability of each ordered outcome with r successes $P^r \cdot Q^{n-r}$	Probability of getting r successes
5	1		.03125	.03125
4	5		.03125	.15625
3	10		.03125	.31250
2	10		.03125	.31250
1	5		.03125	.15625
0	1		.03125	.03125
				1.00000

that do contain r successes in n trials. If we put these two techniques together, we can determine the probability of obtaining r successes in n trials, regardless of which ordered outcome is the specific one to produce the r successes. For example, we have seen that when $n = 5$, there are 5 different ordered outcomes that contain 1 success. When $P = .7$, each of these 5 outcomes has a probability of .00567. Therefore, the probability that we will get 1 success in 5 independent binomial trials when $P = .7$ is $5 \times (.00567) = .02835$. We can summarize this idea in words like this:

$$pr \begin{bmatrix} r \text{ successes} \\ \text{in } n \text{ independent} \\ \text{binomial trials} \end{bmatrix} = \begin{bmatrix} \text{The number of} \\ \text{ordered outcomes} \\ \text{which contain} \\ \text{exactly } r \text{ successes} \end{bmatrix} \times \begin{bmatrix} \text{The probability of} \\ \text{each of these} \\ \text{ordered outcomes} \end{bmatrix}.$$

Alternatively, we can use the simple mathematical notation explained above:

$$pr \begin{bmatrix} r \text{ successes} \\ \text{in } n \text{ independent} \\ \text{binomial trials} \end{bmatrix} = \binom{n}{r} \cdot P^r \cdot Q^{n-r}.$$

This formula may be used to calculate the exact probability of obtaining any specific number of successes on any specific number of binomial trials, provided the probability of a success on one trial is known.

In Tables 6.3 and 6.4, we have presented the probabilities associated with each of the 6 different numbers of successes possible out of 5 trials for the 2 situations when $P = .5$ and when $P = .7$. These tables are referred to as *probability distributions*.

Probability Distributions Represented Graphically. Now we know how to use a few simple mathematical techniques to calculate the probabilities of different numbers of successes in problems that conform to the binomial model. We are almost ready to use the model in the hypothesis testing context. However, we would first ask you to think of the probabilities represented in Tables 6.3 and 6.4 in a slightly different manner, as indicated in our definition of probability given earlier.

You may think of the probabilities as expressing the chance of obtaining each of the 6 different numbers of successes if you perform the experiment once, or you may imagine that the entire experiment is performed an infinite number of times. Then, each probability represents the proportion of times that the experiment produces each of the 6 possible results. If we think of the binomial probabilities in this second way, we can represent them using a discrete graph of the relative frequency distribution of the 6 different results. This has been done for the case in which $P = .5$ in Figure 6.2. It has been done for the case in which $P = .7$ in Figure 6.3.

Comparing these two figures should give you a good idea of how the probability of different numbers of successes changes if the probability of a success on each trial is changed from .5 to .7. If $P = .5$, you would be equally likely to achieve either 2 or 3 successes in 5 trials. If $P = .7$, you would be most likely to obtain 4 successes in 5 trials, although 3 successes would be a fairly likely out-

---

FIGURE 6.2
Probability Distribution of the Number of Successes
in 5 Independent Binomial Trials if $P = .5$

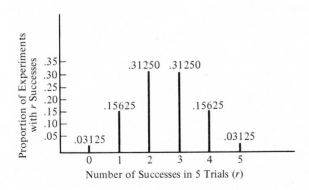

TABLE 6.4
Calculation of Probability of Obtaining r Successes out of $n = 5$ Trials in a Binomial Experiment Where $P = .7$

Number of successes (r)	Number of ordered outcomes containing this number of successes $\binom{n}{r}$	$\times$	Probability of each ordered outcome with r successes $P^r \cdot Q^{n-r}$	Probability of getting r successes
5	1		.16807	.16807
4	5		.07203	.36015
3	10		.03087	.30870
2	10		.01323	.13230
1	5		.00567	.02835
0	1		.00243	.00243
				1.00000

come as well. Notice that when $P = .5$, the probability distribution is symmetrical. That is, 2 successes is as likely as 3, 1 as likely as 4, and 0 as likely as 5. This is what we would expect intuitively. If we have a 50/50 chance of success on each trial, we expect to get 2 or 3 successes in 5 trials, but a deviation in the direction of obtaining more successes is no more likely than an equal deviation in the direction of getting fewer. When $P = .7$, the distribution is no longer symmetrical. If you have a better than 50/50 chance of success on each trial, it stands to reason that you would be more likely to get 5 successes than none in 5 trials.

THE BINOMIAL MODEL
IN HYPOTHESIS TESTING

In applying this model to a real hypothesis testing situation, imagine this situation. Suppose that you have developed what you believe to be an effective method of predicting the winner in tennis matches. You know that your method works, but no one will believe you. Your friends all insist that you're just guessing when you make your predictions. How would you go about convincing them of your ability to predict winners? One way would be to do an experiment. You propose that you and your friends all go to the nearest tennis court and randomly select 5 matches that are about to begin. Each time one of these matches begins, you will predict the winner. The object of the experiment will be to see how many matches out of 5 you can predict correctly.

This experiment conforms to the binomial model. Each prediction has 2 possible outcomes: Your prediction may be correct (a success) or it may be wrong (a failure). Because you make 5 predictions, the experiment has 5 trials. We assume that these trials are independent, and that the probability of a correct prediction does not change from one trial to the next.

Moreover, the experiment constitutes a problem in inferential statistics. When you say that you have developed a method for predicting tennis matches, you are talking about tennis matches in general, not just the 5 matches in the actual experiment. We are really interested in making a statement that will apply to the entire population of tennis matches you might predict, and we will be considering the 5 randomly selected matches to be a sample drawn from that population. Therefore, we would establish a null hypothesis that applies to the population, and test this hypothesis with reference to sample data.

The Hypothesis Testing Procedure

The procedure involved in testing the null hypothesis that your predictions are simply guesses would follow the same sequence of four logical steps that we described earlier in connection with the political argument.

Initial assumption. Because your friends have argued that you are merely guessing, you would adopt this idea as an initial assumption, a null hypothesis that you would like to disprove. If you were merely guessing, then you ought to have a 50/50 chance of guessing right. Another way to say this is to say that the probability of a correct prediction on any match in the population of matches that you might predict is .5. This means that the probability of a cor-

---

FIGURE 6.3
Probability Distribution of the Number of Successes
in 5 Independent Binomial Trials if $P = .7$

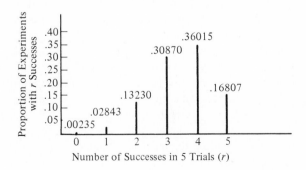

Introduction to Hypothesis Testing

rect prediction on any one of the sample of 5 matches randomly selected for the experiment is .5. Because we have defined a correct prediction as a success in terms of the binomial model, we can just as well say that the probability of a success on any trial is .5. The null hypothesis for the test would be written

$$H_o : P = .5,$$

where H_o reads "null hypothesis," and P refers to the probability of a correct prediction on any match in the population of matches that you might predict. You are trying to show that you are not just guessing, that your prediction technique is effective. But if your technique is effective, your chance of being correct on any prediction is really *greater than .5*. This alternative view, the *alternative hypothesis,* which you are seeking to support in this experiment, is expressed this way:

$$H_a : P > .5,$$

which reads, "The alternative hypothesis is that the probability of a success on any match, and therefore on each trial in our binomial experiment, is greater than .5." To sum up, these two one-line expressions tell us that we are testing the hypothesis that you predict correctly half of the time against the alternative that you do better than that.

Development of an expectation based on the null hypothesis. Once we have set up our null hypothesis, we are in a position to use the appropriate model to determine the most likely results of our 5-trial experiment, assuming that the null hypothesis is a correct statement regarding the entire population of predictions that you might make for all possible matches. Thus, we temporarily assume that it is true that the probability of a success on any prediction is .5, and we use the model to determine how likely it would be to obtain no successes out of the 5 trials in our sample, one success out of the 5 trials, and so on.

We have already worked out the probability distribution for the 6 possible numbers of successes when we have 5 independent binomial trials with $P = .5$. This probability distribution is presented as a relative frequency distribution in Table 6.3 and as a discrete graph in Figure 6.2. On the basis of these probabilities, you and your friends can come to an agreement regarding the number of correct predictions you will have to make in 5 trials in order for them to change their minds and reject the idea that you are simply guessing. Which of the 6 possible results would tend to make them believe that you are doing better than guessing? If you made 2 or 3 correct predictions, this would not make them change their mind about your predicting ability. We can see from the probability distribution that these would be the most likely outcomes if you were guessing. What about 0 correct predictions or 1 correct prediction? These possible results are less likely, under the null hypothesis, but would either of these results help

244

you to prove that you can predict? Clearly, they would not because these results contain even *fewer* correct predictions than we would expect by chance. This leaves us with the results of 4 correct predictions and all 5 correct predictions. Each of these 2 results contains a greater number of correct predictions than we would expect. In this sense, either of these outcomes would provide some evidence against the null hypothesis and in favor of the alternative hypothesis that your predictions are more accurate than chance. You might propose to your friends that if you get either 4 or 5 correct predictions, they should abandon their belief that you are guessing. Do you think they would agree to this rule?

According to our model, if it were true that you were guessing and the null hypothesis were correct, the chance that you would get 4 correct predictions in a sample of 5 predictions is 15.625 percent (see Table 6.3) and the chance that you would get 5 correct predictions in 5 tries is 3.125 percent. This means that if your friends agree to reject the null hypothesis on the basis of either 4 or 5 correct predictions, there would be an 18.75 percent chance of rejecting the null hypothesis, *even if H_o were true*. Your friends would probably not agree to change their minds on this basis. There would be too good a chance of rejecting H_o and concluding you could predict just because you were lucky on these 5 trials. However, they might very well agree to reject the null hypothesis if you succeeded in all 5 predictions. According to the model, if you were just guessing randomly, you would have only a 3.125 percent chance of making 5 correct predictions in 5 attempts. Translating this statement into the language of population and sample, we would state, "If only half of the matches in the population of predicted tennis matches were successful predictions, then the chance that a random sample of 5 such predicted matches would contain 5 successful predictions would be just 3.125 percent."

Thus, if you can actually accomplish the feat of predicting all 5 matches correctly, you would be providing strong evidence that your friends' initial belief (the null hypothesis) was wrong. Of course, we know from the model that even if you do make 5 correct predictions, there is still a chance that the null hypothesis was right. You may have been very lucky on this sample of 5 trials. However, the chance of being this lucky is very small. If your friends insisted on adhering to the null hypothesis in the face of 5 correct predictions, they would be saying that you were lucky enough to come up with a result that occurs only about 3 times in every 100 times the entire experiment is carried out. Clearly, it would make more sense for them to reject the null hypothesis and admit that your ability to predict is better than chance.

Assuming your friends go along with this, you have established what is known in hypothesis testing as a *rejection rule*. In a hypothesis testing situation, the rejection rule is theoretically established prior to the actual experiment. It specifies what result or group of results will cause the null hypothesis to be rejected in favor of the alternative. In this case, the rejection rule would be stated as follows: "We will reject the null hypothesis that you are just guessing if the experiment results in 5 correct predictions out of 5 tries."

Introduction to Hypothesis Testing

Comparison of expectation to factual observation. Now let us assume that you and your friends go to the tennis courts, that you predict the first 5 matches to occur, and that you are correct in all 5 of your predictions. Had you not used the binomial model and established a rejection rule based specifically on this model, you would at this point probably make a statement to your friends like this: "If I couldn't predict the winner of tennis matches, how do you explain the fact that I have just predicted 5 matches in a row correctly?" If you said this, you would be using the logic of hypothesis testing informally, just as the mayoral candidate did in the example presented at the beginning of this chapter. However, because we have employed a mathematical model to arrive at a rejection rule, the comparison of expectation to factual observation comes down to a simple question: "Is the observed result of the experiment the result (or one of the results) that we have already decided will cause us to reject the null hypothesis?" In this case, the observed result, $r = 5$, is one that we have decided will cause us to reject. We know that the observed result in our experiment would be very unlikely if H_o were true, so we reject H_o. That is, if the proportion of correct predictions in the entire population of predictions were really .5, we know that it would be most unlikely that a sample of 5 independent predictions from this population would contain 5 successful predictions. On the basis of our sample results then, we conclude that our initial hypothesis regarding the population is false.

Conclusion. On the basis of the results of the experiment, you and your friends would reject the null hypothesis. You would conclude that when you predict a tennis match, your chance of being correct is not 50 percent, but rather something greater than 50 percent. It is important that you recognize that all you have done is reject the hypothesis that you were guessing. You have not proven that your ability to predict matches is perfect. In fact, you have not really "proven" anything with absolute certainty. What you have done is to decide that it is more reasonable to reject the null hypothesis than it is to persist in adhering to it, given the small probability of obtaining the observed result if the null hypothesis were true.

The Results of a Hypothesis Test

When we establish a null hypothesis, two things are possible: The null hypothesis may really be true, or it may be false. When we reach a decision regarding the null hypothesis, two things are also possible: We may retain our null hypothesis, or we may reject it. Whenever we carry out a hypothesis test, therefore, one of four possible events occurs. These events are depicted in Figure 6.4. Of the four possibilities, two are desirable, and two are undesirable. If the null hypothesis is really true, then of course the correct decision would be *not* to reject this hypothesis. If the null hypothesis were true and we did reject it, then we would have made an error in rejecting the null hypothesis. The error of rejecting a null hypothesis that is in fact correct is referred to as a *Type I error.* If the null

FIGURE 6.4
The Four Possible Outcomes of a Hypothesis Test

Null Hypothesis Is Really

		True	False
Decision Is To	Not reject	Retaining H_o if H_o is true is a correct decision $(1 - \alpha)$	Retaining H_o if H_o is false is a Type II error (β)
	Reject	Rejecting H_o if H_o is true is a Type I error (α or level of significance)	Rejecting H_o if H_o is false is a correct decision ($1 - \beta$ or power)

hypothesis is really false, then the correct decision would be to reject the null hypothesis. If the null hypothesis were false but we did not reject it, then we would have made an error in failing to reject the hypothesis. The error of failing to reject a null hypothesis that is in fact false is referred to as a *Type II error*.

Of course, we never know for sure whether a given null hypothesis is true or false. When we make our decision whether to accept or reject, we are simply making the best decision we can, based on our assumptions, our model, and our observed sample results. We never know for sure if the sample upon which our conclusions are based was typical. However, we do have some control over the risk of Type I and Type II errors.

Type I Error. We control the risk of a Type I error when we select our rejection rule. In the example presented in the previous section, we learned from our model that if H_o: $P = .50$ were true, there would be an 18.75 percent chance of observing either 4 or 5 correct predictions, but only a 3.125 percent chance of observing exactly 5 correct predictions out of 5 trials. In establishing this rejection rule, we were controlling our risk of making a Type I error. We were recognizing that if we chose to reject H_o on the basis of either 4 or 5 correct predictions, our chance of a Type I error would be 18.75 percent. That is, even if H_o were true, we would still have an 18.75 percent chance of observing a result that would lead us to reject. We chose to reduce the risk to 3.125 percent by establishing the stricter rejection rule. The chance of committing a Type I error is also referred to as alpha-risk or *level of significance*.

247

If the results of the hypothesis test cause you to reject a particular null hypothesis, it is better to have done so with a small alpha-risk or level of significance. The smaller alpha was, the more certain you may be that the rejected hypothesis was indeed false. In Figure 6.4, we have placed a symbol α (alpha) and the term "level of significance" in the box referring to a Type I error. This is to reinforce the idea that the symbol α and the term "level of significance" both refer to the probability of a Type I error. In the box that refers to a correct decision not to reject a true null hypothesis we have placed the symbol $1 - \alpha$. Assuming the null hypothesis is true, we will either reject it or not reject it. The probability of doing one of these two things is 100 percent, or 1.00. Therefore, if the probability of rejecting a true hypothesis is α, then the probability of not rejecting it is $1 - \alpha$.

Type II Error. Now let us consider the question of a Type II error. Let us suppose that in the hypothesis test just described we had *not* observed 5 correct predictions in 5 tries. In this case, we would have followed our rejection rule and not rejected the null hypothesis. Would this mean that the null hypothesis was true? Not necessarily. Suppose that the truth of the matter was that your true predicting ability was better than chance, but not perfect. For example, let's say that you really predict matches correctly 70 percent of the time, i.e., $P = .70$. If this were true, then clearly $H_o: P = .50$ would not be true. Yet, if $P = .70$ and you predict 5 matches, there is certainly a good chance that you will not predict 5 out of 5. And unless you predict 5 out of 5, we do not reject H_o. In this case, H_o is false but we do not reject H_o, and we have made a Type II error. The chance of committing a Type II error is referred to as beta-risk. In Figure 6.4, we have placed the symbol β (Greek letter beta) in the box referring to a Type II error. We have placed $1 - \beta$ in the box referring to a correct decision to reject a false null hypothesis. The reasoning here is similar to that for α and $1 - \alpha$. Assuming the null hypothesis is false, we must still either reject it or not reject it. If the chance of not rejecting a false hypothesis is β, then the chance of correctly rejecting it is $1 - \beta$. This chance of correctly rejecting a false null hypothesis is referred to as the *power* of the test, as indicated in Figure 6.4. Obviously, it is desirable that β be small and $1 - \beta$ be large.

How do we determine β and $1 - \beta$? Remember once again that we used a model in this hypothesis test and that we established the rule that we would reject H_o if we saw 5 out of 5 correct predictions. This rejection rule determines our decision, regardless of whether $P = .5$ or $P = .7$ or P equals any other value. Therefore, to find the power of our test, we must determine the chance of obtaining 5 correct predictions. We know that this chance is small if $P = .5$. But now we are saying that we believe $P = .7$. Therefore, to determine the power of this test when $P = .7$, we must determine the chance of obtaining 5 correct predictions when $P = .7$. How do we do this? We use a mathematical model. We use the binomial model again, but now we consider the model where we have 5 independent binomial trials with $P = .7$. Once again, we have already worked out

248

the probability distribution for this situation in our discussion of the binomial model. The probability distribution is presented as a relative frequency distribution in Table 6.4 and as a discrete graph in Figure 6.3.

The probability distribution tells us that if $P = .7$, the chance of observing 5 successes in 5 independent binomial trials is .16807 or 16.807 percent. This, then, is the power of our test of H_o: $P = .5$ against the alternative that $P = .7$. Clearly, this is not a desirable power. If H_o is false and P really is .7, our test has only a 16.807 percent chance of rejecting H_o. This means that if P is really .7 there is a $1 - .16807$ or .83193 percent chance that our test will result in a Type II error.

How can we increase the power of our test? One way would be to change our rejection rule, so that we would reject H_o if either 4 or 5 correct predictions were observed. However, as we noted before, if we chose to reject on the basis of either 4 or 5 correct predictions, there would be a larger chance of a Type I error. Generally, changing the rejection rule to reduce β has the undesirable effect of increasing α. This point is graphically illustrated in Figures 6.5 and 6.6. In Figure 6.5, we have shown the two probability distributions for $P = .5$ and $P = .7$. We have connected them by a dotted line between 4 successes and 5 successes, to distinguish between the outcome that will lead us to reject H_o ($r = 5$ successes) and the outcomes that will lead us not to reject H_o ($r = $ anything less than 5 successes). If H_o is true, the chance of obtaining 5 successes is .03125, so $\alpha = .03125$. The chance of obtaining less than 5 successes is equal to the sum of the chances of obtaining 0 successes, 1 success, 2 successes, 3 successes, or 4 successes. This sum is $.03125 + .15625 + .31250 + .31250 + .15625 = .96875$, so $1 - \alpha = .96875$. On the other hand, if H_o is false, and if the actual value of P is .7, then the chance of obtaining 5 successes is .16807, so $1 - \beta$ is .16807. The chance of obtaining less than 5 successes is $.00243 + .02835 + .13230 + .30870 + .36015 = .83193$. This is the chance of not rejecting H_o when H_o is false, the chance of a Type II error, or β.

In Figure 6.6, we have drawn exactly the same two probability distributions, but we have connected them with a dotted line that separates the outcomes with 4 or 5 successes from those outcomes with less than 4 successes. That is, we have shifted the rejection rule so that we will reject H_o if either 4 or 5 successes are observed. This change in the rejection region makes the test more powerful. If $P = .7$, the chance of getting either 4 or 5 successes is $.36015 + .16807 = .52822$, so $1 - \beta = .52822$. Now the test has a better than 50 percent chance of rejecting H_o: $P = .5$, if P really is .7. However, the shift in the rejection rule has also changed α. If $P = .5$, the chance of getting either 4 or 5 successes is $.15625 + .03125 = .18750$. Thus, if H_o is true, there is a better than 18 percent chance that we will reject H_o anyway. This chance is α, the chance of a Type I error.

Be aware that in determining the power of our test, it was necessary for us to specify a particular alternative value for P, even though the alternative hypothesis that we were attempting to support in our test was simply H_a: $P > .5$, i.e., that P was greater than .5. If you felt that you could predict matches cor-

249

Introduction to Hypothesis Testing

FIGURE 6.5

Interrelation of α, $1 - \alpha$, β, and $1 - \beta$ in Test of $H_o : P = .5$
Against the Alternative, $H_a : P = .7$, in a 5-Trial Binomial Experiment
Where We Reject H_o if $r = 5$

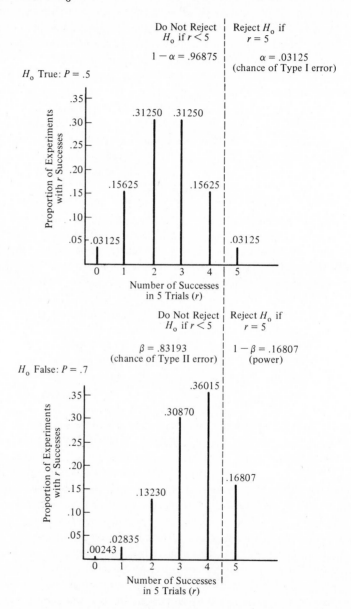

Inferential Statistics

rectly 80 percent of the time, you would use the binomial model with $P = .8$ to determine the chance of obtaining 5 correct predictions in 5 trials. Without calculating this probability, you should be able to see that if $P = .8$, you will have a *greater chance* of obtaining 5 successes in a sample of 5 trials than if $P = .7$. This means that if we consider the power of our test against the alternative that $P = .8$ instead of against the alternative that $P = .7$, the figure we obtain for $1 - \beta$ will be higher. We will be considering power, and the factors that determine the power of various statistical tests, throughout the text. Throughout these discussions, you will find that as we consider alternatives that are further away from the null hypothesis, calculated power increases.

The Binomial Table

We have considered the binomial model and have shown how the model may be used to test hypotheses in certain research situations. At this point, we note that it is not necessary to use the mathematical formulas described previously each time you test a hypothesis in a binomial experiment. It is possible to avoid calculating the probabilities of obtaining each of the various possible numbers of successful trials in a binomial experiment by using the table of binomial distributions contained in Appendix C (Table C.1). This binomial table may be used in binomial experiments having not more than 25 trials.

The table is relatively easy to use. One begins by locating the appropriate number of trials in the extreme left column of the table, headed n. In the experiment just discussed, this number would be 5, because there were 5 trials. Next, look across the top of the table to find the value of P desired. In our example, we tested the hypothesis that $P = .5$. Therefore, we look across the top of the table to find the column headed .50. We follow this column down until we reach the set of numbers beginning opposite the value $n = 5$. There are 6 numbers listed down from this point: 031, 156, 312, 312, 156, 031. These numbers represent the probabilities of the 6 different numbers of successful trials that are possible in a 5-trial binomial experiment with $P = .5$. The number 031 is opposite the value 0 in the column of the table headed r. Thus, the 031 tells us that the probability of obtaining 0 successes in a 5-trial binomial experiment where $P = .5$ is .031, to the nearest 3 decimal places. Compare this figure to the figure we computed earlier and presented in Table 6.3. You will see that the calculated probability, when rounded to 3 decimal places, is .031. Similarly, we find that the next number down in this column is 156. This number is opposite the value 1 in the column headed r. Thus, the 156 tells us that the probability of obtaining 1 success in a 5-trial binomial experiment where $P = .5$ is .156, to 3 decimal places. This listing also agrees with the calculated probability presented in Table 6.3. Thus, the table really presents the entire probability distribution for various binomial experiments.

To check your understanding of the table, locate the probability distribution that would apply to the situation considered earlier, where $P = .7$ rather

FIGURE 6.6

Interrelation of α, $1 - \alpha$, β, and $1 - \beta$ in Test of $H_o : P = .5$
Against the Alternative, $H_a : P = .7$, in a 5-Trial Binomial Experiment
Where We Reject H_o if $r = 4$ or 5

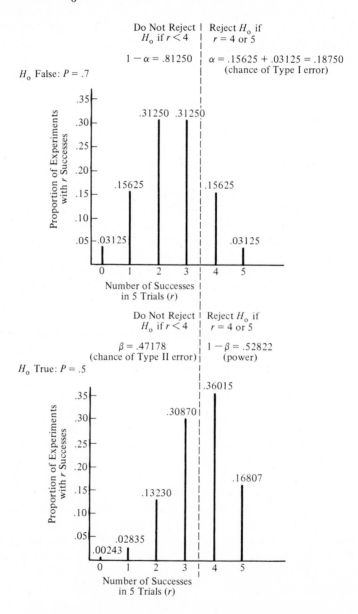

than .5. Can you locate this distribution? To do so, you need to look down the column headed .70. Go down this column until you are opposite $n = 5$ and read the probabilities of the different outcomes by looking across from the various values of r. You should find that the probability of $r = 0$ when $n = 5$ and $P = .7$ is .002, the probability of $r = 1$ when $n = 5$ and $P = .7$ is .028, and so on. You will see that these probabilities are identical to the calculated probabilities presented in Table 6.4, rounded to 3 decimal places. Obviously, the use of the table of binomial distributions will save time. Moreover, there is much less chance of making an error in reading this table than there is in calculating the probabilities.

Two-Tailed Hypothesis Tests

Consider the following research situation. An experimental psychologist is concerned with the effect of a certain drug on long-term memory among the population of college students. On the basis of long experience with this research area, the experimenter knows that when subjects have mastered a particular 20-item learning task, 50 percent of these subjects can still perform the task after a one-week interval from the last training trial. The experimenter believes that the drug in question may affect memory, but he is not sure whether the effect should be to improve memory or to impair it. He selects a random sample of 25 college students. He administers the drug to these students and then allows them to work on the task until they have mastered it. The subjects leave with instructions to return a week later. When they return, the subjects are asked once again to perform the task. When they do, the experimenter records whether they have succeeded or failed.

In this situation, the experimenter is interested in showing that the drug has produced a change in the proportion of subjects who typically succeed in the task after a one-week interval. However, he has no interest in showing that the change is in the direction of decreasing the proportion of subjects who will succeed or increasing the proportion of subjects who will succeed. In this situation, the experimenter's alternative hypothesis is said to be *nondirectional*. In this example, the experimenter would set up the null hypothesis, $H_0: P = .5$, indicating that the drug had not affected long-term memory, so that the chance that any subject would succeed after a week would still be 50 percent, just as it is when no drug is given. The alternative hypothesis would be $H_a: P \neq .5$, indicating that the drug had affected long-term memory in such a way as to alter the proportion of subjects who succeed, but not indicating whether the change is in the direction of decreasing P or increasing P.

Compare this research situation to that of the tennis prediction experiment. In that case, the experimenter's null hypothesis was $H_0: P = .5$, indicating that the chance of being correct in any particular prediction was 50 percent, as we would expect from chance. But the alternative hypothesis was $H_a: P > .5$, indicating that the chance of a correct prediction was *greater* than 50 percent. This alternative hypothesis is *directional*. The experimenter would have no interest

253

whatsoever in showing that the chance of making a correct prediction on any given trial was less than 50 percent, but wants only to show that the chance is greater than 50 percent.

The difference between directional and nondirectional alternative hypotheses is reflected in the type of rejection rules we adopt for the hypothesis tests. In the tennis example, there were 5 trials. If H_o were true, we would have expected to see about 2 or 3 successes in the 5 trials. If H_a were true, however, we would expect to see a *greater* number of successes. Thus, in considering possible rejection rules, we considered only those outcomes containing a high number of successes. Ultimately, we decided to reject H_o if the number of successful trials was 5. In the drug example, there are 25 trials. If H_o were true, we would expect to observe successes on half of these trials. That is, we would expect to see about 12 or 13 successes out of 25 trials. If H_a were true, however, we might see either a small number of successful trials or a large number of successful trials. A small number of successful trials, like 0 or 1, would be an indication that the chance of success among the population of subjects who took the drug was *less* than .5. A large number of successful trials, like 24 or 25, would be an indication that the chance of a success among the population of subjects who took the drug was *greater* than .5. Because H_a includes either of these possibilities, we wish to set up our rejection rule so that both low numbers of successes and high numbers of successes can lead us to reject H_o.

In establishing the actual rejection rule, we may use the table of binomial distributions. Look in this table under $n = 25$ trials and $P = .50$ to find the probability distribution for the experiment under the assumption that the null hypothesis is true. You will notice that for several of the extreme outcomes, such as 0 to 4 successes on the low end of the distribution, the tabled probability is given as 0+. This means that these outcomes are exceedingly unlikely if $P = .5$. As we look further in toward the center of the distribution, we find that the outcomes become more likely. In deciding upon a rejection rule, therefore, the question we need to ask ourselves is, "How far in toward the middle of this distribution do we go?" The answer to the question depends on how large a chance of a Type I error we are willing to take.

It has become accepted procedure in research in psychology and education to keep the chance of a Type I error at or below 5 percent. If we adopt this convention, we must set up our rejection rule in such a way that the total of all the probabilities of all of the outcomes that will cause us to reject H_o will be .05 or less. The easiest way to do this is to consider successively more likely outcomes, until the accumulated probabilities of these outcomes come as close to .05 as possible without exceeding it. We have already seen that the probabilities of the outcomes 0 to 4 successes and 21 to 25 successes are all close to 0. Therefore, we may include all of these in our rejection rule. Moving one success closer to the middle of the distribution on each side, we find that the probabilities of the outcomes 5 and 20 are both equal to .002. If we include 5 and 20 in the rejection region, therefore, the chance of a Type I error would still be very small,

254

just a bit over .004. Therefore, we move in one more step, to the outcomes 6 and 19. These outcomes both have the probability of occurrence of .005. Adding these two .005s to the accumulated probabilities to this point, we find the total chance of a Type I error is still only .004 + .005 + .005 = .014. We are still below .05. If you continue the process in this manner, you will find that the outcomes 7 and 18 may also be included in the rejection area, resulting in a chance of a Type I error of .042. However, no more likely outcomes may be included. If we decided to reject H_o if the number of successes were either 8 or 17 as well, the chance of a Type I error would jump to .106. Because this is greater than .05, we decide not to include these values in the rejection rule. Thus, the rejection rule for this hypothesis test would be to reject H_o if we observe 7 or fewer successes in the 25 trials *or* if we observe 18 or more successes in the 25 trials.

The hypothesis test in question is referred to as a *two-tailed hypothesis test* because both sides of the probability distribution are included in the rejection rule. Hypothesis tests concerning directional alternative hypotheses are *one-tailed tests* because only outcomes in one end of the distribution support the alternative hypothesis in these cases. Hypothesis tests concerning nondirectional alternative hypotheses are two-tailed tests because outcomes at either end of the distribution may support the alternative hypothesis.

Occasionally, we might wish to guard against the possibility of a Type I error even more vigorously than usual. In this case, we can choose to keep the chance of a Type I error below 1 percent rather than below 5 percent. Had that been the case in the example of the drug experiment, what would the rejection rule have been?

THE FORMAL STEPS OF HYPOTHESIS TESTING REVIEWED

In this chapter, we have introduced you to the logic of the hypothesis testing technique. We have illustrated this technique with a particular mathematical model, the binomial model. As you will see, different research situations will require the use of various other mathematical models. In the next chapter, for example, we consider another important model used in inferential statistics, the normal curve model. And as you read on in this book, you will learn about many other statistical tests that utilize still other models. However, regardless of the particular research situation and the particular model, the essential logic involved in testing hypotheses does not change. We have already outlined the elements of this logic in simple, common-sense terms. At this point, we will outline the specific steps of hypothesis testing more formally, with reference to the mathematical model being used. In Table 6.5, we list the steps. In this table, also, we illustrate each step with the tennis prediction example and provide a verbal explanation to link each step to the explanation presented in the chapter.

TABLE 6.5
The Steps of a Hypothesis Test

	Formal step	Example from tennis prediction experiment	Explanation
1	Decide what model to use	The tennis prediction experiment conforms to the binomial model.	Different models are appropriate for different research situations, as you will see.
2	State the null hypothesis	$H_o: P = .5$	Here you are setting up the "straw man," the initial assumption regarding the population that you believe to be incorrect.
3	State the alternative hypothesis	$H_a: P > .5$	Here you are stating what you believe to be true, the alternative to the null hypothesis that you wish to support.
4	Specify the probability distribution of the possible outcomes of the experiment, assuming H_o is true	Binomial distribution for $n = 5$ trials with $P = .5$	When you know the probability distributions for the outcomes under the assumption that H_o is true, you can determine which outcomes are likely and which are not.
5	State a rejection rule	Reject H_o if $r = 5$ successes in 5 trials.	From the probability distribution we see that if H_o is true, the chance of observing 5 successes in a sample of 5 trials is only 3.125 percent. We decide this is a reasonable chance of Type I error, so we decide to reject H_o if $r = 5$.
6	Carry out the experiment and observe the outcome	$r = 5$	In our experiment, we observe 5 correct predictions in our sample of 5 trials.
7	Compare the result to the rejection rule and reach a conclusion	Because $r = 5$, reject H_o.	The observed result of the experiment is a result that we have decided will cause us to reject H_o.

As we study other types of hypotheses tests for other types of statistical problems, we will follow the same logic as we have presented in this chapter, and we will follow the same formal outline of steps as presented in Table 6.5.

256

THE CLASSIFICATION MATRIX

In this chapter, we introduced some of the concepts of statistical inference that we will be applying to different kinds of inferential problems. After we examine a few more concepts of statistical inference, we will be filling in the cells of our matrix with the appropriate statistical tests. We shall see that, for some univariate problems involving categorical data, we can use the binomial test discussed in this chapter. This is indicated in Table 6.6.

TABLE 6.6
Classification Matrix: Inferential Statistics

	Level of Measurement of Variable(s) of Interest		
Type of Problem	Categorical: Nominal or Ordered Categories	Rank-Order	Interval Scale (or Ratio-Scale)
Univariate Problems	Binomial Test		
Bivariate Problems — Correlational Problems			
Experimental and Pseudo-Experimental Problems: Independent Groups			
Dependent Measures: Pretest and Posttest or Matched Groups			

Key Terms

Be sure you can define each of these terms.

hypothesis testing
null hypothesis
probability
binomial variable
binomial experiment
tree diagram
multiplication rule
probability distribution
alternative hypothesis
rejection rule
Type I error
Type II error
level of significance
power of a statistical test
binomial table
nondirectional hypothesis
directional hypothesis
one-tailed hypothesis test
two-tailed hypothesis test

Summary

In Chapter 6, we began our study of inferential statistics. We recalled that an inferential problem is one in which we are trying to learn something about a population by looking at a sample from that population. In this chapter, we pointed out that the two basic approaches to statistical inference are hypothesis testing and interval estimation. Because these methods are based on probability models, we examined some basic concepts of probability and probability distributions. We pointed out that in statistical inference, we make probability statements rather than absolute statements.

We introduced the logic used in hypothesis testing in this chapter. We illustrated this logic with the binomial model, a model in which the variable has only two values, "success" and "failure," and each value has a probability associated with it. The sum of

6

Student
Guide

these probabilities, P and Q, is equal to one. We demonstrated how to use the hypothesis testing technique to test hypotheses about the value of P.

Using a binomial problem, we outlined the steps of a hypothesis test. First, we choose our probability model; in this case, we use the binomial model. Sec-one, we state a null hypothesis; in this case, we specify a value for P. We pointed out that, in fact, we are trying to show our null hypothesis to be so unlikely that we can consider it to be false. Third, we state an alternative hypothesis, either directional, involving a one-tailed test, or nondirectional, involving a two-tailed test. Fourth, we specify the probability distribution of all possible sample outcomes of our experiment. In this case, we specify the probability of different numbers of "successes." Fifth, we establish a rejection rule by indicating which outcomes will cause us to reject our null hypothesis. In this example, we use a table of the binomial distributions to help us establish the rejection rule. Sixth, we observe our result in our sample. Finally, we compare this result to our established rejection rule and decide whether to accept or reject our null hypothesis.

We pointed out that, because we are using probability models, we can never be sure that we have made a correct decision in accepting or rejecting our null hypothesis. In some cases, we will reject a hypothesis that is, in fact, true. We call this a Type I error. If we accept a hypothesis that is really false, we have made a Type II error. We saw that we can control the probability of a Type I error by our choice of the rejection rule. This probability is called the level of significance. We also saw how to calculate the probability of *avoiding* a Type II error when our null hypothesis is really false. This probability, called power, is calculated for specific alternative hypotheses. We saw that power increases when the alternative hypothesis becomes further from the null hypothesis.

The basic concepts presented in this chapter, along with those covered in Chapter 7, will be used throughout the rest of the book. After a thorough introduction to statistical inference, we will be ready to examine the inferential techniques appropriate to the different types of statistical problems we have been considering.

Review Questions

To review the concepts presented in Chapter 6, choose the best answer to each of the following questions.

1 Which of the following is an example of a binomial experiment?
   _____a We select 100 people at random and classify them as male or female.
   _____b We randomly select 50 males from a population of males and 50 females from a population of females.

2 We select 100 people at random and classify them as having brown, blue, or some other eye color. The reason this is not a binomial experiment is because
   _____a each trial of the experiment has more than two outcomes.
   _____b the trials are not independent.

3   People are selected at random and classified as male or female. The selection
    process continues until there are 50 males in the sample. The reason this is
    not a binomial experiment is that
    _____a   the number of trials is not fixed.
    _____b   the P value changes from trial to trial.

4   The variable X has values 0, 1, 2. The probability that $X = 0$ is .3, the prob-
    ability that $X = 1$ is .5, and the probability that $X = 2$ is .4. This informa-
    tion does not define a probability distribution for X because
    _____a   the sum of the probabilities for all values is larger than 1.
    _____b X is not a binomial variable.

5   A binomial experiment is defined as selecting an individual at random from
    a population that consists of 10 percent left-handers and 90 percent non-
    left-handers. The selection process is repeated 20 times and the number of
    left-handers is recorded. In this situation, .10 is called
    _____a   an outcome of the binomial experiment.
    _____b   the P value of the binomial experiment.

6   In Question 5, the probability distribution for the number of left-handers in
    a sample of 20 people
    _____a   is the binomial distribution where $n = 20, P = .1$.
    _____b   cannot be determined without more information.

7   The shape of the probability distribution defined in the previous example
    would be
    _____a   positively skewed.
    _____b   negatively skewed.

8   The binomial probability distribution where $n = 8$ and $P = .4$ is the mirror
    image of the binomial probability distribution where
    _____a $n = 8, P = .6$.
    _____b $n = 2, P = .4$.

9   In hypothesis testing in a binomial experiment, the null hypothesis must
    specify
    _____a   a value for r.
    _____b   a value for P.

10 In hypothesis testing in a binomial experiment, the rejection rule consists of
    values of r that are unlikely outcomes when
    _____a   the alternative hypothesis is true.
    _____b   the null hypothesis is true.

11 In hypothesis testing in a binomial experiment, the probability of the rejec-
    tion rule when H_o is true is called
    _____a   the significance level of the test.
    _____b   the power of the test.

261

12 If we test the hypothesis that $P = .5$ against the alternative hypothesis that $P > .5$ and the results of the experiment are such that we do not reject H_o, we know that this decision is either correct or else we have made

_____a a Type I error.

_____b a Type II error.

13 If we test the hypothesis that $P = .4$ against the alternative hypothesis that $P > .4$, we call this type of test

_____a a one-tailed test.

_____b a two-tailed test.

14 If we test the hypothesis that $P = .4$ against the alternative hypothesis that $P \neq .4$, our rejection rule will most probably contain

_____a only large values of r.

_____b some small values of r and some large values of r.

15 If H_o is true but our experimental results cause us to reject H_o, we have made

_____a a Type I error.

_____b a Type II error.

16 Suppose $P = .5$ is tested against $P > .5$ with $n = 10$. We decide to accept H_o if $r = \{0, 1, \ldots, 8\}$ and to reject H_o if $r = \{9, 10\}$. The probability that $r = \{0, 1, \ldots, 8\}$ when $P = .6$ is called

_____a power.

_____b beta.

17 Suppose we test $P = .5$ against $P > .5$ with $n = 10$. We decide to reject H_o if $r = \{9, 10\}$. The probability that $r = \{9, 10\}$ when $P = .8$ is called

_____a power.

_____b level of significance.

18 In hypothesis testing in a binomial experiment, one reason for using 25 trials rather than 5 trials is that

_____a the test will be more powerful.

_____b the level of significance will be larger.

19 Using a probability model to make statistical inferences, we are able to reach conclusions that are

_____a absolutely correct.

_____b most probably correct.

20 When we call an observed experiment result statistically significant, we are saying

_____a an important event occurred, if the null hypothesis is true.

_____b a rare event occurred, if the null hypothesis is true.

21 When we use a one-tailed hypothesis test, we are implying that

_____a our probability distribution has one tail.

_____b our rejection values all come from one tail of the probability distribution.

Problems for Chapter 6: Set 1

1 A seed company claims that 95 percent of all its corn seeds will germinate. If this claim is correct, what is the probability that a seed selected at random will germinate? What is the probability that a seed selected at random will not germinate?

2 Suppose that according to genetic theory, 10 percent of all people have a certain trait. What is the probability that an individual selected at random will have the trait? What is the probability that the individual selected will not have the trait?

3 Suppose that a test consists of 10 true-false questions. The score on the test is the number of correct answers out of 10 questions. The distribution of possible scores and the probability of getting the score if one is purely guessing follows:

Score	Probability
0	.00
1	.01
2	.04
3	.12
4	.21
5	.24
6	.21
7	.12
8	.04
9	.01
10	.00

If an individual selected at random uses pure guessing while taking the test,
a What is the probability the score will be 0?
b What is the probability the score will be 4?
c What is the probability that the score will be at least 5?
d What is the probability that the score will be no more than 4?
e What is the probability that the score will be 8, 9, or 10?

4 The following distribution shows the percentage of university professors at each rank broken down by sex.

	Male	Female	Totals
Full Professors	.27	.02	.29
Associate Professors	.23	.03	.26
Assistant Professors	.28	.07	.35
Instructors	.06	.04	.10
Totals	.84	.16	1.00

Introduction to Hypothesis Testing

Use this information to figure the probability that a professor selected at random from the population of professors will be the type specified in each of the following statements.

a The selected professor will be a male.

b The selected professor will be an instructor.

c The selected professor will be a full professor.

d The selected professor will be a female associate professor.

e The selected professor will be a male assistant professor.

f The selected professor will be either a full professor or a female.

5 In which ones of Problems 1 through 4 is the variable a binomial variable? Why?

6 For those of Problems 1 through 4 that deal with a binomial variable, what is P and what is Q?

7 To practice using the formulas for calculating the number of outcomes in a binomial experiment and the probabilities of outcomes, do the following exercises.

Calculate the number of ways in which one can get r successes in n trials,

a where $r = 0, 1, 2, 3, 4$ and $n = 4$

b where $r = 0, 1, 2, 3$ and $n = 3$

c where $r = 0, 1, 2, 3, 4, 5$ and $n = 5$

d where $r = 5$ and $n = 10$

e where $r = 2$ and $n = 10$

f where $r = 8$ and $n = 10$

Calculate the probabilities for each of the following outcomes of a binomial experiment:

g Where $r =$ number of successes in 4 trials, $P = .4$, find $pr(r = 0)$, $pr(r = 1)$, and $pr(r = 2)$.

h Where $r =$ number of successes in 3 trials, $P = .95$, find $pr(r = 0)$, $pr(r = 1)$, and $pr(r = 2)$.

i Where $r =$ number of successes in 5 trials, $P = .3$, find $pr(r = 0)$, $pr(r = 1)$, and $pr(r = 2)$.

j Where $r =$ number of successes in 10 trials, $P = .5$, find $pr(r = 2)$, $pr(r = 8)$, and $pr(r = 5)$.

8 Consider Problem 2 in this set. Set up a binomial experiment by selecting 5 people at random. Define the variable "the number of people in 5 who have the trait."

a What is success in this experiment? What is failure?

b What is a trial?

c What is n?

d What are the values of r?

e What is P? What is Q?

f Make a table like Tables 6.3 and 6.4, showing the number of ordered outcomes for each value of r.

g Calculate the probability for each value of r.

264

h Summarize the results into a probability distribution showing the probabilities that *r* people will have the trait in a random sample of 5 people.
i Make a graph of your probability distribution.

9 Refer to Problem 3 in this set. We can use the distribution presented to summarize a binomial experiment.
a What is a trial?
b What is a success?
c What is *P?* What is *Q?*
d What is *n?*
e What are the values of *r?*
f Calculate the probability that *r* = 2. Does your answer agree with the number in Problem 3?
g If pure guessing is used, which values of *r* would be probable outcomes? Which values of *r* would be improbable?
h What outcomes of this binomial experiment might make you wonder whether or not the subject was just guessing? Why?

Set 2

Following is a list of three situations that describe binomial experiments. For each situation, answer the following questions:
a What is the hypothesis to test; that is, what is the null hypothesis?
b What is the hypothesis that will be retained if the null hypothesis is rejected; that is, what is the alternative hypothesis?
c What is the definition of success on one trial?
d What is the variable of the experiment?
e What is the probability distribution of the experimental variable if the null hypothesis is true?
f What is the meaning of a Type I error?
g What is the meaning of a Type II error?
h What is the meaning of the power of the test?

1 A seed company claims that 95 percent of all its corn seeds will germinate. To test this claim an experimenter decides to select a random sample of 100 seeds. He will reject the seed company's claim if the evidence suggests the germination rate is not as high as the company claims.

2 Professor Ann Sharp has been studying the incidence of dreams in adults. Her work leads her to the hypothesis that 60 percent of the adult population have dreams during their sleeping hours. She wonders if this percentage is high enough. She wants to test this under experimental conditions. Her work with rapid eye movement during sleep (REM) has been found to be a good indicator of dreaming. She will use REM as the dependent variable. She selects 25 people at random and gets their cooperation to participate in her experiment. She will keep a record of whether or not subjects have REM during their sleep. If they do, she will assume they dream, if not, she will assume they don't dream.

3 Suppose it is known that 10 percent of a certain kind of white rat develops bladder cancer during their lives. A medical researcher believes that the use of a certain sugar substitute will increase the percentage of rats who get bladder cancer. He randomly selects 25 rats and feeds them a diet that is highly concentrated with the sugar substitute. He records whether or not they develop bladder cancer.

Set 3

1 To practice reading the binomial table (Table C.1), calculate the following probabilities using the table.
a $n = 4, P = .4$ find $pr(r = 0)$, $pr(r = 1)$, $pr(r = 2)$.
b $n = 3, P = .95$ find $pr(r = 0)$, $pr(r = 1)$, $pr(r = 2)$.
c $n = 5, P = .3$ find $pr(r = 0)$, $pr(r = 1)$, $pr(r = 2)$.
d $n = 10, P = .5$ find $pr(r = 2)$, $pr(r = 8)$, $pr(r = 5)$.
e Compare answers a to d with answers g–j in Problem 7, Set 1.
f $n = 15, P = .6$ find $pr(r \leqslant 3)$, $pr(r \geqslant 12)$, $pr(4 \leqslant r \leqslant 10)$
g $n = 25, P = .4$ find $pr(r = 10)$, $pr(r \leqslant 5)$, $pr(r \geqslant 13)$

2 For Problem 2 in Set 2, set up a rejection rule for the experiment so that $\alpha \leqslant .05$. Then calculate the probability for the rule under the conditions that $P = .7, .8, .9$, and $.95$.

3 For Problem 3 in Set 2, set up a rejection rule for the experiment so that $\alpha \leqslant .05$. Then calculate the probability for the rule under the conditions that $P = .2, .3, .4, .5, .6$.

4 For $n = 10$, calculate the probability $pr(r \leqslant 2$ or $r \geqslant 8)$ for $P = .5, .4, .6, .3, .7, .2$, and $.8$.

Set 4

For each of the following three situations, answer the following questions:
a What model will you use?
b What is the null hypothesis?
c What is the alternative hypothesis?
d What is the level of significance?
e Is your test going to be a one- or two-tailed test?
f Define the variable used in making the test.
g What is the probability distribution of the variable if H_o is true?
h State a rejection rule.
i What is the probability of your rejection rule if H_o is true?

1 A manufacturer of machine parts must periodically check the production line to ensure that the parts being made will fit together. Every Wednesday,

Inferential Statistics

he randomly selects 25 parts from the production line and tries them out. If he can believe no more than 1 percent of all parts manufactured are faulty, he continues production. If he concludes that more than 1 percent of the parts manufactured are faulty, he stops production and readjusts the equipment. Because it is costly to stop production, he wishes to set up statistical procedures that will incorrectly stop production just one time in a 100.

2   A cosmetic company has a product on the market that "works" 40 percent of the time. Their research lab has come up with a new product. The company wishes to know how the new product compares to the current product. They find twenty volunteers to test the new product. The company is willing to run a 10 percent risk of incorrectly concluding the products differ.

3   A large company receives its inventory in huge shipments. In accepting the shipments, it is impossible to look through every carton to determine whether or not they are accepting faulty merchandise. They decide to set up procedures to randomly sample the shipment and use the results to make a decision as to whether or not the shipment should be accepted. If they believe that the shipment is at least 90 percent good, they will keep it; otherwise, they will return it to the manufacturer. A random sample of 15 boxes from the shipment will be selected, opened, and inspected. The company wishes to risk returning a good shipment no more than 10 times in 100.

4   In Problem 1 of this set, if the machine is producing 5 percent faulty parts, what is the probability that H_o will be rejected? If the machine is producing 10 percent faulty parts, what is the probability that H_o will be rejected? Based on the experiment set up, is the experimenter apt to stop production if the machine is producing 20 percent faulty parts? Do you think this is a good experiment?

5   In Problem 2 of this set, if the new product works 50 percent of the time, is the company apt to realize it, based on their experiment? If the new product works only 30 percent of the time, is the company apt to realize it, based on their experiment?

6   In Problem 3 of this set, calculate the power for the test against the alternatives $P = .8$, $P = .7$, $P = .6$, $P = .5$. Comment on the likelihood that the company will return a good shipment to the manufacturing company, based on their experiment.

7   If the results of the experiment in Problem 1 are that 15 parts are faulty, what should the manufacturer do?

8   In Problem 2, if 7 people who use the new product find it works, what decision should the drug company make about the new drug?

9   In Problem 3, if the results show that the sampled cartons contain 13 good cartons, what decision should the company make about the shipment?

Introduction to Hypothesis Testing

In Chapter 6, we considered the logic of hypothesis testing. That is, we considered how mathematical models may be used to develop statements regarding the likelihood of various experimental outcomes under the assumptions of a particular null hypothesis. We illustrated this aspect of hypothesis testing with research examples conforming to the binomial model. As we continue our discussion of inferential statistics, you will find that there are many different research situations, and that we must use different mathematical models, depending on the situation. The purpose of the present chapter is to introduce you to a mathematical model that is very important in inferential statistics: the *normal curve model.* You will recall that in binomial problems, we are dealing with the probability distributions of discrete interval variables, such as the number of successes out of a certain number of trials. In normal curve problems, we are dealing with the probability distributions of continuous interval variables. Before proceeding to consider the specific characteristics of the normal curve model, we must consider the way in which we represent the probability distributions of continuous interval variables.

7
The Normal Distribution and Sampling Theory

PROBABILITY DISTRIBUTIONS FOR CONTINUOUS INTERVAL DATA

In Chapter 2, we learned that discrete graphs are typically used to represent the frequency distribution or relative frequency distribution of a set of discrete interval data. In our discussion of the binomial model in the last chapter, we found that the discrete graph may also be employed to represent the probability distribution of a discrete interval variable, such as the number of successes observed in a binomial experiment. By analogy, you might suspect that the techniques used to represent frequency distributions on continuous interval variables would also be used to represent probability distributions for such variables. This is indeed the case. In Chapter 2, we learned that a histogram or frequency polygon is generally used to represent the frequency distribution or relative frequency distribution of a continuous interval variable. We also saw that when we are concerned with large data sets, a smooth curve is sometimes used to represent the frequency distribution of a continuous interval variable. These techniques are also used to represent probability distributions for continuous interval variables.

In illustrating the use of these graphic techniques to represent probability distributions on continuous interval data, we will reconsider the relative frequency distribution of height for a population of 10,000 American males. You will recall that we used this example in Chapter 2 (Figure 2.11) to demonstrate the manner in which a histogram approaches a smooth curve as we progressively reduce the size of the score intervals we use. In Figure 7.1, we have reproduced the four graphs presented in Figure 2.11, but we have shaded a portion of each graph. In shading an area of the graph, we are calling your attention to a certain proportion of the scores that make up the distribution.

Figure 7.1(a), for example, is a histogram representing the distribution of height in which scores have been grouped into intervals 10 centimeters wide. The height of each bar in this histogram represents the proportion of the total group of 10,000 men who had heights that fall within the range specified by the end points of the bar. However, because all the intervals are of equal length, the *area* of any one bar may also be thought of as representing the *proportion* of the total group who had heights within that range. Thus, the area of the bar running horizontally from 155 to 165 cm. is the largest of any bar on the graph, for the proportion of men who had heights in this range was larger than the proportion of men having heights in any of the other 10 cm. ranges. You will note in Figure 7.1(a) that we have shaded the portion of the histogram lying above the point 180 cm. By shading the area of the histogram from the point 180 cm. up to the highest end of the scale, we have represented the approximate proportion of men with heights greater than 180 cm. What proportion is this? First, we would estimate that about 6 percent of the total population have heights between 180 cm. and 185 cm. This is because 12 percent of the group fall in the interval from 175 to 185 cm., and 180 cm. is right in the middle of this interval (remember that we assume scores are spread uniformly over each interval). We also have 8 percent of the total group in the 185 to 195 cm. range, 6 percent in the 195 to 205 cm. range, and 2 percent in the 205 to 215 cm. range. Thus, the percentage of the total group having heights greater than 180 cm. is estimated to be 6 + 8 + 6 + 2 = 22 percent.

Now suppose that one of the 10,000 men in this population was selected at random through a lottery procedure. What would be the chance that this one man would have a height greater than 180 cm.? Logically, if 22 percent of all the men have heights greater than 180 cm., and if the random selection procedure gives each of the 10,000 men an equal chance of being selected, then there

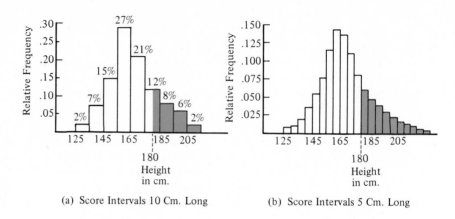

(a) Score Intervals 10 Cm. Long

(b) Score Intervals 5 Cm. Long

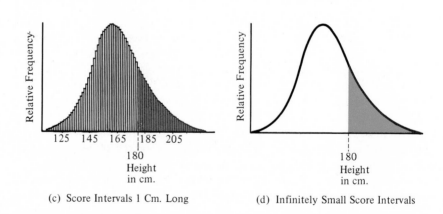

(c) Score Intervals 1 Cm. Long

(d) Infinitely Small Score Intervals

should be a 22 percent chance that *one* man actually selected will have a height greater than 180 cm. Thus, if we were interested in the *probability* that a given score on the variable "height," selected at random from the population of 10,000 scores, would exceed the value 180, we could also represent this probability by the area under the histogram to the right of the value 180. When we think of the histogram of the relative frequency distribution in this manner, we are conceptualizing it as a probability distribution. This distinction may seem trivial to you at this point. However, it is most important, as you will see as you read further in the chapter.

In each of the other three graphs presented in Figure 7.1, the area lying above the point 180 cm. has been shaded. In each case, this area may be thought of as representing either the proportion of all the observations falling in this range or the probability that one subject (observation) selected at random from the population will fall into this range. The only difference between the four figures is the width of the interval considered. As noted in Chapter 2, the shorter the width of the interval, the smoother the graph appears. When the intervals become infinitely small, as in Figure 7.1(d), the graph is a smooth curve.

In this section, then, we have considered two key points. We have shown that if we consider the operation of selecting one score at random from a distribution of scores on a continuously distributed variable, we may choose to represent the distribution as a histogram. Then the probability that the score we select will fall within a specified range of values is represented by that portion of the area of the histogram that lies within this range of values. We have also reviewed the concept, first mentioned in Chapter 2, that when we reduce the size of the class intervals represented in the histogram, the histogram will typically appear more and more like a smooth curve. In this case, we may represent probabilities by areas under sections of a smooth curve.

NORMAL DISTRIBUTIONS

Characteristics of Normally Distributed Variables

Take another look at Figure 7.1(d). Notice the shape of the distribution: It is bell-shaped but not symmetrical. Most of the scores in the distribution of heights cluster around the middle. This stands to reason, for there are relatively many men of average height. As you look away from the middle of the distribution, either to the left (toward shorter heights) or to the right (toward taller heights) the curve drops down toward the horizontal axis. The area is smaller under these sections of the curve, because there are fewer scores in these regions. This also makes sense, because we know there are relatively fewer very short or very tall men than there are men in the average height range. Height, and a great many other natural variables, tend to be distributed in this manner. If you measure a large enough group of individuals on one of these variables, the shape of the distribution will always have this bell-shaped appearance. Variables that do tend to be distributed in this manner are typically referred to as *normally distributed* variables. The reason for this terminology is that the shape of these distributions closely resembles the shape of a theoretically derived family of mathematical

271

curves known as the normal curves. This history of this derivation and the complex equation* that defines a normal curve are not within the scope of this text. However, it is important for you to know the characteristics of normally distributed variables. We can say the following about the characteristics of variables that are truly normally distributed.

Continuity. Normally distributed variables are continuous variables that range from negative infinity to positive infinity. The smooth curve that we use to represent a normal distribution graphically never intersects with the horizontal axis. However, most of the scores in a theoretical normal distribution occur within three standard deviations of the mean of the distribution. Only a very small proportion of the scores in a theoretical normal distribution are extremely low or extremely high. A natural variable such as human height is almost identical to the theoretical model, except of course that there are *no* individuals with infinitely small or infinitely large height. Thus, there is a slight discrepancy between the theoretical model and a natural normally distributed variable. In practice, this discrepancy is of little consequence to us.

Shape and symmetry. Normal distributions are bell-shaped and symmetrical. Because the greatest concentration of scores in a normal distribution occurs at the mean of the distribution, the highest point in the graph of a normally distributed variable occurs at the mean. Moving away from the mean in either the positive or negative direction, we find fewer and fewer scores. Thus, the graph of the distribution falls toward the horizontal axis as we move away from the mean. Because the distribution is symmetrical, the same proportion of the scores in a normal distribution will lie between the mean and the score a given number of units *above* the mean as lie between the mean and the score the same number of units *below* the mean. For example, if we know that height in a particular population is normally distributed with a mean of 160 cm., we know immediately that there are as many members of the population with heights between 150 cm. and 160 cm. as there are with heights between 160 cm. and 170 cm. A second result of the symmetry of normal distributions is that the mean of a normal distribution is always the same score value as the median and the mode.

The normal family of distributions. There is not one "normal distribution." There are many. Normal distributions differ from one another with respect to the mean of the distribution, μ, and the standard deviation of the distribution, σ. We call μ and σ the parameters of the distribution. The mean of a normal distribution determines its position with respect to the horizontal axis. The standard deviation of a normal distribution determines whether the distribution is compact or spread out. To illustrate this point, we present Figure 7.2. In Figure 7.2(a), we have drawn two normal distributions with different means

$$* \ Y = \frac{1}{\sigma\sqrt{2\pi}} e^{-\frac{(X-\mu)^2}{2\sigma^2}}$$

but with the same standard deviation. The two curves are identical in shape, but they are in different positions with respect to the horizontal axis. The scores in the distribution where $\mu = 10$ are lower than the scores in the distribution where $\mu = 20$. If we added 10 to each score in the former distribution, it would be identical to the latter distribution. In Figure 7.2(b), we have sketched three normal curves representing three normal distributions with the same mean but different standard deviations. All three are centered at and symmetrical with respect to their common mean, $\mu = 15$. However, the distribution with the standard deviation of 5 is more peaked than the others, and that with the standard deviation of 20 is flatter, more spread out. You remember, of course, that the standard deviation of a distribution is an index of how spread out the scores are. It should be clear to you that when $\sigma = 5$, the scores are clustered closely around the mean, but when $\sigma = 20$, the scores are more spread out away from the mean.

Area relationships in normal distributions. Regardless of the particular value of the mean and the standard deviation of a given normal distribution, the proportion of scores in that distribution that lie between the mean score and the score one *standard deviation* above the mean will always be the same, 34.13 percent. This is illustrated in Figure 7.3, where we have drawn two normal distributions that have different means and different standard deviations. In distribution A, the mean is 20 and the standard deviation is 10. The area of the curve defined by the mean score ($\mu = 20$) and the score one standard deviation above

---

Figure 7.2
Effect of Changing the Parameters μ and σ of a Normal Distribution

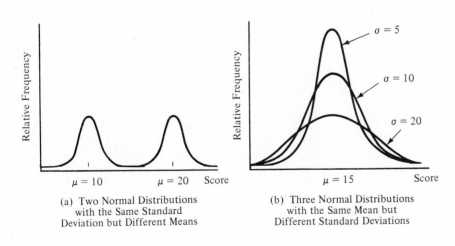

(a) Two Normal Distributions
with the Same Standard
Deviation but Different Means

(b) Three Normal Distributions
with the Same Mean but
Different Standard Deviations

273

the mean $(20 + 10 = 30)$ is equal to 34.13 percent of the total area under normal curve A. Because area is used to represent a proportion of all the scores, this means that 34.13 percent of the scores in this distribution fall between the mean and the score one standard deviation above the mean. In distribution B, the mean is 70 and the standard deviation is 15. However, the area under this normal curve lying between the mean ($\mu = 70$) and the score value one standard deviation above the mean $(70 + 15 = 85)$ is also equal to 34.13 percent of the total area under this normal curve. The same proportion of the total number of scores would lie between the mean and the score one standard deviation above the mean in *any* normal distribution. Because the mean of a distribution corresponds to the z-score of zero and the score one standard deviation above the mean corresponds to the z-score of $+1.00$, we could also say that in *any* normal distribution 34.13 percent of all the scores lie between $z = 0.00$ and $z = +1.00$.

Similar relationships hold for other z-score values. In Figure 7.4, we have graphed a normal distribution in which all the scores have been standardized. That is, all the scores have been converted to z-scores. Such a distribution is generally referred to as a *standard normal distribution*. Referring to Figure 7.4, we find that 13.59 percent of all the scores in a normal distribution lie between $z = +1.00$ and $z = +2.00$, 2.15 percent of the scores lie between $z = +2.00$ and $z = +3.00$, and 0.13 percent of all the scores lie above $z = +3.00$. Furthermore, because the normal distribution is symmetrical, the proportion of scores lying between two positive z scores will be the same as the proportion of scores lying between the corresponding two negative z-scores. That is, just as 13.59 percent

Figure 7.3
Area Relationships in Two Normal Distributions

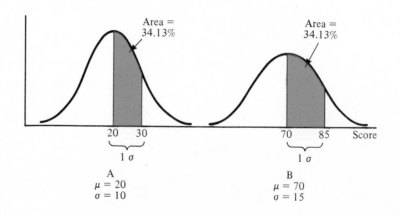

Inferential Statistics

Figure 7.4
The Proportion of the Total Area Under a Normal Curve
Defined by Certain z-score Values

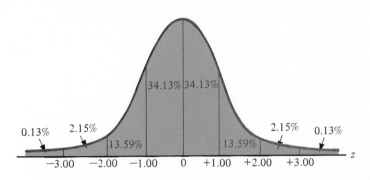

of all the scores lie between $z = +1.00$ and $z = +2.00$, so 13.59 percent of all the scores lie between $z = -1.00$ and $z = -2.00$. In addition, similar relationships hold for z-score values that are *not* whole numbers. For example, in any normal distribution 47.5 percent of the scores will be found between $z = 0$ and $z = +1.96$.

Percentile Ranks in Normal Distributions

The relationship between z-scores in a normal distribution and the proportion of scores located in the areas defined by those z-scores is extremely important to us. If we know that a distribution of scores is normal and if we know the mean and the standard deviation of the distribution, we can find the percentile rank of any score in the distribution without having a complete list of all the scores. Consider once again the way in which we used a cumulative percentage graph to estimate the percentile rank of a score in a distribution of scores (see Chapter 4). We used the frequency distribution to construct a cumulative frequency distribution. Then we used the cumulative frequency distribution to construct a cumulative percentage graph. From this cumulative percentage graph, we could obtain the percentile rank of a particular score or we could find the score that had a particular percentile rank. What we are saying here is that if a distribution is normal, we do not need to see the actual distribution in order to construct a cumulative percentage graph. Because of the special relationship between standard scores and the areas under normal curves cut off by those scores, one cumulative percentage graph describes *all* normal distributions. This is the cumulative percentage graph of the standard normal distribution, presented in Figure 7.5.

275

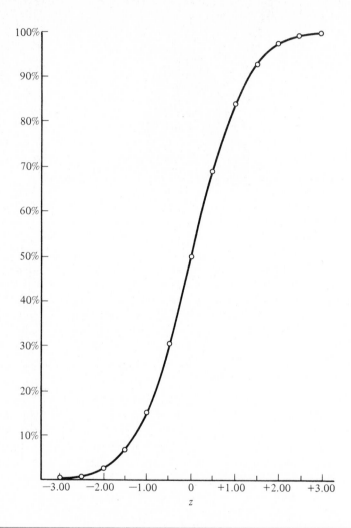

Take a moment to consider this cumulative percentage graph. Standard score values are listed along the horizontal axis and percentiles are listed along the vertical axis. We read this cumulative percentage graph just as we would read any other cumulative percentage graph. To determine the percentile rank of a given z-score, locate that z-score on the horizontal axis, then draw a perpendicular line from the horizontal axis up to the graph, and a horizontal line from the point of intersection with the graph over to the vertical axis. Read the percentile there. To find out what z-score in a normal distribution corresponds to a particular percentile rank, reverse the direction of the lines. To test your understanding, use the cumulative percentage graph of the standard normal distribution to determine the median (fiftieth percentile) z-score in a normal distribution. You should find that the fiftieth percentile z-score is zero. This should make sense to you. Because normal distributions are symmetrical, the mean, median, and mode of a normal distribution all occur at the same score value. Because the mean z-score in a distribution of standard scores is always zero, this score value must be zero.

Now let us compare the cumulative percentage graph in Figure 7.5 to the graph of the normal distribution shown in Figure 7.4. Referring to Figure 7.4, we know that 50 percent of all the scores in the normal distribution fall below the score $z = 0$. We can also see that an additional 34.13 percent of all the scores fall between the scores $z = 0$ and $z = +1.00$. Thus, a total of 50.00 percent plus 34.13 percent, or 84.13 percent, of all the scores in the distribution fall below the z-score of $+1.00$. In other words, the percentile rank of the z-score of $+1.00$ in a normal distribution is 84.13 percent. Now go to Figure 7.5 and use the cumulative percentage graph to determine the percentile rank of the z-score of $+1.00$. You should find that the percentile rank of this score is about 84 here as well. Figures 7.4 and 7.5 are both representations of a normal distribution in standard score form. It is no wonder that we should obtain the same result from each source. As a check of your understanding, find the percentile rank of the z-score of $+2.00$ using both figures. Then do the same for the z-score of -1.00 in both figures.

Of course, the cumulative percentage graph of the standard normal distribution may be used to find the percentile rank of any score in any normal distribution of scores, provided we know the mean and the standard deviation of the distribution. If we have this information, we can use it to convert the raw score we are interested in to a z-score, which will enable us to read the percentile rank from the graph. Furthermore, we may also determine the raw score equivalent of a given percentile rank in a normal distribution. In short, with the cumulative percentage graph of the standard normal distribution and the formulas for converting back and forth from raw scores to z-scores, we can solve two different types of problems. We describe these two types of problems as follows. In the first type, we are given a score and asked to find its percentile rank. This we call the score to area or score to proportion problem because percentile rank corresponds to a proportion of the total area under the standard normal curve. In the second type of problem, we are given a percentile rank and asked to find the score having this percentile rank. This we call the proportion to score problem. The ability to deal with these types of problems is critical in the use of the normal curve model in testing hypotheses. We will now illustrate each type of problem.

277

Score to Proportion Problems.   Suppose that you take a test of mechanical comprehension given by a counselor in your school's counseling center. You score 65 right out of 75 questions. You don't know anyone else who has ever taken the exam, but your counselor tells you that scores on the test are normally distributed, that the average score among students in schools like yours is 50, and that the standard deviation is 10 points. With this information, you can use Figure 7.5 to determine your approximate percentile rank. In this type of problem, we are given a raw score and need to find the proportion of scores in the normal distribution that fall below that score. It is a score to proportion or score to area problem. There are two steps involved in the solution of such problems. First, you need to convert the raw score that you are given into a standard score, using the z-score formula. Here the raw score is 65, so

$$z = \frac{X - \mu}{\sigma} = \frac{65 - 50}{10} = +1.5.$$

Thus, your score of 65 is 1.5 standard deviations above the mean score on the test. Once you have determined the standard score equivalent of your raw score, you can proceed to the second step, which is to use Figure 7.5 to determine the percentile rank of your score. You locate the standard score of +1.5 on the horizontal axis and draw perpendicular lines, first from +1.5 up to the cumulative percentage graph, then from the graph over to the vertical axis. You should find that the percentile rank of a z-score of +1.5 in the standard normal distribution is roughly 93. This means that the percentile rank of your raw score of 65 in the normal distribution of test scores is also 93. Notice that in obtaining the percentile rank, we did not employ any techniques that we have not used before. We used the z-score formula to find the z-score equivalent of a particular raw score, and we read a percentile rank from a cumulative percentage graph. The only new aspect is that the cumulative percentage graph used is the one that describes a normal distribution, rather than some other type of distribution.

Proportion to Score Problems.   We can also use the cumulative percentage graph of the normal distribution to find the score needed to attain a certain percentile rank. Suppose that you know that Law School W will only accept students who scored better than the eightieth percentile on the Law School Admission Test (LSAT). You know that this test is normally distributed with a mean of 500 and a standard deviation of 100, and you want to know what score you will need to obtain on the test in order to reach the eightieth percentile. This problem is just the opposite of the score to proportion problem just discussed. Here you need the eightieth percentile score, so you know you need the score below which 80 percent of all the scores fall. That is, you have a proportion, and you need to find the score that cuts off this proportion. This is a proportion to score or area to score problem. A proportion to score problem is also solved in two steps, but the order of the steps is reversed. First, you use

278

Figure 7.5 to find the eightieth percentile z-score in the standard normal distribution. You should recall that this is done by drawing a line from the 80 percent mark on the vertical axis over to the point where the line intersects the cumulative percentage graph, and then drawing a vertical line straight down from that point to the horizontal axis. You should read that the eightieth percentile score in the standard normal distribution is approximately +0.8. Once you have found the z-score that is the eightieth percentile score in the standard normal distribution, you can proceed to the second step, which is to find the raw score on the LSAT that is equivalent to a z-score of +0.8 in a standard normal distribution. This is done by using the formula for converting z-scores into raw scores:

$$X = z\sigma + \mu = (+.8)(100) + 500 = 580.$$

Thus, you know that you will have to score at least 580 on the Law School Admission Test in order to be considered by Law School W. Once again, this problem did not involve using any new techniques. We read a score from a cumulative percentage graph, and we used a formula to convert a standard score into a raw score.

To review, we have just looked at two types of problems that can be solved using a cumulative percentage graph of the standard normal distribution. In each case, the solution involves two steps: using the graph and using the formula for converting scores from raw scores to z-scores and back again. The difference between the two types of problems is as follows. In the first type of problem, we are given a raw score and we determine the proportion of scores falling below this raw score. That is, we go from score to proportion. In this type of problem, we first transform the raw score to a z-score, then use the graph to determine the desired proportion. In the second problem, we are given a proportion and we determine the score below which this proportion of all the scores falls. In this second type of problem, we go from proportion to score, and the sequence of steps is reversed. First, we use the graph to find the z-score that marks off the desired proportion. Then, we use the formula to transform this z-score to its raw score equivalent. Questions involving the use of the standard normal distribution can be stated in various ways, but in the final analysis they invariably reduce to one of these two procedures. As we move on to consider the use of the normal curve model in hypothesis testing, it will become clear that these procedures are extremely important.

Using a Table of the Standard Normal Distribution

The cumulative percentage graph of the standard normal distribution is a convenient way to find the percentile rank of a score in a normal distribution. There is one drawback to using the graph, however. There is a limit to the accuracy with which percentiles and scores can be read from such a graph. If we are careful, we can read percentiles to the nearest percentage point, and we can read *z*-scores to perhaps the nearest tenth of a point. As you will see, there may be occasions when it is essential that we be able to obtain percentiles and scores with a greater degree of accuracy than is possible with a cumulative percentage graph. For this purpose, we typically employ a table of the standard normal distribution such as that found in Appendix C (Table C.2). This table allows you to read *z*-scores directly to the nearest hundredth of a percentage point, and it provides proportions to the nearest hundredth of a point. Find Table C.2 now. You will note that it runs over several pages. The table is interpreted as follows. In Column 1, we find all the positive *z*-scores from 0.00 to +3.00. In Column 2, we find the proportion of all the scores in the standard normal distribution that fall *below* a particular positive *z*-score. This is indicated by the diagram at the top of Column 2, in which all the area under the normal curve from the extreme left to a particular positive *z*-score is shaded in. Locate the *z*-score of +1.00 in Column 1. You will find that the number in Column 2 immediately adjacent to the *z*-score of +1.00 is .8413. This tells us that the percentage of the total area under the standard normal curve that lies to the left of $z = +1.00$ is 84.13 percent. This means that 84.13 percent of all the scores in a normal distribution fall below the standard score of +1.00. Thus, when you need to find the percentile rank of a particular positive *z*-score, you can obtain this information by looking directly in Column 2 of the table.

What if you need to find the percentile rank of a particular *negative z*-score? The negative *z*-scores are not included in the table. The reason negative *z*-scores are not listed is to save space. Because the normal curve is symmetrical, we can economize on space without giving up any information. The symmetry of the normal distribution means that the percentage of all the scores in a normal distribution that lie *below* a given negative *z*-score will always be the same as the percentage of all the scores that lie *above* the corresponding positive *z*-score. For example, the percentage of scores that lie below the *z*-score of −1.00 is the same as the percentage of scores that lie above the *z*-score of +1.00. If this does not make sense to you immediately, go back and look at Figure 7.4. From the areas given in this figure, you could determine the proportion of scores lying below $z = -1.00$ by adding the 0.13 percent that lie below $z = -3.00$, the 2.15 percent that lie between $z = -3.00$ and $z = -2.00$, and the 13.5 percent that lie between $z = -2.00$ and $z = -1.00$. The result of this addition is 15.87 percent. But this is exactly the same result that we would obtain if we added the 13.59 percent that lie between $z = +1.00$ and $z = +2.00$, the 2.15 percent that lie between $z = +2.00$ and $z = +3.00$, and the 0.13 percent that lie above $z = +3.00$. Because the proportion of scores in a normal distribution that lie below a particular negative *z*-score is always equal to the proportion of scores that lie above the corresponding positive *z*-score, the table includes a third column. Column 3 gives the proportion of scores lying *above* each of the positive *z*-scores listed. This is indicated by the diagram at the top of Column 3, in which all the area under the normal curve from a positive *z*-score to the right has been shaded in. If you need to obtain the percentile rank of a particular negative *z*-score,

280

therefore, you may find it in Column 3. For example, to find the percentile rank of $z = -1.00$, find $z = +1.00$ in Column 1 of the table, then read the entry in Column 3 that is adjacent to the score of $+1.00$. If you do this, you will find that the entry in Column 3 is .1587. This indicates that 15.87 percent of all the scores in a normal distribution lie *above* $z = +1.00$, which is the same thing as saying that 15.87 percent of all the scores lie *below* $z = -1.00$. Therefore, the percentile rank of $z = -1.00$ is 15.87.

The table also contains a fourth column. Column 4 tells the proportion of all the scores in the standard normal distribution that lie between the mean, $z = 0$, and a given positive z-score. This is indicated by the diagram at the top of Column 4, in which the area from the mean of the distribution to a particular positive z-score has been shaded in. This column is particularly useful when you need to determine the proportion of scores in a normal distribution that lie between two particular z-scores, one of which is negative and one of which is positive. For example, suppose you wished to know the proportion of scores falling between $z = -1.96$ and $z = +1.64$. You could find the proportion of scores lying between $z = -1.96$ and $z = 0.00$ by locating the z-score of $+1.96$ in Column 1 and looking at the entry in Column 4 adjacent to this score. This is because the proportion of scores between $z = -1.96$ and $z = 0.00$ is identical to the proportion of scores between $z = 0.00$ and $z = 1.96$ in the symmetrical normal distribution. This proportion is .4750 or 47.5 percent. Similarly, you can find the proportion of scores lying between $z = 0.00$ and $z = +1.64$ by locating the z-score $+1.64$ in Column 1 and looking at the entry in Column 4 adjacent to this score. You will find that this proportion is .4495, or 44.95 percent. You can then add the 47.5 percent of all the scores that fall between $z = -1.96$ and $z = 0.00$ to the 44.95 percent of all the scores that fall between $z = 0.00$ and $z = +1.64$. The result, 92.45 percent, is the proportion of all scores in a normal distribution that lie between $z = -1.96$ and $z = +1.64$.

The table may also be used to find the z-score that marks off a certain proportion of scores in a standard normal distribution. If we wished to find the z-score below which 90 percent of all the scores occurred, we would look in Column 2 of the table for the value .90. Having located .90, we would look adjacent to this number in Column 1 to find the value of the z-score below which 90 percent of all scores fell. Look at the table of the standard normal distribution and find this z-score. You should find that the z-score below which approximately 90 percent of all the scores fall is $+1.28$. Whenever we are dealing with an area to score problem, we will locate the area in the table using Column 2, 3, or 4 and then read the z-score from Column 1.

You should now take some time to practice reading the table of the standard normal distribution. Keep in mind that we use the table in place of the cumulative percentage graph, either to find the proportion of scores marked off by a particular z-score value or values or to find the z-score value or values that mark off a particular portion of the scores in the standard normal distribution. If you are dealing with a problem based on a normal distribution that has not

281

been standardized, it is still necessary to use the formula to transform a raw score into a z-score or to transform a z-score into a raw score. The procedures to be followed in these cases are exactly as outlined earlier. We simply substitute the table for the cumulative percentage graph when it is more convenient or when greater accuracy is required. In using the table to do problems, you will find it helpful to make a sketch of the distribution in which you label the areas and scores you are working with.

SAMPLING THEORY

Although it would be useful to us to use the standard normal distribution solely as a means of determining the percentile rank of a particular score in a normally distributed population of known mean and standard deviation, we use the standard normal distribution in other important ways as well. One extremely important use of the standard normal distribution has to do with the situation in which we are not certain of what the mean of a particular population might be. We might be using the mean of a sample as an estimate of the mean of the population from which that sample was drawn, or we might be using a sample to test the hypothesis that the mean of the population is a particular value.

In these problems of inferential statistics, the key concept is the use of a *sample* to draw an inference about a *population* parameter. In this sense, the use of the normal curve model does not differ from the use of the binomial model, considered in the previous chapter.

The difference lies in the nature of the population being sampled and the nature of the population parameter of interest. In the case of an inferential problem involving the binomial model, the trials of the experiment constitute a random sample from a dichotomous population. The population parameter of interest is P, the proportion of the population falling into one of the two categories on the dichotomous variable. In the case of an inferential problem involving the normal curve model, we draw a random sample of n observations from a population of N measures on a continuous interval variable, and the population parameter of interest is μ, the mean of the population. In this section of the chapter, we continue to lay the foundation for inferential statistics by discussing the essential elements of sampling from a continuously distributed population. In the next section, the material we have already covered on the normal distribution and this new material on sampling from a continuously distributed population will be joined together as we learn how to test hypotheses regarding population means and how to construct interval estimates for population means based on sample data.

A Sampling Problem

Suppose you are the director of a large city's compensatory education program, and you need to have an idea of the average IQ level among all the students in the city who are enrolled in remedial reading classes. Because there would be many students in such classes, you would not want to give each one an IQ test. You would almost certainly select a random sample of students from the population involved in remedial reading. You would test the sample, determine the average IQ for this one sample, and use this average as an estimate of the average IQ of the whole population. That is, you would use a sample mean, $\bar{X}$, as an estimate of a population mean, μ. In using the sample mean this way, you have implicitly assumed that your *sample statistic,* the sample mean, will be reasonably close to the *population parameter* you are really concerned with, the population mean. However, you are aware that the sample mean will not necessarily be *identical* to the population mean. Your random sample may, by chance, have contained proportionately more bright students or proportionately more dull students than the population as a whole. As a result, your sample mean may be somewhat higher or somewhat lower than the population mean. This type of discrepancy is known as *sampling error.*

Why is it that we naturally assume that the mean of a sample will approximate the mean of the population from which the sample is drawn? Given the existence of sampling error, how far off from the true population mean is our sample mean likely to be? To answer these important questions, we need to give some thought to what happens when we select a sample from a population and compute the mean of that sample.

Random Sampling and the Law of Large Numbers

Consider the selection of a random sample of n scores from a large population of N scores on a continuous variable. (Note that we use a lowercase n to indicate sample size and a capital N to indicate the number of scores in our population.) When we select a random sample from a population, we select scores in such a way that any one of the scores in the population sampled has the same chance of being selected as any other score in the population. If a large proportion of the scores in the population occur near a certain value, and if all scores in the population are equally likely to be included in the sample, regardless of their value, then the chances are good that the sample will include some scores that are near this value. Thus, if we are sampling from a normal distribution, where most of the scores lie within a few standard deviations of the mean, then we would expect to find that most of the scores included in the sample would lie within this same range of score values. In fact, the process of random sampling tends to produce samples that are similar to the populations from which they are drawn. They tend to be similar with respect to mean, standard deviation, and shape. Moreover, the larger the sample is, the more similar to the population

283

the sample will tend to be. This is known as the *law of large numbers.* It is perhaps best illustrated by the most extreme case in which the sample drawn from the population is so large that it includes *all* the scores in the population. In this case, the mean, standard deviation, and shape of the sample would be identical to those of the population. Of course, in real sampling problems, we typically do not sample an entire population. However, the law of large numbers tells us that we are more likely to obtain a sample mean close to the mean of the population of interest as we include more observations in our sample. This is the principal reason why you so often hear statisticians asking, "How large was the sample?" On the other hand, you should not get the impression from this that it is always necessary to include hundreds of cases in a sample. Often we can obtain a reasonable estimate of the mean of a population with a sample of size 25 or 30. To see why this is so, we need to take a somewhat different perspective on the sampling process. In real statistical problems, we typically employ a *single sample* in order to estimate a population parameter. In trying to understand sampling theory, however, we need to think in terms of *all* the samples that we might have drawn. If you will, imagine a hypothetical situation in which we have an infinitely large population from which we may draw an infinitely large number of random samples.

Populations, Samples, and the Distribution of Sample Means

Let us first consider an infinitely large, normally distributed population of IQ scores with $\mu = 100$ and $\sigma = 15$. Suppose there are an infinite number of scores in this population. From this population, let us select a random sample of 25 scores. Call this Sample 1. The use of the term *random sample* implies that any one of the infinite number of scores in the population has the same chance of being included in this sample as any other score in the population. Thus, the chances are good that some of the 25 scores included in this sample will lie above the mean of the population, and some will be below. Also, because the population is normally distributed, most of the scores in the population lie relatively close to the mean. Therefore, the chances are also good that many of the scores included in the sample will lie close to the mean of the population. However, because the population does contain a small proportion of scores that lie considerably below or considerably above the mean, there is a good chance that at least a few of the 25 scores in the sample will be rather far away from the population mean. In Figure 7.6, we have represented pictorially the population of IQ scores, and we have shown the 25 scores that were selected by chance from this population to be included in Sample 1. As expected, Sample 1 contains primarily scores that are relatively close to the population mean, $\mu = 100$. Looking at the sample, you can see the scores 86, 89, 103, 100, 95, and so on. However, the sample also contains a few more extreme scores like 75, 78, and 129. What effect will these extreme scores have on the mean of the first sample, $\bar{X}_1$? Computing the sample mean, we find that $\bar{X}_1 = 102.72$. The sample mean

284

Figure 7.6
A Graphic Explanation of the Distribution of Sample Means

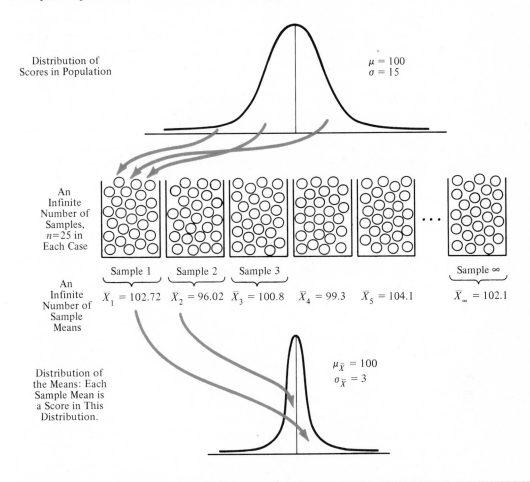

Distribution of
Scores in Population

$\mu = 100$
$\sigma = 15$

An
Infinite
Number of
Samples,
$n=25$ in
Each Case

Sample 1   Sample 2   Sample 3   Sample ∞

An
Infinite
Number of
Sample
Means

$\bar{X}_1 = 102.72$ $\bar{X}_2 = 96.02$ $\bar{X}_3 = 100.8$ $\bar{X}_4 = 99.3$ $\bar{X}_5 = 104.1$ $\bar{X}_\infty = 102.1$

Distribution of
the Means: Each
Sample Mean is
a Score in This
Distribution.

$\mu_{\bar{X}} = 100$
$\sigma_{\bar{X}} = 3$

is just slightly above the mean of the population. Although there were some rather extreme scores in the sample, the high scores and the low scores balanced each other off when we computed the sample mean, with the result that the sample mean was rather close to the population mean. In this case, the balance happened to be somewhat toward the higher scores, and the sample mean $\bar{X}_1$ was a bit greater than $\mu = 100$.

Now, imagine that we go back to the infinitely large population of IQ scores, and that we take a second random sample of 25 scores. Here again, each of the infinite number of scores in the population has an equal chance of being selected.

285

The Normal Distribution and Sampling Theory

Once again, we find that most of the scores selected for the sample lie relatively close to the mean, but that a few more extreme scores are included as well. Once again, a balancing off process occurs. This time, however, the balance of the scores in the sample happens to favor the lower scores. The mean of the second sample, $\bar{X}_2$, turns out to be 96.02.

Now, imagine further that we repeat this process of taking samples an infinite number of times. In each instance, we take a random sample of size 25 from the infinitely large, normally distributed population of IQ scores. Each time we take a sample, we compute the sample mean, $\bar{X}$. The result of this hypothetical procedure would be an infinitely large number of sample means, $\bar{X}_1, \bar{X}_2, \ldots, \bar{X}_\infty$. Taken together, these infinitely many sample means may be thought of as forming a distribution of their own. This hypothetical distribution has been illustrated in Figure 7.6 as well. In this distribution, each of the sample means is a score. It is called, naturally enough, the *distribution of sample means.* This distribution of sample means is most important to us, because it has the following important properties.

The mean of the distribution of sample means. The distribution of sample means has a mean of its own, equal to the mean of the population from which the samples were drawn. If we were to take all of the infinitely many sample means that are the scores in the hypothetical distribution of sample means and if we were to average all these means together, we would find that their average would be exactly equal to the mean of the population from which the samples came. Typically, we denote the mean of the distribution of sample means with the symbol $\mu_{\bar{X}}$, so that the subscript $\bar{X}$ reminds us that we are referring to the "mean of the sample means." Thus we may write:

$$\mu_{\bar{X}} = \mu.$$

If you give it some thought, this statement should make sense to you. Most of the sample means will be close to μ; some will not. Each sample mean has as good a chance of being greater than the population mean as it does of being less than the population mean. Of the sample means actually shown in Figure 7.6, some are greater than $\mu = 100$ and some are less than $\mu = 100$. On balance, those sample means that are greater than μ balance out those that are less. When we consider a distribution of infinitely many sample means, the balancing process is perfect, so $\mu_{\bar{X}} = \mu$. This property of the distribution of sample means makes any single sample mean $\bar{X}$ an *unbiased estimator* of the population mean, μ. Although not every sample mean equals μ, the average of all the sample means does. If we had to guess the value of the mean of a single sample drawn from a population with mean μ, our best guess for this one sample mean would be μ as well. This is an essential property of the distribution of sample means that makes it reasonable for us to use a particular sample mean to estimate a population mean.

286

Of course, the question that immediately comes to mind is "How close to μ will any *particular* $\bar{X}$ be?" You can see from Figure 7.6 that one of the sample means actually listed is 104.1. This is 4.1 points away from μ. If all of the infinitely many sample means could be presented, we would find sample means even further from μ than 4.1 points. In fact, when we take a single sample and compute $\bar{X}$, we can never be sure of exactly how far away from μ this sample mean will lie. This is a matter of chance. However, we can say in this example that it is more likely that a particular sample mean will lie less than 4.1 points from $\mu = 100$ than it is that the sample mean will lie more than 4.1 points from μ. The reason that we can make this general statement is that we not only know something about the relationship between the mean of the population and the mean of the distribution of sample means; we also know something about the relationship between the variability of scores in the population and variability of scores in the distribution of sample means.

The standard deviation of the distribution of sample means. The hypothetical distribution of sample means we have been considering not only has a mean of its own, $\mu_{\bar{X}}$, it also has a standard deviation of its own, $\sigma_{\bar{X}}$. The standard deviation may be thought of as the amount by which a typical sample mean in the distribution of sample means differs from $\mu_{\bar{X}}$ (and therefore from μ, because $\mu_{\bar{X}} = \mu$). Because real problems of statistical inference involve using one of these infinitely many sample means to estimate the population mean, it is important to us to know the amount by which sample means tend to differ from the mean of the population. Obviously, the less the sample means differ from μ, the better. In this regard, statisticians have determined that for samples containing more than one observation, the standard deviation of the distribution of sample means will always be *smaller* than the standard deviation of the original population.

To understand why this is so, let us consider once again what happens when a sample of size 25 is drawn at random. Each time we randomly select a score to be included in our sample, every score in the population has an equal chance of being selected. If the population from which we sample is normally distributed, then each time we select a score, the chance is good that the score will lie within two standard deviations of the mean. Specifically, we know that in a normal distribution only 4.56 percent of all the scores lie more than two standard deviations from μ, 2.28 percent on the low side and 2.28 percent on the high side. Thus, if we randomly select a single score from our normal distribution with $\mu = 100$ and $\sigma = 15$, there is a 4.56 percent chance that *this one score* would be less than 70 or more than 130 ($\mu \pm 2\sigma$). However, we are not really concerned with single scores here, but rather with the means of samples consisting of 25 scores each.

What do you suppose the chance would be that the sample mean, $\bar{X}$, for a given sample of $n = 25$ would be less than 70 or greater than 130? Do you think that this chance would be as great as 4.56 percent? If you cannot answer these questions immediately, think about what a sample of 25 scores would have to be

The Normal Distribution and Sampling Theory

like in order for the mean of the sample to be 130. Either all 25 scores in the sample would have to be right around 130, or else any scores lower than 130 would have to be balanced off by scores even further away from $\mu = 100$ than 130. If we are selecting scores at random from a population whose mean is 100, is it likely that most or all of the scores in the sample would be in the 130 range? Hardly. There might be a few scores in the sample as high as 130, but these scores would almost certainly be balanced off by other scores, perhaps by an equally extreme score on the opposite side of the mean, or by a number of scores closer to the mean, $\mu = 100$. Thus, the mean of a sample of 25 scores is much more likely to be close to the mean of the population than is a single score selected at random from the population. Therefore, the sample means that make up the distribution of sample means will tend to be closer to $\mu_{\bar{X}}$ than the original scores in the population.

This means the standard deviation of the distribution of sample means for samples where $n = 25$ is smaller than the standard deviation of the original population. This is reflected in Figure 7.6, where we note that the hypothetical distribution of sample means appears considerably more compact than the original distribution. Actually, for any sample size greater than $n = 1$, the standard deviation of the distribution of sample means will be smaller than the standard deviation of the population. In addition, the larger the size of the sample, the smaller the standard deviation of the distribution of sample means will be.

Specifically, statisticians have determined that the standard deviation of the distribution of sample means is equal to the standard deviation of the population, divided by the square root of the sample size. That is:

$$\sigma_{\bar{X}} = \frac{\sigma}{\sqrt{n}},$$

where $\sigma_{\bar{X}}$ is the standard deviation of the distribution of sample means, σ is the standard deviation of the original population from which the samples are drawn, and n is the size of each sample. Thus, in our example where $\sigma = 15$ and $n = 25$,

$$\sigma_{\bar{X}} = \frac{15}{\sqrt{25}} = \frac{15}{5} = 3.$$

That is, whereas the standard deviation of the individual scores in the population was 15 points, the standard deviation of the means in this distribution of sample means is only 3 points. The sample means are much more tightly clustered around μ than were the original scores.

Clearly, the researcher who wished to minimize the chance that the mean of the sample would differ very much from the mean of the population would seek a large sample size. In Figure 7.7, we have illustrated the distribution of sample means for several different sample sizes. In each case, the population being

288

Figure 7.7
Three Distributions of Sample Means:
$n = 1$, $n = 25$, and $n = 100$

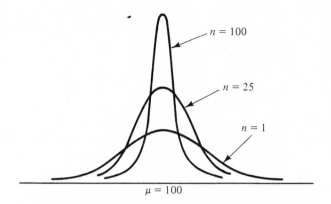

sampled is the same normally distributed population with $\mu = 100$ and $\sigma = 15$. In the first case, we consider the hypothetical situation in which samples of size 1 are selected. Taking a sample of size 1 is the same as selecting an individual score at random from the distribution. According to our equation, the standard deviation of the distribution of sample means in this case will be

$$\sigma_{\bar{X}} = \frac{\sigma}{\sqrt{n}} = \frac{15}{\sqrt{1}} = \frac{15}{1} = 15.$$

That is, when we pick single scores at random, the distribution of sample means is identical to the original population. When the sample size is increased to 25, as we originally considered, the standard deviation of the sampling distribution drops from 15 to 3. Finally, if we were to imagine taking an infinite number of samples with 100 cases in each one, the standard deviation of the distribution of these sample means would be

$$\sigma_{\bar{X}} = \frac{\sigma}{\sqrt{n}} = \frac{15}{\sqrt{100}} = \frac{15}{10} = 1.5.$$

At this point, it should be clear why researchers worry about obtaining an adequate sample size.

The shape of the distribution of sample means. Not only do we know the mean and the standard deviation of a theoretical distribution of sample means,

289

we also have knowledge regarding the shape of the distribution. This knowledge is contained in an important theorem known as the *Central Limit Theorem,* which states: (a) If the population from which samples are drawn is normally distributed, then the distribution of sample means will be normally distributed as well; and (b) even if the population from which samples are drawn is not normally distributed, the distribution of sample means will tend to be normally distributed, as the size of the sample increases. With regard to the second rule, you may wonder just how large the sample size must be in order to obtain a nearly normal distribution of sample means from a non-normal population. This depends on how non-normal the population is. If the population from which the samples are drawn is unimodal and close to being symmetrical, the sample size need not be very large. If the population is bimodal or drastically skewed, a somewhat larger sample size may be required in order that the distribution of sample means be approximately normal. Nearly always, samples of size 30 are large enough that the distribution of sample means is approximately normal. This property of the distribution of sample means is extremely important to us, because it enables us to use a table of the standard normal distribution to determine the percentile rank of a particular sample mean within the hypothetical distribution of sample means. We will see how this allows us to make statements regarding the probability of occurrence of a particular sample mean under various assumptions regarding the mean of the population. This is the basis for the use of the normal curve model in testing hypotheses regarding a population mean.

Probability Statements Regarding Sample Means

Suppose we are concerned with random samples of size 25 drawn from a normally distributed population in which $\mu = 500$ and $\sigma = 100$. Using the ideas presented in the previous section, we know that the distribution of sample means will:

1 have a mean of $\mu_{\bar{X}} = 500$, because $\mu_{\bar{X}} = \mu$;

2 have a standard deviation of $\sigma_{\bar{X}} = 20$, because $\sigma_{\bar{X}} = \dfrac{\sigma}{\sqrt{n}} = \dfrac{100}{\sqrt{25}} = 20$; and

3 be normally distributed.

On the basis of this information, we can use our knowledge of the standard normal distribution to determine what proportion of the sample means in the distribution lie between certain values.

For example, we might wish to know what proportion of the sample means in the distribution are greater than 540. To answer this question, we would proceed in exactly the same manner as we would answer any other normal curve problem. The fact that we are now concerned with a particular type of normal distribution that happens to be the theoretical distribution of sample means has no effect on the steps in the solution. Because this is a score to proportion problem, we first find the z-score equivalent of the raw score of 540. Because we are talking about the distribution of sample means, a score in this case is really a sample mean, the mean of the distribution is the mean of the distribution of sample means, and the standard deviation of the distribution is the standard deviation of the distribution of sample means. Thus, the old formula for finding the z-score equivalent of a raw score,

$$z = \frac{X - \mu}{\sigma},$$

would be written in a somewhat different form:

$$z = \frac{\bar{X} - \mu}{\sigma_{\bar{X}}} = \frac{\bar{X} - \mu}{\sigma/\sqrt{n}} = \frac{540 - 500}{20} = +2.00.$$

That is, in the sampling situation envisioned here, a sample mean of 540 would be located 2 standard deviations above the mean of the distribution of sample means (and therefore 2 standard deviations above the mean of the population, because $\mu_{\bar{X}} = \mu$). This would make the sample mean 540 rather atypical. Referring to the table of the standard normal distribution (Table C.2), we see that only 2.28 percent of all the scores in a normal distribution lie above the value $z = +2.00$. In other words, we happened to select a sample from this population such that only 2.28 percent of all possible samples of size 25 would have means that were higher.

At this point you may be wondering why this is important. What good does it do us to know whether a particular sample mean is relatively typical or relatively rare? If we know the mean of the population, isn't all of this information just academic? The answer to this last question is an emphatic "Yes"! If we know the mean of the population to begin with, as we have been assuming up to now, then this information *is* just academic. But if we really knew the mean of the population to begin with, we would probably not be taking a sample in the first place. It is when we do not know the mean of a population that we take a sample to try to learn about that mean. And when we do not know for sure the mean of the population being sampled, the theorems of sampling theory are very useful indeed. We use these concepts of sampling theory when we test a hypothesis regarding a population mean or when we attempt to estimate a population mean based on a sample mean. At this point, we proceed to consider the use of the normal curve model and sampling theory in inferential statistics. First, we will consider the normal curve in hypothesis testing. Then, we will discuss the idea of interval estimation, again using the normal curve.

The Normal Distribution and Sampling Theory

THE NORMAL CURVE MODEL
IN HYPOTHESIS TESTING

The Normal Curve Model and the Binomial Model Compared

You will recall from Chapter 6 that in hypothesis testing, we follow a specific sequence of logical steps. (1) We adopt an initial assumption regarding the population parameter of interest. This initial assumption is the null hypothesis. (2) We employ an appropriate mathematical model to determine the set of sample results that we would consider relatively likely if the null hypothesis were true. We also use the model to determine a set of results that would be so unlikely if H_o were true that obtaining one of these results in our sample would cause us to reject the notion that H_o was true. This set of results is the rejection region. (3) We examine our observed sample result to see if it falls in the rejection region or not. (4) We draw the appropriate conclusion regarding the truth of the null hypothesis.

In Chapter 6, we illustrated this logic by referring to the binomial model. In the case of the binomial model, the population parameter of interest is P, the proportion of successes; and the sample result is r, the number of successes observed in n binomial trials. In this chapter, we have described a new mathematical model, the normal curve model. In testing hypotheses using the normal curve model, we follow exactly the same sequence of logical steps. However, the normal curve model is used to test hypotheses regarding a different population parameter. Here the population parameter of interest is μ, the population mean; and the sample result is $\bar{X}_{ob}$, the observed mean of a random sample drawn from that population.

We have already noted that the variable of interest in the binomial situation is different from the variable of interest in the normal curve situation. The binomial model applies to categorical variables that have two values, which we arbitrarily denote as success and failure. The normal curve model, on the other hand, applies to interval scale variables. This should be obvious, for the concept of the mean makes sense only when we are dealing with numerical scores. Furthermore, in order to use the normal curve model to test a hypothesis regarding the mean of a population, we not only need to be dealing with an interval scale variable, we also need to know, in advance, the population standard deviation. This is because we use the population standard deviation to compute the standard deviation of the distribution of sample means, which, in turn, we use to generate a probability statement regarding the likelihood of obtaining our observed sample mean, $\bar{X}_{ob}$, under a particular null hypothesis regarding μ. In the following chapter, we will give some attention to the question of *when* we can use the normal curve model and when we must employ an alternative technique. For now, we concentrate on *how* to apply the normal curve model in an appropriate instance in order to test a hypothesis regarding a mean.

The Normal Curve Model in Hypothesis Testing:
A One-Sample z-test for the Mean of a Population

Because the normal curve model is concerned with the mean of an interval scale variable, the null hypothesis of a normal curve test will be stated in terms of a mean or means. The simplest test using the normal curve model is the *one-sample z-test,* which is used to test a hypothesis regarding a population mean on the basis of the mean of a random sample drawn from that population. An example of a null hypothesis for this one-sample z-test would be $H_o: \mu = 500$. The alternative hypothesis in such a test may be nondirectional or directional. In a nondirectional alternative, we would be attempting to determine if the true population mean was most probably different from the value stated in H_o; however we would not be concerned whether the true mean was larger or smaller than the value specified in H_o. In this case, the alternative hypothesis would be written $H_a: \mu \neq 500$. In a directional alternative, we would either seek to show that the true population mean was greater than the value stated in H_o, or we would seek to show that the true population mean was less than the value stated in H_o. In these cases, the alternative hypotheses would be $H_a: \mu > 500$ or $H_a: \mu < 500$, respectively.

Regardless of the alternative hypothesis, in carrying out the test, we would employ a mathematical model built upon the assumption of the null hypothesis and our knowledge of sampling theory. We would use this knowledge to determine what the distribution of sample means would be like, *if H_o were true.* Because the distribution of sample means would be normal or approximately normal, we could refer to the table of the standard normal distribution to determine what sample means would be likely under H_o and what sample means would be unlikely. Then, depending on our alternative hypothesis, we could specify a set of values for $\bar{X}_{ob}$ that would cause us to reject H_o.

A Research Example.   As an illustration of a normal curve test for a population mean, consider the case of Dr. John Baldwin, the academic dean of Ivy College. Dr. Baldwin would like to know whether the average SAT score for students in Ivy College is different from the average SAT score for students in other colleges throughout the state. Dr. Baldwin has read research reports indicating that the average SAT score for the state is 500. He therefore wishes to test the hypothesis $H_o: \mu = 500$ against the alternative $H_a: \mu \neq 500$. He knows that SAT scores in general have a standard deviation of 100 points and is willing to assume that this standard deviation will be the same in his college. He decides to test the hypothesis by selecting from student records a random sample of 25 SAT scores. Temporarily assuming that H_o is true, the dean uses his knowledge of sampling theory to describe the hypothetical distribution of sample means. He knows that if $\mu = 500$, $\mu_{\bar{X}} = 500$ as well. He knows that if $\sigma = 100$ and $n = 25$, then $\sigma_{\bar{X}} = 100/\sqrt{25} = 20$. Thus, if H_o is true, the hypothetical distribution of sample

The Normal Distribution and Sampling Theory

means will have a mean of 500 and a standard deviation of 20. Furthermore, according to the Central Limit Theorem, this distribution will be either normal or close to normal. In Figure 7.8, we have illustrated the distribution of sample means that applies to this situation.

The Rejection Rule. The dean knows that if H_o is true, then the mean of the one sample of size 25 that he will actually obtain in his experiment will be one of the sample means that constitute the scores in this hypothetical distribution. Because the mean of this hypothetical distribution is $\mu_{\bar{X}} = 500$, the mean of his one sample should estimate this value as well. Because of sampling error, the dean knows that even if H_o were true, his sample mean would not necessarily equal exactly 500. However, because the distribution of sample means is normal or approximately normal, the observed sample mean would most likely not differ from $\mu = 500$ by more than a couple of standard deviations. Specifically, if H_o is really true, there is only a 5 percent chance that $\bar{X}_{ob}$ would differ from 500 by more than 1.96 standard deviations in either direction. If this is not clear to you now, go to the table of the standard normal distribution (Table C.2) and look up the proportion of scores in a standard normal distribution that lie below $z = -1.96$ or above $z = +1.96$. You will find that the area in these two "tails" of the standard normal distribution totals 5 percent, 2 1/2 percent on either side.

On the basis of this knowledge, Dean Baldwin decides that if his $\bar{X}_{ob}$ is within 1.96 $\sigma_{\bar{X}}$ of the value of 500, he will accept H_o; but if his $\bar{X}_{ob}$ is farther from 500 than 1.96 $\sigma_{\bar{X}}$, he will reject H_o. In Figure 7.8, we have shaded the area under the normal curve lying farther away from 500 than 1.96 $\sigma_{\bar{X}}$ in either direction. We label this area the "Rejection Region," indicating that any sample mean falling in this area will lead to rejection. In selecting the value 1.96 standard deviations, the dean has adopted the $\alpha = .05$ level of significance for his test. He does so primarily because of tradition. He knows that researchers in the social sciences have traditionally recognized new research findings only when the chance of a Type I error associated with these findings is less than 5 percent. You will recall from the previous chapter that the chance of a Type I error is the chance of rejecting H_o when it is, in fact, true. In this test, we know that if H_o were true, our researcher would still have a 5 percent chance of obtaining an $\bar{X}_{ob}$ farther from $\mu_{\bar{X}} = 500$ than 1.96 $\sigma_{\bar{X}}$, which would lead to rejection. Thus, the chance of a Type I error here is 5 percent. The rejection rule is based on the idea that if the observed sample mean does fall in the rejection area, it is far more likely that H_o is false than it is that H_o is true and an atypical sample mean has been obtained.

In conducting the actual test, the dean needs to determine whether his observed sample mean lies more than 1.96 $\sigma_{\bar{X}}$ away from $\mu = 500$. To do this, he simply transforms his observed sample mean into a z-score, using the formula presented earlier. Because this z-score represents the position of the observed sample mean in the distribution of sample means, we typically refer to this z-score as z-observed (z_{ob}). Furthermore, because the mean of the hy-

FIGURE 7.8

Distribution of Sample Means and Rejection Region in Normal Curve Test of $H_o : \mu = 500$ at $\alpha = .05$, where $\sigma = 100$, $n = 25$

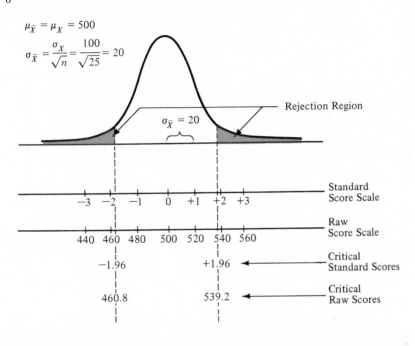

pothetical distribution of sample means is also the hypothesized value of the population mean under H_o, we generally refer to this mean as μ_o to reinforce the idea that this mean is really the hypothesis that we are testing. Thus, the z-score formula is written:

$$ z_{ob} = \frac{\overline{X}_{ob} - \mu_o}{\sigma_{\overline{X}}} = \frac{\overline{X}_{ob} - \mu_o}{\sigma / \sqrt{n}}. $$

This formula is referred to as the z-statistic. By inserting values for the observed sample mean, $\overline{X}_{ob}$, the hypothesized value of the mean of the population, μ_o, the known value of population standard deviation, σ, and the known sample size, n, the z-statistic tells us *how many* standard deviations away from the hypothesized mean our sample mean lies. If z_{ob} falls within the critical region, we reject H_o. In this test, we will reject H_o if z_{ob} is larger than $+1.96$ (which would be the case if $\overline{X}_{ob}$ is more than $1.96 \, \sigma_{\overline{X}}$ above $\mu_o = 500$), or if z_{ob} is less than

295

The Normal Distribution and Sampling Theory

--1.96 (which would be the case if $\bar{X}_{ob}$ is more than 1.96 $\sigma_{\bar{X}}$ below $\mu_o = 500$). In our example, let us assume that Dean Baldwin finds that the mean of his sample of 25 SAT scores is 550. Then

$$z_{ob} = \frac{\bar{X}_{ob} - \mu_o}{\sigma/\sqrt{n}} = \frac{550 - 500}{\dfrac{100}{\sqrt{25}}} = \frac{50}{20} = +2.5.$$

Because +2.5 is larger than +1.96, the dean can reject H_o and conclude that the average SAT score in his college is greater than he had originally hypothesized. In Figure 7.9, we have presented this complete one-sample z-test, according to the seven-step outline of a hypothesis test provided in the previous chapter. At this point, you might find it helpful to compare the seven steps in this one-sample z-test to the seven steps in the binomial test (Table 6.5). Note that although the mathematical models and the parameters of interest differ in the two tests, the logical sequence of steps is identical. In each case, we make an initial statement about a population parameter, the null hypothesis, which we use to generate an expectation regarding likely results in the sample. We then compare our expectation to actual sample results to help us decide if our initial statement is plausible.

The Rejection Rule Stated in Raw Score Terms.　You may already have recognized that the rejection region for the test could be stated in terms of raw scores rather than z-scores. We know that the standard deviation of the distribution of sample means is 20 points. Therefore, the value of the sample mean lying 1.96 standard deviations above $\mu_o = 500$ is $500 + 20(1.96) = 539.2$. Similarly, the value of the sample mean lying 1.96 standard deviations below $\mu_o = 500$ is $500 - 20(1.96) = 460.8$. Thus, the researcher could have stated the rejection rule as follows: reject H_o if $\bar{X}_{ob}$ is greater than 539.2 or less than 460.8. Theoretically, it makes no difference whether we choose to state the rejection rule in terms of z-scores or raw scores. The use of the z-statistic as just presented is traditional. To reinforce the equivalence of the two approaches, we have indicated both standard scores and raw scores in Figure 7.8. You will note that $z = +1.96$ corresponds exactly to $\bar{X}_{ob} = 539.2$, and $z = -1.96$ corresponds exactly to $\bar{X}_{ob} = 460.8$. Had we stated the rejection rule in terms of actual scores, we would simply compare the observed sample mean, $\bar{X}_{ob} = 550$, to the values 539.2 and 460.8. Because $550 > 539.2$, we would again reject H_o.

The Power of the z-test for a Population Mean.　The ability to state the rejection rule as a sample mean rather than as a z-score is helpful to us if we need to determine the power of a particular z-test against a certain alternative. For example, let us suppose that before actually selecting his sample and computing its mean, Dean Baldwin had wondered whether the test he planned would be sufficiently powerful. That is, assuming that the null hypothesis is false and that the

FIGURE 7.9
Illustration of One-Sample z-test for Dr. Baldwin's SAT Scores

Step 1

Because Dean Baldwin is interested in testing a hypothesis regarding the mean of an interval scale variable in a single population and because he knows the standard deviation of the population, he chooses to use the one-sample z-test.

Step 2

$H_o : \mu = 500$.

Step 3

$H_a : \mu \neq 500$.

Step 4

The test statistic is

$$z_{ob} = \frac{\overline{X}_{ob} - \mu_o}{\sigma / \sqrt{n}},$$

which will have the standard normal distribution if H_o is true.

Step 5

$\alpha = .05$: Reject H_o if $z_{ob} < -1.96$ or $z_{ob} > +1.96$.

Step 6

$\overline{X}_{ob} = 550$, $\sigma = 100$, $n = 25$, so:

$$z_{ob} = \frac{550 - 500}{\dfrac{100}{\sqrt{25}}} = +2.5.$$

Because $+2.5 > +1.96$, z_{ob} is in the rejection region.

Step 7

Reject H_o. Conclude that the average SAT score at Ivy College is probably greater than $\mu_o = 500$.

The Normal Distribution and Sampling Theory

true mean of the population of SAT scores at Ivy College is a particular value different from 500, what is the chance that the test will, in fact, result in the rejection of H_o? It is useful for us to determine the power of our test *before* conducting the study, for if we find that the test has a low power, we can alter the procedure to improve the power. For example, increasing the number of observations included in the sample will have the effect of increasing power. How do we determine the power of this one-sample z-test?

Actually, the procedure used to determine the power of the z-test for a population mean is not unlike the procedure for determining the power of the binomial test. In each case, we consider the probability distribution of sample results under the null hypothesis and under the alternative hypothesis. The first step, then, is the selection of the alternative to H_o against which to compute the power of the test. You will recall from our discussion of the power of the binomial test that we can compute power only against a specific alternative to H_o. This alternative will be based on our understanding of the research problem. We might have an idea about what the true mean of the population is. Or, we might have an idea about the size of the difference between μ_o and the true value of μ which would be "important" to detect. For example, let us suppose that Dean Baldwin considers a difference of one standard deviation ($\sigma = 100$ points for SAT) an important difference. If the actual mean SAT score at Ivy College differs from $\mu_o = 500$ by as much as one standard deviation, the dean would like to feel confident that his test will result in the rejection of H_o. Therefore, he selects as an alternative a value of μ that is 100 points away from $\mu_o = 500$. Let us assume he chooses to work with the specific alternative, $\mu_a = 600$. Because he is performing a two-tailed test, he could also have chosen $\mu_a = 400$. The computed value for power would be the same.

Having chosen $\mu_a = 600$ as the specific alternative against which to compute the power of the test, the next step is to draw pictures of the distributions of sample means as they would appear under H_o and under the alternative, $\mu_a = 600$. These pictures are drawn one above the other, in the same manner as the binomial distributions were drawn in Figures 6.5 and 6.6. The two distributions for the Ivy College example are presented in Figure 7.10.

In this figure, the upper curve represents the hypothetical distribution of sample means under H_o. It has a mean $\mu_{\bar{X}} = 500$ and a standard deviation $\sigma_{\bar{X}} = \sigma/\sqrt{n} = 20$. The lower curve represents the hypothetical distribution of sample means under the alternative hypothesis that the mean SAT score in the population was really $\mu_a = 600$. That is, the lower curve supposes that our sample mean, $\bar{X}$, was really one of the many possible sample means that we might obtain by taking samples of sizes 25 from a population where $\mu = 600$ and $\sigma = 100$. This alternative distribution of sample means has $\mu = \mu_a = 600$ and $\sigma_{\bar{X}} = \sigma/\sqrt{n} = 20$. It is, therefore, represented in the figure as shifted 100 points to the right of the distribution of sample means under H_o.

In calculating power, we first consider the distribution under H_o. We note that the values of z_{cr} for our test were ± 1.96, and we convert these critical values

Figure 7.10
The Power of Dean Baldwin's Test Against $H_a : \mu_a = 600$

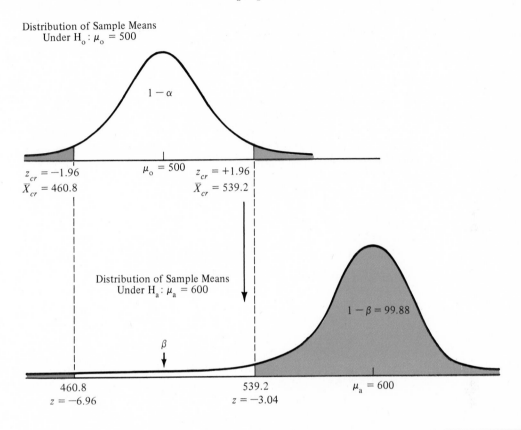

Distribution of Sample Means
Under $H_o : \mu_o = 500$

$1 - \alpha$

$z_{cr} = -1.96$
$\bar{X}_{cr} = 460.8$

$\mu_o = 500$

$z_{cr} = +1.96$
$\bar{X}_{cr} = 539.2$

Distribution of Sample Means
Under $H_a : \mu_a = 600$

$1 - \beta = 99.88$

β

460.8
$z = -6.96$

539.2
$z = -3.04$

$\mu_a = 600$

in terms of z to critical values in terms of raw scores. As noted previously, we do this using the transformation equation $X = z\sigma + \mu$, which here becomes:

$$\bar{X}_{cr} = z_{cr}(\sigma_{\bar{X}}) + \mu_o$$

$$= \pm 1.96(20) + 500$$

$$\bar{X}_{cr} = 539.2 \text{ or } 460.8$$

That is, we reject H_o if $\bar{X}_{ob}$ is either greater than 539.2 or less than 460.8. This conversion of z_{cr} to $\bar{X}_{cr}$ is crucial to the power calculation, because these values

299

of $\bar{X}_{cr}$ may be applied to the distribution of sample means under the alternative that $\mu_a = 600$.

The logic of the procedure is as follows. We have determined to reject H_o if $\bar{X}_{ob} > 539.2$ or if $\bar{X}_{ob} < 460.8$. This rejection rule will be applied, regardless of what the actual value of μ is. If the actual value of μ is really $\mu_a = 600$, then we can determine the power of the test by determining the probability of selecting an $\bar{X}_{ob} > 539.2$ or an $\bar{X}_{ob} < 460.8$ from the distribution of sample means centered at $\mu_a = 600$. We answer this question by referring to the lower curve in Figure 7.10. From this figure, it is obvious that if our sample has really been drawn from a population where $\mu = 600$, there is very little chance that $\bar{X}_{ob}$ would be less than 460.8. However, there seems to be a very good chance that $\bar{X}_{ob}$ would be greater than 539.2. How do we find out exactly what the chance would be? We convert these values of $\bar{X}_{cr}$ into z-scores once again, in terms of the distribution of sample means under the alternative that $\mu_a = 600$. To do this we use the formula, $z = (X - \mu)/\sigma$, written in the form

$$z = \frac{\bar{X}_{cr} - \mu_a}{\dfrac{\sigma}{\sqrt{n}}},$$

which in this problem becomes

$$z = \frac{460.8 - 600}{\dfrac{100}{\sqrt{25}}} = -6.96 \text{ and}$$

$$z = \frac{539.2 - 600}{\dfrac{100}{\sqrt{25}}} = -3.04.$$

That is, when we locate our critical values of $\bar{X}$ with respect to the specific alternative that $\mu_a = 600$, we find that the lower critical value lies 6.96 standard deviation units below μ_a, and the upper critical value lies 3.04 standard deviation units below μ_a. This means that there is virtually no chance that a sample mean drawn from the population where $\mu_a = 600$ will fall below 460.8, but there is an excellent chance that such a mean will fall above 539.2. Specifically, the table of the standard normal distribution tells us that 99.88 percent of all the scores in a normal distribution lie *above* the value $z = -3.04$. Thus, drawing a sample of size 25 from a population where $\sigma = 100$ and $\mu = 600$, we have a 99.88 percent chance of obtaining a sample mean greater than 539.2. This in turn means that if our original $H_o : \mu_o = 500$ was wrong and the specific alternative $H_a : \mu_a = 600$ was right, we have a 99.88 percent chance that $\bar{X}_{ob}$ will exceed 539.2, resulting in the rejection of H_o. That is, the power of Dean Baldwin's test against the

300

alternative that $\mu_a = 600$ is 99.88 percent. This is obviously an acceptable level for power.

We may summarize the steps involved in finding the power of a one-sample z-test for a population mean as follows:

Step 1   State the value (one-tailed test) or values (two-tailed test) of z_{cr}.

Step 2   Convert z_{cr} to $\bar{X}_{cr}$ using

$$\bar{X}_{cr} = z_{cr}(\sigma_{\bar{X}}) + \mu_o.$$

Step 3   Reconvert $\bar{X}_{cr}$ to a z-score (one-tailed test) or z-scores (two-tailed test) in terms of the specific alternative to H_o, using

$$z = \frac{\bar{X}_{cr} - \mu_a}{\dfrac{\sigma}{\sqrt{n}}}.$$

Step 4   Use the table of the standard normal distribution to find the probability associated with the z or z's found in Step 3.

In carrying out these steps, it is always helpful to use a diagram relating the distribution of sample means under H_o to the distribution of sample means under the specific alternative to H_o.

Exact Significance Probability.   Up to this point, we have followed the convention of employing specific levels of significance like $\alpha = .05$ and $\alpha = .01$ to obtain values of z_{cr} for our hypothesis test. You should understand that this is purely a matter of tradition. In some cases, we might wish to determine the *exact significance probability* of the observed result. Significance probability refers to the chance of obtaining a result as extreme as or more extreme than the observed result, under the assumption that H_o is true. In the example just provided, Dean Baldwin obtained the observed sample result $\bar{X}_{ob} = 550$. Because this $\bar{X}_{ob}$ yielded a z_{ob} of $+2.5$, which falls in the rejection region for this test, we know that the significance probability of the observed result was less than 5 percent. We know this because the test was conducted at the $\alpha = .05$ level of significance. But why stop with the statement that the significance probability is less than 5 percent? Why not determine exactly the probability of the observed result? To obtain the exact significance probability, we have only to use z_{ob} and refer to the table of the standard normal distribution. In the case of Dean Baldwin's problem, we found z_{ob} to be $+2.5$. Referring to the table of the standard normal distribution, we find that only 0.62 percent of the scores in a normal distribution lie *above* $z = 2.5$. Because Dean Baldwin's test was a nondirectional, two-tailed test, we want to know the proportion of scores lying this far from the hypothesized mean in either direction. Therefore, we double the 0.62 percent to take into account equally extreme scores on the other side of μ_o. This yields the *two-tailed significance probability* of 1.24 percent. In other words, if H_o: $\mu = 500$ were true, then the chance that Dean Baldwin's sample of size 25 would have a mean as far away from 500 as 50 points in either direction would be exactly 1.24 percent. Because 1.24 is less than 5 percent, Dean Baldwin's test resulted in the rejection of H_o at the $\alpha = .05$ level. However, if Dean Baldwin had chosen to conduct his test at the $\alpha = .01$ level of significance, the test would not have resulted in the rejection of H_o.

THE NORMAL CURVE MODEL IN INTERVAL ESTIMATION

Hypothesis Testing and Interval Estimation Compared

You will recall that in our introduction to inferential statistics in the previous chapter, we noted two different approaches to inferential statistics: hypothesis testing and interval estimation. The two approaches are both inferential in that they involve the use of sample data to make a statement about a parameter of the population from which the sample has been drawn. The two approaches are also similar in that both involve the use of appropriate mathematical models in arriving at statements about population parameters. However, the two approaches differ with respect to the logical sequence of steps and with respect to the nature of the resulting statement. We have seen that the logic of hypothesis testing involves the establishment of an initial (null) hypothesis regarding the relevant population parameter. We use a mathematical model to determine the sample results that would be likely if this null hypothesis were true. Then, we look at our actual sample result to see if it is a likely or unlikely result under H_o. If it is sufficiently unlikely, we reject H_o. The result of the hypothesis test is, therefore, a yes or no decision. We either reject H_o, or we do not.

The *interval estimation* approach is different. There is no initial assumption, no null hypothesis regarding the parameter of interest. We do not employ any such assumption to build a model specifying likely sample results. Instead, we begin with the sample result itself, and we use this result as an *estimate* of the population parameter of interest. Because sample results are subject to sampling error, we know that our sample estimate may not be exactly equal to the parameter, so we allow a little room on either side of our sample estimate for error. In deciding how much room to allow, we employ a mathematical model like the one we would have employed in the hypothesis test. The result is a statement that the parameter of interest has a value falling into a certain range of values. This range of values is generally referred to as a *confidence interval*. We use the term confidence interval because our mathematical model allows us to state that the relevant parameter lies between two values with a specific degree of con-

302

fidence that this statement is true. Note that the result of interval estimation is not a yes/no decision on whether the population parameter is a specific value. Rather, it is a statement that the value of the parameter falls within a certain range of values. We illustrate interval estimation with the normal curve model by constructing a confidence interval for the mean of a single population. To enable you to compare hypothesis testing and interval estimation, we will again use the example of Dean Baldwin and the average SAT score at Ivy College.

The Confidence Interval for a Population Mean

You will recall that Dean Baldwin selected a random sample of 25 SAT scores from student records to test his hypothesis regarding the average SAT score at Ivy College. He used the null hypothesis H_o: $\mu = 500$ and his knowledge of sampling theory to establish a rejection region, as illustrated in Figure 7.8. Note that the rejection region consists of all sample means further below $\mu_o = 500$ than $1.96\sigma_{\overline{X}}$ and all sample means further above $\mu_o = 500$ than $1.96\sigma_{\overline{X}}$. Between the points $z = -1.96$ and $z = +1.96$ lie all the values of sample means that will cause us to accept H_o rather than reject it. We could refer to this region as the "acceptance" region. This acceptance region is $2(1.96)\sigma_{\overline{X}}$ in length, and it is centered about $\mu_o = 500$. We know that the region is $2(1.96)\sigma_{\overline{X}}$ in length because 95 percent of all sample means of samples of size 25 will fall within $1.96\sigma_{\overline{X}}$ of the population mean, and because Dean Baldwin chose to use $\alpha = .05$. Now consider the possibility that the mean of the population is not 500, but 510. In that case, 95 percent of all the sample means of samples of size 25 would fall within $1.96\sigma_{\overline{X}}$ of the population mean of 510. In fact, regardless of the true value of the population mean, sampling theory and our knowledge of the normal curve tell us that 95 percent of all the sample means of samples of size 25 would fall within $1.96\sigma_{\overline{X}}$ of the true population mean. Moreover, this statement implies that the true population mean of a population would be located within $1.96\sigma_{\overline{X}}$ of 95 percent of all the sample means of samples of size 25 drawn from that population. This concept provides the theoretical basis for the calculation of a 95 percent confidence interval for μ, based on the $\overline{X}_{ob}$ of a single random sample. It enables us to take the $\overline{X}_{ob}$ and allow $1.96\sigma_{\overline{X}}$ in either direction for error, and then make the statement that the true population mean lies between $\overline{X}_{ob} - 1.96\sigma_{\overline{X}}$ and $\overline{X}_{ob} + 1.96\sigma_{\overline{X}}$, with 95 percent confidence that we are correct in this statement. This is an important concept. We illustrate it in Figure 7.11.

The Theory of Confidence Intervals Illustrated. In Figure 7.11, we have drawn the hypothetical distribution of sample means for samples of size 25 drawn from a population with $\sigma = 100$ and mean $= \mu$. We have purposely not specified an exact value for μ, to reinforce the notion that there is no initial assumption regarding μ when we construct a confidence interval. Regardless of the value of μ, 95 percent of all the sample means in the distribution of sample means of

Figure 7.11
Illustration of Confidence Intervals for μ Based on Three Samples

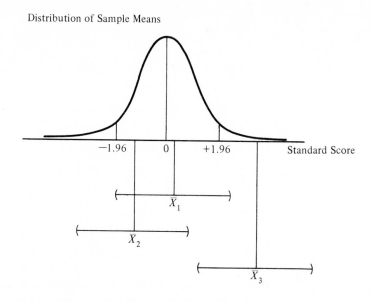

Distribution of Sample Means

samples of size 25 will fall within $1.96\sigma_{\bar{X}}$ of μ. In the figure, we have noted 3 possible sample means. One is denoted $\bar{X}_1$. It is close to μ, as most of the sample means would be. The second is denoted $\bar{X}_2$. It is not as close to μ as $\bar{X}_1$, but it is still one of the 95 percent of all the sample means falling within $1.96\sigma_{\bar{X}}$ of μ. The third, $\bar{X}_3$, is one of the extreme cases in the distribution of sample means. It is one of the 5 percent of the sample means in the distribution lying more than $1.96\sigma_{\bar{X}}$ from μ. Around each of these 3 sample means, we have sketched a band 2 times $1.96\sigma_{\bar{X}}$ in width.

You will note that the bands constructed around $\bar{X}_1$ and $\bar{X}_2$ overlap the mean of the distribution, μ. Because neither of these 2 sample means lies further away from μ than $1.96\sigma_{\bar{X}}$, an interval extending $1.96\sigma_{\bar{X}}$ on either side of these sample means contains the value μ. Furthermore, because 95 percent of all the sample means in the distribution of sample means do lie within $1.96\sigma_{\bar{X}}$ of μ, 95 percent of all the intervals of this size constructed about sample means would contain the value μ. On the other hand, the band constructed about $\bar{X}_3$ does not overlap the mean of the distribution. This sample mean lies further away from μ than $1.96\sigma_{\bar{X}}$, so an interval extending just $1.96\sigma_{\bar{X}}$ on either side of $\bar{X}_3$ will *not* contain μ. Because 5 percent of the sample means in the distribution of sample means do not lie within $1.96\sigma_{\bar{X}}$ of μ, 5 percent of all the intervals of this size would not contain the value μ.

Thus, if we select a single sample, compute $\bar{X}_{ob}$, and construct an interval $2(1.96\sigma_{\bar{X}})$ wide around this $\bar{X}_{ob}$, we will have a 95 percent chance that this interval contains μ. If our $\bar{X}_{ob}$ was one of the 95 percent of all sample means within $1.96\sigma_{\bar{X}}$ of μ, our interval will contain μ. If our $\bar{X}_{ob}$ was one of the 5 percent of all sample means lying farther than $1.96\sigma_{\bar{X}}$ from μ, our interval will not contain μ. Of course, because we do not actually know μ, we will simply construct the interval. We will make a statement that μ lies between two values. We will not know for sure whether this statement is true or false; we will only know that there is a 95 percent chance that it is true.

Calculating the Confidence Interval.   We illustrate the actual calculation of the interval by returning to Dean Baldwin. He had selected a sample of 25 from the population with $\sigma = 100$, and he obtained an $\bar{X}_{ob}$ of 550. To construct a 95 percent confidence interval for μ, he would start with 550 and allow $1.96\sigma_{\bar{X}}$ on either side for error. As we have seen, $\sigma_{\bar{X}} = \sigma/\sqrt{n} = 20$, so the range of Dean Baldwin's 95 percent confidence interval would be from a lower limit of

$$550 - 1.96(20) = 510.8$$

to an upper limit of

$$550 + 1.96(20) = 589.2.$$

He would then say, "The mean SAT score in my population lies between 510.8 and 589.2, and I am 95 percent sure that this is a correct statement."

If Dean Baldwin wanted to be more certain, he could allow a larger margin for sampling error. He could, for example, compute a 99 percent confidence interval for μ. Then, instead of using the value 1.96 for the number of standard deviations to allow, he would use the value 2.58. Do you know why? What are the values of the z-scores in a normal distribution that cut off the most extreme 1 percent of all scores?

The procedure for computing a confidence interval for a population mean may be summarized in general terms as follows:

$$\text{upper and lower limits of interval} = \bar{X}_{ob} \pm z_{cr}\left(\frac{\sigma}{\sqrt{n}}\right),$$

where z_{cr} is the value of the score in the standard normal distribution appropriate for the desired level of confidence. Usually this value is 1.96, for a 95 percent interval, or 2.58, for a 99 percent interval. However, in theory the researcher may compute a confidence interval at any desired level. Obviously, when greater confidence is required, we must allow more room for error, so the size of the interval increases.

305

THE CLASSIFICATION MATRIX

In this chapter, we continued our introduction to the basic concepts of statistical inference. We are now ready to return to our classification matrix and fill in the remaining cells with the appropriate inferential techniques. We shall examine these techniques in the next four chapters. Up to this point, we have looked at the binomial test, the one-sample z-test, and the z-method for finding a confidence interval for a population mean. These are indicated in Table 7.1.

TABLE 7.1
Classification Matrix: Inferential Statistics

Level of Measurement of Variable(s) of Interest

Type of Problem		Categorical: Nominal or Ordered Categories	Rank-Order	Interval Scale (or Ratio-Scale)
Univariate Problems		Binomial Test		One-Sample z-Test Confidence Interval for Population Mean (z-Method)
Bivariate Problems	Correlational Problems			
	Experimental and Pseudo-Experimental Problems: Independent Groups			
	Dependent Measures: Pretest and Posttest or Matched Groups			

Key Terms

Be sure you can define each of these terms.

normal curve model
normally distributed variable
standard normal distribution
sample statistic
population parameter
sampling error
law of large numbers
random sample
distribution of sample means
unbiased estimator
Central Limit Theorem
one-sample z-test
exact significance probability
interval estimation
confidence interval (z-method)

Summary

In Chapter 7, we continued our discussion of the basic concepts of statistical inference. We introduced the normal curve model and showed how it could be used as a probability model for continuous interval variables. We also looked at some basic principles of sampling theory and showed how to use the normal curve for hypothesis testing and interval estimation with continuous variables.

We examined the characteristics of the family of normal distributions and showed how we could interpret the area under the curve between two scores as the probability of selecting a score at random that lies between those two scores. We also pointed out that if we know the parameters of a normal distribution, μ and σ, we can find the percentile rank of any score by converting that score to a z-score and using a table of the standard normal distribution. We also saw how to make use of this property of normal curves in hypothesis testing and interval estimation.

To develop the concepts of sampling theory, we introduced the idea of a sampling distribution of sample means—a theoretical distribution consisting of an infinite number of means of samples of the same size drawn at random from a population of scores on a continuous variable. We saw that, because of sampling error, not all the sample means would be equal to the true value of the population mean, but that the mean of all the sample means would be equal to the population mean. We also pointed out that the standard deviation of the sampling distribution would be equal to the standard deviation of the original distribution divided by the square root of the sample size. Thus, we saw mathematically that larger samples result in smaller average sampling errors. The Central Limit Theorem told us that, with large enough sample sizes, any sampling distribution of sample means would be approximately normally distributed.

We then showed how the normal curve model could be used to test hypotheses regarding the value of a population mean of a continuous variable. We called this type of test the one-sample z-test. We saw how we could use our seven steps of a hypothesis test developed in Chapter 6 with normal curve problems as well as binomial problems. We used our knowledge of sampling distributions and the normal curve to establish a rejection rule for a specific null hypothesis about a population mean. As in our binomial example, a sample result (in this case, a mean) that would be highly unlikely under the null hypothesis would cause us to reject the null hypothesis. We pointed out that the concepts of level of significance, power, and Type I and Type II errors were all applicable to hypothesis tests using the normal curve model, just as they were with the binomial model.

Finally, we saw how we could use our knowledge of the sampling distribution of sample means to estimate a population mean from a sample mean. We set up a range of scores, called a confidence interval, in which we stated that our true population mean fell, and we indicated how confident we were that this was true. In the case of confidence intervals, we make no initial hypothesis about the population mean. Instead, we try to estimate the population mean on the basis of our sample mean.

Review Questions

To review the concepts presented in Chapter 7, choose the best answer to each of the following questions.

1 In an inference problem, if our statistic is the number of successes in a sample of 20, the distribution of the statistic over all possible samples of 20 will be

_____a a binomial distribution.

_____b a normal curve.

2 The use of area to represent the notion of probability implies that we are looking at the distribution of

_____a a continuous variable.

_____b a discrete variable.

309

3 If we are studying a continuous variable and we talk about the probability of getting a score at or below 52, we are talking about

_____a the standard score for 52.

_____b the percentile rank for 52.

4 Which one of the following is a parameter of a normal curve?

_____a μ

_____b $\bar{X}$

5 When you know the mean and standard deviation of a normal distribution, the distribution is completely defined. For this reason, the mean and standard deviation are called

_____a variables of the distribution.

_____b parameters of the distribution.

6 If we have a normal distribution and we wish to find out which score is the seventy-fifth percentile, we have

_____a a score to proportion problem.

_____b a proportion to score problem.

7 If we use the sample mean as an estimate for a population mean, we know that the value of the sample mean will depend on which sample we draw. For this reason, the sample mean is called a

_____a statistic.

_____b parameter.

8 The fact that the mean of a single sample is not exactly the same as the mean of the population illustrates the notion of

_____a an unbiased estimator.

_____b sampling error.

9 If we look at the distribution of all sample means of random samples of size 3, we are looking at

_____a a sample distribution.

_____b a sampling distribution of the mean.

10 If we select a random sample of 50 observations and make a frequency distribution of the results, we are looking at

_____a a sample distribution.

_____b a sampling distribution.

11 When we say that $\bar{X}$ is an unbiased estimator of μ, we mean that

_____a $\mu_{\bar{X}} = \mu$.

_____b $\bar{X} = \mu$.

12 To say that the distribution of sample means will be approximated by a normal distribution when n is large is a statement of
_____ a the Central Limit Theorem.
_____ b the law of large numbers.

13 One advantage of a large sample is that there is high probability that our sample mean will be close to μ. This fact is based on the equation
_____ a $\bar{X} = \mu$.

_____ b $\sigma_{\bar{X}} = \dfrac{\sigma}{\sqrt{n}}$.

14 If we take 25 random observations from a normal distribution with mean $= 100$ and standard deviation $= 15$, then
_____ a we know that the standard deviation of our sample will be 3.
_____ b our sample mean is likely to be between 97 and 103.

15 We can say that σ is a parameter of a population and $\sigma_{\bar{X}}$ is the standard deviation of
_____ a a single sample.
_____ b the distribution of sample means.

16 To test $H_o: \mu = 50$ against $H_a: \mu \neq 50$, our statistic is

_____ a $\dfrac{\bar{X} - \mu}{\sigma_{\bar{X}}}$.

_____ b $\dfrac{\bar{X} - \mu}{\sigma}$.

17 In using a normal curve to establish a rejection rule to test $H_o: \mu = 50$ vs. $H_a: \mu > 50$, we will
_____ a use the upper tail of the curve.
_____ b use both tails of the curve.

18 If we want a rejection rule to test $\mu = 50$ against $\mu \neq 50$ at a 5 percent level of significance using a z-test, we need to know which z-score is
_____ a the ninety-fifth percentile.
_____ b the 97.5 percentile.

19 In using the one-sample z-test to test $H_o: \mu = 50$ against $H_a: \mu \neq 50$, we will use a
_____ a one-tailed test.
_____ b two-tailed test.

20 In testing the hypothesis that $\mu = 50$ using an upper tail test, the alternate hypothesis will be
_____ a $\mu < 50$.
_____ b $\mu > 50$.

The Normal Distribution and Sampling Theory

21 If we test $H_o: \mu = 50$ vs. $H_a: \mu \neq 50$ and our z_{ob} causes us to reject H_o, we know that either $\mu \neq 50$ or

_____a we made a Type I error.

_____b we made a Type II error.

22 If we use $\dfrac{\bar{X} - 50}{\sigma/\sqrt{n}}$ as our statistic to test $H_o: \mu = 50$ vs. $H_a: \mu > 50$ at a 5 percent level of significance, the probability that $z_{ob} > 1.645$ when $\mu = 60$ is called

_____a the power of the test.

_____b a Type I error.

23 If we perform a two-tailed z-test and our z_{ob} is 1.9, we see from Table C.2 that the probability is .0574 that a z-score selected at random would be more extreme than 1.9. This probability is called

_____a the power of the test.

_____b the exact significance probability of the result.

24 A 95 percent confidence interval for μ can be interpreted as saying

_____a 95 times out of 100, μ is within the confidence limits reported.

_____b I have 95 percent confidence that my interval contains the true value of μ.

25 Using the same number of observations, a 95 percent confidence interval for μ will be

_____a shorter than a 99 percent confidence interval.

_____b longer than a 99 percent confidence interval.

Problems for Chapter 7: Set 1

1 Suppose we have a continuous variable, z, whose distribution is a normal curve. The mean of z is 0 and the standard deviation of z is 1. For each of the following statements, draw a picture of the normal curve, designate the range of z-scores specified, and shade the area under the curve that falls over the specified range.

a $0 \leqslant z \leqslant .5$ i $-1.96 \leqslant z \leqslant -1.645$

b $-.5 \leqslant z \leqslant 0$ j $-1 \leqslant z \leqslant 1$

c $-.5 \leqslant z \leqslant .5$ k $z \geqslant 1.28$

d $z \leqslant 1.5$ l $-2.33 \leqslant z \leqslant 2.33$

e $z \geqslant 1.5$ m $-3 \leqslant z \leqslant 3$

f $z \leqslant -.65$ n $z \geqslant 1.28$ or $z \leqslant -1.28$

g $z \geqslant .65$ o $z \leqslant -1.645$ or $z \geqslant 2.33$

h $-1.5 \leqslant z \leqslant .32$

2 Use Figure 7.5 to estimate standard score percentiles for the following percentile ranks:

a 5   d 77

b 1   e 50

c 97.5   f 39

312

3 Use Figure 7.5 to estimate the percentile rank for the following standard scores:

a 1.25   e −2
b 0   f −.75
c .9   g −.67
d 3

4 Use Figure 7.5 to estimate the area shaded in each situation in Problem 1.

Set 2

1 Use Table C.2 in Appendix C to calculate answers to questions 2, 3, and 4 of Set 1. Compare the calculated answers to the graph estimates.

2 Suppose a normal distribution of scores has $\mu = 100$ and $\sigma = 15$.
  a Which score is the tenth percentile?
  b Which score is the eightieth percentile?
  c What percentage of scores is above 120?
  d What is the percentile rank for a score of 135?
  e What is the percentile rank for a score of 90?
  f What is the range for the middle 50 percent of scores?
  g What percentage of all scores falls between 90 and 120?
  h Beyond which two scores will the extreme 5 percent of all scores fall? (half in either direction)

3 Answer the questions a–h of Problem 2 about a normal curve where $\mu = 100$, but $\sigma = 10$.

4 Answer the questions a–h of Problem 2 about a normal curve where $\mu = 110$ and $\sigma = 15$.

5 A high-school student has taken a series of tests and goes to her guidance counselor to discuss the results. The tests taken have all been standardized on large populations of students similar to the high-school student. Each test is assumed to have an approximately normal distribution with the mean and standard deviation as follows.

Test	Student's score	National mean	National standard deviation
Intelligence	145	100	15
Math Aptitude	730	500	100
Verbal Aptitude	510	500	100
Artistic Aptitude	15	20	3
Vocational Preference			
Math-Science	50	50	8
Fine Arts	49	40	5
Social Service	50	55	9
Business	55	45	7

The Normal Distribution and Sampling Theory

a What is the percentile rank for the student's score on each test?

b Does it appear that the student's interests match her aptitudes?

Problems 6 and 7 deal with the normal curve as a probability distribution.

6 Suppose scores on an IQ test are distributed normally over a population of adults with $\mu = 100$ and $\sigma = 15$. If one person is selected at random from the population, what is the probability that the person selected will have an IQ

a between 90 and 100?

b greater than 135?

c less than 85?

d between 85 and 115?

e between 88 and 120?

f more than 140 or less than 60?

g less than or equal to 100?

7 Suppose the mean mentioned in Problem 6 is 110 instead of 100. What will be the answers to questions a–g?

Set 3

1 If the annual average family income in the United States is $12,000 per year with a standard deviation of $2,000, what would be the mean and standard deviation of the distribution of sample mean incomes for samples of size n? Assume the samples are random.

a $n = 4$

b $n = 25$

c $n = 100$

d $n = 400$

e $n = 1600$

2 Assume that infinitely many random samples of size n are drawn from a normal distribution with $\mu = 500$ and $\sigma = 100$. For the values of n listed below, find the range for the middle 50 percent of sample means and the range for the middle 95 percent of sample means.

a $n = 16$

b $n = 64$

c $n = 100$

d $n = 400$

3 In Problem 2, what is the effect on the distribution of sample means of increasing the sample size?

4 a If a random sample of size 25 is selected from the population described in Problem 6 of Set 2, would you be surprised if the sample had a mean IQ of at least 105?

b Suppose the sample contained 400 observations. Should you be surprised if the sample mean IQ is more than 105?

1 The average IQ score for a certain IQ test is 100 and the standard deviation is 15. The test is normed on a large population of United States children. It has been 20 years since the test was normed, and the author of the test wonders whether or not the average is still 100. He selects a random sample of 400 children and administers the test.

    a Follow the hypothesis test outline and design statistical procedures to help the author decide whether or not the mean is 100. Use $\alpha = .05$.

    b What does it mean to make a Type I error?

    c What does it mean to make a Type II error?

    d What conclusion should be made if the results show a sample mean IQ of 103.7?

    e What is the significance probability for the observed result in d?

2 A sociologist is interested in determining whether or not her random sample of 50 is typical of the population at large on the variable "family income." For the population, family income is known to have a mean of $12,000 with a standard deviation of $2,000.

    a Use the hypothesis test outline to set up statistical procedures to help the sociologist decide whether or not her sample is typical. Use $\alpha = .10$.

    b What is the consequence of making a Type I error?

    c What is the consequence of making a Type II error?

    d Suppose the results yield a sample mean income of $10,800. What should the sociologist conclude?

    e What is the significance probability for the observed result in d?

3 In the early 1900s, the average height of adult males was found to be 174 cm. with a standard deviation of 6.5 cm. To determine whether on the average adult males are taller today, an investigator selects a random sample of 64 adult males and measures their height.

    a Use the hypothesis test outline to set up statistical procedures to help decide whether adult males are taller on the average than they used to be. Use $\alpha = .001$.

    b What is the consequence of a Type I error?

    c What is the consequence of a Type II error?

    d Suppose the random sample yields a mean height of 179.4 cm. Based on the procedures in a, what conclusion should be drawn concerning average height of adult males?

    e Calculate the significance probability for the observed result in d.

1 An investigator wishes to estimate the average IQ for a population in which she is interested. She selects a random sample of 100 subjects from her population of interest and administers the IQ test. The sample mean IQ score turns out to be 105. The standard deviation of the population of IQ scores is assumed to be 15.

a Calculate a 95 percent confidence interval for the population mean based on her sample mean.

b Interpret the result calculated in a.

c Calculate a 99 percent confidence interval for the population mean. What effect does the change in confidence level have on the precision of her estimate?

2 Suppose family income is known to have a standard deviation of $2,000. A random sample of 64 families is selected. They have an average income of $14,200. Use this information to estimate the population average family income with 90 percent confidence.

3 Suppose family income is known to have a standard deviation of $2,000. An investigator wishes to estimate the population average income. He wants his estimate to be off by no more than $500 with 95 percent confidence. How many families should he have in his random sample?

4 a How does the level of confidence affect the width of a confidence interval?

b How does the sample size affect the width of a confidence interval?

Set 6

1 Suppose an investigator is testing $H_o: \mu = 32$ vs. $H_a: \mu > 32$. She chooses $\alpha = .05$ and σ is known to be 8. She uses a random sample of 16 observations.

a Write the rejection rule for her test in terms of z and in terms of $\bar{X}$.

b Find the power of her test against the alternatives that $\mu = 36$, $\mu = 40$, and $\mu = 38$.

2 Suppose the investigator in Problem 1 uses a sample of 64 observations instead of 16.

a Write the rejection rule for her test in terms of z and in terms of $\bar{X}$.

b Calculate the power for the alternatives $\mu = 36$, $\mu = 40$, and $\mu = 38$.

c What seems to be the effect of increasing the sample size on the power of the test?

3 a Refer to Problem 1 of Section 4. Calculate the power of the test against the alternatives $\mu = 97, 98, 99, 101, 102, 103$.

b Summarize these results in a chart.

8

Inferential
Statistics
for Univariate
Problems

In the last two chapters, we have considered the nature of statistical inference. We have illustrated the manner in which a mathematical model may be employed to enable us to make a probability statement regarding a population parameter, based on data observed in a random sample drawn from the population. The two models that we have considered thus far are the binomial model and the normal curve model. As you will see, these two models are used very frequently in research in psychology and education, because many research studies involve variables that fit one of these two models. However, there are other research situations in which different mathematical models are appropriate. In the next four chapters, we will be considering various types of research situations and the techniques of inferential statistics that are appropriate to use in these situations. These chapters are organized according to our classification scheme. We will consider univariate problems, bivariate correlational problems, problems involving the comparison of two or more independent groups, and problems involving dependent measures. As different research situations are considered, new mathematical models will be employed. It is not necessary to describe each new model in the same detail as we described the binomial and normal curve models because, regardless of the mathematical details of a particular model, the nature of the process of statistical inference remains the same. The model is the tool required in order for us to use a sample to generate a probability statement about a population.

In this chapter, we will consider the techniques of inferential statistics that are appropriate for univariate research situations. Both the binomial model and the normal curve models will be reconsidered here, and other models will be presented as well. We will consider techniques appropriate in univariate research situations involving categorical data and interval scale data. With respect to categorical data, we will consider tests that may be employed with nominal scale data as well as a test appropriate for data in the form of ordered categories. We will not consider univariate problems involving rank-order data because there are no inferential statistics applicable to a single set of ranks.

INFERENTIAL STATISTICS FOR UNIVARIATE PROBLEMS: CATEGORICAL DATA

The Binomial Test

One of the most basic research situations occurs when we are concerned with a single variable that is categorical and has only two values. In this situation, we are interested in the proportion of cases within the population of interest that fall into each of the two categories. For example, if our population of interest were the population of adult education students in Texas preparing for the high-school equivalency examination, the variable of interest might be "result of examination." This variable would be a categorical variable with the values "pass" and "fail." As a second example, if our population of interest were the population of deep sea fishermen sailing from Key West, Florida, the variable of interest might be "result of trip." This variable could be a categorical variable with the values "caught fish" or "did not catch fish." This type of research situation is generally approached using the binomial model.

The use of the binomial model has been discussed in detail in Chapter 6. In this chapter, we will be considering a technique known as the *normal approximation to the binomial.* The normal approximation to the binomial is used in research situations that conform to the binomial model but involve such a large number of trials that the table of the binomial distributions (Table C.1) does not contain the necessary probabilities. Before presenting the normal approximation, we will review briefly the essential characteristics of a research situation conforming to the binomial model.

Suppose that past experience indicates that half of all juvenile offenders discharged from the State Training School for boys will be incarcerated again within 2 years. Suppose further that social worker Smith has developed a group therapy technique that he believes will reduce the rate of recidivism. He employs this technique with 36 subjects prior to their release. He follows this group carefully for 2 years following their release, noting which subjects again become inmates. He finds that 5 subjects are once again incarcerated before the end of the 2-year period.

This research situation conforms to the binomial model for the following reasons. Smith is concerned with a single variable in a single population. The variable of interest is a categorical variable having two values: (1) incarcerated again within 2 years and (2) not incarcerated again within 2 years. The population of interest is the population of all boys at the State Training School who might receive the Smith group therapy program. You will recall from Chapter 6 that in a binomial experiment of this nature, we typically designate one of the two categories of the variable of interest as a "success." Furthermore, we designate the proportion of cases in the population falling into this category as P. This proportion, P, is the population parameter about which our inferences are made. Because there are only two categories on the variable of interest and because each case in the population must fall into one of these two categories, knowledge of P also automatically tells us the proportion of cases that do not fall into the "success" category. This proportion is generally designated Q, and $Q = 1 - P$. In this example, we shall arbitrarily define P as the proportion of boys who are incarcerated again within two years. Then Q is the proportion of boys in the population who are not incarcerated again.

You will also recall from Chapter 6 that in a binomial experiment, we draw inferences regarding P on the basis of the evidence contained in a sample of n independent binomial trials, each of which may fall into the success or failure category. In the research situation being considered here, we regard each boy in the sample of 36 as a trial. We assume that these trials are independent of each other. The term independent means that the chance that any one trial, in this case, the selection of a boy, will result in a success, P, does not change from one observation to another.

Because past experience has shown that the rate of recidivism is 50 percent, the null hypothesis for our binomial test should be $H_o : P = .50$. Because Smith wishes to show that his therapy program reduces recidivism, the alternative hypothesis will be $H_a : P < .50$.

In considering the results of our study, we will count the number of successes observed in our sample of $n = 36$ trials or observations. We denote the number of successes observed as r_{ob}, so r_{ob}/n equals the *proportion* of successes that we observe in our sample. This sample proportion is denoted p and represents our best estimate of the population parameter P. As indicated in Chapter 6, there are mathematical formulas that enable us to determine the exact probability of obtaining a specific number of successes, r, in a sample of n trials, under a particular assumption regarding P. In practice, these formulas are not the most convenient approach to testing hypotheses in binomial situations. For experiments in which the sample size is 25 or fewer, we can obtain the required probabilities from the table of the binomial distributions. This table is contained in Appendix C (Table C.1). A description of the use of the table was presented in Chapter 6, and it is most important that you understand this material. In the case of our example of social worker Smith's research, however, the table of the binomial distributions will not help us. Mr. Smith's experiment has 36 trials, and the probabilities for experiments with 36 trials are not included in Table C.1. There are tables that provide the exact probabilities for the outcomes of binomial experiments where the sample size is greater than 25. However, these tables obviously become quite large in size as n increases, and they are not included in statistics textbooks. When $n > 25$, we typically employ a different approach to statistical inference in binomial experiments. This technique is the normal approximation to the binomial. At this point, we will consider some of the logic behind this alternative technique. Then we will return to Mr. Smith's research to illustrate its use.

The Normal Approximation to the Binomial

In Chapter 7, we noted that many variables in nature are distributed approximately normally. In the case of a binomial experiment, as the number of trials or observations becomes large, the variable "number of observed successes" tends to approximate a normal distribution. That is, as the sample size, n, increases, the distribution of r becomes approximately normal. Of course, the variable "number of successes" is a discrete interval variable, whereas a true normal distribution requires a continuous interval variable. However, under certain conditions the distribution of r is close enough to a normal distribution for us to employ the normal curve model to make inferences regarding the population parameter P.

Figure 8.1 illustrates the distribution of r when $P = .5$ for binomial experiments having 5, 10, and 25 trials. You will note that by the time $n = 25$, the distribution fits the normal curve quite well. The binomial distribution in which $P = .5$ and $n = 25$ is most closely approximated by a normal curve with a mean of 12.5 and a standard deviation of 6.25. The value 12.5 is easily understood if we ask, "How many successes would we expect to observe in a binomial experiment where $n = 25$ and $P = .5$?" Clearly, we would expect to find successes on about half of the trials, and half of 25 is 12.5. Of course, we know that it is not possible to observe 12.5 successes in any binomial experiment. We could observe either 12 successes or 13 successes, but not 12.5 successes. In fact, the binomial table indicates that the two most likely actual outcomes would be $r = 12$ and $r = 13$. If the experiment were repeated an infinite number of times, however, the *average* number of successes observed over all experiments would be 12.5. Thus, the normal distribution that would most nearly approximate the distribution of r for our 25-trial binomial experiment where $P = .5$ would have a mean of 12.5.

In general, the mean of the normal distribution that most closely approximates the distribution of r is given by the formula

$$\mu_r = nP,$$

where μ_r denotes the average number of successes observed when n is the number of trials or sample size and P is the population proportion of successes. Furthermore, the standard deviation of the normal distribution that most closely approximates the distribution of r is given by

$$\sigma_r = \sqrt{nPQ}.$$

These are important formulas because if we can compute the mean and standard deviation of the normal distribution that closely approximates the distribution of r in a particular binomial experiment, then we can use the normal curve model to perform hypothesis tests for the parameter P. You will recall from the previous chapter that the formula for the one-sample z-statistic used to test hypotheses regarding a population mean is

$$z_{ob} = \frac{\bar{X}_{ob} - \mu_o}{\dfrac{\sigma}{\sqrt{n}}},$$

where $\bar{X}_{ob}$ is the observed sample mean, μ_o is the hypothesized population mean, and $\sigma/\sqrt{n}$ is the standard deviation of the distribution of sample means. We can apply this same formula to the binomial problems with large numbers of trials by making appropriate substitutions. In place of the sample mean $\bar{X}_{ob}$, we substitute r_{ob}, the actual number of successes observed in our sample of n binomial trials. In place of the hypothesized population mean μ_o, we substitute nP_o, where n is the number of trials in our experiment and P_o is the hypothesized value for P, the proportion of successes in the population. If the proportion of successes in the population is really P_o, the expected number of successes for our experiment is nP_o. Finally, in place of the standard deviation of the dis-

---

Figure 8.1
The Binomial Distribution Approximates the Normal as n Increases

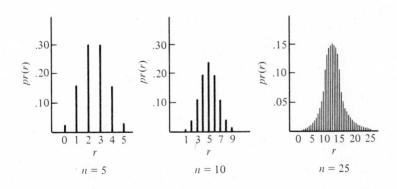

tribution of sample means, $\sigma/\sqrt{n}$, we substitute $\sqrt{nP_oQ_o}$, the standard deviation of the distribution of r under the null hypothesis. Thus, the z-statistic becomes:

$$z_{ob} = \frac{r_{ob} - nP_o}{\sqrt{nP_oQ_o}}.$$

This statistic will have approximately the standard normal distribution with large values of n.

How large n must be depends on our hypothesized value of P. When P_o is near .50, increasing the size of n brings the distribution of r close to the normal rather quickly. When P_o is close to 0 or close to 1.0, n must be quite large for the distribution of r to be close to normal. You will understand this difference if you remember that normal distributions are symmetrical. A binomial distribution with $P = .5$ is also symmetrical, even when n is small. A binomial distribution with $P = .1$ or $P = .9$ is not symmetrical. Nevertheless, if n is increased sufficiently, the normal approximation to the binomial may be used, even if P_o is not close to .5. Generally, the normal approximation to the binomial will be quite accurate with $n = 25$, if P_o is close to .5. As a rule of thumb, when P_o is near 0 or 1, however, the sample size should be large enough that nP_oQ_o is at least 9. For example, if $P_o = .2$, then $Q_o = .8$, and we would require a sample size of 57 for an accurate test:

$$nP_oQ_o \geqslant 9$$

$$n(.2)(.8) \geqslant 9$$

$$n(.16) \geqslant 9$$

$$n \geqslant 56.25$$

If it were not possible to increase our sample size to meet the criterion that nP_oQ_o equal at least 9, then it would be advisable to use the exact probabilities. This might involve finding a binomial table for $n > 25$, or it might involve the calculation of exact probabilities using the formulas presented in Chapter 6.

You may be wondering at this point how you can know for certain whether the sample size is large enough because we do not really know the value of the parameter P when we perform a hypothesis test. Remember that when hypothesis tests are performed, the parameters of the mathematical model employed are dictated by the assumptions embodied in the null hypothesis. Thus, the $P_o = .2$ just referred to would be derived from the null hypothesis of the test.

The Continuity Correction. The accuracy of the normal approximation to the binomial is improved by the addition of a small correction factor, which makes the formula for z_{ob} appear as follows:

$$z_{ob} = \frac{(r_{ob} \pm .5) - nP_{o}}{\sqrt{nP_{o}Q_{o}}}$$

This *continuity correction factor* improves accuracy because the binomial distribution with which we are really concerned is discrete, whereas the normal distribution being employed as our model is continuous. The correction works as follows. If r_{ob} is greater than nP_{o}, we subtract .5 from r_{ob} before finding the difference between r_{ob} and nP_{o}; but, if r_{ob} is smaller than nP_{o}, we add .5 to r_{ob} before finding the difference. Thus, the effect of the continuity correction will always be to reduce the absolute value of the difference between the observed sample value of r_{ob} and the expectation under the null hypothesis, nP_{o}. This will always increase the accuracy of the normal approximation. As n becomes very large, however, the effect of the continuity correction becomes negligible.

Social Worker Smith's Problem. As an illustration of the use of the normal approximation to the binomial, let us return to social worker Smith. You will recall that Mr. Smith was testing the hypothesis that P, the proportion of recidivists, was .50, against the alternative that P was less than .50. He had a sample of size 36, and of this number he observed that 5 subjects were incarcerated again within two years. The hypothesis test for this research is presented in Figure 8.2, following the same step-by-step outline for hypothesis tests that was presented in Chapter 6. As you follow the example through, check to see if you recall why the alternative hypothesis is directional (Step 3) and why the rejection region is one-tailed (Step 5). Also, check to be sure that you understand why the continuity correction was, in this case, to add .5 to r_{ob}, rather than to subtract it.

The result of the test is $z_{ob} = -4.2$. This means that the observed result of just 5 subjects returning within two years differs from the expected result by 4.2 standard deviation units. The critical value for z_{ob} was -1.645. Because -4.2 is less than -1.645, we reject H_{o}. Social worker Smith could conclude that the proportion of juvenile offenders in the population receiving his therapy is really less than 50 percent. He will, therefore, consider his program a success.

An Alternative Formula for the Normal Approximation to the Binomial. The formula for the z-statistic used in the normal approximation to the binomial may be written in several different ways. Specifically, we may express the formula not only in terms of r_{ob} and nP_{o}, the observed and expected *numbers* of successes, but also in terms of p_{ob} and P_{o}, the observed and expected *proportions* of successes in n trials. Because the proportion of successes is equal to the

FIGURE 8.2
Example of the Normal Approximation to the Binomial

Step 1:

Because we have a univariate problem where the variable of interest is categorical and has only 2 values, we consider the use of the binomial model. We assume observations are independent. Because $n > 25$, we consider using the normal approximation to the binomial.

Step 2:

$H_o: P = .50$ (Because H_o states $P = .50$, we know a sample size of 36 is adequate.)

Step 3:

$H_a: P < .50$ (A directional alternative)

Step 4:

The test statistic is: $z_{ob} = \dfrac{(r_{ob} \pm .5) - nP_o}{\sqrt{nP_o Q_o}}$.

The statistic will have approximately a standard normal distribution if H_o is true.

Step 5:

$\alpha = .05$. Reject H_o if $z_{ob} < -1.645$. (The rejection region is on the left side of the curve because values of r_{ob} less than $nP_o = 36(.5) = 18$ will tend to support the idea that P is really less than .50.)

Step 6:

$z_{ob} = \dfrac{(5 \pm .5) - 36(.50)}{\sqrt{36(.50)(.50)}}$

$z_{ob} = \dfrac{(5 + .5) - 18}{\sqrt{9}}$ (Here the continuity correction requires us to add .5 to r_{ob} because $r_{ob} < nP_o$.)

$z_{ob} = \dfrac{-12.5}{3}$

$z_{ob} = -4.2$

Step 7:

Because $z_{ob} < -1.645$, the observed statistic is in the rejection region. We reject H_o and we conclude that P is really less than .50.

number of successes divided by the number of trials, we simply divide each term in the formula presented earlier by n. Thus,

$$z_{ob} = \frac{r_{ob} - nP_o}{\sqrt{nP_oQ_o}} = \frac{\dfrac{r_{ob}}{n} - \dfrac{nP_o}{n}}{\dfrac{\sqrt{nP_oQ_o}}{n}} = \frac{p_{ob} - P_o}{\sqrt{\dfrac{nP_oQ_o}{n^2}}} = \frac{p_{ob} - P_o}{\sqrt{\dfrac{P_oQ_o}{n}}},$$

where p_{ob} represents the proportion of successes actually observed in the sample and $\sqrt{P_oQ_o/n}$ represents the standard deviation of the distribution of sample proportions under the null hypothesis that $P = P_o$. Just as we employ a continuity correction with the formula written in terms of r_{ob}, so we use a continuity correction with the formula when it is written in terms of p_{ob}. With the continuity correction, the formula becomes:

$$z_{ob} = \frac{\left(p_{ob} \pm \dfrac{1}{2n}\right) - P_o}{\sqrt{\dfrac{P_oQ_o}{n}}}.$$

We subtract $1/2n$ from p_{ob} before subtracting P_o when p_{ob} is greater than P_o. We add $1/2n$ to p_{ob} before subtracting P_o when p_{ob} is less than P_o. Thus, the effect of the continuity correction is always to reduce the absolute value of the difference between p_{ob} and P_o. In a hypothesis testing situation, it makes no difference which form of the normal approximation one chooses to employ. The results will be identical. You can verify this for yourself by using the formula in terms of p_{ob} to recalculate the z-statistic for the research example presented in Figure 8.2. We present the alternative formula for the normal approximation because it is useful in demonstrating the method used to obtain *confidence intervals for population proportions.*

Confidence Intervals for P.   You will recall that there are two approaches to inferential statistics, the hypothesis testing approach and the interval estimation approach. We have just seen how to test the hypothesis that the proportion of successes in a particular binomial population is a specific value. But suppose our research situation is such that we do not have an initial hypothesis regarding P. Suppose, in our example of social worker Smith's group therapy program, that there were no figures available on the present rate of recidivism among juvenile offenders. In this case, Smith would not have a baseline to which he could compare his own program. That is, he would not have the null hypothesis that $P = .50$ that he would seek to disprove. In this situation, Smith might employ the interval estimation approach. Rather than attempting to show that the rate of recidivism for participants in his program is lower than a particular value, he might simply establish a confidence interval for the rate of recidivism among the population of juvenile offenders who might be included in the program. To do this, Smith would begin with p_{ob}, the proportion of recidivists among the sample of 36 juveniles actually participating, and he would allow for sampling error by establishing a band extending a fixed number of standard deviation units on either side of this estimate.

At this point, you should review the section on the confidence interval for a population mean, presented in Chapter 7. The principles involved in the confidence interval for a population proportion are identical. In the confidence interval for μ, we begin with the sample estimate $\bar{X}_{ob}$ and allow an amount equal to $z_{cr}(\sigma/\sqrt{n})$ for error on either side of this estimate. You will recall that z_{cr} is the factor that determines the number of standard deviations to be allowed for error on either side of $\bar{X}_{ob}$, 1.96 for a 95 percent confidence interval and 2.58 for a 99 percent confidence interval. This factor is unchanged when we construct a confidence interval for P. In the formula for the confidence interval for the mean, the term $\sigma/\sqrt{n}$ represents the size of the standard deviation of the sampling distribution of $\bar{X}$. When we compute a confidence interval for a population proportion, we must use a factor representing the standard deviation of the distribution of sample proportions. We have seen that this factor is $\sqrt{PQ/n}$. This factor appears as the denominator of the z-statistic used to test a hypothesis regarding P as $\sqrt{P_0 Q_0/n}$. The notation P_0 is used because the null hypothesis specifies a particular value of P to be used in our model. In the interval estimation situation, however, we have no initial assumption that $P = P_0$. Therefore, in estimating the standard deviation of the distribution of sample proportions, we again return to the best estimate of P that we have, p_{ob}. Substituting p_{ob} for P_0 and $(1 - p_{ob})$ for Q_0, we have

$$\sqrt{\frac{p_{ob}(1 - p_{ob})}{n}}$$

as an estimate of the standard deviation of the distribution of sample proportions. Thus, our allowance for sampling error will be

$$\pm z_{cr} \sqrt{\frac{p_{ob}(1 - p_{ob})}{n}}.$$

In computing confidence intervals for population proportions, we also include a continuity correction of $1/2n$. When the continuity correction is included, the formulas for the lower and upper limits of the confidence interval become:

$$\text{lower confidence limit for } P = \left(p_{ob} - \frac{1}{2n}\right) - z_{cr}\sqrt{\frac{p_{ob}(1 - p_{ob})}{n}}$$

and

326

upper confidence limit for $P = \left(p_{ob} + \dfrac{1}{2n} \right) + z_{cr} \sqrt{\dfrac{p_{ob}\,(1 - p_{ob})}{n}}$.

In Figure 8.3, we have applied these formulas to the data on social worker Smith's program. Our calculations indicate that among the population of juvenile offenders given Smith's treatment, the proportion of recidivism is somewhere between .012 and .266. Because the confidence interval was a 95 percent interval, we are 95 percent sure that the true rate of recidivism is in this interval.

A Chi-Square One-Variable Test

Now, let us consider the research situation in which we are concerned with a single variable that is categorical, but that has more than two values. An example of such a variable would be political party affiliation, defined as having three

---

FIGURE 8.3
A 95 Percent Confidence Interval for the Proportion of Recidivists Among the Population of Juvenile Offenders Given the Smith Group Therapy Program

$r_{ob} = 5, p_{ob} = \dfrac{r_{ob}}{n} = \dfrac{5}{36} = .1389.$

Because level of confidence = 95 percent, $z_{cr} = 1.96.$

Lower limit $= \left(p_{ob} - \dfrac{1}{2n} \right) - z_{cr} \sqrt{\dfrac{p_{ob}\,(1 - p_{ob})}{n}}$

$= \left(.1389 - \dfrac{1}{2(36)} \right) - 1.96 \sqrt{\dfrac{.1389\,(1 - .1389)}{36}}.$

Lower limit = .012.

Upper limit $= \left(p_{ob} + \dfrac{1}{2n} \right) + z_{cr} \sqrt{\dfrac{p_{ob}\,(1 - p_{ob})}{n}}$

$= \left(.1389 + \dfrac{1}{2(36)} \right) + 1.96 \sqrt{\dfrac{.1389\,(1 - .1389)}{36}}.$

Upper limit = .266.

We say that the true proportion of recidivists in the population lies between the values .012 and .266, and we can be 95 percent certain that this statement is true.

categories—Republican, Democrat, or Independent. Another example would be academic title, with four values—instructor, assistant professor, associate professor, and full professor. In these situations, we are interested in the proportion of cases in the population of interest that fall into each of the several categories. Because there are more than two categories, the binomial model is not appropriate. We typically test hypotheses regarding the distributions of such variables using a *chi-square* (χ^2) *one-variable* test, sometimes referred to as a *goodness-of-fit test*. This χ^2 one-variable test may also be used in the situation where we have a single categorical variable that has only two values. However, the binomial or the normal approximation to the binomial is more frequently used in this situation.

In this χ^2 one-variable test, our starting point is a null hypothesis that specifies the proportion of cases in the population of interest that falls into each of the categories of the variable of interest. These proportions are typically based on past experience or records. For example, a political scientist might be concerned with determining whether there has been any shift in political party affiliation in a given county since the last election. In this situation, she might consult voter registration records from the time of the last election to generate a null hypothesis. If records indicated that 30 percent of registered voters were Republican, 40 percent Democrat, and 30 percent Independent at that time, she might establish as a null hypothesis

$$H_o : P_R = .3, P_D = .4, P_I = .3.$$

Of course, what she would be interested in showing was any change in these proportions since that time. That is, she would be interested in determining whether the population being sampled *now* was distributed as indicated in the null hypothesis. Thus, the alternative hypothesis would be

$$H_a : H_o \text{ is not true.}$$

In order to test H_o, the political scientist would select a random sample of n independent observations from the population of registered voters from the county. She would examine her sample to determine how many subjects fell into each of the three categories. Then, she would compare the number of subjects *actually observed* to fall into each of the categories to the number of subjects she would *expect* to find in that category, under the assumption that the population sampled is really distributed as specified in H_o. In this sense, she is testing to see how well the data in her sample fit the distribution specified in the null hypothesis, thus, the name goodness-of-fit test.

We have already considered the idea of expected frequencies in another context (see Chapter 5), but we will review the concept as it applies here. Suppose that our political scientist selected a random sample of 100 registered voters for

328

this research study. If H_o were a true description of the distribution of party affiliation at the time of the survey, how many of these 100 subjects would we expect to be Republicans? Clearly, if 30 percent of the population sampled are Republicans and if we select a sample of size $n = 100$, we would expect 30 percent of these 100 subjects, or 30 subjects, to be Republicans. Similarly, we would expect 40 of the 100 subjects to be Democrats and 30 to be Independents. Of course, we recognize that these are only expected values. Because of sampling error, even if H_o accurately describes the population sampled, we will not necessarily observe exactly 30 Republicans, 40 Democrats, and 30 Independents in our sample of 100. However, it is fair to say that the greater the differences are between the expected frequencies and the actual observed frequencies, the more evidence we have that H_o may not really describe the population we have sampled. This χ^2 one-variable test enables us to decide whether the expected frequencies and the observed frequencies in a particular example are sufficiently different for us to reject H_o with a reasonable degree of confidence that we have made the correct decision. In making this decision, we employ a statistic called the Pearson χ^2 statistic.

Requirements for Use of χ^2 One-Variable Test.   In order to use the χ^2 statistic to test a hypothesis such as $H_o: P_R = .3, P_D = .4, P_I = .3$, certain conditions must be satisfied. Some of these conditions concern the nature of the research problem and sampling and have already been noted. The first condition is that there is a single, categorical variable of interest; the second condition, that all observations on this variable are sampled randomly and independently of any other observations; and the third, that all observations must fall into one and only one category on the variable of interest. We have seen that the problem considered here meets these conditions.

However, there are additional requirements for the use of the χ^2 statistic. These requirements concern the expected frequencies for the several categories of the variable of interest. They may be summarized as follows. If we wish to apply the χ^2 test to the situation where the variable of interest has two categories, it is essential that the *expected* frequency in each category is not less than 5. If this criterion is not met, the χ^2 statistic will be inaccurate and should not be used. If we wish to apply the χ^2 test to the situation where the variable of interest has more than two categories, we would still hope to find that the expected frequency in each category was not less than 5. However, the χ^2 statistic will be fairly accurate if not more than 20 percent of the categories have expected frequencies less than 5, provided no one category has an expected frequency of less than 1. In cases where the variable of interest has several categories and these constraints are not met, we sometimes combine two or more categories together in order to raise the expected frequency to the required size. This process, often referred to as "collapsing" categories, was considered in our discussion of frequency distributions in Chapter 3.

329

The Calculation of the Observed Statistic in the χ^2 One-Variable Test. In conducting this chi-square one-variable test, we typically organize our data as shown in Table 8.1. You will note that we have calculated an expected frequency for each of the 3 categories of the variable of interest. We note that all 3 expected frequencies are greater than 5. Below each expected frequency we have tabled the actual number of subjects observed to fall into that category, out of the 100 subjects sampled. Organizing the data in this fashion facilitates the calculation of the chi-square statistic, which we compute according to this formula:

$$\chi^2_{ob} = \sum^{\substack{\text{all} \\ \text{categories}}} \frac{(Ob - Ex)^2}{Ex}.$$

The formula directs us to do the following calculations:

Step 1 For *each category,* find the difference between the observed frequency and the expected frequency.

Step 2 For each category, square the result of Step 1.

Step 3 Again for each category, divide the result of Step 2 by the expected frequency for that category. (At this point we have computed a quantity for each category known as the contribution to χ^2 of that category.)

Step 4 The capital sigma tells us to add up the results of Step 3 for all the categories to obtain a single number, χ^2_{ob}, our observed statistic.

Obviously, the larger the differences between the observed frequencies and the expected frequencies in the various categories, the larger the value of χ^2_{ob} will be. If χ^2_{ob} is big enough, we will reject H_o and conclude that the differences between the observed frequencies and the expected frequencies derived from H_o are too great for us to believe that our sample came from a population distributed as specified in H_o. But how big is big enough?

Determining the Significance of χ^2_{ob}. As in the other statistical tests we have considered so far, the observed statistic must exceed a critical value, which we denote χ^2_{cr}. The value of χ^2_{cr} for this goodness-of-fit test will depend upon two things: the level of significance we wish to employ in the test and the number of *degrees of freedom.* You are already familiar with the idea of level of significance from our previous discussion of hypothesis tests employing the binomial and the normal curve models. There is nothing different about this χ^2 goodness-of-fit test in this regard. In choosing α, we are selecting the risk of a Type I error. By tradition, we typically set this risk at 5 percent ($\alpha = .05$) or 1 percent ($\alpha = .01$).

The concept of degrees of freedom is new. A theoretical discussion of the χ^2 statistic is beyond the scope of an elementary text. We note here simply that

Inferential Statistics

TABLE 8.1

Expected and Observed Frequencies in Survey of Voting Registration

	Republican	Democrat	Independent	Total
Expected frequency under H_o	$Ex_R = P_R(n)$ $Ex_R = (.30)(100)$ $Ex_R = 30$	$Ex_D = P_D(n)$ $Ex_D = (.40)(100)$ $Ex_D = 40$	$Ex_I = P_I(n)$ $Ex_I = (.30)(100)$ $Ex_I = 30$	$n = 100$
Observed frequency in sample	$Ob_R = 21$	$Ob_D = 42$	$Ob_I = 37$	$n = 100$

there are, in fact, a large number of χ^2 *distributions* and that only one of these distributions is applicable to any χ^2 test. To know which distribution is applicable, we must know the number of degrees of freedom applicable to the test. This is easy to determine because the number of degrees of freedom for this χ^2 goodness-of-fit test is always one fewer than the number of categories of the variable of interest. In our example, there were 3 categories for the variable political party affiliation, so the number of degrees of freedom for this test will be $3 - 1$ or 2. In general, we use the letter K to represent the number of categories, so the number of degrees of freedom is $(K - 1)$.

When we have decided upon a level of significance and determined the number of degrees of freedom, we can locate the value of χ^2_{cr} in a table of critical values of the χ^2 distributions, such as Table C.3 in Appendix C. In a chi-square goodness-of-fit test where $K > 2$, we use the table as follows. First, we locate the desired level of significance among the values .01, .05, and .10 listed along the top of the table. Whichever α-risk we select, we find our χ^2_{cr} in the column of figures below this heading. Having located the appropriate column for alpha, we go as far down the column as we need to go in order to reach the correct number of degrees of freedom for our test. These degrees of freedom are listed in the column on the far right of the table. In our example, we would need to go down to the second row of the table because we have 2 degrees of freedom. If we chose $\alpha = .01$ for the test, χ^2_{cr} would be at the intersection of the column headed .01 and the row labeled 2. There we read the value 9.2. If we had chosen $\alpha = .05$, χ^2_{cr} would lie at the intersection of the column headed .05 and the row labeled 2. There we read the value 6.0. In either case, we would reject H_o if the value of χ^2_{ob} computed on the basis of our data exceeded the tabled value of χ^2_{cr}.

Note that we have not mentioned the question of the directionality or nondirectionality of this χ^2 test. When $K > 2$, the χ^2 one-variable test is always nondirectional. This is because the observed value of χ^2 will increase as observed frequencies differ from expected frequencies over all values of the variable of interest. The alternative hypothesis in the test does not specify which categories will show the largest differences between expected and observed frequencies, nor does it specify in which categories the observed will exceed or fall short of the expected frequency. The table of the χ^2 distributions (Table C.3) provides critical values for nondirectional tests. Therefore, when $K > 2$ in the χ^2 one-variable test, we simply locate our α-level along the top and look down to find χ^2_{cr}.

In the case where the variable of interest has just 2 categories, it is possible to conduct a χ^2 one-variable test that is directional. That is, we may state as an alternative hypothesis that the proportion of cases falling into one of the two categories is *greater than* the value specified in H_o, rather than just different from it. To locate the value of χ^2_{cr} for such a test, we would look under the column of the table of χ^2 distributions with the level of significance twice that actually desired for the test. That is, to find χ^2_{cr} for a directional test at the $\alpha = .05$ level of significance, we would look under the column of the table headed $\alpha = .10$. To find the value of χ^2_{cr} for a directional test at the $\alpha = .01$ level of significance, we would look under the column of the table headed $\alpha = .02$. Keep in mind that it is possible to specify a directional alternative *only* when $K = 2$. Therefore, we would always be seeking χ^2_{cr} for a directional test in the row of the table for $df = K - 1 = 2 - 1 = 1$.

The χ^2 one-variable test corresponding to the data presented in Table 8.1 has been carried out in Figure 8.4, following our standard outline of a hypothesis test. As you follow through the outline, check to make sure you understand how to obtain expected frequencies, how to determine χ^2_{cr}, and how to compute χ^2_{ob}.

You will note that the value of χ^2_{ob} in our example is 4.43. Because this observed value does not exceed the critical value of χ^2 for 2 degrees of freedom, we cannot reject H_o at the specified level of significance, $\alpha = .05$. This means that the distribution of political party preference within the sample was not sufficiently different from the distribution specified in the null hypothesis to allow us to conclude with confidence that the distribution specified in H_o was not a correct description of the population.

The Kolmogorov-Smirnov One-Variable Test

We now consider the research situation in which the experimenter has obtained independent observations on a single categorical variable where the categories have a clear order. As an example of such a situation, consider the case of master winemaker Nappy Sonoma, who has been called in as a consultant by a major California vineyard to determine the optimum method of processing a particular variety of grape. Mr. Sonoma believes that the grape in question is best when used in making a very dry wine. However, the owners of the vineyard know that sweeter wines are often more popular and they will not accept the winemaker's judgment until they are convinced that American wine drinkers prefer the drier products of this grape to the sweeter products. Mr. Sonoma prepares four

332

FIGURE 8.4
Example of χ^2 One-Variable Test

Step 1:

Because we have a univariate problem where the variable of interest (political party preference) is categorical and has more than 2 values, we consider using the χ^2 one-variable test. Because a random sample has been drawn from a large population of voters, we may assume all observations are independent. We compute the expected frequencies for each category and find that none is less than 5. We therefore determine that the χ^2 one-variable test is appropriate.

Step 2:

$H_o: P_R = .30, P_D = .40, P_I = .30.$

Step 3:

$H_a: H_o$ is not true.

Step 4:

The test statistic is the χ^2 one-variable test or goodness-of-fit test. Because the variable of interest has 3 categories, our degrees of freedom are $(K - 1) = (3 - 1) = 2$. So

$$\chi^2_{ob} = \sum \frac{(Ob - Ex)^2}{Ex}, \text{ which will have 2 df.}$$

Step 5:

$\alpha = .05$. Using the table of the χ^2 distributions, we find that χ^2_{cr} for $\alpha = .05$ in a test with 2 df is 6.0. We will reject H_o if $\chi^2_{ob} > 6.0$.

Step 6:

$$\chi^2_{ob} = \frac{(21 - 30)^2}{30} + \frac{(42 - 40)^2}{40} + \frac{(37 - 30)^2}{30} = 4.43.$$
(For calculation of expected frequencies, see Table 8.1.)

Step 7:

Because χ^2_{ob} is not greater than 6.0, we cannot reject H_o. Although there were differences between observed and expected frequencies in our sample, these differences were not large enough for us to conclude that the population is not distributed as specified in H_o.

333

batches of wine from the variety of grape in question. These batches are prepared in such a way that wine drinkers can easily and reliably rank order the wines in terms of the underlying dimension of sweetness. Using these batches of wine, Mr. Sonoma will seek to demonstrate that wine drinkers consistently prefer the drier batches. To do this, he selects a random sample of 20 wine drinkers. Each subject is asked to taste all four wines. To guard against possible contamination of results because of order of presentation, the wines are tasted in random order by each taster. Between tastes, each subject thoroughly rinses his or her mouth. After tasting all four wines, all of the 20 subjects are asked which wine they like *best*. Mr. Sonoma reasons that if wine drinkers have no preference in terms of dryness, his subjects should select each of the four wines in approximately equal numbers. However, if they really have a preference for the drier wine, as he believes, more of his subjects should select the drier wines than the sweeter wines.

χ^2 One-Variable Test and Kolmogorov-Smirnov One-Sample Test Compared.

At this point, you may be thinking that the χ^2 one-variable test could be employed in this situation. Mr. Sonoma could consider each of the four varieties of wine as a category of the variable "wine type preferred." He could establish as the null hypothesis the idea that equal proportions of wine drinkers preferred the four types of wine, i.e., $H_o: P_1 = .25, P_2 = .25, P_3 = .25, P_4 = .25$. He could compute expected frequencies for each category and employ the χ^2 one-variable test to see how well his sample of 20 wine tasters' preferences fits this hypothetical distribution. This is good reasoning; we could use the χ^2 one-variable test here. We have already noted that the statistical procedures that are employed with purely nominal categories may also be used with ordered categories. However, to do so may not always be the optimum course of action. In the wine-making problem, we must realize that the order of the four categories of wine is important. When subjects tell us their preference for one of the four wines, they are giving us more information than they would be giving us if the four wines were simply different from each other. In this situation, the *Kolmogorov-Smirnov one-variable test* is used in preference to the χ^2 one-variable test because it takes advantage of this additional information to provide us with a *more powerful* test. (For a review of the concept of power, see Chapter 6.)

The Kolmogorov-Smirnov test takes advantage of the additional information inherent in the ordering of the categories because it focuses not simply on the hypothetical and observed frequency distributions, but on the hypothetical and observed *cumulative frequency distributions*. Whereas the χ^2 one-variable test compares expected to observed frequencies for each category taken by itself, the Kolmogorov-Smirnov test compares the expected *cumulative* relative frequency for each category to the observed cumulative relative frequency for that category. Thus, the null hypothesis that there is no preference in terms of dryness would not be expressed as in the χ^2 test: that the percentage of subjects indicating a preference for a given category would be 25 percent for each category.

334

Rather it would be stated that 25 percent would prefer the driest wine, that 50 percent would prefer either the driest or the next driest, that 75 percent would prefer one of the three driest, and of course that 100 percent would prefer one of the four. This is illustrated in Table 8.2, in the row headed "Hypothetical cumulative relative distribution under H_o." Obviously, it is possible to state the null hypothesis of no preference in this manner only when we have ordered categories, because it does not make sense to speak of cumulative frequency distributions in the absence of a clear order.

Requirements for Use of the Kolmogorov-Smirnov One-Variable Test.   The Kolmogorov-Smirnov one-variable test is like the χ^2 one-variable test, except that the categories of the variable of interest must be ordered. The requirements are therefore that there must be a single categorical variable of interest having categories that are clearly ordered; that all observations on this variable must be sampled randomly and be independent of any other observations; and that each observation must fall into one and only one of the ordered categories of the variable of interest. The Sonoma wine experiment fits these conditions.

TABLE 8.2
Expected and Observed Cumulative Relative Distributions on Wine Preferences

$n = 20$

	Rank of Wine on Dryness			
	Most dry	Second most dry	Third most dry	Least dry
Hypothetical cumulative relative distribution under H_o	.25	.50	.75	1.00
Observed distribution	12	6	2	0
Observed cumulative relative distribution	$\frac{12}{20} = .60$	$\frac{18}{20} = .90$	$\frac{20}{20} = 1.00$	$\frac{20}{20} = 1.00$
D	.35	.40	.25	0.00

Carrying Out the Kolmogorov-Smirnov One-Variable Test. Once we have determined that our data fit the requirements for the Kolmogorov-Smirnov one-variable test, the procedure in carrying out the test is quite simple. The first step is to specify the hypothetical cumulative relative frequency distribution of the variable of interest. As already indicated, the null hypothesis in the wine tasting example would be that 25 percent of wine drinkers prefer the driest wine, 50 percent prefer one of the two driest, and so on. We write this hypothetical distribution in a table as shown in Table 8.2. Then we examine the observed results of the experiment and construct the cumulative relative frequency distribution for these observed results. In Table 8.2, we have presented the distribution of responses and the cumulative relative frequency distribution of these responses. The results show that 12 subjects preferred the driest wine, 6 the next driest, 2 the third driest, and none the sweetest. Thus, 12/60 or 60 percent preferred the driest, $(12 + 6)/20 = 18/20 = 90$ percent preferred either the driest or the next driest, and 100 percent preferred one of the three driest. In Figure 8.2, these figures are presented as proportions in the row titled "Observed cumulative relative distribution."

When we have labeled both the expected and observed cumulative relative frequencies, we find the difference between the expected cumulative relative frequency and the observed cumulative relative frequency for each category of the variable of interest. This step is shown in Table 8.2 in the row labeled D, for difference. The observed statistic D_{ob} in the Kolmogorov-Smirnov one-variable test is simply the absolute value of the largest of these differences. In our experiment, the largest of these differences occurred at the point of the category pertaining to the second driest wine. Here the difference between the observed and expected cumulative relative frequencies was 40 percent, or .40. Thus, D_{ob} in this example is .40.

Determining the Significance of the D-Statistic. As in other types of hypothesis tests, we decide whether the Kolmogorov-Smirnov one-variable test is significant by comparing the observed statistic to a critical value based on a mathematical probability model. Also, as in the case of other types of hypothesis tests, we obtain the specific critical value from a table. The table of critical values for D_{ob} in the Kolmogorov-Smirnov one-variable test is found in Appendix C, Table C.4, To use this table, we locate the row of the table corresponding to the sample size in the column on the left, and then we look across this row until we reach the column corresponding to the chosen level of significance. In this example, our sample size is 20, so we look down the left hand column of the table until we find the value 20. Looking across the row for $n = 20$, we find that the critical value for $\alpha = .05$ would be .294, and the critical value for $\alpha = .01$ would be .356. You should verify these values to check your understanding of how to read the table. In every case, the critical values presented in the table are for two-tailed tests of nondirectional hypotheses. The critical values for one-tailed Kolmogorov-Smirnov one-variable tests have not been fully worked out.

336

Therefore, on a directional hypothesis test such as that in our example, the critical values will be conservatively large.

In our example, we note that our D_{ob} of .40 is larger than D_{cr} for $\alpha = .01$, which is .356. Thus, Mr. Sonoma will reject the null hypothesis at the .01 level of significance. He will conclude that wine drinkers really prefer the drier wines made from this grape. Mr. Sonoma's test is illustrated in Figure 8.5.

You will note that the table of critical values for the Kolmogorov-Smirnov one-variable test (Table C.4) contains entries for sample sizes as large as 35. For sample sizes larger than 35, critical values for the test are found by dividing the constants tabled in the last row by the square root of the sample size. Thus, if our n in this example had been 100, the critical value of D for $\alpha = .01$ would have been $1.63/\sqrt{100} = 1.63$.

INFERENTIAL STATISTICS FOR UNIVARIATE PROBLEMS: INTERVAL DATA

The One-Sample z-Test for a Population Mean and the z-Method of Confidence Intervals

In Chapter 7, we considered the normal family of probability distributions. We considered the characteristics of these distributions in some detail in order to help you understand how the normal curve model could be used in drawing inferences about the mean of a population on the basis of a sample. At the conclusion of that chapter, we outlined the one-sample z-test for a population mean and the z-method of computing a confidence interval for a population mean. In terms of our classification scheme, the one-sample z-test and the z-method of computing a confidence interval for a mean are among those techniques of inferential statistics that are appropriate for univariate problems with interval scale data.

However, it is not possible to use the one-sample z-test or the z-method of confidence intervals for *every* univariate research situation with interval data. You will recall from Chapter 7 that in order to compute a z-statistic or a z-confidence interval, we must either have a random sample from a normally distributed population, or a random sample from a non-normal distribution large enough that the sampling distribution of the mean will be normal. Furthermore, we must know σ, the standard deviation of the sampled population.

We need to know σ in order to calculate the standard deviation of the distribution of sample means, $\sigma_{\bar{X}} = \sigma/\sqrt{n}$. There are occasions in research where σ is known. For example, if the variable of interest is a score on a standardized instrument, such as an intelligence test or a graduate school entrance examination, we can sometimes obtain the exact value of σ from the test manual. However, there are also occasions in research where σ is not known. Often, the variable

337

FIGURE 8.5
Example of the Kolmogorov-Smirnov One-Variable Test

Step 1:

Because we have a univariate problem where the variable of interest (dryness of preferred wine) is an ordered categorical variable, we use the more powerful Kolmogorov-Smirnov one-variable test in preference to the equally appropriate χ^2 one-variable test.

Step 2:

H_o: There is no preference in terms of dryness, or
H_o: $P_1 = .25$, $P_1 + P_2 = .50$, $P_1 + P_2 + P_3 = .75$, $P_1 + P_2 + P_3 + P_4 = 1.00$.

Step 3:

H_a: H_o is not true.

Step 4:

The test statistic is D_{ob}, the absolute value of the largest difference between the expected and observed relative cumulative frequencies for any value of the variable.

Step 5:

$\alpha = .01$. Using the table of critical values for the Kolmogorov-Smirnov one-variable test, we find that for $n = 20$ and $\alpha = .01$, $D_{cr} = .356$.

Step 6:

$D_{ob} = |.50 - .90| = .40$ where .90 represents the 18 out of 20 subjects who preferred either the driest or next driest wines.

Step 7:

Because $D_{ob} > D_{cr}$, we reject H_o and conclude that drier wines are preferred.

---

of interest in a study is a score on an instrument that is not standardized. Perhaps the researchers have developed an original instrument specifically for their study, or perhaps they are using an instrument that has been used before, but not often enough to provide an exact value for σ. Moreover, even if the variable of interest in a study is a score on a standardized instrument, there may be occasions when the researchers will not want to assume that the σ reported in the test manual applies to the specific group they have tested. For example, if researchers in special education were to employ a standardized intelligence test in an experiment where subjects were mentally retarded individuals, they would probably not want to assume that the σ for the test reported in the manual, based on a large norming sample drawn from the population in general, would apply to the population of retarded individuals. It might be that the population of retarded individuals differs from the general population not only with respect to the mean score on the intelligence test, but with respect to the variability of test scores as well. Whenever researchers are unwilling to assume that they know the *exact* value of the standard deviation of the sampled population, they cannot use the z-test or the z-method of computing confidence intervals. However, they do not necessarily need to revert to the χ^2 or Kolmogorov-Smirnov one-variable tests. In this situation, they may choose to employ another model: one of the *t-distributions*. We now proceed to a consideration of this method.

One-Sample *t*-Test for a Population Mean

You will recall that the z-statistic measures the number of standard deviation units by which an observed sample mean differs from the value expected under the null hypothesis. The z-statistic requires that we know σ because we need it to find $\sigma_{\bar{X}}$, the standard deviation of the distribution of sample means. The *t-statistic* attempts to provide the same type of measure in the absence of definite knowledge of σ. In doing this, the t-statistic employs an estimate of σ. This estimate of the population parameter σ is the sample statistic s, the *sample standard deviation*. Just as we can compute a sample mean as an estimate of the mean of the population, we can compute a sample standard deviation as an estimate of the standard deviation of the population. We can then use this estimate in place of the actual value of σ to carry out a hypothesis test for the population mean σ. The result is the *one-sample t-test*. This test may be interpreted in the same way as a z-test, except that the use of s as an estimate of σ results in a slight change in the shape of the distribution of sample means. The nature of this change depends on the size of the sample employed in the test. Because of the change, the significance of the test is no longer determined by reference to the standard normal distribution, but rather to one of a series of rather similar distributions known as the *t*-family of probability distributions. Before we consider the procedures involved in carrying out the one-sample t-test, however, we must consider the manner in which we use sample data to estimate the standard deviation of the sampled population.

The Sample Standard Deviation.   You will recall from our discussion of descriptive statistics in Chapter 4 that we calculate the descriptive or population standard deviation, σ, according to one of these formulas:

$$\sigma = \sqrt{\frac{\Sigma X^2}{N} - (\mu)^2} = \sqrt{\frac{\Sigma X^2 - \frac{(\Sigma X)^2}{N}}{N}}.$$

We reemphasize here that these formulas are appropriate when we have observed *every score* in a population of scores and are seeking to describe the variability of this population alone. But now we have shifted our focus from descriptive statistics to inferential statistics. Now we are concerned with generalizing from a sample to the larger population from which the sample was drawn. Specifically, when we compute a sample standard deviation, s, we are using sample data to estimate the standard deviation of the population, σ.

Statisticians have determined that in this situation, it is not correct to apply the formulas just presented directly to sample data to obtain an estimate of σ. To do so would result in estimates of σ that tend to be slightly too small on the average. The correct formula to use when using sample data to estimate a population standard deviation is:

$$s = \sqrt{\frac{\Sigma X^2 - \frac{(\Sigma X)^2}{n}}{n-1}}.$$

You will note that this formula differs from the descriptive formula in that we now use a lowercase n to represent the number of observations in our *sample*, rather than a capital N to represent the number of observations in an entire population. Also, what was simply N in the denominator of the formula for the population standard deviation is now $n - 1$ in the formula for the sample standard deviation. Dividing by $n - 1$ rather than by n has the effect of making the computed value of s slightly larger. Statisticians have determined that the use of $n - 1$ makes s the best available estimate of the population standard deviation, σ.

In problems of inferential statistics, we always use the formula for s to obtain an estimate of σ. When we conduct a one-sample t-test for a population mean, we use the data in our sample to compute s, which we then plug into the formula for the t-statistic.

One-Sample z-Test and One-Sample t-Test Compared. As you might suspect from the foregoing discussion, the one-sample z-test and the one-sample t-test for population means are analogous. In each case, we are testing a hypothesis regarding the mean of a population. As far as the actual test statistic is concerned, the two differ only in that the σ used in the one-sample z-test is replaced by s in the one-sample t-test:

z-test

$$z_{ob} = \frac{\bar{X}_{ob} - \mu_{o}}{\frac{\sigma}{\sqrt{n}}}$$

t-test

$$t_{ob} = \frac{\bar{X}_{ob} - \mu_{o}}{\frac{s}{\sqrt{n}}}$$

340

Inferential Statistics

As far as interpretation is concerned, we can say that the z-test is a measure of the number of standard deviation units by which a sample mean differs from the hypothesized mean of the distribution of sample means, and the t-test is an *estimate* of this same measure.

Requirements for One-Sample t-Test. The one-sample t-test assumes that we have n independent random observations on an interval scale variable drawn from a population that is normally distributed. In practice, it is often impossible to know whether the population from which our sample is drawn is really normally distributed. However, we must at least be willing to make the assumption that the distribution is normal. In general, we will check the distribution of the sample to make sure that it is unimodal and not highly skewed.

The t-Distributions. Another difference between the t-test and the z-test for a population mean is that the process of estimating σ with s results in a change in the distribution of the test statistic. Whereas the one-sample z-test employs the standard normal distribution, the one-sample t-test employs one of the family of t-distributions. There are many different t-distributions. Which distribution we employ in a given one-sample t-test for a population mean depends on the number of degrees of freedom associated with the particular test. In this sense, the t-distributions are like the χ^2 distributions, considered in connection with the χ^2 goodness-of-fit test described previously. The number of degrees of freedom associated with a one-sample t-test for a population mean is equal to one less than the number of observations included in the sample.

The t-distributions are rather similar in shape to the standard normal distribution, in that they are unimodal and symmetrical. They differ from the normal in that they are somewhat less peaked at the mode, somewhat less flat at the tails. Thus, a bit more of the area in a t-distribution lies out toward the tails. In Figure 8.6, we have sketched a standard normal distribution and two t-distributions. You will note that because the t-distributions have more area out in the tails, we need a higher t-score to cut off the upper 2.5 percent of a t-distribution than the z-score we need to cut off the upper 2.5 percent of the standard normal distribution. In Figure 8.6, we have illustrated this point for the t-distribution with df = 10. In this distribution, the score $t = +2.23$ cuts off the upper 2.5 percent of all observations. This is written in shorthand notation as $t_{10;.975} = +2.23$, which says that in the t-distribution with 10 degrees of freedom the 97.5 percentile score is $+2.23$. But we already know that in a standard normal distribution, the z-score of $+1.96$ is the 97.5 percentile score that cuts off the uppermost 2.5 percent of all observations in the distribution.

As the number of degrees of freedom associated with a t-distribution increases, the distribution becomes more and more like a normal distribution. In Figure 8.6, we see that the t-distribution with df = 25 is more similar in shape to the standard normal distribution than is the t-distribution with df = 10. This is reflected in the fact that the 97.5 percentile score in the t-distribution with 25

Figure 8.6
The Standard Normal Distribution Compared to *t*-Distributions
With 10 and 25 Degrees of Freedom

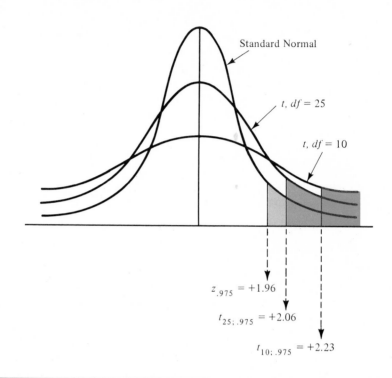

degrees of freedom, $t_{25;.975}$, is 2.06. This is closer to the 97.5 percentile score in the standard normal distribution than was the corresponding score in the *t*-distribution with only 10 degrees of freedom, $t_{10;.975} = 2.23$. When the degrees of freedom associated with a *t*-distribution increase still further, the distribution becomes closer and closer to the standard normal distribution. When df = ∞ (infinity), the *t*-distribution is identical to the standard normal distribution. If you stop and think about it, this should make sense. A *t*-distribution with df = ∞ applies to a situation where the sample size is infinitely large. In the case where we have an infinite sample size, an estimate of σ based on *s* would be perfect. In this case, we would really have the exact value of σ, so our *t*-test would in effect become a *z*-test.

Reading the Table of the *t*-Distributions. Because there are many *t*-distributions, it is not practical to provide a complete table for each distribution as we have done for the table of the standard normal distribution. Instead, we provide a table of the t-distributions that contains only the most commonly used values of the t-distributions for one- and two-tailed hypothesis tests at various levels of significance. This is presented in Table C.5 of Appendix C. To use the table, we first look down the extreme left column to find the appropriate number of degrees of freedom. When we have located the degrees of freedom, we have determined which row of the table to use in finding our critical value of t. Next, we look at the top of the table. If our test is to be a two-tailed test, we consult the row of figures labeled "Level of significance for two-tailed test." If our test is to be a one-tailed test, we check the row of figures labeled "Level of significance for one-tailed test." Here we locate the α-level chosen for our test. This α-level will be at the heading of a column of figures in the table. We look down this column until we reach the row of the table corresponding to our degrees of freedom. At the intersection of this row and column lies the critical value of t for the t-test. You will note that all the values in the table are positive. Of course, it is possible to have negative values for t_{cr}. In the case of a two-tailed test, there will always be one positive and one negative critical value. In this case, we use both the positive and the negative of the value tabled. In the case of a one-tailed test, the critical value will be either positive or negative. If the alternative hypothesis of the test is that the true mean is larger than the value specified in H_o, then we will be hoping to obtain a large positive value for t_{ob}, and t_{cr} will be a positive number. However, if the alternative hypothesis is that the true mean is less than the value specified in H_o, then we will be hoping to obtain an extreme negative value for t_{ob}, and t_{cr} will be a negative number. In this event, we simply take the negative of the value we find in the table of the t-distributions.

Example of a One-Sample t-Test. As an example of a one-sample t-test, consider the following hypothetical research problem. Counseling psychologist Weber is the director of a large vocational rehabilitation program for prisoners in the state correctional facility. Dr. Weber would like to secure funding for a remedial reading expert to join the program. However, state law requires that funds for such a professional staff member may be authorized only if it can be shown that the intended client group is performing below par on a standardized reading achievement test. Because it is not feasible to test the entire population of prisoners, Dr. Weber selects a random sample of 20 inmates. She administers to this sample the Blackstone Reading Achievement Test. From the test manual, Dr. Weber learns that the average score on this achievement test among the general population is 80, and that the standard deviation of the distribution of test scores in the norming group is 14 points. Dr. Weber therefore needs to show that the population of inmates has an average score in this test of less than 80. That is, she will want to test $H_o: \mu = 80$ against the alternative $H_a: \mu < 80$. In conducting this test, Dr. Weber might assume that the standard deviation of 14 points reported in the test manual was also the standard deviation for her population of inmates. However, the manual indicates that the group used in norming the test was a cross section of twelfth-grade high-school students. Because these

343

twelfth-graders constitute a population very dissimilar to the inmate population in terms of age, sex, ethnic origin, and education, Dr. Weber wisely decides that it would *not* be reasonable to assume that the standard deviations of the two groups are identical. For this reason, Dr. Weber determines not to use a z-test, but rather a *t*-test. She decides to conduct the test at the $\alpha = .05$ level.

Because Dr. Weber's alternative hypothesis here is a directional hypothesis, her test will be a one-tailed test. Because she proposes to take a sample of 20 inmates, the degrees of freedom associated with the test will be $20 - 1$ or 19. To find the value of t_{cr} to be used in her study, therefore, Dr. Weber looks in the row of the table of the *t*-distributions adjacent to the value 19 in the left-hand column. She looks in the column of the table headed $\alpha = .05$ for a one-tailed test. At the intersection of this row and column, she finds the value 1.73. Because her alternative hypothesis states that the true population mean is *less than* the values specified in H_o, she knows that negative values of t_{ob} will support her case. Therefore, she assigns a negative value to the tabled value. t_{cr} will be -1.73. She will reject H_o if t_{ob} is less than -1.73.

Computing t_{ob} in a One-Sample *t*-Test. On the basis of her testing, Dr. Weber obtains the data presented in Figure 8.7. She uses these data to compute the sample mean, $\bar{X}_{ob}$, and sample standard deviation, s. These calculations are carried out in Figure 8.7, to illustrate the calculation of the sample standard deviation. Having calculated $\bar{X}_{ob}$ and s, Dr. Weber plugs these values into the formula for *t*-observed and obtains the value -6.03. Because t_{ob} here is less than -1.73, the statistic is in the rejection region. Dr. Weber rejects H_o and concludes that the average reading score for her inmate population is less than 80, the mean for the general population represented in the norm group. The one-sample *t*-test is presented according to our hypothesis testing outline in Figure 8.8.

The *t*-Confidence Interval for a Population Mean

Just as we can construct a confidence interval for a mean using the normal curve model and the known value of σ, we can construct a *confidence interval for a mean using the t-distributions* and an estimated value for σ, the sample standard deviation s. The formulas for z-method confidence intervals and for *t*-method confidence intervals can be compared as follows:

<u>z-Method</u>   <u>*t*-Method</u>

Upper confidence limit for μ $= \bar{X}_{ob} + z_{cr}\left(\dfrac{\sigma}{\sqrt{n}}\right)$ Upper confidence limit for μ $= \bar{X}_{ob} + t_{cr}\left(\dfrac{s}{\sqrt{n}}\right)$

Lower confidence limit for μ $= \bar{X}_{ob} - z_{cr}\left(\dfrac{\sigma}{\sqrt{n}}\right)$ Lower confidence limit for μ $= \bar{X}_{ob} - t_{cr}\left(\dfrac{s}{\sqrt{n}}\right)$

FIGURE 8.7
Calculations of Sample Mean, $\bar{X}_{ob}$, and Sample Standard Deviation, s, for Use in One-Sample t-Test

Subject	Score (X)	X^2
1	60	3600
2	79	6241
3	81	6561
4	48	2304
5	51	2601
6	27	729
7	89	7921
8	58	3364
9	55	3025
10	59	3481
11	48	2304
12	40	1600
13	56	3136
14	51	2601
15	44	1936
16	72	5184
17	84	7056
18	63	3969
19	60	3600
20	54	2916
	$\Sigma X = 1179$	$\Sigma X^2 = 74{,}129$

$$\bar{X}_{ob} = \frac{\Sigma X}{n} = 58.95$$

$$s = \sqrt{\frac{\Sigma X^2 - \frac{(\Sigma X)^2}{n}}{n-1}} = \sqrt{\frac{74{,}129 - \frac{(1179)^2}{20}}{19}} = 15.61$$

The two formulas are analogous. The differences that distinguish z-method from t-method confidence intervals are the same as the differences that distinguish z-tests from t-tests. In the z-method interval, we know σ and we employ a value, z_{cr}, taken from the standard normal distribution. In the t-method interval, we use s as an estimate of σ and employ a value, t_{cr}, taken from one of the t-distributions. The number of degrees of freedom for the t-method confidence interval for μ is the same as the number of degrees of freedom for the corresponding one-sample t-test, i.e., $n - 1$. The value of t_{cr} will depend on the level of confidence desired. For a 95 percent confidence interval, we would find the appropriate

FIGURE 8.8
Example of One-Sample t-Test

Step 1

We have drawn a random sample of independent observations from a population of measures on a single interval scale variable that is either normally distributed or nearly normally distributed. We cannot assume knowledge of the standard deviation of this population. Therefore, the one-sample t-test is appropriate.

Step 2

$H_o: \mu = 80$.

Step 3

$H_a: \mu < 80$.

Step 4

The test statistic is the one-sample t-test. Because n in this case is 20, the number of degrees of freedom is $n - 1 = 19$. We will therefore be employing the t-distribution with 19 df.

Step 5

$\alpha = .05$. The alternative hypothesis is directional, so the test will be one-tailed. Because values of $\bar{X}_{ob} < 80$ support the alternative hypothesis, the rejection region will be on the left (lower) tail of the distribution. It will be the area lying below the fifth percentile of that distribution, $t_{cr} = t_{19;.95} = -1.73$.

Step 6

$$t_{ob} = \frac{\bar{X}_{ob} - \mu_o}{\dfrac{s}{\sqrt{n}}} = \frac{58.95 - 80}{\dfrac{15.61}{\sqrt{20}}} = -6.03.$$

Step 7

Because $t_{ob} < -1.73$, we reject H_o and conclude that the mean of the sampled population is less than $\mu_o = 80$.

346

Inferential Statistics

value of t in the column headed $\alpha = .05$ in a two-tailed test. For a 99 percent confidence interval, we would look in the column headed $\alpha = .01$ for a two-tailed test.

In the example of Dr. Weber and the reading scores, the research situation lent itself to a hypothesis test. However, if we were to assume that Dr. Weber was simply interested in obtaining an interval estimate for the mean reading score of the inmate population, then it would be appropriate to calculate a confidence interval for this parameter. Let us assume that a 99 percent confidence interval is desired. The sample size of 20 still means that we will be interested in the t-distribution with 19 df. We look for the 99.5 percentile in this distribution, which we find by looking in the column $\alpha = .01$ for a two-tailed test. We find this value to be 2.86. We plug this value into the formulas given earlier, along with the values of $\bar{X}_{ob}$ and s calculated from sample data. Thus,

$$\text{Lower confidence limit} = \bar{X}_{ob} - t_{cr}\left(\frac{s}{\sqrt{n}}\right)$$

$$= 58.95 - 2.86\left(\frac{15.61}{\sqrt{20}}\right)$$

Lower confidence limit = 48.97

and

$$\text{Upper confidence limit} = \bar{X}_{ob} + t_{cr}\left(\frac{s}{\sqrt{n}}\right)$$

$$= 58.95 + 2.86\left(\frac{15.61}{\sqrt{20}}\right)$$

Upper confidence limit = 68.93,

which leads us to state: The mean reading score for this population lies between 48.97 and 68.93, and we are 99 percent certain that this statement is true. Because the value of t_{cr} here is somewhat larger than the corresponding percentile in the z-distribution, the confidence interval computed using the t-method will tend to be larger than the equivalent interval using the z-method. However, the interpretation of the interval is identical in each case.

THE CLASSIFICATION MATRIX

In this chapter, we returned to our classification matrix to fill in the inferential techniques in the appropriate cells. We looked at inferential techniques for univariate problems and added the normal approximation to the binomial, the chi-square one-variable test, the Kolmogorov-Smirnov one-variable test, the one-sample t-test, and the t-method of confidence intervals to those techniques we have already examined.

TABLE 8.3
Classification Matrix: Inferential Statistics

Level of Measurement of Variable(s) of Interest

Type of Problem		Categorical: Nominal or Ordered Categories	Rank-Order	Interval Scale (or Ratio-Scale)
Univariate Problems		Binomial Test Normal Approximation to Binomial χ^2 One-Variable Test *With Ordered Categories:* Kolmogorov-Smirnov One-Variable Test		
Bivariate Problems	Correlational Problems			
	Experimental and Pseudo-Experimental Problems: Independent Groups			
	Dependent Measures: Pretest and Posttest or Matched Groups			

Key Terms

Be sure you can define each of these terms.

normal approximation to the binomial
continuity correction factor
confidence interval for P
χ^2 one-variable test
goodness-of-fit test
degrees of freedom
χ^2 distribution
Kolmogorov-Smirnov one-variable test
t-distribution
t-statistic
sample standard deviation
one-sample t-test
confidence interval (t-method)

Summary

In Chapter 8, we returned to our classification matrix and examined inferential techniques for univariate problems. We looked at techniques for nominal scale variables, ordered categories, and interval scale variables. We pointed out that there are no inferential techniques applicable to a single set of ranks.

We reviewed the use of the binomial model to test hypotheses about population proportions with categorical variables having only two values. We went on to show how the normal curve can be used to approximate the binomial distribution with sufficiently large sample sizes. With the normal approximation to the binomial, we use a one-sample z-test with a correction factor. We can also use the normal approximation to set up confidence intervals for the population proportion based on our sample proportion.

We then discussed a χ^2 one-variable test that can be used to test hypotheses regarding population proportions for categorical variables with two or more values. In this case, we use the difference between expected frequencies and observed frequencies to compute the χ^2 statistic and then compare this value to a critical value obtained from a table of the χ^2

probability distribution. If our observed value is higher than the critical value, we reject our hypothesis regarding the various population proportions. We briefly introduced the idea of degrees of freedom in this context, because to find our critical value of χ^2, we need to know our number of degrees of freedom as well as our level of significance. For this test, the number of degrees of freedom is equal to one less than the number of values of our variable.

We saw that if our categories are ordered, it is preferable to use the Kolmogorov-Smirnov one-variable test because it is more powerful than the χ^2 one-variable test. With the Kolmogorov-Smirnov one-variable test, we test hypotheses regarding cumulative proportions rather than simple proportions. This is possible because the values of our variable have a clear order in this case. With the Kolmogorov-Smirnov one-variable test, we compare the largest difference between expected and observed cumulative proportions to a critical value obtained from a table. If our difference is greater than the critical value, we reject our hypothesis regarding cumulative proportions.

We also reviewed the one-sample z-test for the population mean of a continuous interval variable. We pointed out that this test cannot be used unless we know the population standard deviation. Because we often do not know the population standard deviation, we must estimate it with the sample standard deviation, which has a slightly modified formula. In this case, we use one of the t-distributions in place of the normal curve and the t-statistic in place of the z-statistic. Which t-distribution we use depends on our number of degrees of freedom; in this case, this number is one less than the sample size. We noted that as the number of degrees of freedom increases, the t-distribution becomes more and more like the normal curve.

The one-sample t-test and the t-method of confidence intervals for a population mean are analogous to the one-sample z-test and the z-method of confidence intervals. In both cases, we test hypotheses about or construct confidence intervals for population means. However, we must use the t-methods in the absence of knowledge of the population standard deviation.

Review Questions

To review the concepts presented in Chapter 8, choose the best answer to each of the following questions.

1 The reason we correct for continuity when we use the normal curve to estimate binomial probabilities is that
_____a r is a discrete variable as a binomial variable but a continuous variable as a normal variable.
_____b the binomial distribution is less precise than a normal curve.

2 To evaluate the probability that we would see at least 160 heads in 300 tosses of a fair coin, we may use the normal curve where
_____a $\mu = 150$ and $\sigma = 8.66$.
_____b $\mu = 160.5$ and $\sigma = 8.66$.

3 The reason we use the standard normal distribution to make probability statements about the variable $\dfrac{r - np}{\sqrt{nPQ}}$ is that

_____a it is a standard score and standard scores are normal.

_____b it is a standard score for r, and r has an approximate normal distribution when n is large.

4 In using a normal curve to estimate binomial probabilities, we know this method will be good provided

_____a $n > 30$.

_____b $nPQ \geqslant 9$.

5 The distribution of the t-statistic approaches the standard normal distribution as

_____a n approaches ∞.

_____b μ approaches ∞.

6 In comparing t-distributions to the standard normal distribution, we note that

_____a the standard normal distribution is bell-shaped but t-distributions are not.

_____b both kinds of curves are bell-shaped but the t-distributions are "fatter" in the tails.

7 In using a normal curve to establish a one-tailed rejection rule, we pick a percentile value and reject H_o for any z_{ob} that is further from zero. The name given to the chosen percentile value is

_____a a critical value.

_____b the level of significance.

8 The degrees of freedom for the χ^2 distribution in the goodness-of-fit test depend on

_____a the number of observations.

_____b the number of categories.

9 One requirement for the appropriateness of the chi-square goodness-of-fit test is that

_____a the hypothetical proportion times the sample size be at least 5 for every cell.

_____b no cell have less than 5 observations.

10 In the one-variable χ^2 test, the null hypothesis specifies the proportion of the population in each category and the alternate hypothesis says that at least one of the proportions is not as specified. We say that the alternate hypothesis is

_____a nondirectional.

_____b directional.

352

11 The number of degrees of freedom for the t-statistic in the one-sample t-test depends on

    _____a the number of observations in the sample.

    _____b the number of values of the variable.

12 The difference in the formulas for the t- and z-statistics is that

    _____a t has two variables in its formula.

    _____b z has two variables in its formula.

13 In hypothesis testing situations concerning the mean of a normal distribution, t will be used instead of z when

    _____a we don't know about the population sampled.

    _____b we don't know the population standard deviation of the sampled normal distribution.

14 Suppose we use a normal curve to establish the proportion of the population that falls into a low IQ category, an average IQ category, and a high IQ category. A random sample is classified into the three categories. To see whether we have hypothesized the correct normal distribution we could

    _____a use the Kolmogorov-Smirnov one-variable test.

    _____b use the one-sample t-test.

15 Suppose we are interested in a color preference task. Each of 50 subjects selects his or her favorite color (blue, red, yellow, or green) from the list. Recent studies have shown that the proportion of the population favoring blue is .40, favoring red is .30, favoring yellow is .10, and favoring green is .20. To test the hypothesis that these proportions describe the population sampled in this study, we would use

    _____a the D statistic.

    _____b the χ^2 statistic.

16 If we are interested in the proportion of a population that falls into one of two categories and we have no idea as to what this proportion might be, we could use the information from a sample most appropriately to

    _____a calculate a confidence interval for P.

    _____b test some hypothesis about P.

17 The proportion of the vote Candidate A will receive is estimated as $.51 < P_A < .53$ with 99 percent confidence. This estimate is based on a random sample of 1,000 voters. We can say

    _____a the probability is .99 that another sample of 1,000 will say that A receives from 51 to 53 percent of the vote.

    _____b with 99 percent confidence that A will get more than 50 percent of the vote because .5 is not a value in the confidence interval.

18 To estimate the average SAT score for seniors in a large metropolitan community, 400 seniors are selected at random and their average SAT score is

calculated. This information is used to construct a 95 percent confidence interval for μ. The calculation places the community average SAT score between 470 and 490. This tells us that

_____a $H_o: \mu = 500$ would be rejected in favor of $\mu \neq 500$ at $\alpha = .05$.
_____b 95 times out of 100, μ is between 470 and 490.

Problems for Chapter 8: Set 1

1 Draw the graphs of the following binomial distributions using the binomial tables in Appendix C (Table C.1).
   a $n = 20, P = .5$
   b $n = 20, P = .8$
   c $n = 25, P = .5$
   d $n = 25, P = .8$

2 a Describe the shape of each distribution in Problem 1.
   b How does the shape change when P stays the same, but n increases?
   c How does the shape change when n is the same, but P increases?

3 Calculate the mean and standard deviation for each of the four distributions in Problem 1.

4 a Use the tables of binomial probabilities to evaluate the following probabilities:
   i   that $r \leqslant 7$ if $n = 20$ and $P = .5$.
   ii that $r \geqslant 8$ if $n = 20$ and $P = .5$.
   iii that $r = 7$ if $n = 20$ and $P = .5$.
   b Next, we will use a normal curve to estimate the exact probabilities given in a. We will use the normal curve with $\mu = 10$ and $\sigma = 2.24$. We will try two ways, without a continuity correction and with a continuity correction, to see which way gives us the best estimate.

	Without correction	With correction
i	$r \leqslant 7$	$r \leqslant 7.5$
ii	$r \geqslant 8$	$r \geqslant 7.5$
iii	$r = 7$	$6.5 \leqslant r \leqslant 7.5$

   c Compare normal curve estimates to exact binomial probabilities. In b, which gives the closest answer to the corresponding probabilities in a: without continuity correction or with continuity correction?
   d Describe how a normal curve can be used to estimate binomial probabilities.

5 a Use the tables of binomial probabilities to estimate the following binomial probabilities:
   i $r \geqslant 19$ if $n = 20$ and $P = .8$

354

ii $r \leqslant 18$ if $n = 20$ and $P = .8$

iii $r = 19$ if $n = 20$ and $P = .8$

b Use a normal curve with $\mu = 16$ and $\sigma = 1.79$ to attempt to evaluate probabilities for the events of a. Try without continuity correction and with continuity correction.

c Does the normal curve give good estimates for the binomial probabilities either with or without the continuity correction?

d Why does the normal curve not work in this example?

6 A seed company claims that at least 90 percent of all its seeds will germinate. To test this claim, a farmer selects a random sample of 100 seeds and records the proportion of his sample that germinate. What should he conclude about the company's claim if 85 percent of his sample germinate? Use $\alpha = .05$ and follow our seven-step outline for hypothesis testing.

7 To predict the winner of an election that features two candidates, A and B, a pollster selects a random sample of 500 voters and asks whom they will vote for. Calculate a 99 percent confidence interval for the proportion of the total vote going to candidate A based on the pollster's results that 260 prefer A and 240 prefer B. Is the pollster justified in predicting a winner?

8 A geneticist claims that 40 percent of the entire population has a certain trait. To test this claim an experimenter selects a random sample of 25 people. She believes the geneticist is underestimating the proportion of the population who have the trait. Using $\alpha = .05$, set up procedures to test the geneticist's claim using:

a an exact binomial test.

b the normal curve to approximate the binomial test.

c What should the experimenter conclude if her sample yields 16 people having the trait?

Set 2

1 Use the chi-square tables in Appendix C (Table C.3) to answer the following questions.

a What is the ninety-fifth percentile in the χ^2 distribution with 1 degree of freedom?

b In the χ^2 distribution with 7 degrees of freedom, what value has percentile rank 99?

c What is the ninetieth percentile in the χ^2 distribution with 4 degrees of freedom?

d If an investigator has four categories and is using the chi-square one-variable test with $\alpha = .05$, what should the critical value be?

2 Use a chi-square one-variable test to test the hypothesis mentioned in Problem 8, Set 1.

3 An experimenter writes a story that has a moral. He has written the story so that a reader might possibly think the moral is any one of four possibilities. He wonders whether or not the members of the population are equally likely to choose any one of the four as the moral to the story. To explore this hypothesis, he selects a random sample of 80 subjects from his population; then, he has them read the story and choose what they think the moral is.

a Set up a statistical test for his hypothesis. Use $\alpha = .01$.

b Would the test set up in a be appropriate if he had selected 10 subjects? Explain.

c Suppose he selected 20 people and asked each to do the task four times, giving 80 observations. Would the procedures established in a be appropriate? Explain.

d Carry out the statistical test designed in a if the observed results show that 40 subjects chose moral 1, 10 chose moral 2, 10 chose moral 3, and 20 chose moral 4. What conclusions can be made?

4 In a survey of students' attitudes toward their teachers, a random sample of 900 students was asked to rate the extent of their agreement with the statement "I get tired of hearing my teachers talk all the time" on a rating scale from "strongly disagree" to "strongly agree." The investigator uses the sample results that follow to test the hypothesis that the student responses will be proportionately the same in each of the response categories. She plans to use the χ^2 one-variable test with $\alpha = .05$.

a Set up the χ^2 test for the investigator.

b Carry out the test using the observed results.

c What conclusion can be made?

Category	Number of students choosing the category
Strongly Agree	230
Agree	410
Undecided	105
Disagree	130
Strongly Disagree	25
Total	900

Set 3

1 A teacher of statistics wonders if there is reason to believe that students feel very anxious while taking statistics examinations. At the end of a recent exam, she has 30 randomly selected students rate themselves on the extent of their agreement with the statement "I felt anxious while I was taking the test." Use the null hypothesis that the proportion of students at each of five levels of agreement is the same to do the following.

a Set up a Kolmogorov-Smirnov one-variable test using $\alpha = .01$.

b Carry out the test if the results show that 5 strongly agree, 15 agree mildly, 7 feel neutral, 2 disagree mildly, and 1 strongly disagrees.

c Draw a conclusion.

2 Previous research has shown that student attitudes towards a statistics class are generally hostile. A teacher of statistics believes that if the class is made relevant, comprehensible, and nonthreatening, then students will enjoy learning statistics. She develops such a course. She selects 30 students at random to take the course. At the end of the course, they rate their feelings about the statistics class. A rating scale is used that was also used in previous research. How do the results of her sample compare to the results of previous research? Use the Kolmogorov-Smirnov test with $\alpha = .05$.

Category	Proportions from previous research	Sample frequency
Statistics is my favorite class.	.05	15
Statistics was enjoyable.	.15	10
Statistics was tolerable.	.50	5
Statistics was horrible.	.30	0
	1.00	30

Set 4

1 What is the effect of the increase in degrees of freedom on the shape of the t-distributions?

2 Assume a normal distribution is to be sampled. Suppose you wish to test $H_o: \mu = 10$ vs. $H_a: \mu > 10$. You want $\alpha = .01$.
a Write down critical values of t for your test if you have a random sample of observations of $n = 5$.
b Do the same thing for $n = 10, 20, 30, 60, \infty$.
c What do you notice about the critical value as n increases?
d What sample size will give you the lowest critical value? Why?

3 When could a normal curve be used to approximate t-distribution probabilities?

4 A random sample of 15 people were measured on the variable "weight." To the nearest pound, the measurements are as follows:
115, 100, 95, 130, 127, 114, 150, 180, 127, 190, 140, 150, 192, 125, 134.
Calculate the sample mean and the standard deviation.

Set 5

1 A sample of 20 students are selected at random and measured on an IQ test that has a mean of 100 and a standard deviation of 15. The test is assumed to have a normal distribution on the normed population. The results are as follows:

357

104, 123, 141, 121, 94, 103, 70, 123, 112, 127,
111, 109, 110, 102, 99, 105, 113, 106, 104, 105.

a   Test the hypothesis that the 20 constitute a random sample from the normed population. Use a z-test with $\alpha = .05$.

b   Test the same hypothesis using a t-test with $\alpha = .05$.

2   Use the data in Problem 1 to calculate a 95 percent confidence interval for μ using

a   the z-method.

b   the t-method.

c   Compare the results.

3   A random sample of 10 adult males yields the following heights: 63.2 in., 80.7 in., 69.5 in., 68.2 in., 67.9 in., 70.5 in., 72.3 in., 69.4 in., 75.2 in., and 73.4 in. Is it reasonable to believe that this sample is from a normal distribution of heights with $\mu = 68.5$? Use $\alpha = .01$.

4   A gasoline company uses an additive in its product in the hope that the additive will increase mileage. For a certain car, without the additive, the average miles per gallon is 13.2. A random sample of 36 runs is made with the additive and the results show an average of 15.1 m.p.g. for the 36 trials with $s = 3.2$ m.p.g. At a .01 level of significance, what can the company conclude?

9

Inferential Statistics for Bivariate Correlational Problems

You will recall from Chapter 1 that we have a bivariate correlational problem when we obtain information regarding two variables from a set of subjects without manipulating or controlling either of these variables. In that chapter, we noted that correlational studies can be used to establish whether two variables are related to each other but not to establish a definite cause and effect relationship. Bivariate correlational problems may be either descriptive or inferential in nature. A bivariate correlational problem is descriptive if we have actually measured the two variables for the entire population of interest. In Chapter 5, we presented the descriptive statistics employed with bivariate correlational data. These techniques included Cramér's index of contingency, for use when the two variables measured are categorical variables; the Spearman rank-order correlation, for use when the two variables measured are rank-order variables; and the Pearson product-moment correlation, for use when the two variables measured are interval variables. Each of these statistics provides a numerical index of the strength of the relationship between the two variables measured within the specific sets of data used in the calculation. If the problem is descriptive and we have no desire to generalize the results to any group other than the group actually used to calculate the statistic, then we need only to report the appropriate descriptive statistic.

However, bivariate correlational problems may also be inferential in nature. We may select a sample of subjects from some specified population. We may measure each of these subjects on two variables, not to see if a relationship exists in the sample, but to determine whether a relationship exists between these two variables in the larger population from which the sample was drawn. When we have a bivariate correlational problem that is inferential rather than descriptive, we will need to use some different or additional statistical techniques. If our data are categorical, we will employ a statistical test known as the chi-square test of association. If our data are ordinal or interval in nature, we will probably begin by computing the Spearman rank-order correlation or the Pearson product-moment correlation for the data in our sample. Then, we will perform a test to determine whether this correlation is large enough, given our sample size, to infer that a relationship exists in the population. Let us now consider the specific techniques employed with each of the three types of data.

INFERENTIAL TECHNIQUES FOR BIVARIATE CORRELATIONAL PROBLEMS: CATEGORICAL DATA

The Chi-Square Test of Association

You will recall our study of the relationship between sex and opinion on the ERA among residents of Tranquility, New York. Remember that this was a descriptive problem because our concern was only with the group of subjects actually contacted. Let us now assume that an executive of a major women's organization learns of our study, and commissions us to carry out a research project in which the same questions will be asked, but this time with respect to the entire state of New York. Because it is not practical to attempt to survey all of the registered voters in New York, we will have to rely on a random sample of voters to answer this question. Of course, this means that we now have an inferential problem. The question is the same; we still have a bivariate correlational problem. The two variables are the same; we still have categorical data. But now the problem is inferential rather than descriptive.

In this situation, we would employ a *chi-square test of association.* In this test, the null hypothesis is that the two variables of interest are unrelated in the population of interest. The alternative hypothesis is that they are related. In drawing a random sample, our assumption is that whatever relationship that exists between the two variables in the population will be approximated within the sample. Of course, we know that, because of sampling error, the relationship in the sample will not always be identical to the relationship in the population. Nevertheless, we know that if we have a true random sample of sufficient size, the sample should resemble the population most of the time. For this reason, if we find a rather substantial relationship between the two variables within our sample, we will most likely conclude that some relationship exists between the two variables in the population. We use the chi-square statistic as a tool in making this decision. It enables us to decide whether to reject the null hypothesis with knowledge of the chance of making a Type I error in this rejection. In this regard, it is similar to the statistical tests considered in the previous chapter, and, in fact, to all inferential statistics.

The Computation of the Chi-Square Test of Association. The computation of the chi-square statistic for the chi-square test of association begins with the preparation of a crosstabulation table for the sample data. The statistic is based on the comparison of observed and expected cell frequencies. In fact, the chi-square statistic for a given crosstabulation table is actually computed as an intermediate step in the calculation of the index of contingency for that table. If you recall the discussion of the index of contingency in Chapter 5, you already know how to compute the chi-square statistic for the test of association. If you feel uncertain regarding the meaning of expected and observed frequencies, take a moment to review these concepts in Chapter 5. Otherwise, read on from here as we illustrate the calculation of chi-square with data from our hypothetical study.

Let us assume that we have obtained a random sample of $n = 500$ voters from New York State, and that we have ascertained their sex and their opinion on the ERA. The crosstabulation table for this hypothetical sample of 500 voters is presented in Table 9.1. You will note that our random sample of New York voters contains 230 males and 270 females (column marginals). It has 170 voters in favor, 190 opposed, and 140 with no opinion (row marginals). The percentages of the 500 subjects who fell into each of the 3 different categories on the variable "opinion" are presented in parentheses below the number of subjects in that category. You will recall from Chapter 5 that these percentages are

---

TABLE 9.1
Crosstabulation of Opinion on ERA by Sex for a Random Sample
of 500 New York State Voters

		Sex		
		Males	Females	Total
Opinion	In favor	$Ex = 78.2$ $Ob = 50$	$Ex = 91.8$ $Ob = 120$	170 (34.0%)
	Opposed	$Ex = 87.4$ $Ob = 90$	$Ex = 102.6$ $Ob = 100$	190 (38.0%)
	No opinion	$Ex = 64.4$ $Ob = 90$	$Ex = 75.6$ $Ob = 50$	140 (28.0%)
	Total	230	270	500 (100.00%)

useful in the calculation of expected cell frequencies. Under the null hypothesis that the two variables are unrelated, we would expect that if 34.0 percent of all voters were in favor, then 34.0 percent of voters in each sex would have been in favor. Thus, 34.0 percent of the 230 males in our sample would be expected to be in favor, and 34.0 percent of the 270 females in our sample would be expected to be in favor. We find that 34.0 percent of 230 is 78.2, which is the expected frequency for the cell corresponding to males who were in favor; and 34.0 percent of 270 is 91.8, the expected frequency for the cell corresponding to females who were in favor. Expected frequencies for the other 4 cells are computed similarly. The expected frequencies are presented in Table 9.1 for all 6 cells in the shaded upper left portion of each cell, labeled *Ex*. The actual or observed number of subjects falling into each of the 6 cells is presented in the lower right portion of the cell, labeled *Ob*. All of this should be familiar to you from Chapter 5.

Having calculated the expected frequencies for each cell and tabled them near the corresponding observed frequencies, we proceed to calculate the observed value of the chi-square statistic using the formula:

$$\chi^2_{ob} = \sum\sum \frac{(Ob - Ex)^2}{Ex}.$$

A step-by-step guide to the calculation of χ^2_{ob} has been provided in Chapter 5, in Steps 1 through 4 of the guide to computing the index of contingency. The calculation of χ^2_{ob} for the present example is presented in Figure 9.1. We find that the value of χ^2_{ob} here is 37.83. As you may already have guessed, this value of the observed statistic must be compared to a critical value in order to determine whether or not our sample results are significant.

The Critical Value of χ^2 in the Test of Association. In the χ^2 test of association, as in the χ^2 goodness-of-fit test described in Chapter 8, the critical value of χ^2 depends upon the number of degrees of freedom associated with the problem. In a chi-square test of association, the number of degrees of freedom for the test is found using the following formula:

$$\text{degrees of freedom} = (R - 1)(C - 1),$$

where R is the number of rows in the crosstabulation table, and C is the number of columns in the table. The row and column for the marginal totals do not count. Only the rows and columns representing the values of the variables of interest are counted. In the problem at hand, we have 3 rows, one for each of the 3 values of the variable "opinion," and 2 columns, one for each of the 2 values of the variable "sex." Thus, in this problem, the number of degrees of freedom is:

$$df = (3 - 1)(2 - 1) = 2.$$

FIGURE 9.1
Calculation of χ^2_{ob} for Crosstabulation of Opinion on ERA by Sex
for a Random Sample of 500 New York Voters

	Step 1	Step 2	Step 3	
Cell	$Ob - Ex$	$(Ob - Ex)^2$	$\dfrac{(Ob - Ex)^2}{Ex}$	
1, 1	$50 - 78.2 = -28.2$	$(-28.2)^2 = 795.24$	$\dfrac{795.24}{78.2} = 10.17$	Add up all these terms to obtain χ^2_{ob}
1, 2	$120 - 91.8 = +28.2$	$(+28.2)^2 = 795.24$	$\dfrac{795.24}{91.8} = 8.66$	
2, 1	$90 - 87.4 = +2.6$	$(+2.6)^2 = 6.76$	$\dfrac{6.76}{87.4} = 0.08$	
2, 2	$100 - 102.6 = -2.6$	$(-2.6)^2 = 6.76$	$\dfrac{6.76}{102.6} = 0.07$	
3, 1	$90 - 64.4 = +25.6$	$(+25.6)^2 = 655.36$	$\dfrac{655.36}{64.4} = 10.18$	
3, 2	$50 - 75.6 = -25.6$	$(-25.6)^2 = 655.36$	$\dfrac{655.36}{75.6} = 8.67$	

Step 4

$$\chi^2_{ob} = \sum\sum \frac{(Ob - Ex)^2}{Ex} = 37.83$$

Having obtained the number of degrees of freedom associated with our test, we can obtain a critical value of χ^2 using the χ^2 table contained in Appendix C (Table C.3). The use of this table has already been explained in connection with the χ^2 goodness-of-fit test in Chapter 8. To recapitulate briefly, we locate the appropriate number of degrees of freedom in the right-hand column of the table, and note the row of the table adjacent to that number of degrees of freedom. We locate the desired significance level of our test in the top row of the table, and note the column of the table that runs down from this level of significance. We find our χ^2_{cr} at the intersection of the noted row and column. You should verify that for a test with 2 df at the $\alpha = .05$ level of significance, χ^2_{cr} is 6.0. For a test with 2 df at $\alpha = .01$, χ^2_{cr} is 9.2.

363

We compare χ^2_{ob} to χ^2_{cr} to determine whether our test is significant. In the problem at hand, χ^2_{ob} was found to be 37.83, greater than either 6.0 or 9.2. Thus, this test is significant beyond the .01 level. In fact, this test is significant beyond the .001 level. (What is χ^2_{cr} for $\alpha = .001$?) Accordingly, we decide to reject H_o, which stated that the two variables were unrelated in the population of New York State voters. We conclude instead that there is a relationship between sex and opinion on the ERA among the voters of New York. The results of our sample are strong enough for us to infer a relationship in the larger population. The entire χ^2 test of association for our sample of 500 voters has been summarized, according to our hypothesis testing outline, in Figure 9.2. As you consider this summary, you will notice in Step 1 that there are certain conditions that must be met in order to use the χ^2 test of association. We consider these constraints on the use of the χ^2 test of association next.

Constraints on the Use of the χ^2 Test of Association.   As in the case of any inferential statistic, the χ^2 test of association employs a mathematical model. The mathematical model is one of a family of probability distributions known as the chi-square distributions. Exactly which chi-square distribution is appropriate for a given problem depends upon the number of degrees of freedom associated with that problem. If we were to obtain our critical value for χ^2 from a χ^2 distribution with the wrong number of degrees of freedom, our test would not be accurate.

But it is not enough for us to be sure to use the χ^2 distribution with the correct number of degrees of freedom. In order for the mathematical model to provide accurate results, certain other criteria must be met as well. These criteria are similar to those noted in connection with the χ^2 goodness-of-fit test. Specifically, in order to use the chi-square model with confidence, the following assumptions must be met.

1   *Observations must be independent.* In terms of the problem with which we have been working, this means that each one of the 500 subjects included in our sample must be selected randomly, in such a manner that the inclusion of any one voter in the sample will have no effect on the chance that any other voter in the population will be included.

2   *Cells in the crosstabulation table must be mutually exclusive.* This means that we must be able to place each one of our 500 subjects in one and only one of the cells of the table. In order for this to be true, we must be able to place each subject in one and only one category on each of the two variables we have measured.

3   *The expected frequencies for the cells must be large enough.* If the χ^2 approximation is to be accurate, it is advisable that the expected frequency for each cell in the crosstabulation table be at least 5.

FIGURE 9.2
Example of χ^2 Test of Association: Is There a Relationship Between
Sex and Opinion on the ERA Among Voters in New York State

Step 1:

Because we have a bivariate correlational problem with categorical variables, we consider the use of the χ^2 test of association. Because we have independent observations, mutually exclusive cells in the crosstabulation table, and sufficiently large cell frequencies, we know that our data are appropriate for this statistic.

Step 2:

H_o: There is no relationship between sex and opinion on the ERA in the sampled population.

Step 3:

H_a: There is a relationship in the population.

Step 4:

The test statistic is the χ^2 statistic with $(R-1)(C-1) = 2$ degrees of freedom, so

$$\chi^2 = \sum\sum \frac{(Ob - Ex)^2}{Ex} \text{ will have 2 df.}$$

Step 5:

$\alpha = .01$. Using the table of the χ^2 distributions, we find that χ^2_{cr} for $\alpha = .01$ in a test with 2 df. is 9.2. We will reject H_o if χ^2_{ob} is greater than 9.2.

Step 6:

$\chi^2_{ob} = 37.83$.

Step 7:

Because $\chi^2_{ob} > \chi^2_{cr}$, we reject H_o at the .01 level of significance. We conclude that there is a relationship between sex and opinion on the ERA among the population of New York State voters.

Bivariate Correlational Problems

Be sure to note that this last rule applies to *expected* frequencies, not *observed* frequencies. Observed frequencies may be any value. Indeed, crosstabulation tables for strongly related variables will often have observed frequencies of 0 or close to 0. However, in order for the χ^2 model to provide an accurate significance test, expected cell frequencies must not be too small.

In those cases where there are a relatively large number of cells in the crosstabulation table, it is not absolutely essential that *every* expected frequency be at least 5. Here, the general rule of thumb is that not more than 20 percent of the cells should have expected frequencies of less than 5, and no cell should have an expected frequency of less than 1.

If the cell frequencies are not large enough for a particular table, we sometimes combine two or more categories on one or both of the variables. For example, consider the crosstabulation presented in Table 9.2. This table presents the results of a hypothetical random sample of 300 Minneapolis heads-of-household conducted by a researcher seeking to investigate the relationship between level of education and annual income among city residents. In obtaining observations on these two variables for the 300 subjects included in the sample, the researcher employed 5 levels of education and 8 levels of annual income. Although there is certainly nothing inherently wrong with employing this many values for the variables measured, this decision does present a problem with respect to the calculation of a chi-square statistic for the chi-square test of association. You will note that 21 of the 40 cells represented in the table have expected frequencies of less than 5. This is because of the combination of: (1) the large number of cells; (2) the relatively small number of subjects for this number of cells; and (3) the nature of the marginal distributions of the two variables, which are such that large proportions of observations fall in some categories, and very small proportions fall in other categories. With respect to this third point, we simply mean that many subjects have incomes in the $10,000–$14,999 and $15,000–$19,999 ranges, but few subjects have incomes in the categories of $25,000 and over. Similarly, there are many subjects with high-school diplomas and bachelor's degrees, but few with doctoral degrees. With the expected cell frequencies we find in Table 9.2, it would not be appropriate to conduct a chi-square test. To do so would produce a wildly inflated statistic that would have no meaning.

However, by combining or collapsing several of the categories, it is possible to rectify this situation. Furthermore, in this problem it makes good sense to combine certain categories. For example, the 2 lowest income categories may be grouped into a category called "less than $10,000," and the 3 highest income categories may be grouped into a category called "$25,000 or more." Similarly, we could combine the 2 highest education categories. In each case, we are combining categories that have small marginal frequencies because these small marginals are producing the small expected frequencies. Note that we are combining only adjacent categories. It would not make sense, for example, to combine the lowest income category with the highest. The collapsed crosstabulation resulting from

366

TABLE 9.2
Crosstabulation of Yearly Income by Level of Education
for 300 Minneapolis Heads-of-Household

Income	Education					Total
	No high-school diploma	High-school diploma, not Bachelor's degree	Bachelor's degree, but no graduate degree	Master's level degree	Doctoral level degree	
Less than $5,000	Ex = 5.38 Ob = 9	Ex = 5.76 Ob = 6	Ex = 4.69 Ob = 3	Ex = 2.47 Ob = 1	Ex = 0.70 Ob = 0	19
$5,000–$9,999	Ex = 7.08 Ob = 12	Ex = 7.58 Ob = 8	Ex = 6.17 Ob = 5	Ex = 3.25 Ob = 0	Ex = 0.92 Ob = 0	25
$10,000–$14,000	Ex = 19.55 Ob = 21	Ex = 20.93 Ob = 24	Ex = 17.02 Ob = 19	Ex = 8.97 Ob = 5	Ex = 2.53 Ob = 0	69
$15,000–$19,999	Ex = 28.62 Ob = 31	Ex = 30.64 Ob = 30	Ex = 24.91 Ob = 27	Ex = 13.13 Ob = 12	Ex = 3.70 Ob = 1	101
$20,000–$24,999	Ex = 13.60 Ob = 12	Ex = 14.56 Ob = 16	Ex = 11.84 Ob = 10	Ex = 6.24 Ob = 8	Ex = 1.76 Ob = 2	48
$25,000–$29,999	Ex = 5.38 Ob = 0	Ex = 5.76 Ob = 2	Ex = 4.69 Ob = 6	Ex = 2.47 Ob = 9	Ex = 0.70 Ob = 1	19
$30,000–$34,999	Ex = 2.83 Ob = 0	Ex = 3.03 Ob = 4	Ex = 2.47 Ob = 1	Ex = 1.30 Ob = 2	Ex = 0.37 Ob = 3	10
$35,000 or more	Ex = 2.55 Ob = 0	Ex = 2.73 Ob = 1	Ex = 2.22 Ob = 3	Ex = 1.17 Ob = 2	Ex = 0.33 Ob = 3	9
Total	85	91	74	39	11	300

these combinations is presented in Table 9.3. You will note that none of the
cells in this collapsed table has an expected frequency less than 5. This table
meets the criteria necessary for the use of the chi-square model. We could now

Bivariate Correlational Problems

proceed to test the hypothesis that the two variables measured are unrelated in the population of Minneapolis heads-of-household. In determining our critical value for χ^2, we would use the number of rows and the number of columns in the collapsed table. Degrees of freedom would be $(5-1)(4-1) = 12$. You may wish to check your understanding of the χ^2 test of association at this point by calculating the observed value of χ^2 for Table 9.3.

INFERENTIAL TECHNIQUES FOR BIVARIATE CORRELATIONAL PROBLEMS: RANK-ORDER DATA

Test for the Significance of the Spearman Correlation

In Chapter 5, we discussed the use of the Spearman rank-order correlation coefficient in describing the relationship between two sets of ranks. As long as we are dealing with a purely descriptive problem, such as the problem of the rank in class of 15 students in math and English, we need to do nothing more than compute the Spearman correlation. However, a bivariate correlational problem involving ordinal data may be inferential as well as descriptive. We may want to *test the significance of the Spearman correlation.* For example, suppose Psychology Professor Falk would like to test his belief that there is a positive relationship between the personality variables "popularity" and "assertiveness" in the population of college-age males. Dr. Falk selects a random sample of 12 college-age males from the population of Ann Arbor, Michigan. He then arranges to have these 12 subjects spend several 3-hour periods in an entertainment room at the university, where they are free to play games, talk to each other, listen to music, or just relax. During these sessions, the subjects are observed by 2 groups of judges. One group of judges is instructed to rank the 12 subjects from most popular to least popular. The other set of judges is instructed to rank the subjects from most assertive to least assertive. The two groups of judges are physically separate from each other and are not informed of what the other group is doing. Within each group, judges rank subjects independently. Then the ranks given to each subject by the judges are averaged. Thus, the final rankings on popularity and assertiveness are based on the average rank given subjects by several different judges. The purpose of this procedure is to make the rank ordering of subjects as reliable as possible.

This situation is clearly an inferential problem. Professor Falk is certainly not concerned with the relationship between these two variables among the 12 subjects actually observed. Rather, he is concerned with the relationship between the two variables in the larger population. He would like to be able to perform a test to determine whether the relationship present in his sample of 12 is sufficiently strong to conclude that there is a relationship in the population. If the

368

TABLE 9.3
Collapsed Crosstabulation of Yearly Income by Level of Education for 300 Minneapolis Heads-of-Household

Income	Education: No high-school diploma	High-school not Bachelor's degree	Bachelor's degree, no graduate degrees	Has graduate degree	Total
Less than $10,000	$Ex = 12.47$ / $Ob = 21$	$Ex = 13.35$ / $Ob = 14$	$Ex = 10.85$ / $Ob = 8$	$Ex = 7.33$ / $Ob = 1$	44
$10,000-$14,000	$Ex = 19.55$ / $Ob = 21$	$Ex = 20.93$ / $Ob = 24$	$Ex = 17.02$ / $Ob = 19$	$Ex = 11.50$ / $Ob = 5$	69
$15,000-$19,000	$Ex = 28.62$ / $Ob = 31$	$Ex = 30.64$ / $Ob = 30$	$Ex = 24.91$ / $Ob = 27$	$Ex = 16.83$ / $Ob = 13$	101
$20,000-$24,999	$Ex = 13.6$ / $Ob = 12$	$Ex = 14.56$ / $Ob = 16$	$Ex = 11.84$ / $Ob = 10$	$Ex = 8.00$ / $Ob = 10$	48
$25,000 or more	$Ex = 10.77$ / $Ob = 0$	$Ex = 11.53$ / $Ob = 7$	$Ex = 9.37$ / $Ob = 10$	$Ex = 6.33$ / $Ob = 21$	38
Total	85	91	74	50	300

variables "popularity" and "assertiveness" are not related in the population, then the value of the Spearman correlation coefficient for the population, ρ_s, would be 0. Thus, Professor Falk will seek to use the data obtained from his sample of 12 to test the null hypothesis that the Spearman correlation between the two variables is zero in the population sampled, i.e., $H_o: \rho_s = 0$.

Carrying Out the Test of H_o: $\rho_s = 0$. In conducting this test, the first step is always the calculation of the Spearman correlation for the sample data. This is done exactly as described in Chapter 5, just as if the problem were descriptive. The only difference to be noted is in the symbol we employ to denote the Spearman correlation. Because the correlation here is used as a sample statistic estimating the value of a population parameter, we now employ the symbol r_s. It is common practice to use a Greek letter to represent a population parameter and the corresponding Latin letter to represent the sample statistic that estimates the parameter. You will recall from Chapter 7 that we use the Greek σ to represent the population standard deviation, and the Latin s to represent the sample estimate of σ. Here, the sample Spearman correlation r_s is used to estimate the population Spearman correlation, ρ_s.

The hypothetical data for the experiment just described are presented in Figure 9.3, along with the calculation of the rank-order correlation for the data. If you have any question regarding the steps involved in this calculation, you should review Chapter 5. Once we have computed the Spearman correlation for the data in the sample, one of two methods is employed to determine whether that correlation is strong enough in the sample to infer a relationship in the population. Which of the two methods we use depends on the size of the sample.

Sample Size Less than 25. When the sample size is less than 25, the significance of the rank-order correlation may be determined by reference to the table of values of the Spearman correlation significant at $\alpha = .05$ and $\alpha = .01$ in Appendix C (Table C.6). This table presents the values of r_s that are significant at the .05 and .01 levels in a one-tailed hypothesis test of $H_o : \rho_s = 0$. To use the table, you need only know the size of the sample and the value of r_s calculated from sample data. You look in the table for the value of n appropriate to your problem. Adjacent to this n you will find the critical values of r_s for the .05 and .01 levels. If the value of r_s calculated from your sample data exceeds the tabled critical value for $\alpha = .05$, then the result is significant at the .05 level. If the value of r_s calculated from your sample data exceeds the tabled critical value for .01 as well, then the result is significant at the .01 level as well.

Note that the tabled critical values are for one-tailed tests. They are to be used to test the hypothesis $\rho_s = 0$ against the alternative $\rho_s > 0$ or $\rho_s < 0$. Should you be seeking a negative relationship, the observed r_s obtained from sample data would have to be less than the negative of the tabled critical values for a significant result. The table may also be used for two-tailed hypothesis tests, by doubling the level of significance, alpha, of the result. That is, the critical value of r_s tabled under the .05 column would be the critical value in a two-tailed test at the $\alpha = .10$ level; and the value of r_s tabled under the .01 column would be the critical value in a two-tailed test at the $\alpha = .02$ level.

Note also that the table does not contain every sample size from 1 to 30. The values from 1 to 4 are missing because sample sizes of this size are too small to ever yield a significant result. Also, odd-numbered sample sizes are missing. If you have an odd-numbered sample size, you can use the next lowest sample size contained in the table to obtain critical values.

As an example of the use of the table, let us consider Professor Falk's study. We note that the critical values provided in Table C.6 for a sample size of 12 are .506 for $\alpha = .05$ and .712 for $\alpha = .01$. Because Professor Falk has hypothesized a positive relationship between popularity and assertiveness, he will be testing the null hypothesis that the Spearman correlation between the two variables measured is zero in the population against the alternative that the Spearman correlation in the population is greater than zero. That is, his test will be a one-tailed test of $H_o : \rho_s = 0$ vs. $H_a : \rho_s > 0$. Therefore, the tabled critical values are correct

FIGURE 9.3
Computation of r_s for Professor Falk's Experiment

$$n = 12$$

Subject	Rank on popularity	Rank on assertiveness	D	D^2
1	1	3	−2	4
2	2	4	−2	4
3	3	6	+3	9
4	4	5	−1	1
5	5	7	−2	4
6	6	8	−2	4
7	7	1	+6	36
8	8	2	+6	36
9	9	12	−3	9
10	10	9	+1	1
11	11	10	+1	1
12	12	11	+1	1
				$\Sigma D^2 = 110$

$$r_s = 1 - \frac{6\Sigma D^2}{n(n^2 - 1)} = .62$$

as they are for the .05 and .01 levels. We note that the observed value of r_s based on the sample data is .61. Thus, the observed statistic exceeds the critical value for $\alpha = .05$, but not for $\alpha = .01$. Professor Falk would, therefore, reject H_o at the .05 level. He would conclude that most likely there is a positive relationship between popularity and assertiveness in the population of college-age males. The test has been summarized according to the seven-step outline of a hypothesis test in Figure 9.4.

Sample Size Greater Than 25. When the sample size is greater than 25, the significance of the rank-order correlation may be estimated by means of the following statistic:

$$z_s = r_s \cdot \sqrt{n - 1},$$

which is approximately normally distributed when $\rho_s = 0$ and $n \geqslant 25$. Critical regions for this statistic are obtained by reference to the table of the standard normal distribution, just as they were obtained in the case of the one-sample z-test for a population mean. Thus, for a one-tailed test of $H_o: \rho_s = 0$ against the

Bivariate Correlational Problems

FIGURE 9.4
Example of a Test for the Significance of Spearman Correlation

Step 1:

Because we have a bivariate correlational problem with rank-order variables of interest, we consider the test for the significance of the Spearman correlation.

Step 2:

$H_o: \rho_s = 0$.

Step 3:

$H_a: \rho_s > 0$.

Step 4:

Because $n = 12$ in Professor Falk's study, we evaluate the significance of r_s, the sample statistic, by reference to the table of values of the Spearman correlation significant at $\alpha = .05$ and $\alpha = .01$.

Step 5:

Using the table, we find that when $n = 12$, the critical values of r_s are .506 for $\alpha = .05$ and .712 for $\alpha = .01$.

Step 6:

The observed value of r_s here is .62.

Step 7:

Because $.62 > .506$ but $.62 < .712$, we can reject H_o at the .05 level of significance. We conclude that there is a positive relationship between popularity and assertiveness among the population of college-age males.

---

alternative $H_a: \rho_s > 0$, the critical values would be $z = +1.645$ at $\alpha = .05$ and $z = +2.33$ at $\alpha = .01$. For a two-tailed test of the same null hypothesis against the alternative $H_a: \rho_s \neq 0$, the critical values would be $z = \pm1.96$ at $\alpha = .05$ and $z = \pm2.575$ at $\alpha = .01$.

To illustrate this *normal approximation* approach to testing the significance of the Spearman rank-order correlation, assume that Professor Falk had conducted his study using 60 subjects rather than 12, and that he obtained the same observed r_s of .62. Because the hypothesis was directional, he would still use a one-tailed test. Thus, his critical values for z_s would be +1.645 at $\alpha = .05$ and +2.33 at $\alpha = .01$. The calculation of the observed statistic would be as follows:

$$z_s = r_s \sqrt{n-1} = .62 \sqrt{60-1} = +4.76.$$

Because the observed value of z_s exceeds +2.33, Professor Falk would reject H_o at the .01 level. His conclusion would again be that there is a relationship between popularity and assertiveness among the population of college-age males.

Note that with the increase in sample size from the original $n = 12$ to the present $n = 60$, the significance of the observed sample result of $r_s = .62$ has changed from beyond the .05 level to beyond the .01 level. This should make good sense to you, based on the discussion of sampling in Chapter 7.

INFERENTIAL STATISTICS FOR BIVARIATE CORRELATIONAL PROBLEMS: INTERVAL DATA

When we move into bivariate correlational problems involving interval scale data, we move from the Spearman rank-order correlation to the Pearson product-moment correlation. The procedures involved in making inferences regarding Pearson correlations have something in common with the procedures involved in making inferences regarding Spearman correlations. In each type of problem, we begin by computing the correlation coefficient for the data in the sample, just as if the problem were descriptive. Then, we use the observed correlation for the sample to make an inference regarding the correlation in the population of interest. At this point then, you should review your understanding of the calculation and interpretation of the Pearson product-moment correlation. This material is presented in Chapter 5.

With respect to Pearson correlations, the range of techniques of inferential statistics available to us is wider than with respect to Spearman correlations. This is because mathematical statisticians have done considerably more work on the Pearson correlation than on the Spearman. Thus, more is known about the distributions of the sample values of the Pearson r when random samples are drawn from populations in which the population correlation coefficient, ρ, has different values. Most of this work was done by R. A. Fisher. It is not necessary for you to become immersed in the sampling theory of Pearson correlations at this time. It is important, however, that you become aware of the greater variety of inferential techniques that may be employed with Pearson correlations, in contrast to the Spearman correlations.

373

The only inferential technique open to us when dealing with Spearman correlations is the test of the hypothesis that the population correlation is zero. This is because the sampling distribution for r_s has been worked out only for the case where $\rho_s = 0$. Because of Fisher's work, however, we are not so limited when it comes to Pearson correlations. We can not only test the hypothesis that $\rho = 0$, but we can also test the hypothesis that $\rho = C$, where C is a constant lying between -1.00 and $+1.00$. In addition, it is possible for us to use a sample value of the Pearson correlation to obtain a confidence interval for the value of the Pearson correlation in the population from which the sample was drawn. We will now consider how to use these techniques.

Test of the Hypothesis That $\rho = 0$

A Research Example. Take a moment to reconsider the example of Professor Falk's study of the relationship between popularity and assertiveness among college-age males. Note that Dr. Falk developed his own technique for measuring these personality variables, and that this technique resulted in measurement in terms of rank order. It is sometimes the case in psychological research that personality variables can be measured through procedures yielding interval scale data. This often occurs when such variables are measured using intensively researched instruments such as commercially produced paper-and-pencil personality inventories. Suppose that Dr. Falk had chosen to examine the relationship between "outgoingness," measured by the Cattell Sixteen Personality Factor Questionnaire, and "ascendance," measured by the Guilford-Zimmerman Temperament Survey. The term "ascendance" refers to leadership or a tendency to take the initiative in conversation. It is also associated with an enjoyment of situations that involve influencing or persuading others. It certainly seems reasonable to hypothesize a relationship between outgoingness and ascendance.

The use of these instruments would simplify the procedure involved in Dr. Falk's study. He would select a random sample of subjects from the population of interest and administer each of the two instruments to each subject in the sample. To control for the possibility that the experience of taking one test might influence a subject's score on the other test, he would randomly divide his sample into two groups. One group of subjects would complete the Sixteen Personality Factor Questionnaire first; the other group would complete the Guilford-Zimmerman Temperament Survey first. In addition, Dr. Falk would probably provide for a space of a week or two between the two tests. Each of these instruments would yield a numerical score for the personality trait being measured. Because the instruments are the products of extensive research on the items that form the outgoingness and ascendance scales, it is reasonable to assume that these numerical scores do indeed form an interval scale. Accordingly, in this situation Professor Falk would employ the Pearson product-moment correlation to measure the relationship between the two variables in his sample of college-age men.

374

Let us assume that Professor Falk has theoretical reasons for believing that outgoingness and ascendance are positively related. The null hypothesis of his test would be H_o: $\rho = 0$, where ρ signifies the value of the Pearson correlation between the two variables in the population of college-age men. The alternative hypothesis would be H_a: $\rho > 0$, i.e., that there is a positive Pearson correlation between the two variables in the population. Thus, Professor Falk would be conducting a one-tailed test of a directional hypothesis. Let us further assume that Professor Falk tested 60 subjects, and that the correlation he observed in the sample was $r = .35$. Note the use of the Latin letter r for the Pearson correlation when it is used as a sample statistic estimating the population parameter, ρ. This is analogous to the use of r_s to denote a sample Spearman correlation. On the basis of the sample Pearson correlation, r, and the number of pairs of scores, n, we can carry out the test.

Calculation of the Test of H_o: $\rho = 0$.   The mathematical model employed in the test will involve one of the t-distributions. R. A. Fisher determined that when two normally distributed* variables have a Pearson correlation of $\rho = 0$, the distribution of the statistic,

$$t = \frac{r\sqrt{n-2}}{\sqrt{1-r^2}},$$

is a t-distribution with $n - 2$ degrees of freedom. This means that we can plug in the relevant values of r and n in this formula and obtain an observed statistic that can be compared to critical values obtained from the table of the t-distributions. You will recall that the table of the t-distributions was considered in Chapter 8 in connection with one-sample t-tests.

In the example at hand, the calculated value of t_{ob} for $r = .35$ and $n = 60$ would be:

$$t_{ob} = \frac{r\sqrt{n-2}}{\sqrt{1-r^2}} = \frac{.35\sqrt{60-2}}{\sqrt{1-(.35)^2}} = +2.85.$$

The significance of this observed statistic would be evaluated by reference to the t-distribution with $(60 - 2) = 58$ degrees of freedom. Turning to Table C.5 in Appendix C, we note that there is no listing for 58 df. We therefore check the next lowest number of degrees of freedom tabled, i.e., 40 df. We find that for a one-tailed test, the value of t_{cr} for $\alpha = .05$ is $+1.68$, and for $\alpha = .01$ it is $+2.42$. Comparing the observed value of $+2.58$ to these critical values, we see that the

---

* Technically, the model is appropriate only when the joint distribution of the two variables is bivariate normal. This means, in part, that for any value of either variable, the conditional distribution of the second variable is normal.

375

result is significant beyond the $\alpha = .01$ level. Prof. Falk would reject $H_o: \rho = 0$. He would conclude that within the population of college-age men there is a relationship between "outgoingness," as measured by the Cattell Sixteen Personality Factor Questionnaire, and "ascendance," as measured by the Guilford-Zimmerman Temperament Inventory.

Test of the Hypothesis That $\rho = C$.

A Research Example.   For the last 10 years, the American Telegraph Company has been using a commercially produced test of mechanical aptitude as a means of selecting trainees for the position of equipment technician. Studies have shown that the correlation between scores on the mechanical aptitude test and grades in the technician training course is .40. Industrial psychologist Watson has developed a new test that she feels will have a higher correlation with grades in the technician training course. This is important to the company, because it is expensive to train a technician, and candidates who do not complete the course satisfactorily must be fired or placed in less skilled jobs. Dr. Watson knows that if her new test does indeed correlate more highly with success in training than the old test, then it can be used more effectively than the old test as a predictor in a regression equation for predicting grades in the technician training course from mechanical aptitude test scores. (If the idea of the regression equation seems unfamiliar at this point, you should review Chapter 5.) The use of a better predictor could save the company a great deal of money and perhaps save some unqualified candidates from failure and frustration in the training program.

To determine the relationship between scores on her new test and grades in training, Dr. Watson performs the following procedures. She selects a random sample of 60 applicants for the job of technician trainee. She gives the new test to these 60 applicants. She then arranges to have all 60 applicants go through the training program, and at the conclusion of the training she records their grades. Dr. Watson calculates the Pearson product-moment correlation between test score and grade in training for this sample of 60. She finds that the correlation is .66.

The question faced by Dr. Watson at this point is not simply whether there is a relationship between the two variables of interest in the population of technician trainee applicants. It would not help her to know that her observed correlation of .66 is significantly different from zero. If her new test is to be adopted as the selection criterion for this position, Dr. Watson must show evidence that the correlation between the new test and grade in training is higher than the correlation between the old test and grade in training, which we know to be .40. Thus, Dr. Watson will not seek to test the hypothesis $H_o: \rho = 0$ against $H_a: \rho > 0$. Rather, she will seek to test the hypothesis $H_o: \rho = .40$ against $H_a: \rho > .40$. The value .40 is a constant, chosen because previous knowledge indicated that this value was a relevant point of comparison. Because of the work of Fisher, we may test hypotheses of the form $H_o: \rho = C$, where C is any constant between -1.00 and $+1.00$.

376

Carrying Out the Test of H$_0$: $\rho = C$. The hypothesis H$_0$: $\rho = C$ cannot be tested in the same manner as the hypothesis H$_0$: $\rho = 0$, for the following reason. When $\rho = 0$, the sampling distribution of r is symmetrical, but when $\rho = C$, where C is a value other than 0, the sampling distribution of r is skewed. Because the t-distributions used in the test of H$_0$: $\rho = 0$ are symmetrical, they cannot be used to test hypotheses for nonzero values of the correlation coefficient. However, Fisher discovered that it was possible to employ a mathematical transformation to convert the skewed sampling distribution of r into an approximately normal distribution that could be used to perform statistical tests. This transformation is known as the *Fisher z_r transformation*. The mathematical formula for this transformation may look strange to those without a strong background in math,* but fortunately a table has been developed that makes the actual job of transformation an easy one. This table for transforming r to z_r is presented in Table C.7. The steps involved in carrying out the hypothesis test are presented here.

First, use Table C.7 in Appendix C to convert both the value of ρ stated in the null hypothesis and the observed value of r in the sample into transformed values. This is quite easily done. The first decimal place in the correlation coefficient is presented in the left-hand column of the table, headed r. The second decimal place of the correlation coefficient is presented along the top row of the table.

Thus, to determine the transformed value for the correlation stated in the null hypothesis, $\rho_0 = .40$, we proceed as follows. We locate the first decimal place of the correlation, .4, in the left-hand column of the table headed .00 because the second decimal place in the correlation .40 is 0. At the intersection of this row and column, we find the value .42365. We call this transformed score z_{r_0}, to indicate that it is the transformed score corresponding to the correlation specified in the null hypothesis, r_0.

Similarly, we determine the transformed value for the correlation actually observed in the sample, r_{ob}. Because r_{ob} in Dr. Watson's sample was .66, we look in the row of the table adjacent to the value .6 in the left-hand column and in the column of the table headed .06. At the intersection of this row and column, we find the value .79281. We call this transformed score $z_{r_{ob}}$, to indicate that it is the transformed score corresponding to the correlation observed in the sample, r_{ob}.

Then, having used the table to obtain both z_{r_0} and $z_{r_{ob}}$, we plug these values into the following formula:

$$z_{ob} = \frac{z_{r_{ob}} - z_{r_0}}{\dfrac{1}{\sqrt{n-3}}} = (z_{r_{ob}} - z_{r_0})(\sqrt{n-3}),$$

---

* The actual transformation formula involves natural logarithms and appears as follows:

$$z = 1/2 \log_e \left(\frac{1+r}{1-r} \right).$$

377

where n is the sample size and $1/\sqrt{n-3}$ is the standard deviation of the distribution of sample values of $z_{r_{ob}}$. The statistic z_{ob} will have approximately the standard normal distribution if H_o is true, so we can test the significance of the observed result by reference to critical values obtained from the table of the standard normal distribution. In the case of Dr. Watson's study, the test was $H_o: \rho = .40$ vs. $H_a: \rho > .40$. It is therefore a one-tailed test of a directional hypothesis. The critical values would be $z_{cr} = +1.645$ for $\alpha = .05$ and $z_{cr} = +2.33$ for $\alpha = .01$. Carrying out the calculations, we find

$$z_{ob} = (z_{r_{ob}} - z_{r_o})(\sqrt{n-3})$$

$$z_{ob} = (.79281 - .42365)(\sqrt{60-3})$$

$$z_{ob} = 2.79.$$

Because z_{ob} exceeds the critical value of $+2.33$, Dr. Watson will reject H_o at the .01 level of significance. She will conclude that the Pearson correlation between scores on her new selection examination and grades in the technician trainee course is greater than .40 among the entire population of technician candidates. This will probably lead her to recommend that the new test be adopted as a selection device in place of the old test.

Confidence Intervals for ρ

As indicated in Chapter 6 and again in Chapter 7, there are two approaches to statistical inference: the hypothesis testing approach and the interval estimation approach. Because much is known about the sampling distribution of r for various values of ρ, it is possible for us to employ the interval estimation approach when we are dealing with Pearson correlations. That is, given a sample value of r, we might construct a *confidence interval for the value of ρ* in the population from which the sample was drawn.

A Research Example. Suppose Dr. Watson had been asked to develop a test to be used in selecting candidates to be trained to perform a completely new job at American Telegraph. Because the job is brand-new, no tests in existence have ever been used to select candidates for it. Accordingly, Dr. Watson has no beforehand knowledge of the size of the correlation she is likely to find between scores on the test she develops and grades in the job training program. Under these circumstances, Dr. Watson's research would be of an exploratory nature. She would not be concerned with showing that the correlation between her test and grades in training was greater than any specific value. Rather, she would be interested in finding out approximately what the correlation was, so that she could seek improvement in later versions of the test. In this research situation, Dr. Watson might well choose to calculate a confidence interval for ρ.

378

Let us suppose that Dr. Watson has made a careful analysis of the skills required to perform the new job and that she has developed a test based on this analysis. Let us further assume that she administered her test to a group of 100 applicants who then completed the training program. Finally, assume that Dr. Watson correlated her test scores with grades in training and obtained a sample r of +.37 for her group of 100 subjects. These data, the observed sample correlation and the sample size, are sufficient to obtain a confidence interval for the correlation in the greater population of all applicants who might undergo this testing and training program. There are two methods available to us to obtain the interval.

The Fisher z-Method for Confidence Intervals for ρ.   The most accurate way to obtain a confidence interval for a population correlation is to use the Fisher z-transformation to calculate the confidence interval. The procedure involves transforming r_{ob} into a Fisher z, calculating a confidence interval for the population value of z, and then transforming the limits of this interval back into correlations. The transformation is necessary because the sampling distribution of r is not symmetrical for values of ρ other than zero. We illustrate the procedure with Dr. Watson's data.

First, we transform r_{ob} into a Fisher z. This is done using the r to z_r table (Table C.7) as described already. Because r_{ob} here is +.37, we look in the row adjacent to .3 and the column below .07. At the intersection of this row and column we find .38842. This is the transformed value of r_{ob}.

Then, we calculate a confidence interval for the population value of the Fisher z. Because transformed values of r_{ob} have approximately a normal distribution with a standard deviation of $1/\sqrt{n-3}$, it is possible to construct a confidence interval for the true population value of the transformed value of ρ based on the transformed value of r_{ob}. Once again, we will use a Latin letter to represent a sample statistic and the corresponding Greek letter to represent the population parameter represented by that statistic. Because we refer to the transformed value of the statistic r_{ob} as $z_{r_{ob}}$, we refer to the transformed value of the parameter ρ as ζ (zeta, the Greek letter z). Thus, we are computing a confidence interval for ζ, the transformed value of ρ, based on $z_{r_{ob}}$, the transformed value of r_{ob}. This is done in exactly the same manner that we compute a confidence interval for a population mean. That is, we begin with the sample estimate and allow a certain number of standard deviation units on either side of this estimate to allow for sampling error. For a 95 percent confidence interval, we allow 1.96 standard deviation units, and for a 99 percent confidence interval, we allow 2.58. The formulas for the upper and lower confidence limits for ζ are therefore:

$$\text{Upper limit for } \zeta = z_{r_{ob}} + z_{cr}\left(\frac{1}{\sqrt{n-3}}\right)$$

and

$$\text{lower limit for } \zeta = z_{r_{ob}} - z_{cr}\left(\frac{1}{\sqrt{n-3}}\right),$$

where z_{cr} is 1.96 for a 95 percent interval or 2.58 for a 99 percent interval. Let us calculate a 95 percent confidence interval for Dr. Watson's problem:

$$\text{Upper limit for } \zeta = .38842 + 1.96\left(\frac{1}{\sqrt{100-3}}\right) = +.59$$

and

$$\text{lower limit for } \zeta = .38842 - 1.96\left(\frac{1}{\sqrt{100-3}}\right) = +.19.$$

Thus, we say that the population value of ζ lies between $+.59$ and $+.19$, and we are 95 percent certain that this statement is correct.

Finally, we transform the upper and lower limits for ζ into upper and lower limits for ρ. We do this because we are not really interested in the upper and lower limits for the transformed value of ρ; we are really interested in the upper and lower limits of ρ itself. Accordingly, our final step will be to transform the values of ζ just obtained back into correlation coefficients. The transformation of the upper and lower limits for ζ back into upper and lower limits for ρ is accomplished by use of a second table, contained in Appendix C (Table C.8). This new table is the table for transforming ζ to ρ. In this table, values of ζ are shown along the left-hand column and across the top, and values of ρ are tabled in the body. We locate the first decimal place of a ζ in the left-hand column and the second decimal place across the top.

Thus, to use Table C.8 to determine the value of ρ corresponding to the lower limit for ζ, .19, we find the row of the table adjacent to .1 and the column of the table headed by .09. At the intersection of this row and column we find the desired value of ρ, .1878. Similarly, to find the value of ρ corresponding to the upper limit for ζ, .59, we find the row of the table adjacent to .5 and the column headed by .09. At the intersection of this row and column we find the upper limit of ρ, .5299. Thus, we can now state that the population value of ρ lies between .1878 and .5229, and we can be 95 percent sure that this is a true statement. This is useful information to Dr. Watson. She knows that a test of significance at the .05 level of the form $H_0: \rho = C$ would result in the rejection of the null hypothesis for all values of C less than .1878 or greater than .5299. In a sense, the calculation of the confidence interval for ρ takes the place of many hypothesis tests.

A Table for Confidence Intervals for ρ.   Although the calculation of a confidence interval for a population correlation coefficient is not difficult, there are several steps involved. There is a one-step method that can be used to obtain approximate confidence intervals for ρ very quickly. Tables C.9 and C.10 contain charts that enable us to read 90 percent and 95 percent confidence limits for ρ directly. To use either chart, we locate the value of the observed sample r along the horizontal scale at the top or the bottom of the chart. Using a straight edge, we proceed into the chart at this point until we cross the sets of curved lines corresponding to our sample size. There will be two curved lines for each sample size. At the points where the straight edge crosses the line corresponding to the sample size, we draw horizontal lines extending to the vertical scales on the left. The points where these horizontal lines cross the vertical scale are the lower and upper confidence limits. Using the observed sample r of $+.37$ and the sample size 100 for Dr. Watson's study, you will find that the chart for 95 percent confidence limits provides estimates of the upper and lower 95 percent confidence limits quite close to those obtained using the Fisher z-transformation method. If we take care in reading the chart, it is usually accurate enough for most purposes.

THE CLASSIFICATION MATRIX

In this chapter, we continued our discussion of inferential techniques by looking at those techniques applicable to bivariate correlational problems. To our matrix we added the chi-square test of association, tests for the significance of Spearman and Pearson correlations, and techniques for setting up confidence intervals for the population Pearson correlation.

TABLE 9.4
Classification Matrix: Inferential Statistics

Type of Problem		Level of Measurement of Variable(s) of Interest		
		Categorical: Nominal or Ordered Categories	Rank-Order	Interval Scale (or Ratio-Scale)
Univariate Problems		Binomial Test Normal Approximation to Binomial χ^2 One-Variable Test *With Ordered Categories:* Kolmogorov-Smirnov One-Variable Test		One-Sample z-Test One-Sample t-Test Confidence Interval for Population Mean (z- or t-Method)
Bivariate Problems	Correlational Problems	χ^2 Test of Association	Test for Significance of Spearman r_s	t-Test for Significance of Pearson r Fisher z_r Transformation for Testing Significance of Pearson r Confidence Interval for ρ
	Experimental and Pseudo-Experimental Problems: Independent Groups			
	Dependent Measures: Pretest and Posttest or Matched Groups			

Key Terms

Be sure you can define each of these terms.

chi-square test of association
test for significance of Spearman correlation
normal approximation to test for significance
   of Spearman correlation
test of significance for Pearson correlation
Fisher z_r transformation
confidence interval for Pearson correlation

Summary

In Chapter 9, we examined inferential techniques for bivariate correlational problems.
We looked at techniques for problems with
categorical variables, rank-order variables, and
interval variables. We pointed out that in any
correlational problem, we use inferential techniques if we want to make an inference about
a relationship in a population based on our
observation of a sample.

To examine the relationship between two
categorical variables, we looked at the chi-
square test of association. In this case, we test
the null hypothesis that there is no relationship between our two variables. As in the chi-
square one-variable test, we compute a statistic based on the difference between observed
and expected frequencies. If these differences
are large enough, then our statistic will be
larger than critical value and we will reject our
null hypothesis and conclude that our variables are related in the population. As with the
chi-square one-variable test, we need to know
the number of degrees of freedom and the
level of significance to determine our critical
value. In this test, the number of degrees of
freedom is equal to one less than the number
of values of one variable times one less than
the number of values of the other.

If our two variables are rank-order, we
can test whether or not the sample Spearman
correlation is statistically significant. In this
case, our null hypothesis is that the Spearman
correlation in the population is zero. We saw

that with small sample sizes, we can use a table to obtain critical values for r_s. If our sample correlation is larger than the critical value, we reject our null hypothesis and conclude that our variables have some relationship in the population. With large sample sizes, we saw that we can test the same hypothesis, $p_s = 0$, by using a z-statistic and the standard normal curve.

We also saw that with interval variables, we can use a wider variety of inferential techniques than with rank-order or categorical data. We saw that we can not only test the hypothesis that the Pearson correlation in the population is equal to zero, but we can also test the hypothesis that the correlation equals any other value between -1.00 and $+1.00$. In the case of the test with $H_0: \rho = 0$, we saw that we can use the t-distribution with the number of degrees of freedom equal to one less than the number of pairs in our sample. In this case, we compute a t-statistic derived from our sample r and n and compare it to a critical value of t. To test the hypothesis that the correlation in the population is some value other than zero, we use the Fisher z_r transformation. We use a table to obtain a value of z derived from our sample r and then use the normal curve to see if our value of z is beyond the critical value. If so, we reject our null hypothesis and conclude that the population correlation is different from our hypothesized value.

We also saw how to use the Fisher z_r transformation to set up confidence intervals for population values of ρ. In this case, we use one table to transform our sample r to a derived z-score, determine our confidence interval for this derived value, and then use another table to transform this confidence interval into a confidence interval for ρ.

Review Questions

To review the concepts presented in Chapter 9, choose the best answer to each of the following questions.

1 If we have obtained information on two variables for a single sample of subjects without manipilating or controlling either variable, we have an inferential problem that is
_____a correlational.
_____b experimental.

2 If we have a problem in which we are attempting to establish a definite cause and effect relationship between two variables based on information from two independent samples, our inferential problem is
_____a correlational.
_____b experimental.

3 If we have two nominal variables and we wish to use a sample to determine whether a relationship exists between the two variables in our population of interest, we will test the null hypothesis that
_____a there is no relationship in the population between these two variables.
_____b the sample data will show no relationship.

4 The probability distribution for $\sum\sum \dfrac{(Ob-Ex)^2}{Ex}$ is discrete but will be
well approximated by a continuous χ^2 distribution when H_o is true provided
_____a we have at least 5 observations in each cell.
_____b the expected frequency in each cell is at least 5.

5 The degrees of freedom for the chi-square statistic in the test for association
between two nominal variables depends on
_____a the number of observations.
_____b the number of values of each variable.

6 If we have rejected the null hypothesis that there is no relationship between
two nominal variables, then we can say that in the population there is a relationship between the two variables. If we wish to estimate the size of the
relationship, we should calculate
_____a ϕ'.
_____b r.

7 The sampling distribution for the Spearman r may be approximated by the
standard normal distribution
_____a when n is large.
_____b when n is large and H_o is true.

8 In testing H_o: $\rho_s = 0$ vs. H_a: $\rho_s \neq 0$, if r_s is observed to be .5, the significance of the result
_____a will depend on n.
_____b will be the same for any value of n.

9 We know that if two variables have a relationship in a certain population, a
sample from the population will not have exactly the same amount of relationship. The reason for this is
_____a sampling error.
_____b that the sample isn't random.

10 If the null hypothesis that two variables are unrelated is tested and accepted,
we can
_____a be sure that there is no relationship between the two variables.
_____b say that either our decision is a Type II error or there is no rela-
tionship between the two variables.

11 One assumption that is common to all the statistical tests presented in
Chapter 9 is that
_____a the sample is a random sample.
_____b both variables have normal distributions.

12 If a 95 percent confidence interval places ρ between $-.1$ and .3, we can say
_____a there is a 95 percent chance that $\rho = 0$.
_____b the hypothesis that $\rho = 0$ cannot be rejected.

385

13 The sampling distribution of r depends on the true value of ρ in the population. If $\rho = 0$, then the sampling distribution of r is symmetrical. If $\rho = .7$, the sampling distribution of r would be

_____a skewed positively.

_____b skewed negatively.

14 When we talk about the sampling distribution of r, we are talking about

_____a the value of r in our sample.

_____b the distribution of all possible r's calculated on all possible random samples of size n.

15 When we talk about ρ as a measure of relationship in a bivariate population, we are

_____a speaking of a linear relationship.

_____b speaking of any kind of relationship.

16 If $\rho = -.7$ in a bivariate population, then in a very large random sample from that population, we expect that

_____a $r = -.7$.

_____b r will be close to $-.7$.

17 The reason we use the Fisher z_r transformation to either test $H_o: \rho = C$ or to construct a confidence interval for ρ is that

_____a the transformed distribution is approximately a normal distribution.

_____b the computations are easier.

18 A 95 percent confidence interval for ρ based on $n = 20$ will differ from one based on $n = 100$ in that

_____a the one based on $n = 20$ will be shorter.

_____b the one based on $n = 20$ will be longer.

19 If $H_o: \rho = 0$ is tested against $H_a: \rho \neq 0$ and H_o is rejected, we can say that

_____a ρ has a significant nonzero value.

_____b either ρ is not zero or we made a Type I error.

20 In using r to make an inference about ρ, both of the variables must be

_____a interval variables.

_____b rank-order or interval variables.

Problems for Chapter 9: Set 1

1 A random sample of 100 people took part in a study of the relationship between dreams and Rapid Eye Movement (REM) while sleeping. The results showed:

		Dreams		
		Report having	Report not having	
Rapid Eye Movement	REM	25	25	50
	No REM	5	45	50
		30	70	100

Using $\alpha = .01$, test the independence of the variables "report having dreams" and "rapid eye movement." Give H_o, the expected frequency table under H_o, the critical region, and your conclusion. Estimate the size of the relationship using ϕ'.

2   At a large university, 125 faculty persons were selected and cross classified by sex and rank. The results are:

		Sex		
		Male	Female	
Rank	Full Professor	40	10	50
	Associate Professor	20	15	35
	Assistant Professor	15	25	40
		75	50	125

Test the hypothesis that rank and sex are independent in the university. Use $\alpha = .05$.

3   In a large metropolitan area, an investigator wishes to study the relationship between extent of smoking cigarettes and cause of death. He goes to the death records and selects a random sample of 1,000 and cross classifies subjects according to cause of death and extent of smoking. The results are:

		Extent of Smoking				
		Heavy smoker	Moderate smoker	Light smoker	Never smoked	
Cause of Death	Lung cancer	100	50	50	100	300
	Heart ailments	100	100	50	150	400
	Respiratory ailments	50	0	50	0	100
	Other causes	50	50	50	50	200
		300	200	200	300	1,000

Are the variables "extent of smoking" and "cause of death" related? Use $\alpha = .01$.

4 As part of a large survey, an investigator looks at the possible relationship between the types of television programs watched and family membership. A sample of 800 families is selected and cross-classified into the following cells:

		Type of Television Program					
		Soaps	Police or detective shows	Sit-coms	Sports	News	Other
Family Membership	Mother						
	Father						
	Child						

Why is a chi-square test inappropriate to this situation?

5 An investigator goes to the racetrack and asks a random sample of 100 subjects whether they would bet on a 2:1 shot, a 15:1 shot, or not bet. She also asks the 100 subjects their birth order. The results are:

		Betting Preference			
		2:1	15:1	Wouldn't bet	
Birth Order	1	25	5	5	35
	2	10	10	10	30
	Lower	5	25	5	35
		40	40	20	100

a What is the population sampled?
b Test the hypothesis that betting preference is independent of birth order. Use $\alpha = .05$.
c What conclusion can you make?

Set 2

1 Use Table C.6 to test the hypothesis that $\rho_s = 0$ vs. $\rho_s \neq 0$ for each of the following. Use $\alpha = .10$.
a $n = 10, r_s = .3$
b $n = 15, r_s = .7$
c $n = 20, r_s = -.5$

Inferential Statistics

2 Use the normal approximation to test the hypothesis that $\rho_s = 0$ vs. $\rho_s \neq 0$ for each of the following. Use $\alpha = .01$.

a $n = 30, r_s = -.6$
b $n = 28, r_s = .3$
c $n = 50, r_s = .15$

3 A random sample of 10 overweight people were measured on the variables "weight loss" and "weeks on diet." Because weight loss and weeks on diet are not normally distributed, an investigator ranked the 10 subjects on each variable and calculated Spearman r for his sample. The sample data shows $r_s = .62$. Test the hypothesis that weight loss and weeks on diet are unrelated variables. Use $\alpha = .02$.

4 To investigate whether or not math aptitude and artistic aptitude are independent variables, an investigator selected a random sample of 40 subjects and administered both a math aptitude test and an artistic aptitude test. Because normed scores were given in percentile ranks, she decided to use the Spearman rank-order correlation coefficient to test her hypothesis. The observed data yielded $r_s = .47$. Using $\alpha = .02$, test her hypothesis. What can she conclude?

5 A random sample of 20 college football teams were selected from the population of college football teams. Two different polling groups, a coaches' group and a sports writers' group, were asked to rank the teams from 1 to 20 according to a scale on which 1 was best and 20 was worst. The results of the two polls were correlated using Spearman r. A correlation of .81 was computed. Test the hypothesis that the two polling groups were unrelated in their judgment of football teams. Use $\alpha = .02$.

Set 3

1 To practice using Table C.7, use the Fisher z_r transformation to test H_0: $\rho = .7$ vs. $H_a: \rho \neq .7$. Use $\alpha = .05$.

a $n = 20, r_{ob} = .3$
b $n = 50, r_{ob} = .5$
c $n = 100, r_{ob} = .6$
d $n = 100, r_{ob} = .8$

2 To practice using Tables C.7 and C.8, use the Fisher z_r transformation to calculate a 95 percent confidence interval for ρ based on the following information:

a $n = 20, r_{ob} = .3$
b $n = 50, r_{ob} = .3$
c $n = 100, r_{ob} = .3$
d Comment on the rule of thumb that you should have at least 100 observations to reliably estimate a correlation coefficient.

Bivariate Correlational Problems

3 Use Table C.9 to get 95 percent confidence intervals for ρ if:
   a $n = 20, r_{ob} = .3$.
   b $n = 50, r_{ob} = -.6$.
   c $n = 100, r_{ob} = .54$.

4 A random sample of 50 children are given a standardized reading test at the beginning of the year and again at the end of the year. An investigator wants to know whether or not the pretest scores are related to gains made over the year. She calculates r for the two variables and observes that for her sample, $r = -.3$. Using $\alpha = .05$, what conclusion can she make about the two variables?

5 In a large metropolitan community, a random sample of 500 students took a verbal aptitude test and a mathematics aptitude test. The correlation coefficient between these two variables was observed to be $r = .55$. What conclusion can be made about the relationship between verbal aptitude and mathematics aptitude for the community of students? Use a 99 percent confidence interval.

6 To study the relationship between test anxiety and performance in mathematics, a random sample of 200 college students from a large university were measured on test anxiety and mathematics achievement. The observed r was $-.4$. What conclusion can be made about these two variables for the university population? Use $\alpha = .01$.

10

Inferential Statistics for Experimental and Pseudo-Experimental Problems

Up to this point, we have considered two classes of inferential statistics: those appropriate for use with univariate problems and those appropriate for use with bivariate correlational problems. In this chapter, we consider those inferential techniques that are appropriate for experimental and pseudo-experimental problems. You will recall that experimental and pseudo-experimental problems are alike in that they involve the comparison of two or more independent groups on some variable. You should also recall the distinction between experimental problems and pseudo-experimental problems. In experimental problems, group membership is determined by the experimenter, as in the case where subjects are randomly assigned to one of two or more treatment groups. In pseudo-experimental problems, on the other hand, group membership is determined naturally, as in the case where a sample of women is compared to a sample of men on some variable. The distinction between natural groups and experimentally determined groups is very important. Only when group membership has been experimentally determined do we have a true experimental study, and only when we have a true experimental study can we speak with certainty regarding the issue of causation. If you have any questions regarding this distinction, it would be useful to review Chapter 1 at this point.

As far as the selection of appropriate inferential statistical techniques is concerned, however, the distinction between experimentally determined groups and naturally determined groups is not a critical issue. The techniques that will be presented in this chapter are appropriate for both types of problems. The key requirements for the use of the techniques presented here is that some variable has been measured in two or more different groups of subjects, and that the subjects included in the different groups have been selected independently from each other.

Thus, a study in which a group of randomly selected men was to be compared to a group of randomly selected women would be appropriate as well as a study in which a group of subjects randomly assigned to an experimental treatment was to be compared to a group of subjects randomly assigned to a control treatment. As you read through this chapter, you will find research examples of each type.

Following the pattern of previous chapters, the techniques of inferential statistics will be presented here in the following order. First, we will consider techniques appropriate for group comparisons in which the variable of interest is categorical; next, we consider techniques for group comparisons in which the variable of interest is a rank-order variable; and finally, we consider techniques for group comparisons in which the variable of interest is an interval or ratio scale variable.

INFERENTIAL STATISTICS FOR THE COMPARISON OF INDEPENDENT GROUPS ON A CATEGORICAL VARIABLE

The Comparison of Two Independent Groups on a Dichotomous Variable

A common inferential problem is one in which the researcher seeks to compare two populations on a categorical variable that has only two values (a dichotomous variable). We might seek to determine if Virginians differ from Californians on whether they favor state aid to parochial schools; we might seek to determine if Republicans differ from Democrats on whether they support or oppose strategic arms limitation; or we might seek to determine if subjects assigned to Remedial Reading Program I differ from subjects assigned to Remedial Reading Program II on whether they pass or fail the high-school equivalency examination.

All of these problems are similar in that they are inferential in nature. In each case, we will take samples from the two populations to determine whether the populations differ. The problems are also similar in that they may be thought of as involving two variables. In each case one of the variables may be thought of as "population" or "group": Virginians vs. Californians, Republicans vs. Democrats, or subjects assigned to Reading Program I vs. subjects assigned to Reading Program II. Finally, all these problems are similar in that in each case, both the group variable and the variable on which the groups are to be compared have two values. Any subject in this type of study may have one of four combinations of values on the two variables. In the remedial reading program example, there may be Program I readers who pass, Program I readers who fail, Program II readers who pass, and Program II readers who fail. Sample data for such studies are typically presented in a crosstabulation table with four cells, often referred to as a 2 × 2 contingency table. A hypothetical example of such a table, corresponding to the remedial reading example, is presented in Table 10.1. Let us take some time to consider the procedure that yielded these data.

392

TABLE 10.1

Performance on High-School Equivalency Examination Among 8 Program I Readers and 8 Program II Readers

		Program		
		I	II	
Performance	Pass	8	2	10
	Fail	0	6	6
		8	8	16

Suppose Dr. Williams, a professor of reading, believes that her program, Program I, is superior to the existing program, Program II. She therefore seeks to perform an experiment to test the null hypothesis that there is no difference in performance among Program I readers and Program II readers against the directional alternative that Program I readers have a higher pass rate. To do so, she randomly assigns 8 students identified as poor readers to Program I and another 8 poor readers to Program II. On completion of the reading programs, all 16 students take the reading section of the high-school equivalency examination.

At this point, it is useful to reiterate the distinction between bivariate correlational problems and group comparison problems. We have already noted that bivariate correlational problems involving categorical data lead to the use of crosstabulation tables that look like Table 10.1. The similar appearance of these tables can be confusing. To distinguish between the two classes of problems, remember the manner in which different subjects are selected. In a bivariate correlational problem, a single population of interest is identified and sampled. Then, subjects within the sample are measured on each of two variables. The researchers in this situation can determine only the total number of subjects selected. They cannot determine how many subjects will fall into any one value on either of the two variables. In the group comparison problem, two separate populations are identified and sampled. Then subjects are measured on one variable. Here the researchers may determine the number of subjects to be selected from each of these populations. Thus, Dr. Williams determines not only the total number of subjects, but also the marginal values of the "group" variable. Note that there is no requirement in these group comparison studies that the two groups be represented equally. This is up to the experimenter and will often depend on the availability of subjects within each of the two populations. The point is that whatever number of subjects appears in each of the two groups, that number has been chosen by the experimenter.

393

Once Dr. Williams has determined how many subjects will be sampled from each reading program, she checks the results of their examinations. Let us assume that of the 16 students, 10 pass and 6 fail. These data establish the values of the other set of marginals of the 2 × 2 contingency table presented in Table 10.1. Obviously, Dr. Williams does not have any control over these marginals. They are determined by the performance of the subjects.

Given that 10 of the 16 subjects passed, the question that is relevant for Dr. Williams is "How many of the students who passed were from the Program I group, and how many were from the Program II group?" Obviously the larger the number of students from the Program I group who passed, the more evidence there is that Program I is more effective than Program II. When we consider the question of how many students from a particular group passed, we are considering the cell frequencies of the 2 × 2 table. For a moment, let us forget about the observed values for the cell frequencies that we have actually observed in Dr. Williams' study. Let us consider instead the various possibilities for these frequencies, given that Dr. Williams selected a sample of 8 subjects from each group and given that 10 of the 16 students passed. Actually, with these marginals established, there are only 7 different configurations of cell frequencies possible. These different possible outcomes are presented in Table 10.2. Let us consider these 7 configurations to determine which support and which do not support Dr. Williams' alternative hypothesis that Program I readers are more likely to pass the exam than Program II readers.

In Table 10.2(a), we see the outcome that we would expect to observe, if the null hypothesis were true. If the populations of Program I and Program II readers do not differ with respect to passing the exam, we would expect to see this reflected in our samples. In this table, the two samples show identical performance on the test. Of the 8 subjects in each program, 5 passed. Clearly, this supports the null hypothesis rather than the alternative.

In Tables 10.2(b) through 10.2(d), we see 3 possible outcomes in which subjects assigned to Program II actually performed better than subjects assigned to Program I. Although these results are not what we would expect if the two populations were identical with respect to the exam, they do not support Dr. Williams' alternative hypothesis. In fact, they are in the opposite direction from what we would expect if Program I readers are superior.

In Tables 10.2(e) through 10.2(g), we see the 3 possible cell configurations that lend some support to Dr. Williams' alternative hypothesis. In Table 10.2(e), we see that 6 out of 8 Program I readers have passed the exam, compared to only 4 out of 8 Program II readers. In Tables 10.2(f) and 10.2(g), the superiority of the Program I subjects is even more clear. The table that supports Dr. Williams' alternative most strongly is Table 10.2(g). Here, all 8 of the Program I readers passed, and only 2 of the 8 Program II readers passed. Given that 10 of 16 subjects passed overall, this is the most extreme observed result possible in the direction suggesting the superiority of Program I.

394

TABLE 10.2
The Seven Different Possible Configurations of Cell Frequencies
in Dr. Williams' Experiment, Given That 10 Subjects Pass and 6 Fail

(a)   Program

		I	II	
Performance	Pass	5	5	10
	Fail	3	3	6
		8	8	16

(b)

	I	II	
Pass	4	6	10
Fail	4	2	6
	8	8	16

(c)

	I	II	
Pass	3	7	10
Fail	5	1	6
	8	8	16

(d)

	I	II	
Pass	2	8	10
Fail	6	0	6
	8	8	16

(e)

	I	II	
Pass	6	4	10
Fail	2	4	6
	8	8	16

(f)

	I	II	
Pass	7	3	10
Fail	1	5	6
	8	8	16

(g)

	I	II	
Pass	8	2	10
Fail	0	6	6
	8	8	16

Of course, understanding the kinds of outcomes that favor the null hypothesis and the kinds of outcomes that favor the alternative hypothesis is not enough to enable us to carry out a hypothesis test. We have seen that in order to establish a rejection rule for a statistical test, we must be able to determine *how likely* these various outcomes are under the assumption that the null hypothesis is true. In problems of this nature, where we are making an inference regarding two independently selected samples with respect to a dichotomous variable, one of two statistical tests can be employed. In cases such as Dr. Williams' experiment, where the sample size is small, we generally employ the *Fisher exact probability test,* which is based on a mathematical model known as the *hypergeometric model.* As sample sizes increase, the Fisher exact test becomes rather difficult to compute. However, when samples are sufficiently large, we may test the hypothesis of no difference between two populations using a chi-square test of homogeneity. We now consider these two techniques.

Experimental and Pseudo-Experimental Problems

The Fisher Exact Probability Test

The Exact Probability of a Particular Outcome. Once we have determined the marginal frequencies for a 2 × 2 table, we use a formula to determine the exact probability of obtaining a particular configuration of cell frequencies under the null hypothesis of no difference between the sampled populations. This formula is based on the marginals and on the cell frequencies of the particular outcome. We label the cells of the 2 × 2 table as follows:

A	C	$A + C$
B	D	$B + D$
$A + B$	$C + D$	

Then, the probability of observing any particular outcome is given by

$$pr(\text{outcome}) = \frac{(A+B)!\,(C+D)!\,(A+C)!\,(B+D)!}{n!\,A!\,B!\,C!\,D!}.$$

This formula may appear a bit imposing at first, but it is actually quite simple to use. The terms in the numerator are simply the factorials of the marginal frequencies of the 2 × 2 table, and the terms in the denominator are the factorials of the total number of subjects (n) and the cell frequencies. The factorial notation should be familiar to you from our discussion of the binomial distribution in Chapters 6 and 8. As an example of the use of the formula, let us consider the result in which equal numbers of subjects in the two groups passed:

$$pr(A = 5) = \frac{10!\,6!\,8!\,8!}{16!\,5!\,5!\,3!\,3!} = .3916.$$

Note that in this equation, we have defined the outcome in terms of a single cell frequency. This is possible because once we have accepted the marginal frequencies of a 2 × 2 table as given, specifying the frequency of a single cell automatically determines the frequencies of the other 3 cells as well. In the same way, we can find the probability that all 8 of the subjects from Program I passed. Then, the frequency of cell A is 8, and

$$pr(A = 8) = \frac{10!\,6!\,8!\,8!}{16!\,8!\,2!\,0!\,6!} = .0035.*$$

---

* Don't forget that $0! = 1$.

It is possible to continue in this manner until we determine the exact probability of obtaining each of the 7 possible outcomes given the marginals in Dr. Williams' study. At this point, you might wish to calculate the probability of some other outcomes, to be certain that the procedure is clear. The probabilities of each of the 7 different possible cell configurations are presented in Table 10.3.

Given the observed marginal totals of 10 subjects passing and 6 subjects failing, Table 10.3 constitutes the probability distribution of the possible outcomes of the experiment, under the null hypothesis that Program I readers do not differ from Program II readers. With the knowledge of the probability distribution, we may establish a rejection region with an acceptable level for alpha. In Figure 10.1, we have represented the probability distribution graphically. We have drawn a dotted line between the values 7 and 8 on the horizontal axis to indicate that we have chosen to reject H_o if we observe 8 subjects from Program I passing. We have established a one-tailed rejection region because the hypothesis is directional. Only large values for the variable "number of subjects from Program I who pass" support Dr. Williams' alternative hypothesis. We have chosen to reject if and only if we find 8 subjects from Program I who pass, because we wish to keep our chance of a Type I error below 5 percent. If H_o is true, the probability distribution tells us that there is only a .0035 or 0.35 percent chance that 8 subjects from Program I will pass. Thus, in adopting 8 as a critical value, we are allowing only a 0.35 percent chance of a Type I error. This is less than 5 percent by a good deal.

What if we had included in the outcome that 7 subjects from Program I passed in the rejection region? In this case, we would reject H_o if either 7 or 8 subjects from Program I passed. From the probability distribution, we see that

---

TABLE 10.3
Probabilities of Each of the 7 Different Possible Configurations of Cell Frequencies in Dr. Williams' Experiment, Given That 10 Subjects Pass and 6 Fail

Outcome in terms of number from Program I who pass (A)	Probability of outcome
$A = 2$	.0035
$A = 3$	.0559
$A = 4$	.2448
$A = 5$	.3916
$A = 6$	.2448
$A = 7$	.0559
$A = 8$	.0035
	1.0000

Experimental and Pseudo-Experimental Problems

FIGURE 10.1

Probability Distribution of Outcomes of Dr. Williams' Experiment, Given That 10 Subjects Pass and 6 Subjects Fail

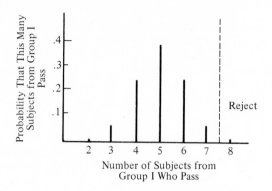

if H_0 were true, the chance of observing either 7 or 8 subjects from this group passing is $(.0559) + (.0035) = .0594$. This means that if H_0 were true, we would have about a 6 percent chance of rejecting H_0 by chance. That is, the chance of a Type I error would be .0594 or about 6 percent. Because this exceeds the traditional value of 5 percent for the level of significance, we chose to reject only if the observed result is 8.

It is possible to employ the Fisher exact probability test in this fashion to test any hypothesis involving the comparison of two independently selected groups on a dichotomous variable. Once we have determined the marginal distribution on this dichotomous variable, we can calculate the probability distribution for the various possible outcomes in terms of cell frequencies. We can use this distribution to determine a rejection region (if the test is one-tailed) or rejection regions (if the test is two-tailed). We can then check to see if the actual observed outcome lies in the rejection region. However, it must be noted that as the total number of subjects increases, both the number of possibilities for different combinations of marginal frequencies and the number of possible cell configurations for any particular set of marginal frequencies increases rather rapidly. This can make the job of calculating the probability distribution for a given set of marginals quite tedious. Of course, you may have already figured out that it is not essential to calculate the exact probability of *every* possible outcome for a given set of marginals. We could begin by calculating the probability of the most extreme outcome favoring the alternative hypothesis and work inward until the sum of the accumulated probabilities of the most extreme outcomes exceeded the desired level of significance of the test. Nevertheless,

Inferential Statistics

even this procedure could involve a great deal of work when n was 20 or more. For this reason, techniques have been developed that enable us to determine the significance of 2 X 2 tables without actually calculating the probabilities of individual outcomes. Tables exist that provide probabilities for all the outcomes in such problems for total sample sizes up to 100.* Unfortunately, these tables are so large that they alone fill a volume as large as this text. In this text, we have included a table of a more manageable size that may be used to determine the significance of 2 X 2 tables in which neither sample is larger than 15. Thus, the table may be used with a total n as large as 30 in the case of two samples of 15 observations each. When the sample sizes are such that the table cannot be used, we employ instead an alternative chi-square test. We consider first the use of the table, then the chi-square test.

Table of Critical Values for the Fisher Exact Probability Test. The table of critical values for the Fisher exact probability test is presented in Appendix C (Table C.11). In using this table, we denote the marginal and cell frequencies of the particular 2 X 2 contingency table with the A, B, C, D notation presented earlier. The quantities $(A + B)$ and $(C + D)$ correspond to the marginal frequencies of the variable "group." That is, $(A + B)$ will be the size of one of the two independently selected samples, and $(C + D)$ will be the size of the other independently selected sample. For convenience, we call the *larger* of the two sample sizes $(A + B)$. With this scheme in mind, we proceed as follows.

Step 1   Find the appropriate value of $(A + B)$ in the table. Values of $(A + B)$ appear in the left-hand column of the table. The table runs over several pages, with smaller values of $(A + B)$ at the beginning and larger values of $(A + B)$ at the end. The minimum value of $(A + B)$ is 3, because 2 X 2 tables having their smallest marginal frequency less than 3 cannot be statistically significant, regardless of how the cell frequencies appear. The maximum value of $(A + B)$ is 15.

Step 2   Once we have found the correct value of $(A + B)$, we look in the next column to the right to find the appropriate value of $(C + D)$. This will be equal to or smaller than $(A + B)$, because there is no requirement that sample sizes be equal in this test.

Step 3   After locating both $(A + B)$ and $(C + D)$, we shift our attention one more column to the right, to the column headed "B (or A)." Here, we locate either the value of cell frequency A or the value of cell frequency B, *whichever is larger*. Once we have located all 3 of the quantities as described in Steps 1–3, we have located the row of the table of critical values in the Fisher test that will tell us whether our 2 X 2 table is significant.

* Lieberman, G. I. and Owen, D. B., *Tables of the Hypergeometric Probability Distribution,* Stanford, Calif.: Stanford University Press, 1961.

Experimental and Pseudo-Experimental Problems

Step 4   To determine if the 2 × 2 table is significant, we use the critical values listed under the heading "Level of significance." If A was the larger of the cell frequencies that we located in Step 3, then these are critical values for C. In this case, any value of C that is *smaller* than the tabled value for a given significance level means that our result is significant at that level. If B was the larger of the cell frequencies that we located in Step 3, then the tabled values are critical values for D. In this case, if D is smaller than the tabled value for a given significance level, then our particular 2 × 2 table is significant at that level. These critical values are for one-tailed tests of directional hypotheses. When we test a nondirectional hypothesis, we must double the significance level indicated in the table.

To illustrate the use of the table of critical values for the Fisher exact probability test, let us reconsider Dr. Williams' experiment. Let us suppose that Dr. Williams had carried out her experiment as already described and that her results were those presented in Table 10.2(g). To determine the significance of this result using the table just described, she would first find the values for $(A + B) = 8$ and $(C + D) = 8$, because her two samples each had 8 subjects. Next, she would look in the column headed A or B for $A = 8$, because here cell frequency A is larger than cell frequency B. Because she has found A rather than B in the third column, Dr. Williams knows that the critical values given in the table apply to the value of C. Looking in the row of the table corresponding to $(A + B) = 8$, $(C + D) = 8$, and $A = 8$, we find that the observed frequency for cell C, 2, is significant beyond $\alpha = .005$ in a one-tailed test. This agrees with our result using the hypergeometric formula, where we found that the exact significance probability of this result was .0035.

Note how much simpler it is to use the table of critical values than to calculate actual significance probabilities. It is not necessary to consider the actual marginal distribution of the variable of interest at all. We need only consider the sample sizes, $(A + B)$ and $(C + D)$, and two cell frequencies (either A and C or B and D) to determine the significance of the result. The only disadvantage in the use of the table is that we do not obtain exact significance probabilities.

One word of caution is in order with respect to tests of directional hypotheses. Before using the table of critical values, you are advised to be sure that the observed result is in the direction that supports your alternative hypothesis. Table C.11 tells us whether a particular 2 × 2 crosstabulation table is significantly different from what we would expect if the two populations being compared did not differ on the variable of interest. Users of the table must determine the direction of the deviation from expectation themselves. The Fisher exact probability test for Dr. Williams' study has been summarized, according to our seven-step outline for a hypothesis test, in Figure 10.2.

FIGURE 10.2
Fisher Exact Probability Test for Dr. Williams' Reading Study

Step 1

Because we have two independent samples that are to be compared on a dichotomous categorical variable, we consider either the Fisher exact probability test or a χ^2 test of homogeneity. Because neither sample in Dr. Williams' study has more than 15 observations, we may conduct the Fisher exact probability test using the table of critical values (Table C.11).

Step 2

H_o: The population of Program I readers and the population of Program II readers are distributed identically on the variable "performance on high-school equivalency exam."

Step 3

H_a: The populations are not distributed identically. Program I readers are more likely to pass than Program II readers.

Step 4

The test is the Fisher exact probability test. Critical values for observed cell frequencies are obtained from Table C.11.

Step 5

Each sample size in the problem is 8. The larger of the cell frequencies for cells A and B is that of cell A, which is 8. Therefore, the critical values provided in the table will be for cell C. Looking under $(A + B) = 8$, $(C + D) = 8$, and $A = 8$, we find that the test would be significant at $\alpha = .05$ if $C = 4$ and significant at $\alpha = .01$ if $C = 2$.

Step 6

In our problem, $C = 2$.

Step 7

Because $C = 2$, the test is significant at beyond the $\alpha = .01$ level. We reject H_o and conclude that Program I readers are more likely to pass than Program II readers.

Experimental and Pseudo-Experimental Problems

The χ^2 Test of Homogeneity for 2 X 2 Tables

As noted already, the table of critical values for the Fisher exact probability test may be used only for problems in which neither sample size exceeds 15. Of course, it is always possible to employ the formula presented earlier to calculate the probability distribution for the outcomes that are possible with a particular set of marginals, but this becomes a tedious task when the sample sizes become large. Fortunately, it is often possible to test the hypothesis of no difference between the sampled populations using a χ^2 *test of homogeneity*. This test may be used to test the hypothesis that the two populations sampled are homogeneous with respect to the variable of interest. The essential requirement for the use of the χ^2 statistic with 2 X 2 contingency tables is that the expected frequency in all 4 cells of the table must be at least 5. You will recall from our discussion of the χ^2 test of association (see Chapter 9) that the expected frequency of a cell may be found by multiplying the marginal frequencies of the row and column occupied by the cell and dividing the result by n.

The value of χ^2_{ob} may be computed using the regular formula for the χ^2 statistic, based on the comparison of observed frequencies to expected frequencies. However, in the case of the 2 X 2 contingency table, there is a special formula that is easier to use and that actually provides a more precise significance test. This formula is typically written in terms of the *A, B, C*, and *D* notation for cell frequencies, as follows:

$$\chi^2_{ob} = \frac{n(|AD-BC|-n/2)^2}{(A+B)\,(C+D)\,(A+C)\,(B+D)}.$$

The significance of the observed result may be determined by reference to the χ^2 table with 1 degree of freedom. You will recall that the levels of significance presented in the table of the χ^2 distributions (Table C.3) refer to tests of nondirectional hypotheses. This is because most χ^2 tests are inherently nondirectional. In the case of the 2 X 2 contingency table, however, it is possible to state a directional alternative. That is, we may state as an alternative not only that the two populations sampled differ on the dichotomous variable of interest, but that one of the two populations has a higher proportion of observations falling into one of the two categories on the variable of interest. This was the case in Dr. William's study, where the directional alternative was that a larger proportion of Program I readers would fall in the "pass" category than the proportion of Program II readers. When we are testing a directional hypothesis in a 2 X 2 table, we seek χ^2_{cr} in the table under the column for the level of significance twice that actually desired for the test. Thus, if $\alpha = .05$ in a *directional* test, we would find the value of χ^2_{cr} under the column headed .10. Keep in mind that it is possible to specify a directional hypothesis in the χ^2 test of homogeneity only in the situation where there are 2 populations compared on a dichotomous variable. As you will learn shortly, there are also χ^2 tests of homogeneity for contingency tables that are larger than 2 X 2. Such tests are always nondirectional.

As an example of a χ^2 test of homogeneity for a 2 × 2 table, let us return to a research question posed at the start of this chapter: Do Virginians differ from Californians on whether they favor state aid to parochial schools? Suppose a researcher selects independent random samples of 100 Virginians and 100 Californians. He asks each subject in each sample to answer the question, "Do you favor or oppose state aid to parochial schools?" The results of the study are shown in Table 10.4.

It is clear from the table that the sample sizes involved here are too large for us to test the hypothesis of no difference between the sampled populations using the table of critical values for the Fisher exact probability test. It is also clear that the expected cell frequency for each of the 4 cells in the 2 × 2 table is greater than or equal to 5. (Actually, the smallest expected frequency for this table is [90/200] × 100 = 45.) Thus, the researcher would be justified in using the χ^2 test of homogeneity. The test has been carried out, according to the seven-step hypothesis testing outline, in Figure 10.3.

Two cautions are warranted in using the formula for obtaining χ^2_{ob} in 2 × 2 tables. First, in calculating χ^2_{ob}, note that we use the absolute value of the difference between AD and BC. Thus, if BC is larger than AD, so that the difference is a negative number, we ignore the negative sign. Second, in the case where we are dealing with a directional alternative hypothesis, we should again check the 2 × 2 table itself to be sure that the observed results are in the direction supporting this alternative hypothesis. If the observed results are in the opposite direction, we should not compute the statistic.

The Comparison of Two Independent Groups on a Categorical Variable Having More Than Two Values

In research in psychology, education, and the social sciences, we frequently encounter research situations where two independent groups must be compared on a categorical variable that has more than two values. In this situation, we test the

---

TABLE 10.4
Opinion on State Aid to Parochial Schools Among
100 Virginians and 100 Californians

	Virginia	California	
Favor	70	40	110
Oppose	30	60	90
	100	100	200

Experimental and Pseudo-Experimental Problems

FIGURE 10.3
χ^2 Test of Homogeneity for a 2 × 2 Contingency Table

Step 1

Because we have two independent samples that are being compared on a dichotomous categorical variable, we consider either the Fisher exact probability test or the χ^2 test of homogeneity. Because the sample sizes are large, we rule out the Fisher exact probability test. Because none of the 4 cells has an expected frequency of less than 5, the χ^2 test of homogeneity is appropriate.

Step 2

H_o: The two populations sampled do not differ with respect to the variable of interest.

Step 3

H_a: The two populations sampled do differ (a nondirectional alternative).

Step 4

The test is the χ^2 test of homogeneity. The statistic has df = 1, because the contingency table is a 2 × 2 table.

Step 5

$\alpha = .05$. $\chi^2_{cr} = \chi^2_{1;.95} = 3.84$. We will reject H_o if $\chi^2_{ob} > 3.84$.

Step 6

$$\chi^2_{ob} = \frac{n\left(|AD - BC| - \dfrac{n}{2}\right)^2}{(A + B)(C + D)(A + C)(B + D)}$$

$$= \frac{200\left(|(70)(60) - (40)(30)| - \dfrac{200}{2}\right)^2}{(100)(100)(110)(90)} = 16.99.$$

Step 7

Because $\chi^2_{ob} > \chi^2_{cr}$, we reject H_o. We conclude that the populations of Virginians and Californians do differ with respect to the variable "opinion on state aid to parochial schools."

null hypothesis of no difference between the groups using the χ^2 test of homogeneity. The statistic employed in this test is calculated in exactly the same manner as the χ^2 statistic for the χ^2 test of association, presented in Chapter 9. The two tests are distinguishable from each other, however, in that the χ^2 test of association is used with bivariate correlational problems, and the χ^2 test of homogeneity is used in the comparison of independent groups. To review this distinction briefly, the χ^2 test of association applies to research in which a sample has been drawn from a single population, and each subject in the sample is measured on two variables. The χ^2 test of association enables us to come to a conclusion on whether these two variables are related to each other in this one population. The χ^2 test of homogeneity, on the other hand, applies to research in which independent random samples have been drawn from two (or more) populations, and each subject in each sample is measured on a single variable. The χ^2 test of homogeneity enables us to come to a conclusion on whether the sampled populations differ with respect to the measured variable of interest. If the distinction between bivariate correlational studies and studies involving the comparison of independent groups is unclear, you should review the discussion of these topics in Chapter 1. We now consider a research problem involving the comparison of two groups.

A Research Example.   Dr. Hillman is a special education expert in charge of a program designed to teach trainable mentally retarded individuals (TMRs) to travel to and from work independently. Participants in this program achieve one of three levels of success: (1) They may be graduated from the program and certified as capable of independent travel; (2) they may complete the program but not receive such certification; and (3) they may be dropped from the program. In selecting students for the program, there has been some question regarding the chance of success of participants with IQ scores in the 40 and above range vs. the chance of success of participants with IQ scores below 40. Because resources available for training are limited, it is important to know whether the relatively high IQ TMRs really do differ from the relatively low IQ TMRs with respect to success in travel training.

Dr. Hillman arranges the following experiment. He randomly selects a group of 30 TMRs with IQs in the below 40 category, and he randomly selects another group of 30 TMRs with IQs in the 40 and above category. Without informing any of the program instructors about the measured IQ of these subjects, Dr. Hillman allows all 60 to undergo travel training. He carefully monitors the progress of each participant, recording the final decision in each case. The results are presented in Table 10.5. To determine whether the two IQ groups differ with respect to final decision in travel training, Dr. Hillman will use the χ^2 test of homogeneity. The null hypothesis of this test is that the two populations sampled do not differ with respect to final decision in travel training. The test is based on the idea that if this null hypothesis is true, then the proportion of subjects in each sample falling into each of the categories on final decision should

405

TABLE 10.5
Final Decision in Travel Training for Two Groups
of Trainable Mentally Retarded Individuals ($n = 60$)

| Decision | | Group | | |
		Group I IQ > 40	Group II IQ < 40	Total
	Dropped from program	4	8	12
	Completed program but not certified	3	12	15
	Certified for independent travel	23	10	33
		30	30	60

tend to be equal. Deviations from this expectation under the null hypothesis
are measured by the difference between observed and expected cell frequencies.

Calculation of the χ^2 Test of Homogeneity.   The actual calculation in the χ^2
test of homogeneity is identical to the calculation in the χ^2 test of association.
The two tests are mathematically identical. We differentiate them here because
they are different from the point of view of the research design. Because they
are mathematically equivalent, we perform the identical steps in each case. We
compute expected frequencies as we explained in Chapter 5. We compare expected and observed frequencies over the entire table using the formula:

$$\chi^2_{ob} = \sum \sum^{\substack{\text{all} \\ \text{cells}}} \frac{(Ob - Ex)^2}{Ex}.$$

We determine the appropriate number of degrees of freedom for our test using
the formula

$$df = (R - 1)(C - 1),$$

where R equals the number of rows in the 2 × 2 table, and C equals the number
of columns. The χ^2 test of homogeneity has been laid out according to our outline for a hypothesis test in Figure 10.4.

406

Inferential Statistics

FIGURE 10.4
χ^2 Test of Homogeneity Comparing the Two Groups in Dr. Hillman's Study
on Final Decision in Travel Training

Step 1

Because we have independent samples that are to be compared on a categorical
variable, we consider the use of the χ^2 test of homogeneity. We check expected
cell frequencies to be sure the minimum requirements are met.

Step 2

H_o: The two populations sampled (TMRs with IQ $>$ 40 and TMRs with IQ $<$ 40)
do not differ with respect to the variable of interest (final decision in travel
training).

Step 3

H_a: They do differ.

Step 4

The test is the χ^2 test of homogeneity. The test statistic will have approximately
a χ^2 distribution with $(R-1)(C-1) = (2)(1) = 2$ df if H_o is true.

Step 5

$\alpha = .05$. $\chi^2_{cr} = \chi^2_{2;.95} = 6.0$. We will reject H_o at $\alpha = .05$ if $\chi^2_{ob} > 6.0$.

Step 6

$$\chi^2_{ob} = \sum \sum \frac{(Ob - Ex)^2}{Ex} = \frac{(4-6)^2}{6} + \frac{(8-6)^2}{6} + \frac{(3-7.5)^2}{7.5}$$

$$+ \frac{(12-7.5)^2}{7.5} + \frac{(23-16.5)^2}{16.5} + \frac{(10-16.5)^2}{16.5} = 11.85.$$

Step 7

Because $\chi^2_{ob} > \chi^2_{cr}$, we reject H_o at the .05 level of significance and we conclude
that TMRs with IQs below 40 differ from TMRs with IQs of 40 and above with
respect to final decision in travel training.

Constraints on the Use of the χ^2 Test of Homogeneity. The requirements for the use of the χ^2 test of homogeneity are the same as the requirements for the use of the χ^2 test of association: (1) all observations must be independent; (2) the cells of the crosstabulation table must be mutually exclusive; and (3) the expected frequencies for the cells must be large enough. With respect to this last requirement, the ideal situation would be that in which no cell in the table would have an expected frequency of less than 5. As before, however, in larger tables this rule is relaxed somewhat. In these cases, not more than 20 percent of the cells should have expected cell frequencies of less than 5 and none should have expected cell frequencies of less than 1.

Extensions of the χ^2 Test of Homogeneity. You may already have thought ahead to consider other types of problems involving group comparisons on a categorical variable. Clearly, it would be possible to consider whether 3 or more independently selected groups differ from each other with respect to a categorical variable. Furthermore, the variable of interest in this situation might have 2 values or more than 2 values. The χ^2 test of homogeneity may be extended to cover these situations as well. We need only adjust the number of degrees of freedom associated with the test when we obtain our critical value for χ^2. The requirements relating to minimum expected cell frequencies remain in effect for tables having larger numbers of cells. As pointed out before, it is quite common in situations involving large tables to group categories in order to meet the requirements for minimum expected cell frequencies.

One word of caution is in order regarding the results of a test in which 3 or more independent groups are compared. It should be noted that obtaining a significant χ^2 in this situation allows us to conclude only that not all of the groups being considered are identical with respect to the variable of interest. The test does not tell us specifically which groups are different from which other groups. To judge the nature of the differences between the groups, we must examine the actual percentages of subjects in the different groups falling into each of the different categories on the variable of interest.

INFERENTIAL STATISTICS FOR THE COMPARISON OF INDEPENDENT GROUPS ON A RANK-ORDER VARIABLE

Two Group Comparisons: The Mann-Whitney *U*-Test

When subjects from two independent groups can be rank ordered in terms of a variable of interest, we typically employ the *Mann-Whitney U-test* to test the hypothesis that the two groups do not differ with respect to the variable of interest. This test is also employed when we have measured the variable of in-

408

terest with numerical scores, but we are not confident that we have met all the assumptions required to use the two-sample t-test (to be discussed later in this chapter).

A Research Example.   Suppose Clinical Psychologist Shafer has developed a postdoctoral training workshop that she believes will help psychotherapists become more accepting of their clients' beliefs and attitudes. To test the effectiveness of this workshop, Dr. Shafer arranges an experiment. She randomly selects a group of 8 therapists to participate in the program. She also randomly selects a second group of 8 therapists to serve as a control group. Subjects in the second group do not receive the new training workshop, but, instead, receive an equivalent amount of more traditional supervision. At the conclusion of these treatments, subjects from both the experimental and control groups are videotaped during several therapy sessions. Dr. Shafer arranges for an experienced training therapist to view these tapes and rank all of the subjects from most accepting to least accepting. The judge used to rank the subjects has no knowledge of whether a given subject was an experimental or control subject. Dr. Shafer keeps a record of this, however. The data obtained in this hypothetical study are presented in Table 10.6. The rank 1 is assigned to the therapist judged most accepting. You will note that more of the experimental group subjects have been ranked high on acceptance. Within these samples, it appears that therapists participating in Dr. Shafer's workshop are more accepting than those who do not participate in the program. However, we need to employ a statistical test to determine whether

---

TABLE 10.6
Rank on "Acceptance" and Group Membership for 16 Subjects
in Dr. Shafer's Study of Two Training Methods

Rank	Training group		Rank	Training group
1	E		9	C
2	E		10	C
3	E		11	E
4	C		12	E
5	E		13	C
6	C		14	C
7	C		15	E
8	E		16	C

E = experimental
C = control

Experimental and Pseudo-Experimental Problems

these results are sufficient to warrant the conclusion that the population of therapists who might undergo the Shafer program will differ in the variable "acceptance" from the population of therapists who do not undergo the program. The test we employ here is the Mann-Whitney U-test.

The Calculation of the U-Statistic. The concept behind the U-statistic is that the more different the two groups are, the more subjects from one of the groups will be ranked ahead of the subjects from the other group. Specifically, the U-statistic can be defined as the number of scores from one group that *precede* each of the scores from the other group. In determining U for small sets of data, we can actually count scores to find U. For example, in Professor Shafer's data, we might look at each subject in the experimental group and count the number of control group subjects who precede him or her in the rankings. When we consider each of the first three experimental subjects (the subjects ranked 1, 2, and 3 on "acceptance"), we find that no control group subjects precede any of these experimental subjects in the rankings. The next experimental subject has the rank 5. This subject is preceded by one control group subject, the subject who has the rank 4. The next experimental subject has the rank 8 and is preceded by 3 control group subjects, those subjects ranked 4, 6, and 7. We proceed in this manner until we have accounted for each of the experimental group subjects. Thus,

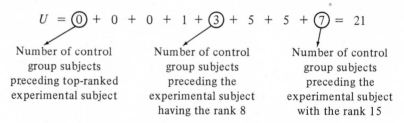

You may be wondering why we chose to consider the number of control group subjects preceding each of the experimental subjects. Obviously, we could have considered the reverse—the number of experimental subjects preceding each of the control subjects—which would have yielded a different value for U. This procedure would have yielded a larger value for U because the experimental subjects tended to be ranked higher, so more of them would precede control group subjects. In fact, whenever we consider the U-statistic, there are always two possible values, depending on which group we use in counting. We generally refer to the two possible values as U and U'. These are inversely related to each other. The larger U is, the smaller U' will be, and the more evidence we have that the two populations truly differ. In evaluating the significance of the U-statistic, we typically focus on the smaller of the two values. In Dr. Shafer's data, it was easy to see which method of counting would produce the smaller value for U. This is not always the case. It is sometimes unclear which manner of counting will yield the smaller value of U, especially with larger sets of data. In fact, in such cases, we typically obtain U not by counting, but by using a formula that yields the same result more conveniently.

When the number of subjects is large, we typically find U by using the formula

$$U_{ob} = n_1 n_2 + \frac{n_1(n_1 + 1)}{2} - R_1,$$

where n_1 and n_2 refer to the number of subjects in each of the two groups and R_1 is the sum of the ranks assigned to the subjects in group number 1. In the case of Dr. Shafer's data, if we regard the control group as group 1, we find

$$U_{ob} = n_1 n_2 + \frac{n_1(n_1 + 1)}{2} - R_1 = (8 \times 8) + \frac{8 \times 9}{2} - 79 = 21.$$

Of course, we have the same problem here that we had when counting, i.e., how do we know which group to regard as group 1? When we counted directly, we could see that we would have to concentrate on the number of controls preceding each of the experimentals to get the smaller value for U. But when we use the formula, we must count the sum of the ranks occupied by control group subjects to get the same answer. How do we know whether to count the sum of the ranks occupied by the control group subjects or the sum of the ranks occupied by the experimental group subjects?

The answer is that we do not immediately know, so we must test to be sure that the value we have obtained for U is the smaller of the two possible values. We do this as follows. The *maximum* possible value for U would be $n_1 \cdot n_2$, which would occur when all the scores in one group were ranked ahead of all the scores in the other group. Consider the values we would have obtained for U in Dr. Shafer's example if all 8 experimental subjects had been ranked ahead of all 8 control subjects. If we counted the number of experimental subjects preceding each control subject, we would see that all 8 experimental subjects preceded each of the 8 control subjects, so that U would be $8 \cdot 8 = 64$. But if we counted the number of control subjects preceding each experimental subject, we would see that no control subject ever preceded an experimental subject, so that U would be 0. In all problems where it is possible to use the Mann-Whitney U-test, the sum of the two different possible values of U will be $n_1 \cdot n_2$, or:

$$n_1 \cdot n_2 = U + U'.$$

Thus, we can always determine whether we have obtained the smaller value of U by checking to see if the computed value is less than or equal to $\dfrac{n_1 \cdot n_2}{2}$. If we

411

compute a value for U using the formula and this value is greater than $\dfrac{n_1 \cdot n_2}{2}$, then we know that we have *not* obtained the smaller of the two possible values for U. That is, we have calculated U' rather than U. In this case, we can determine the appropriate value for U very easily by subtracting the calculated value of U' from $n_1 \cdot n_2$:

$$U = (n_1 \cdot n_2) - U'.$$

We can illustrate this using the data obtained in Dr. Shafer's study. In using the formula to calculate U, we might have chosen to concentrate on the experimental group rather than the control group. Had we counted the sum of the ranks assigned to experimental group subjects, we would have obtained $R_1 = 57$ instead of the 79 we found when we considered the control group as group 1. Then

$$U = n_1 \cdot n_2 + \frac{n_1(n_1 + 1)}{2} - R_1 = (8 \cdot 8) + \frac{(8 \cdot 9)}{2} - 57 = 43.$$

When we obtained the value 43 for U, we would see that 43 is larger than the quantity $n_1 \cdot n_2 / 2$, which in this case equals $64/2 = 32$. Thus, we would have known that our 43 was really U', the larger of the possible values for U. Then, we could have found the smaller value by the formula

$$U = n_1 \cdot n_2 - U' = 64 - 43 = 21.$$

Once we have determined the smaller of the possible values for U, we use this smaller value to determine the significance of our test. U is then an observed statistic that must be compared to a critical value based on a probability distribution. In most cases, we can obtain critical values of U from a table. When the sample sizes are especially large, we employ an approximation technique.

Tables of Critical Values for U. When the smaller of the two sample sizes is 20 or less and the larger of the two sample sizes is 40 or less, we can obtain critical values for U from a table of critical values for the Mann-Whitney U (see Table C.12 in Appendix C). There are actually four separate tables, for both one- and two-tailed tests at the $\alpha = .05$ and the $\alpha = .01$ levels of significance. To use any one of these tables, we need only find the number of subjects in the smaller sample among the column headings at the top of the table, and the number of subjects in the larger sample along the row headings on the left-hand side of the table. In Dr. Shafer's example, the alternative hypothesis would be directional, because she was seeking to show that her experimental group would rank *higher* than the control group on "acceptance." This means that her test would be a

412

one-tailed test. Let us assume she chose the $\alpha = .05$ level of significance. Then, she would turn to Table C.12 and find the table of critical values for U in a one tailed test at $\alpha = .05$. She would locate $n_1 = 8$ among the column headings and $n_2 = 8$ among the row headings. At the intersection of this row and column, she would find the critical value 15. In this case, U_{ob} must be $\leq U_{cr}$ in order for the test to be significant. This should make sense to you if you recall that we use the smaller of the possible values of U as the statistic. In Dr. Shafer's study, U_{ob} was found to be 21. Because U_{ob} is not smaller than U_{cr}, Dr. Shafer could not reject the null hypothesis that experimentals do not differ from controls at the .05 level. Although the subjects in the experimental sample tended to outperform the subjects in the control sample, their performance was not so much better that we can safely generalize to the larger populations. Dr. Shafer might have to consider replicating her study with larger samples. The hypothesis test for Dr. Shafer's study has been presented according to our hypothesis testing outline in Figure 10.5.

A Normal Approximation for Large Samples. If both samples involved are larger than 20, or if either sample exceeds 40, the tables of critical values for the Mann-Whitney U cannot be used. Fortunately, as the size of the two samples increases, the sampling distribution of U becomes very nearly normal. If the null hypothesis of no difference between the sampled populations is true, then the mean of the sampling distribution is $\mu_U = n_1 n_2 / 2$, and the standard deviation is

$$\sigma_U = \sqrt{\frac{(n_1)(n_2)(n_1 + n_2 + 1)}{12}}.$$

We can, therefore, construct a z-statistic by substituting this mean and standard deviation into the formula for a z-score, $z = \frac{X - \mu}{\sigma}$. Here the formula becomes:

$$z_{ob} = \frac{U_{ob} - \dfrac{n_1 n_2}{2}}{\sqrt{\dfrac{(n_1)(n_2)(n_1 + n_2 + 1)}{12}}}.$$

When we use this formula, it does not matter whether the U_{ob} is the larger or the smaller of the two possible Us. The absolute value of the difference between U_{ob} and the expected value of U, $n_1 \cdot n_2 / 2$, will be identical in either case. The significance of the z-statistic is determined by reference to the table of the standard normal distribution (Table C.2).

Treatment of Tied Scores. When tied scores occur, the procedure employed with the Mann-Whitney U-test is to give each of the tied observations the average

Experimental and Pseudo-Experimental Problems

FIGURE 10.5
Mann-Whitney U-Test Comparing the Two Groups in Dr. Shafer's Study
of Therapist Acceptance

Step 1

Because we have two independent samples that are to be compared on a rank-order variable, we consider the Mann-Whitney U-test to be appropriate.

Step 2

H_o: The two populations sampled, experimental therapists and control therapists, do not differ with respect to the variable of interest, "acceptance."

Step 3

H_a: The two populations do differ, with experimental therapists having higher ratings.

Step 4

The test statistic is the Mann-Whitney U. Because the sample sizes in this problem are small, we can obtain a critical value for U from the table of critical values (Table C.12).

Step 5

$\alpha = .05$. Because the test is directional, we locate the critical value for U when $\alpha = .05$ in a one-tailed test. Referring to the table, we find $U_{cr} = 15$. We will reject H_o if U_{ob} is *smaller* than 15.

Step 6

$U_{ob} = n_1 n_2 + \dfrac{n_1(n_1 + 1)}{2} - R_1 = 64 + \dfrac{8 \cdot 9}{2} - 79 = 21$. Comparing U_{ob} to
$\dfrac{n_1 n_2}{2} = 32$, we see $U_{ob} < \dfrac{n_1 n_2}{2}$, so our calculated U is the smaller of the two possible values. Thus, there is no need to find $U = (n_1 n_2) - U'$.

Step 7

Because $U_{ob} = 21$ is not smaller than $U_{cr} = 15$, we cannot reject H_o at the $\alpha = .05$ level of significance. We cannot conclude that Dr. Shafer's workshop produces more accepting therapists.

of the ranks they would have held if no ties had occurred. Clearly, if tied scores are from the same group, the value of the U-statistic does not change. If tied scores come from different groups, the value of U may change. The effect on the test is generally inconsequential. However, in the case of the normal approximation, there is an alternative formula for obtaining z_{ob} that corrects for tied observations. This formula is

$$z_{ob} = \frac{U - \dfrac{n_1 n_2}{2}}{\sqrt{\left(\dfrac{n_1 n_2}{N(N-1)}\right)\left(\dfrac{N^3 - N}{12} - \Sigma T_U\right)}}$$

In this formula, $N = n_1 + n_2$, or the total number of observations in both groups in the study; and ΣT_U is defined as follows:

$$\Sigma T_U = \Sigma \frac{(t^3 - t)}{12},$$

where t is the number of observations that are tied for any one rank, and ΣT_U instructs us to sum the T_Us across all such ranks. In most cases, the effect of the correction for ties is small. In all cases, the effect is to increase the value of z_{ob}. Unless there are large numbers of ties in the data, the correction is generally inconsequential.

Comparisons of More Than 2 Independent Groups on a Rank-Order Variable: The Kruskal-Wallis Analysis of Variance by Ranks

When we are comparing 3 or more independent groups on a rank-order variable, we typically employ the *Kruskal-Wallis analysis of variance by ranks* to test the hypothesis that these groups do not differ with respect to the variable of interest. The Kruskal-Wallis test is also frequently used when we have measured the variable of interest with numerical scores, but we are not confident that we have met all the assumptions required to use the standard analysis of variance (to be discussed later in this chapter).

A Research Example. Suppose that Clinical Psychologist Shafer wished to compare 3 postdoctoral psychotherapy training workshops with respect to the level of acceptance of client beliefs and attitudes achieved by participating therapists. Dr. Shafer randomly selects 10 therapists to participate in each of the 3 workshops. At the conclusion of the workshops, subjects from each group are videotaped during several therapy sessions. An experienced training therapist views these tapes and rates all the subjects from most accepting to least accepting. Obviously, the problem described here is identical to that described to illustrate the use of the Mann-Whitney U-test, except that in this case subjects from 3 independent populations are being compared, rather than subjects from 2 populations. In cases where more than 2 groups are to be compared, we typically employ the Kruskal-Wallis analysis of variance by ranks. The data corresponding to Dr. Shafer's hypothetical 3-group study are presented in Table 10.7. The therapist judged to be most accepting has been assigned the rank 1. To reinforce the idea that this test does not require equal sample sizes, let us suppose that one subject in Group A became ill and dropped out of the study, leaving two groups of 10 subjects each and one group of 9. Also, to illustrate the handling of ties, let us say that there were 2 occasions on which the judge was unable to differentiate between a pair of subjects. This means that there are 2 sets of 2 subjects each in which scores are tied.

The Calculation of the Kruskal-Wallis H-Statistic. In the event that there are no tied observations, the observed statistic, the H-statistic, in the Kruskal-Wallis analysis of variance by ranks is computed according to the following formula:

$$H_{ob} = \left[\frac{12}{N(N+1)} \cdot \sum_{}^{\substack{\text{all} \\ \text{samples}}} \frac{(R_j)^2}{n_j} \right] - 3(N+1).$$

In this formula, the terms have the following meaning:

N: The total number of subjects in all the samples
R_j: The sum of the ranks assigned to the subjects in the jth sample
n_j: The number of subjects in the jth sample

The formula directs us to perform the following steps.

Step 1   For each of the samples in the study, add up the ranks assigned to subjects in the sample. The result of this step for each sample is R_j. The j is a subscript that identifies the sample. If there are 3 samples, j might be A for the first sample, B for the second, and C for the third.
Step 2   Again for each sample, square the result of Step 1. This is $(R_j)^2$.
Step 3   And again for each sample, divide the result of Step 2 by n_j, the number of subjects contained in that sample.
Step 4   Sum up the results of Step 3 across all samples.
Step 5   Multiply the result of Step 4 by $12/N(N+1)$, where N is the total number of subjects in *all* the groups.
Step 6   From the result of Step 5, subtract $3(N+1)$.

If there are no tied observations in the sample data, these 6 steps are all that are

416

TABLE 10.7
Rank on "Acceptance" and Group Membership for 29 Subjects
in Dr. Shafer's Study of 3 Training Programs

Rank	Group	Rank	Group
1	A	16	B
2	A	17	B
3	B	18	C
4	A	19.5	B
5	B	19.5	C
6	A	21	B
7.5	A	22	B
7.5	C	23	C
9	A	24	C
10	A	25	C
11	B	26	C
12	A	27	C
13	A	28	C
14	B	29	C
15	B		

required to compute the *H-statistic*. However, when there are ties in the data, the formula just presented is adjusted slightly, and becomes

$$
H_{ob} = \frac{\left[\frac{12}{N(N+1)} \cdot \sum_{}^{\substack{\text{all} \\ \text{samples}}} \frac{(R_j)^2}{n_j} \right] - 3(N+1)}{1 - \frac{\Sigma T_k}{N^3 - N}}.
$$

You will note that the only change is the addition of the denominator, which has the effect of increasing slightly the value of H_{ob} when there are ties. The new term, T_k, is equal to $t_k^3 - t_k$, where t is the number of observations tied at a particular rank and k is the number of *sets* of ties. The formula for the *H*-statistic adjusted for ties, therefore, directs us to perform the following additional steps in calculating H_{ob}:

Step 7 For each instance in which tied observations occur, compute $T = t^3 - t$. If 2 scores are tied at a particular rank, $T = (2)^3 - 2 = 6$. If 3 are tied, $T = (3)^3 - 3 = 24$.

417

Step 8   Sum up the Ts for all the cases in which ties occurred.

Step 9   Divide the result of Step 8 by $N^3 - N$, where N still equals the total number of observations in the study.

Step 10 Subtract the result of Step 9 from 1.

Step 11 Divide the result of Step 6 by the result of Step 10.

The computational procedure for obtaining H_{ob} has been illustrated in Figure 10.6. You will note that the change that takes place in the computed value of H as a result of correcting for ties is very small. It may be somewhat greater when a large proportion of the scores are ties.

Once we have calculated the value of the observed statistic H, we determine its significance in one of two ways, depending on the number of groups and the size of these groups.

Determining the Significance of H With Small Samples.   When the problem is a comparison of 3 groups, and none of the samples is larger than 5, we can determine the significance of H_{ob} by reference to a table of probabilities associated with various values of H_{ob} such as Table C.13. To use the table, we first locate the portion of the table that matches the sizes of our 3 samples. The table then provides us with the exact significance probability of various observed values of H. That is, the table tells us what the chance would be of obtaining an H_{ob} as large as the tabled values, under the assumption of the null hypothesis that the 3 sampled populations really do not differ with respect to the variable of interest. If our H_{ob} is as large as the particular tabled value of H for a particular combination of sample sizes, then the significance probability of our observed result is equal to the probability listed in the table alongside that tabled value of H. For example, let us suppose that we had calculated the Kruskal-Wallis statistic for an experiment with 3 samples of size 5, 4, and 4. Let us further suppose that the calculated value for H_{ob} in our study was 7.744. Referring to Table C.13, we see that the significance probability of this observed result is .011. That is, if H_o is true and the 3 samples have been drawn from populations that are identical with respect to the variable of interest, the chance of obtaining H_{ob} as large as the one actually obtained is just over 1 percent. The result of the test, therefore, would be significant beyond $\alpha = .05$, but it would not quite reach the $\alpha = .01$ level.

Determining the Significance of H_{ob} with Larger Samples.   If our experiment involves more than 3 samples, or if any one of the samples contains more than 5 observations, then Table C.13 cannot be used. In this case, we rely on the fact that the sampling distribution of H approximates a χ^2 distribution when the null hypothesis of no differences between the populations is true. The number of degrees of freedom associated with the χ^2 distribution is $K - 1$, where K is the number of samples.

To find the critical value for H_{ob}, therefore, we use the table of the χ^2 distribution (Table C.3). For Dr. Shafer's 3-group study, we look in the table under

FIGURE 10.6
Calculation of Kruskal-Wallis H-Statistic in Dr. Shafer's 3-Group Comparison

Step 1

Sum of ranks for Group A $= 1 + 2 + 4 + 6 + 7.5 + 9 + 10 + 12 + 13 = 64.5 = R_A$.
Sum of ranks for Group B $= 3 + 5 + 11 + 14 + 15 + 16 + 17 + 19.5 + 21 + 22 = 143.5 = R_B$.
Sum of ranks for Group C $= 7.5 + 18 + 19.5 + 23 + 24 + 25 + 26 + 27 + 28 + 29 = 227 = R_C$.

Step 2

$(R_A)^2 = (64.5)^2 = 4,160.25$

$(R_B)^2 = (143.5)^2 = 20,592.25$

$(R_C)^2 = (227)^2 = 51,529.0$

Step 3

$$\frac{(R_A)^2}{n_A} = \frac{(64.5)^2}{9} = \frac{4,160.25}{9} = 462.25$$

$$\frac{(R_B)^2}{n_B} = \frac{(143.5)^2}{10} = \frac{20,592.25}{10} = 2,059.23$$

$$\frac{(R_C)^2}{n_C} = \frac{(227)^2}{10} = \frac{51,529}{10} = 5,152.9$$

Step 4

$$\sum \frac{(R_j)^2}{n_j} = \left[\frac{(R_A)^2}{n_A} + \frac{(R_B)^2}{n_B} + \frac{(R_C)^2}{n_C} \right] = [462.25 + 2,059.23 + 5,152.9] = 7,674.38$$

Step 5

$$\frac{12}{N(N+1)} \cdot \sum \frac{(R_j)^2}{n_j} = \frac{12}{29(29+1)} \cdot 7,674.38 = (.01379)(7,674.38) = 105.83$$

Step 6

$$\left[\frac{12}{N(N+1)} \cdot \sum \frac{(R_j)^2}{n_j} \right] - 3(N+1) = 105.83 - 3(29+1) = 15.83$$

Step 7

For tied scores at rank 7.5, $T_k = t^3 - t = (2)^3 - 2 = 6$.
For tied scores at rank 19.5, $T_k = t^3 - t = (2)^3 - 2 = 6$.

Step 8

$\Sigma T_k = 6 + 6 = 12$

Step 9

$$\frac{\Sigma T_k}{N^3 - N} = \frac{12}{(29)^3 - 29} = \frac{12}{24,389 - 29} = \frac{12}{24,360} = .0004926$$

Step 10

$1 - .0004926 = .9995074$

Step 11

$$H_{ob} = \frac{\dfrac{12}{N(N+1)} \cdot \sum \dfrac{(R_j)^2}{n_j} - 3(N+1)}{1 - \dfrac{\Sigma T_k}{N^3 - N}} = \frac{15.83}{.9995074} = 15.84$$

$K - 1 = 3 - 1 = 2$ degrees of freedom. We find χ^2_{cr} for $\alpha = .05$ is 6.0 and for $\alpha = .01$ is 9.2. Because H_{ob} was found to be 15.84, the null hypothesis may be rejected at the .01 level of significance. In this case, Dr. Shafer may conclude that the 3 populations examined are not identical with respect to the ordinal variable "acceptance." The Kruskal-Wallis test that applies to Dr. Shafer's 3-group experiment has been laid out according to our hypothesis testing outline in Figure 10.7.

Notice that the significant result obtained in this case allows only the conclusion that not all of the 3 populations under investigation are identical on the variable of interest. It does not allow us to state positively *which* populations differ from the others. That is, we cannot say that Population A ranks higher than Population B and that Population B ranks higher than Population C. We can get a feeling for the relative positions of the 3 populations by examining the sample data. However, this is not the same as performing a significance test. If you wish to specify that the ranks assigned to 2 samples are significantly different from each other, we recommend the use of the Mann-Whitney *U*-test on the data from these 2 samples alone. Another way of looking at the same issue is to note that the alternative hypothesis in the Kruskal-Wallis analysis of variance by ranks is always nondirectional. It states that the populations do differ, but not which are high and which are low.

INFERENTIAL STATISTICS FOR THE COMPARISON OF INDEPENDENT GROUPS ON AN INTERVAL SCALE VARIABLE

Two Independent Groups: The *t*-Test for Two Independent Samples

When we wish to compare two populations on an interval scale variable, we can sometimes employ the *t-test for two independent samples* to test the null hypothesis that the two populations have the same mean. We use the word "sometimes" because there are certain assumptions made in the use of the *t*-test and we must believe that our data come reasonably close to meeting these assumptions in order to use the *t*-test with confidence.

Assumptions of *t*-Test for Two Independent Samples. The assumptions underlying the *t*-test for two independent samples may be summarized as follows: (1) We have randomly selected n_1 observations from a population with a mean μ_1 and a variance σ^2_1 (you will recall that the variance of a population is simply the square of the standard deviation). Also, we have randomly selected n_2 observations from another population with a mean μ_2 and a variance σ^2_2. These two samples are selected independently; that is, the selection of a particular subject or observation from one of the two populations of interest will have no effect whatsoever on the chance of selection of any other subject, either within that population or in the other population sampled. (2) We further assume that each

420

FIGURE 10.7
Kruskal-Wallis Analysis of Variance by Ranks Comparing 3 Groups
in Dr. Shafer's Second Study of Therapist "Acceptance"

Step 1

Because we have 3 independent samples that are to be compared on a rank-order variable, we consider the Kruskal-Wallis analysis of variance by ranks to be appropriate.

Step 2

H_o: The 3 populations sampled, therapists from 3 different training programs, do not differ with respect to the variable of interest, "acceptance."

Step 3

H_a: The 3 populations do differ with respect to this variable.

Step 4

The test statistic will be the Kruskal-Wallis H-statistic. Because the number of subjects in this study is too large for us to determine significance from the table of probabilities associated with various values of H_{ob}, we use the χ^2 approximation. Here df $= (K - 1) = (3 - 1) = 2$.

Step 5

$\alpha = .01$. χ^2_{cr} for df $= 2$ and $\alpha = .01$ is 9.2. We will reject H_o at the .01 level of significance if $H_{ob} > 9.2$.

Step 6

$$H_{ob} = \frac{\left[\dfrac{12}{N(N+1)} \cdot \sum \dfrac{(R_j)^2}{n_j} \right] - 3(N+1)}{1 - \dfrac{\Sigma T_k}{(N^3 - N)}} = 15.84 \qquad \text{(See Figure 10.4 for details of computing } H_{ob}.)$$

Step 7

Because $H_{ob} > \chi^2_{cr}$, we reject H_o at the $\alpha = .01$ level of significance. Dr. Shafer can conclude that the 3 populations of therapists sampled do differ with respect to "acceptance."

421

of the two populations sampled is normally distributed; and (3) that the variability of scores in the two populations is identical.

For the third assumption, we sometimes write: $\sigma_1^2 = \sigma_2^2 = \sigma^2$, where σ^2 refers to the common variance of the two populations. These assumptions underlie the mathematical model employed in the t-test for independent samples. They are rather rigorous assumptions, and you may be wondering if there are any research situations where they are actually met. The answer to this question is that we can make sure that the first assumption is met by employing careful sampling procedures, but we cannot make sure that the second and third assumptions are met. These assumptions concern the nature of the populations sampled, and they are beyond our control. Fortunately, it has been found that the t-test for two independent samples, like the one-sample t-test, will still produce accurate results even if the populations sampled do not conform exactly to the assumptions. In truth, the populations need not be exactly normal. If they are unimodal and not highly skewed, the t-test will still work well. Similarly, it is not absolutely essential that the variances of the populations be identical. This is particularly true when the two samples are the same size or close to the same size. The assumption of equal variability in the two populations is more critical when the two samples are of very different sizes. The statistician would say that the t-test is rather "robust" to violations of its underlying assumptions. That is, the test will provide reasonably accurate results even if the assumptions are not fully justified.

Of course, we will most often lack knowledge of the true shape and variance of the sampled populations. For this reason, when we have interval scale data from two independent samples, we will typically consider the shape and variability of these *samples* in deciding whether it would be appropriate to use the two-sample t-test. If we find that both samples are unimodal and not highly skewed, we will be on firm ground. If the two samples are of equal size, the sample variances s_1^2 and s_2^2 may be in a ratio as large as 4 to 1, and the t-test should provide accurate results. If the two samples are of unequal size, more care must be taken, for unequal population variances will have the effect of increasing the actual chance of a Type I error over the level specified in the test. If we have unequal sample sizes and find that the variances in our two samples are in a ratio of 4 to 1 or larger, we might wish to transform our numerical scores into ranks and employ the Mann-Whitney U-test as a substitute.

A Research Example.   Measurement Specialist Wilson is interested in the effect of subject motivation on the results of standardized personality inventories. Dr. Wilson randomly selects two groups of college students. Group 1, consisting of 16 students, is asked to complete the Hartman Social Responsiveness Inventory. They are instructed to work very carefully because their score on the instrument will become a permanent part of their student records. Group 2, consisting of 15 students, is also asked to complete the Hartman Social Responsiveness Inventory. However, they are instructed to leave their names off the inventory altogether, because the experimenter seeks only to gain information about students in general, rather than any information on specific students. Dr. Wilson makes no prediction regarding the question of which group will score higher on the Hartman Social Responsiveness Inventory. She will seek to test the null hypothesis that the average score on the Inventory will be the same in the population given personalized motivating instructions as in the population instructed to remain anonymous. If we designate the mean of the first population as μ_1 and the mean of the second population as μ_2, the null hypothesis may be written $H_o: \mu_1 = \mu_2$ or $H_o: \mu_1 - \mu_2 = 0$. The nondirectional alternative hypothesis would be written $H_a: \mu_1 \neq \mu_2$ or $\mu_1 - \mu_2 \neq 0$.

Dr. Wilson reads in the manual for the test that scores on the Inventory are normally distributed. The manual also provides the mean and standard deviation of the instrument for the large group of subjects used in establishing test norms. Dr. Wilson feels that the new instructions to be given to her own subjects may very well cause a change in the average score on the scale. She also feels that the instructions may cause a change in the variability of the test scores. However, she strongly believes that the distribution of test scores among the populations represented in her samples will continue to be normal or at least close to normal. She therefore feels that the t-test for two independent samples may be the appropriate inferential statistic to test her hypothesis. On inspecting the data obtained in her two samples, Dr. Wilson finds that the sample standard deviation for the first group, s_1, is 5.82. The sample standard deviation for the second group, s_2, is 6.34. She also finds that the distributions of scores in the two samples are unimodal and symmetrical. These observations satisfy Dr. Wilson that the t-test for two independent samples is appropriate.

Calculating the Pooled Sample Variance.   You will recall that one assumption underlying the test is that the variances of the sampled populations are identical. If this assumption is true, then each of the two sample variances, $s_1{}^2$ and $s_2{}^2$, must be an estimate of the common population variance, σ^2. For this reason, the first step in calculating the t-statistic for the t-test for two independent samples is to obtain a weighted average of the two sample variances that will be used to obtain the best available estimate of σ^2. This estimate of σ^2 is known as the *pooled sample variance* and is designated $s_p{}^2$. The formula for finding $s_p{}^2$ is:

$$s_p{}^2 = \frac{(n_1 - 1) s_1{}^2 + (n_2 - 1) s_2{}^2}{n_1 + n_2 - 2}$$

where n_1 is the number of subjects in the first group, n_2 is the number of subjects in the second group, $s_1{}^2$ is the square of the sample standard deviation for the first group (the sample variance), and $s_2{}^2$ is the square of the sample standard deviation for the second group. You will note that this formula is simply a weighted average. If there are more subjects in one sample than the other, we count the larger sample's variance more heavily in arriving at our estimate of the pooled variance. In the case of Dr. Wilson's data, we find

$$s_p^2 = \frac{(16-1)\,(5.82)^2 + (15-1)\,(6.34)^2}{16 + 15 - 2} = 36.92.$$

You will note that the value calculated for s_p^2 lies between s_1^2 and s_2^2. This will always be the case, provided the calculations are performed correctly.

Calculating the Observed Value of t.   Once the appropriate value of s_p^2 has been obtained, we proceed to calculate the value of the observed statistic. The formula for t_{ob} in the t-test for two independent samples is

$$t_{ob} = \frac{(\bar{X}_1 - \bar{X}_2) - (\mu_{1_o} - \mu_{2_o})}{\sqrt{s_p^2\left(\dfrac{1}{n_1} + \dfrac{1}{n_2}\right)}}$$

In this formula, $(\bar{X}_1 - \bar{X}_2)$ represents the observed difference between the two sample means. The term $(\mu_{1_o} - \mu_{2_o})$ represents the difference between the means of the two populations sampled, as stated in the null hypothesis. The use of zeros as second subscripts refers to the null hypothesis. Finally, the denominator of the statistic represents the estimated standard deviation of the hypothetical distribution of differences between the sample means. This terminology should sound somewhat familiar to you, based on the material on one-sample t-tests presented in Chapter 8.

In fact, the formula for obtaining t_{ob} in this two-sample test is exactly analogous to the formula for obtaining t_{ob} in the one-sample t-test considered in Chapter 8. You will recall that in the one-sample t-test, we envisioned a hypothetical distribution of sample means having a mean equal to the value specified in the null hypothesis and a standard deviation estimated to be equal to $s/\sqrt{n}$. The observed statistic in the one-sample t-test is, therefore, an estimate of the number of standard deviations by which the observed sample mean differs from the expectations embodied in the null hypothesis. In the two-sample t-test, we also envision a hypothetical distribution derived from the following procedure. We imagine drawing two samples, one from each of two infinitely large normal distributions having the same variance. We imagine calculating the mean of each sample and recording the *difference* between these means. We further imagine repeating this process an infinite number of times so that we have recorded an infinite number of these differences. Finally, we imagine that each of these differences is a score in a new distribution, the distribution of differences between sample means. Mathematical statisticians have shown that the mean of this distribution of differences will be equal to the difference between the means

424

of the two populations from which the samples are drawn. They have further shown that the standard deviation of this distribution of differences is estimated by

$$\sqrt{s_p^2 \left(\frac{1}{n_1} + \frac{1}{n_2} \right)}.$$

Thus, the t-statistic in the t-test for two independent samples is an estimate of the number of standard deviations by which an observed difference between the means of two independent samples differs from the expectation for this difference embodied in the null hypothesis. The one-sample t-test and the t-test for two independent samples are compared in Figure 10.8.

In the case of Dr. Wilson's study, the null hypothesis was that $\mu_1 - \mu_2 = 0$, so the expected value for the difference between the two sample means was also 0. Let us suppose that Dr. Wilson obtains an observed sample mean for Group 1 of 43.64, and an observed sample mean of 36.20 for Group 2. In this case, the computed value of t_{ob} would be:

$$t_{ob} = \frac{(43.64 - 36.20) - (0)}{\sqrt{36.92 \left(\frac{1}{16} + \frac{1}{15} \right)}} = 3.41.$$

That is, in terms of the hypothetical distribution of differences between sample means, the observed difference of $(43.64 - 36.20) = 7.44$ is estimated to be 3.41 standard deviations larger than what we would have expected, if H_o were true.

Determining the Significance of t_{ob} in the t-Test for Two Independent Samples. The significance of the observed statistic in the t-test for two independent samples is determined by reference to the table of the t-distributions (Table C.5). The number of degrees of freedom associated with the t-test for two independent samples is always $n_1 + n_2 - 2$. In the case of Dr. Wilson's study, df = 16 + 15 − 2 = 29. We therefore refer to the table of the t-distributions under the row for df = 29. We find that for a two-tailed test of a nondirectional hypothesis, t_{cr} is 2.045 at the .05 level of significance, and 2.756 at the .01 level of significance. Because t_{ob} in Dr. Wilson's study is 3.41, she would reject H_o at the .01 level. She would conclude that subjects receiving the different motivating instructions employed in her experiment have different mean scores on the Hartman Social Responsiveness Inventory. The two sample means tell her that subjects in Group 1 scored higher than subjects in Group 2 on the Inventory. On the basis of the significance test, Dr. Wilson would conclude that subjects receiving motivating instructions have a higher mean score than subjects not receiving such instructions. The entire t-test for two independent samples for

425

FIGURE 10.8
One-Sample t-Test and t-Test for Two Independent Samples Compared

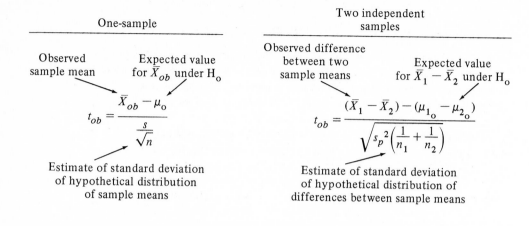

Dr. Wilson's data has been laid out, according to our outline for a hypothesis test, in Figure 10.9.

A Confidence Interval for the Difference Between Two Population Means.
You will recall that when we considered the one-sample t-test, we immediately noted that it was also possible to apply the interval estimation approach. That is, we could not only test the hypothesis that the mean of a population was equal to some specified value, we could also calculate a confidence interval for the actual value of the population mean. Now, having considered the t-test for two independent samples, we again note that the interval estimation approach may be used. That is, rather than test the null hypothesis that the difference between two population means is some specified value, we may also calculate a *confidence interval for the difference between two population means.*

In calculating a confidence interval for the difference between two population means, we employ the standard interval estimation technique. We start with an estimate of the difference derived from our sample data, and we allow an amount on either side of this sample estimate to take into consideration sampling error. In the case of the confidence interval for the difference between two population means, we obtain our confidence limits using these formulas:

$$\text{lower limit} = (\bar{X}_1 - \bar{X}_2) - t_{cr} \sqrt{s_p^2 \left(\frac{1}{n_1} + \frac{1}{n_2}\right)}$$

426   and

$$\text{upper limit} = (\bar{X}_1 - \bar{X}_2) + t_{cr} \sqrt{s_p^2\left(\frac{1}{n_2} + \frac{1}{n_2}\right)},$$

where $(\bar{X}_1 - \bar{X}_2)$ is the estimate of the difference between the two population means, t_{cr} is the factor that determines the level of confidence for the confidence interval, and $\sqrt{s_p^2\left(\frac{1}{n_1} + \frac{1}{n_2}\right)}$ is the estimated standard deviation of the hypothetical distribution of differences between sample means. To obtain t_{cr}, we look for the t-distribution with $n_1 + n_2 - 2$ df. For a 95 percent confidence interval, we find t_{cr} for a two-tailed test at $\alpha = .05$. For a 99 percent confidence interval, we find t_{cr} for a two-tailed test at $\alpha = .01$. You will note that the formulas for the confidence interval for the difference between two population means are exactly analogous to the formulas for the confidence interval for a single population mean in that we begin with the sample estimate as a base and allow a certain number of estimated standard deviation units on either side of this estimate for sampling error.

In the case of Dr. Wilson's data, we find that the 99 percent confidence interval is:

$$\text{lower limit} = (43.64 - 36.20) - 2.756 \sqrt{36.92\left(\frac{1}{16} + \frac{1}{15}\right)} = +1.42$$

and

$$\text{upper limit} = (43.64 - 36.20) + 2.756 \sqrt{36.92\left(\frac{1}{16} + \frac{1}{15}\right)} = +13.46.$$

That is, the difference between the means of these two populations lies between $+1.42$ and $+13.46$, and we can be 99 percent confident in making this statement. Notice that the 99 percent confidence interval does not contain the value 0. This means that it is highly unlikely that the true difference between the two population means is 0. This finding agrees with the result of our hypothesis test for Dr. Wilson's data, in which we rejected the null hypothesis $H_o: \mu_1 - \mu_2 = 0$.

More Than Two Independent Groups: One-Way Analysis of Variance

The final research situation that we will be considering in this chapter is that in which more than two independent populations are compared on an interval scale variable. In such situations, we may sometimes employ the statistical technique known as the *one-way analysis of variance* (ANOVA) to test the hypothesis that the several populations of interest have the same mean on the variable of interest. The one-way analysis of variance may be viewed as an extension of the t-test for two independent samples. Alternatively, the t-test for two independent samples may be viewed as the special two-sample case of the one-way analysis of

FIGURE 10.9
t-Test for Two Independent Samples for Dr. Wilson's Study
of the Effect of Motivating Instructions on Test Scores

Step 1

Because Dr. Wilson has drawn independent random samples from each of two populations, and because she is willing to assume that these two populations are normally distributed with a common variance, the *t*-test for two independent samples is appropriate.

Step 2

$H_o: \mu_1 - \mu_2 = 0$

Step 3

$H_a: \mu_1 - \mu_2 \neq 0$

Step 4

The test is the *t*-test for two independent samples. Because there are 16 subjects in one sample and 15 in the other, the degrees of freedom for the test will be $df = n_1 + n_2 - 2 = (16 + 15 - 2) = 29$.

Step 5

$\alpha = .01$. t_{cr} in a nondirectional test at $\alpha = .01$ is $t_{29;.995} = 2.756$. We will reject H_o if t_{ob} is greater than $+2.756$ or less than -2.756.

Step 6

$$s_p{}^2 = \frac{(n_1 - 1)s_1{}^2 + (n_2 - 1)s_2{}^2}{n_1 + n_2 - 2} = \frac{(16 - 1)(5.82)^2 + (15 - 1)(6.34)^2}{16 + 15 - 2} = 36.92$$

$$t_{ob} = \frac{(\bar{X}_1 - \bar{X}_2) - (\mu_{1_0} - \mu_{2_0})}{\sqrt{s_p{}^2\left(\frac{1}{n_1} + \frac{1}{n_2}\right)}} = \frac{(43.64 - 36.20) - (0)}{\sqrt{36.92\left(\frac{1}{16} + \frac{1}{15}\right)}} = +3.41$$

Step 7

Because $t_{ob} > +2.756$, we reject H_o. We conclude that the two populations have different means.

Inferential Statistics

variance. Regardless of how we choose to relate the two techniques to each other, there are many similarities. A most important area of similarity concerns the assumptions underlying the use of the techniques.

Assumptions of One-Way Analysis of Variance. The assumptions underlying the one-way ANOVA may be summarized as follows: (1) We have randomly and independently selected a sample of n_j observations from each of K populations. Here the subscript j designates a particular sample and K is the number of populations sampled. Thus, if our research problem dealt with 4 populations, $K = 4$. For the subscript j, we may employ numbers from 1 to 4, designating Sample 1, Sample 2, Sample 3, and Sample 4.

Alternatively, we might use the letters A, B, C, and D to designate samples. If we use letters to designate our samples, then n_A would represent the number of subjects in Sample A, n_B the number of subjects in Sample B, and so on. There is no requirement that all the samples be the same size. The requirement is that all observations be independent, both within and across samples. This means that the selection of a particular subject or observation from any of the sampled populations will have no effect whatsoever on the chance of selection of any other subject, either in the same population or in a different population. (2) We also assume that each of the K populations sampled is normally distributed, and (3) that the variability of scores in all K populations is identical.

As in the case of the t-test for two independent samples, we have some latitude with regard to the last two assumptions. The one-way analysis of variance is also "robust" to violations of these assumptions. In determining whether or not to employ the one-way analysis of variance in a specific research situation, we consider the shape and variability of the samples. We make sure that the samples are unimodal and not highly skewed. We compare the largest sample variance to the smallest sample variance. With respect to the assumption of equal variances, we keep in mind our sample sizes. When the sizes of the samples employed are equal or nearly equal, we can feel confident in proceeding with the ANOVA, even if the largest sample variance is as much as 4 times the smallest variance. If the sample sizes are quite different, however, a difference in sample variances of this nature would lead us to decide not to use the ANOVA because the results could be inaccurate.* In this event, we might employ instead the Kruskal-Wallis analysis of variance by ranks, which has only the assumption of random independent samples. This would involve translating the interval scale scores into ranks.

A Research Example. Educational Evaluation expert Edison has been asked to compare the efficacy of 3 different first-grade mathematics curricula, which we will call curricula A, B, and C. Dr. Edison begins by locating 3 suburban school

---

* The effect of various deviations from the assumption of equal variances is considered in some detail in Scheffé H., *The Analysis of Variance*. New York: John Wiley & Sons, 1959.

Experimental and Pseudo-Experimental Problems

districts that are similar in terms of socioeconomic status and past student performance on standardized achievement tests. The 3 districts differ, however, in that each has adopted a different one of the 3 mathematics curricula for use in its schools. From each of these districts, Dr. Edison selects a random sample of 20 first-grade students. In order to control for the possible effects of different teachers and schools, Dr. Edison draws his random sample in each case from a list of *all* the first-graders in the district regardless of which particular school the child is in or which particular class the child has been assigned to within that school. By sampling randomly across schools and teachers, Dr. Edison hopes that the 3 samples from the 3 different districts will turn out to be roughly equivalent in the sense that there will be some good and some bad schools and some good and some bad teachers in each case. At the conclusion of the year, Dr. Edison administers the standardized Indiana Test of Elementary Mathematics Achievement to each of the subjects. Two subjects from the district employing curriculum B become ill on the day of the test so Dr. Edison's final sample sizes are 20, 18, and 20 respectively. These scores are presented in Figure 10.10, along with some computations to be described shortly. Dr. Edison knows that scores on the Indiana Test are normally distributed across the country, and he believes that they will be normally distributed among the populations of students using curricula A, B, and C as well. He consults the sample distributions to see that they are unimodal and not seriously skewed. He compares the sample variances and finds that they are all rather close to each other. He therefore concludes that it would be appropriate to use the one-way analysis of variance to test the hypothesis that the mean score on the Indiana Test is identical among the populations of first-graders using curricula A, B, and C. This null hypothesis is generally written $H_o: \mu_A = \mu_B = \mu_C$. The alternative hypothesis for the test is H_a: Not all the means are the same.

The Idea Behind Analysis of Variance. We will not consider in detail the derivation of the mathematical model that underlies the one-way analysis of variance. However, it may be useful to you to have a general intuitive notion of what is involved in the analysis of variance. We hope to provide such a notion by reference to Figure 10.11. In this figure, we have depicted the scores obtained by the subjects in Dr. Edison's study along a common axis. Scores achieved by subjects using Curriculum A are represented with an A, scores achieved by subjects using Curriculum B with a B, and scores achieved by subjects using Curriculum C with a C. You will note that the scores of subjects in Group A seem to be lower than the scores of subjects in Groups B or C. This is not what we would expect to see in our 3 samples if the null hypothesis, $\mu_A = \mu_B = \mu_C$, were true. If the null hypothesis were true, we would expect to see that the scores in our 3 samples would be similar to each other. Therefore, we might say that just looking at Figure 10.11 provides us with some evidence that the null hypothesis of our test is not true. However, we have seen that in hypothesis testing, we employ

430

FIGURE 10.10
The Calculation of Sums of Squares in the One-Way ANOVA
Comparing 3 Elementary Mathematics Curricula

Step 1

Curriculum A	Curriculum B	Curriculum C
39	54	59
41	56	60
43	57	61
43	58	64
45	60	64
45	60	66
45	61	66
46	61	67
46	61	67
46	61	67
46	61	67
47	62	68
48	63	68
48	63	68
49	63	68
49	64	69
51	69	71
53	72	72
55		73
57		74

Totals

	Curriculum A	Curriculum B	Curriculum C	
Step 2 ΣX_j:	$\Sigma X_A = 942$	$\Sigma X_B = 1{,}106$	$\Sigma X_C = 1{,}339$	**Step 5** $\Sigma\Sigma X = 3{,}387$
Step 3 n_j:	$n_A = 20$	$n_B = 18$	$n_C = 20$	**Step 6** $n_{TOT} = 58$
$\bar{X}_j$:	$\bar{X}_A = 47.10$	$\bar{X}_B = 61.44$	$\bar{X}_C = 66.95$	$\bar{X}_G = 58.40$
Step 4 ΣX_j^2:	$\Sigma X_A^2 = 44{,}742$	$\Sigma X_B^2 = 68{,}262$	$\Sigma X_C^2 = 89{,}949$	**Step 7** $\Sigma\Sigma X^2 = 202{,}953$
s_j:	$s_A = 4.44$	$s_B = 4.23$	$s_C = 3.99$	

Step 8 $SST = \Sigma\Sigma X^2 - \dfrac{(\Sigma\Sigma X)^2}{n_{TOT}} = (202{,}953) - \dfrac{(3{,}387)^2}{58} = 5{,}163.88$

Step 9 $SSB = \left[\Sigma \dfrac{(\Sigma X_j)^2}{n_j} \right] - \dfrac{(\Sigma\Sigma X)^2}{n_{TOT}} = \left[\dfrac{(942)^2}{20} + \dfrac{(1{,}106)^2}{18} + \dfrac{(1{,}339)^2}{20} \right] - \dfrac{(3{,}387)^2}{58} = 4{,}182.68$

Step 10 $SSW = SST - SSB = 5{,}163.88 - 4{,}182.68 = 981.20$

Experimental and Pseudo-Experimental Problems

a mathematical model that enables us to make a statement regarding the probability of observing a particular set of sample results under the assumption that H_o is true. In the one-way analysis of variance, we use a mathematical model known as an *F-distribution* to make such a statement. We compute a statistic called F_{ob}. The manner in which the *F-statistic* works is as follows. The further apart the actual means of the sampled populations are, the further apart the means of our samples will tend to be. The further apart our sample means are, however, the larger F_{ob} tends to be. We compare the value of F_{ob} to a critical value obtained from a table. If F_{ob} exceeds the critical value for a given level of significance, we reject the null hypothesis that the population means are all the same.

What we would like to accomplish in this section is to provide you with an idea of why F_{ob} tends to be small when the means of the several samples are similar to each other, and why F_{ob} tends to be large when the means of the several samples are very different from each other. The statistic F_{ob} is a ratio that compares the extent to which the means of the samples differ from each other to the extent to which scores within the samples differ from the means of those samples. The ratio is generally written

$$F_{ob} = \frac{MSB}{MSW},$$

where *MSB* is the *mean square between groups,* our measure of the extent to which the sample means differ from each other; and *MSW* is the *mean square within groups,* our measure of the extent to which scores within the several samples spread out around the means of those samples. Take a moment to consider the meaning of these phrases. What are we talking about when we say "the extent to which the means of samples differ from each other" and "the extent to which scores within samples differ from the means of those samples"? First of all, we are talking about variability. Thus, the name "analysis of variance." But if the null hypothesis concerns population means, why are we concerned with variance? The reasoning is as follows.

If the null hypothesis is true, the sampled populations have the same mean. But if the sampled populations have the same mean, then the means of the samples we draw from these populations are all estimating the same value, $\mu_A = \mu_B = \mu_C$. And if this is the case, then the sample means $\bar{X}_A$, $\bar{X}_B$, and $\bar{X}_C$ should differ from each other by no more than an amount expected because of sampling error. In the *F-statistic*, the term *MSB* is our measure of the extent to which the sample means differ from each other. We shall provide a formal definition of *MSB* shortly. For now, it is sufficient to state that we average all the scores in all the samples to obtain an overall figure $\bar{X}_G$, known as the grand mean. If all the sample means were equal to each other, they would all equal the grand mean as well. The more the sample means differ from each other, the more they will differ from the grand mean. The differences between sample

432

FIGURE 10.11
Within and Between Groups Variability in a One-Way ANOVA

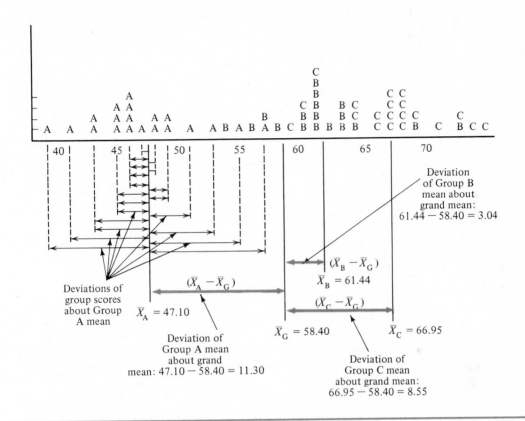

means and the grand mean are referred to as deviations. The 3 deviations of sample means from the grand mean for Dr. Edison's study are shown in Figure 10.11. If these deviations are not too large, *MSB* will not be too large. Moreover, if *MSB* is reasonably close to the value we would expect from sampling error alone, we will not have evidence that H_o is false. However, if *MSB* is much larger than we would expect from sampling error alone, we do have evidence that H_o is false. The next question, then, is "How do we know how much variability we should expect from sampling error alone?"

The answer to this question is that we estimate sampling error from a pooled estimate of the common population variance, σ^2. This pooled estimate is known as *MSW* and it depends upon the extent to which the individual scores within each sample differ from the means of their respective samples. It is analogous to

433

$s_p{}^2$ in the *t*-test for 2 independent samples. We will provide a formal definition shortly. For now, it is enough to know that *MSW* is the best estimate we have of sampling error, and that *MSW* is large when the individual scores within samples differ greatly from the means of those samples. These differences are referred to as deviations of individual scores about their sample means. These deviations have been sketched in Figure 10.11 for the scores in Group A only. In the actual ANOVA, the deviations of all scores from the means of their respective groups contribute to the value of *MSW*.

   The key point to be made at this time is that if the null hypothesis is true, both *MSB* and *MSW* are estimating the same thing: sampling error. But, if this is true, *MSB* and *MSW* should be about equal, and the *F*-ratio *MSB/MSW* should be about 1.00. On the other hand, if H_o is false and the sampled populations have different means, then *MSB* will tend to be larger than *MSW*. This is because the means of the samples will tend to differ from each other not only because of sampling error but because they are actually estimating different population means. This would appear to be the case in Figure 10.11, where the typical deviation of a group mean about the grand mean would seem to be larger than the typical deviation of a Group A score from the mean of Group A. As *MSB* increases, the ratio of *MSB* to *MSW* increases. Thus, *F* will tend to rise above the value of 1.00 that we expect, if H_o is true. This should become more clear to you as we define *MSW* and *MSB* more precisely.

Definition of *MSW*.   You will recall that one of the assumptions of the ANOVA is that all sampled populations have a common variance, σ^2. If this is true, then the sample variance in each of the several samples should provide an estimate of σ^2. *MSW* is a pooled estimate of this common population variance based on the variability of individual scores about the means of their respective samples. It is defined as follows:

$$MSW = \frac{\Sigma\Sigma(X - \bar{X}_j)^2}{n_{TOT} - K}.$$

Do not allow the summation notation here to confuse you. Were we using this formula to compute *MSW*, we would simply perform these steps:

Step 1   Measure the deviation of each score from the mean of the sample of which it is a member.

Step 2   Square all these deviations.

Step 3   Add up all the squared deviations for all groups.

Step 4   Divide the result of Step 3 by $n_{TOT} - K$, where n_{TOT} is the total number of subjects and K is the number of groups.

   In fact, we do not use this formula to compute *MSW*. We use a more convenient method to be described. The preceding formula is presented to give you

434

an idea of what *MSW* really means. It is simply a measure of the average squared deviation of scores from their respective group means. Thus, we call it mean square within groups. In Figure 10.11, we have used the data from Dr. Edison's study to illustrate the deviations of the individual scores in Group A from the mean of that group, $\bar{X}_A = 47.10$. In calculating *MSW*, we consider the deviations of every score from the mean of the particular sample to which that score belongs. We have sketched in the deviations of the individual scores in Group A from $\bar{X}_A$.

Definition of *MSB*. *MSB* is a measure of how close the sample means are to each other. It is computed by measuring the deviations of the sample means from the overall mean for all the scores in all the groups. As already indicated, the overall mean is generally referred to as the grand mean and is designated $\bar{X}_G$. Then, *MSB* is defined as follows:

$$MSB = \frac{\Sigma n_j (\bar{X}_j - \bar{X}_G)^2}{K - 1}.$$

This formula would instruct us to perform the following steps:

Step 1   Measure the deviation of each of the sample means from the grand mean, $\bar{X}_G$.
Step 2   Square each of these deviations.
Step 3   Multiply each squared deviation by the number of subjects in the group.
Step 4   Add up all the results of Step 3.
Step 5   Divide the result of Step 4 by $K - 1$, where K is the number of groups.

Thus, *MSB* is the weighted average of the squared deviations of the sample means around the grand mean. In Figure 10.11, we have used the data from Dr. Edison's study to illustrate the deviations of the 3 group means from the grand mean. These deviations are sketched in color.

In viewing Figure 10.11, it should be clear to you that the deviations of the sample means about the grand mean are, on the average, larger than the deviations of individual scores about their sample means. This is reflected in an F_{ob} in Dr. Edison's study of 117.23. This is much larger than the F_{ob} of 1.00 that we would expect to obtain if H_o were true. As we shall see shortly, this F_{ob} is highly significant. However, before we turn to the question of determining the significance of a calculated F_{ob}, we present formulas that may be used to compute F_{ob} conveniently.

Calculating F_{ob}.   The actual calculation of the *F*-statistic in the one-way ANOVA is accomplished with the computational formulas illustrated in Figures 10.10 and

10.12. The first steps in obtaining F_{ob} are those required to obtain the "sums of squares," SSB and SSW, which lead ultimately to MSB and MSW. These steps are outlined as follows and illustrated using Dr. Edison's data in Figure 10.10.

Step 1   Lay out the data so that all the scores in each group are listed in a column, with additional room at the bottom and along the right margin.

Step 2   For each group add up all the scores to obtain ΣX_j, the sum of all scores in the group. List these sums beneath the appropriate column of scores.

Step 3   For each group, count how many scores there are in the group, n_j. List these numbers beneath the appropriate column of scores.

Step 4   Again for each group, square each score and add up the squared scores to obtain ΣX_j^2, the sum of the squared scores in the group. List the sum of the squared scores for each group in the appropriate column. (ΣX_j, n_j, and ΣX_j^2 for each group can be obtained easily from calculators that have standard deviation functions and accessible memories.)

Step 5   Add the results of Step 2 for all the groups to obtain $\Sigma\Sigma X$, the sum of all the scores in all the groups.

Step 6   Add the results of Step 3 for all the groups to obtain n_{TOT}, the total number of subjects in all the groups.

Step 7   Add the results of Step 4 for all the groups to obtain $\Sigma\Sigma X^2$, the sum of all the squared scores in all the groups.

Step 8   At this point, we obtain the value SST using the formula:

$$SST = \Sigma\Sigma X^2 - \frac{(\Sigma\Sigma X)^2}{n_{TOT}}.$$

The formula instructs us to perform these operations:
a   Square the result of Step 5.
b   Divide the result of Step 8a by n_{TOT}.
c   Subtract the result of Step 8b from the result of Step 7. This yields SST, the total sum of squares.

Step 9   Next, we obtain SSB using the formula:

$$SSB = \left[\Sigma \frac{(\Sigma X_j)^2}{n_j} \right] - \frac{(\Sigma\Sigma X)^2}{n_{TOT}}.$$

This second formula instructs us to perform these operations:
a   For each group, square ΣX_j and divide the result by n_j. That is, square the sum of the scores in each group and divide the squared sum by the number of scores in that group.

436

b   Add the results of Step 9a for all the groups, yielding

$$\left[\sum \frac{(\Sigma X_j)^2}{n_j} \right].$$

c   Subtract the result of Step 8b from the result of Step 9b. This yields *SSB*, the between groups sum of squares.

Step 10   Subtract the result of Step 9c from the result of Step 8c: *SST* − *SSB* = *SSW*. This yields the within groups sum of squares.

Once these sums of squares have been obtained, we proceed to calculate the observed statistic with the aid of the ANOVA table. The ANOVA table for Dr. Edison's data is presented in Figure 10.12. The ANOVA table typically has the 5 columns shown in Figure 10.12. In the first column, we list the sources of variability in the ANOVA. For a one-way ANOVA, these sources will be between groups variability, within groups variability, and their sum, total variability. Adjacent to each source we table the calculated sum of squares corresponding to that source. These are the figures obtained using the formulas presented earlier and illustrated in Figure 10.10, i.e., *SSB, SSW,* and *SST.* Next, we table the degrees of freedom corresponding to *SSB* and *SSW.* In a one-way ANOVA, the df corresponding to the between groups term is always equal to one less than the number of samples being compared, i.e., $K − 1$. In the case of Dr. Edison's study, $K − 1 = 3 − 1 = 2$. The df corresponding to the within groups term in a one-way ANOVA is always equal to the total number of subjects included in the experiment minus the number of groups, i.e., $n_{TOT} − K$. In Dr. Edison's study, there were 58 subjects and 3 groups, so $n_{TOT} − K = 58 − 3 = 55$. To check to make sure we have found the correct number of degrees of freedom for the between group and within group terms, we add the two degrees of freedom together. The result should equal $n_{TOT} − 1$. We proceed in our calculation by dividing the terms *SSB* and *SSW* by the appropriate number of degrees of freedom. In each case, this yields a "mean square." Here, $\dfrac{SSB}{K − 1} = \dfrac{4,182.68}{2} = 2,091,34$, the mean square between groups (*MSB*). Similarly, $\dfrac{SSW}{n_{TOT} − K} = \dfrac{981.20}{55} = 17.84$, the mean square within groups (*MSW*). We enter the terms *MSB* and *MSW* in the column of the ANOVA table headed "Mean square." Finally, we divide *MSB* by *MSW* to obtain F_{ob}. F_{ob} here is $\dfrac{MSB}{MSW} = \dfrac{2,091.34}{17.84} = 117.23$.

Determining the Significance of F_{ob}.   When we have obtained F_{ob}, we determine the significance of the observed result by reference to a table of the *F*-distributions such as Table C.14 in Appendix C. This table contains the critical values of *F* for the $\alpha = .05$ and $\alpha = .01$ levels of significance. In order to use the

FIGURE 10.12
Analysis of Variance Table for Dr. Edison's Comparison
of 3 Elementary Mathematics Curricula

Source	Sum of squares	df	Mean square	F
Between groups	$SSB = 4{,}182.68$	$K - 1 = 2$	$MSB = \dfrac{SSB}{K-1} = 2{,}091.34$	117.23
Within groups	$SSW = 981.20$	$n_{TOT} - K = 55$	$MSW = \dfrac{SSW}{n_{TOT} - K} = 17.84$	
Total	5,163.88	$n_{TOT} - 1 = 57$		

table, we must first find the appropriate distribution. Every F-distribution is defined by *two* sets of degrees of freedom, one for the numerator of the F-ratio and the other for the denominator of the F-ratio. For the one-way ANOVA, we have seen that the df for the numerator is $K - 1$, and the df for the denominator is $n_{TOT} - K$. To use the table, we locate the degrees of freedom for the numerator across the top of the table. Then, we locate the degrees of freedom for the denominator along the left-hand column. In this case, we find 2 df for the numerator and 55 df for the denominator. At the intersection of the column headed 2 and the row headed 55, we find two numbers entered in the table, 3.17 and 5.01. The number tabled on the top, 3.17, is F_{cr} for the .05 level of significance. The number tabled on the bottom, 5.01, is F_{cr} for the .01 level of significance. You will note that as the number of degrees of freedom increases for both the numerator and denominator, not every single possible value for the df is tabled. If you need a value of F_{cr} for a df that is not tabled, you should substitute the next smallest number of degrees of freedom contained in the table.

In the case of Dr. Edison's data, our F_{ob} was 117.23, far in excess of F_{cr} for the .01 level of significance. Accordingly, Dr. Edison would reject H_0 and conclude that not all of the means of the 3 populations of interest are equal. Note that in rejecting H_0, Dr. Edison can only conclude that not all of the population means are the same. He cannot, on the basis of the significant ANOVA, conclude that any particular population has a mean that is significantly different from any other population. There is, however, a technique that does allow us to pinpoint the significant differences that exist within several independent samples. The technique is an interval estimation technique known as the *Scheffé contrast*. We proceed now to a consideration of this technique.

438

Scheffé Contrasts

When our one-way ANOVA has led us to reject the null hypothesis that the means of the sampled populations are all the same, the method of Scheffé contrasts may be used to determine specifically where the significant differences lie. For example, in Dr. Edison's study, we may wonder whether the mean of Population A differs from the mean of Population B, or whether the mean of Population B differs from the mean of Population C. The Scheffé contrasts may be used to help us in answering questions such as these. In the case of the question whether the mean of Population A differs from the mean of Population B, the contrast would appear as follows:

$$\theta_1 = \mu_A - \mu_B,$$

where θ_1 simply stands for contrast number one, and $\mu_A - \mu_B$ is the difference between the two population means. The Scheffé method enables us to compute a confidence interval for the difference between the two means. If the confidence interval contains the value 0, then we have shown that 0 is one of the values for the difference that is a reasonable value. That is, the difference between μ_A and μ_B could be 0. Then, the contrast would not be significant and we would conclude μ_A may equal μ_B. On the other hand, if the confidence interval does not contain 0, then we have shown that 0 is not a reasonable value for the difference between the two means. Then the contrast would be significant and we would conclude that μ_A does not equal μ_B. All of this should sound familiar to you, for it is analogous to our discussion of confidence intervals for the difference between two population means considered earlier in connection with the t-test for two independent samples. In fact, we could use the data from any two of the samples employed in our ANOVA to compute a confidence interval for the difference between the two population means. However, when we are considering not just two populations but more than two populations, the Scheffé method is the more desirable technique. There are two reasons why this is the case. One reason has to do with the fact that in an ANOVA there are more than two populations, which makes it possible to establish multiple comparisons of pairs of means. In Dr. Edison's study, for example, we could compare μ_A to μ_B, μ_B to μ_C, or μ_A to μ_C. In ANOVAs with more than three groups, still more pairwise comparisons would be possible. The other reason is that the Scheffé method enables us to include more than two population means in a single contrast. For example, we can use the Scheffé method to compare the mean of Group A to the average of the means of Groups B and C. This kind of comparison is often highly illuminating to researchers. We now consider each of these two aspects of the Scheffé method.

439

Accumulating Probability of Type I Error.   With respect to the possibility of multiple pairwise comparisons of means in the ANOVA, the Scheffé method is superior to the use of several separate *t*-method confidence intervals because the use of the Scheffé method reduces the risk of making a Type I error. You will recall that when we compute a 95 percent confidence interval for a mean, or a 95 percent confidence interval for the difference between two means, we are allowing a 5 percent chance that, because of sampling error, the computed confidence interval will not really contain the true difference between the means. In the case where the true difference is 0, this means that there is a 5 percent chance that the confidence interval will not contain 0, causing us to conclude incorrectly that the two means differ. This would be a Type I error, because it would amount to rejecting the hypothesis that the two means were equal when, in fact, they were equal. Now, as long as we calculate just a single confidence interval, this is generally considered an acceptable risk. However, as just noted, in the ANOVA we are typically concerned with at least three population means, and very often with more. This means that there are several possible comparisons that we might be interested in. In a three-group problem such as Dr. Edison's example, we might want to compare μ_A to μ_B, μ_A to μ_C, and/or μ_B to μ_C. That is, there are three separate pairwise comparisons that we might wish to make. When we have four groups in the ANOVA, there are six such comparisons; and as the number of groups increases, the number of comparisons that might be of interest increases rapidly. Each time we compute a new confidence interval, we have the same 5 percent chance that the interval is wrong. But this means that if we compute several such intervals, there is a *greater than 5 percent* chance that *at least one* of the intervals so computed will be wrong. This can become a serious problem when numerous intervals are to be computed. This is what we mean by the accumulating probability of a Type I error. The Scheffé contrast technique, to be described here, has the advantage that we can compute as many 95 percent confidence intervals as we need following a given ANOVA, and the *overall* level of confidence for the intervals will be 95 percent. That is, even if we compute 3 or 6 or 10 such intervals, the chance that *any one* of the intervals will be incorrect is just 5 percent.

Contrasts Involving More Than Two Means.   The Scheffé method has the additional advantage that more complicated comparisons can be made. For example, it might be of interest to us to know if the mean of Group A is significantly different from the average of the means of Groups B and C. Such a contrast can be computed using the Scheffé method. A contrast is defined as a linear combination of means such that the sum of the coefficients for all the means is equal to zero. Thus, in Dr. Edison's study, the direct comparison of Population A to Population B is a contrast:

$$\theta_1 = \mu_A - \mu_B,$$

and so is the comparison of the mean of Population A to the average of the means of Populations B and C:

$$\theta_2 = \mu_A - \frac{\mu_B + \mu_C}{2}.$$

The first, θ_1, is a contrast because the coefficient of μ_A is $+1$ and the coefficient of μ_B is -1. These two coefficients add up to 0. The second, θ_2, is a contrast because the coefficient of μ_A is $+1$ and the coefficients of μ_B and μ_C are each $-\frac{1}{2}$. Thus, these three coefficients also add up to 0. The second contrast, as we pointed out, would be used to compare the mean of Population A to the average of the means of Populations B and C. A contrast of this nature would be of interest if we felt that the mean of Population A was quite different from the means of Populations B and C, which were quite close to each other. A contrast of this nature is also of interest in the event that the ANOVA yields a significant result, but none of the possible direct pairwise comparisons of means are significant. If the ANOVA for a given problem yields a significant F_{ob}, there is at least one contrast that will be significant. However, there may be times when none of the direct pairwise contrasts like $\theta_1 = \mu_A - \mu_B$ will be significant. In this case, a more complicated contrast may be used to find the nature of the significant differences between sample means.

Computing Scheffé Contrasts. The formulas for computing the confidence interval for a particular contrast are as follows:

$$\text{lower limit} = \hat{\theta} - \sqrt{(K-1)F_{cr}} \cdot \sqrt{MSW\left(\frac{C_1^{\,2}}{n_1} + \frac{C_2^{\,2}}{n_2} + \ldots + \frac{C_K^{\,2}}{n_K}\right)}$$

and

$$\text{upper limit} = \hat{\theta} + \sqrt{(K-1)F_{cr}} \cdot \sqrt{MSW\left(\frac{C_1^{\,2}}{n_1} + \frac{C_2^{\,2}}{n_2} + \ldots + \frac{C_K^{\,2}}{n_K}\right)}.$$

The terms in the formulas have the following meanings:

$\hat{\theta}$: This is the sample estimate of the desired contrast. Thus, if $\theta_1 = \mu_A - \mu_B$, $\hat{\theta}_1 = \bar{X}_A - \bar{X}_B$. And if $\theta_2 = \mu_A - \frac{\mu_B + \mu_C}{2}$, $\hat{\theta}_2 = \bar{X}_A - \frac{\bar{X}_B + \bar{X}_C}{2}$.

441

K: This is the number of groups involved in the ANOVA on which the contrast is based.

F_{cr}: This is the value of F_{cr} in the ANOVA on which the contrast is based. If we wish a 95 percent confidence interval, it is F_{cr} for the .05 level of significance. If we wish a 99 percent confidence interval, it is F_{cr} for the .01 level of significance. In either case, the df for the F_{cr} are the same as in the ANOVA, $K - 1$ for the numerator and $n_{TOT} - K$ for the denominator.

MSW: This is the MSW from the ANOVA.

$C_1, C_2, \ldots, C_K$: These are the coefficients corresponding to each of the K means examined in the ANOVA. In Dr. Edison's study, if $\theta_1 = \mu_A - \mu_B$, then the coefficients would be as follows:

$$C_A = 1, C_B = -1, C_C = 0.$$

Group C would not be involved in the contrast at all. If $\theta_2 = \mu_A - \dfrac{\mu_B + \mu_C}{2}$, then $C_A = 1$, $C_B = -\dfrac{1}{2}$, and $C_C = -\dfrac{1}{2}$.

$n_1, n_2, \ldots, n_K$: These refer to the number of subjects in each of the groups.

In Figure 10.13, we have calculated 99 percent confidence intervals for θ_1 and θ_2 using the data from Dr. Edison's study.

Interpreting Confidence Intervals for Scheffé Contrasts.   The interpretation of confidence intervals for Scheffé contrasts is similar to the interpretation of confidence intervals in general. Any value contained in the interval is considered a reasonable possible value for the contrast, and values not contained in the interval are not considered reasonable possible values. Thus, with regard to θ_1 in Figure 10.13, we conclude that the contrast is significant because the value 0 is not contained in the confidence interval for the contrast. This means that zero is not a reasonable value for the difference between μ_A and μ_B. In other words, μ_A and μ_B are almost certainly different. With regard to θ_2 in Figure 10.13, we conclude that this contrast is also significant, for the same reason. Because 0 is not included in the confidence interval for θ_2, we conclude that 0 is not a reasonable value for the difference between μ_A and the average of μ_B and μ_C.

The special aspect of the Scheffé method that may be restated here is that the level of confidence applies not to individual intervals, but to all such intervals calculated. That is, when we say we are 99 percent confident, we are 99 percent confident that *all* the intervals are true statements.

FIGURE 10.13
Calculation of 99 Percent Confidence Intervals for 2 Scheffé Contrasts
for Dr. Edison's Study of 3 Math Curricula

$$\theta_1 = \mu_A - \mu_B$$

$$\hat{\theta}_1 = \bar{X}_A - \bar{X}_B = (47.10 - 61.44) = -14.34$$

lower limit $= \hat{\theta}_1 - \sqrt{(K-1)F_{cr}} \cdot \sqrt{MSW\left[\dfrac{C_A^2}{n_A} + \dfrac{C_B^2}{n_B} + \dfrac{C_C^2}{n_C}\right]}$

$= (-14.34) - \sqrt{(3-1)(5.01)} \cdot \sqrt{17.84\left[\dfrac{(1)^2}{20} + \dfrac{(-1)^2}{18} + \dfrac{(0)^2}{20}\right]}$

$= (-14.34) - 4.34$

lower limit $= -18.68$

upper limit $= (-14.34) + 4.34$

upper limit $= -10.00$

Thus, $-15.64 < \theta_1 < -10.00$ with 99 percent confidence.

- -

$\theta_2 \qquad = \mu_A - \dfrac{\mu_B + \mu_C}{2}$

$\hat{\theta}_2 \qquad = 47.10 - \dfrac{61.44 + 66.95}{2} = -17.10$

lower limit $= \hat{\theta}_2 - \sqrt{(K-1)F_{cr}} \cdot \sqrt{MSW\left[\dfrac{C_A^2}{n_A} + \dfrac{C_B^2}{n_B} + \dfrac{C_C^2}{n_C}\right]}$

$= (-17.10) - \sqrt{(3-1)(5.01)} \cdot \sqrt{17.84\left[\dfrac{(1)^2}{20} + \dfrac{(-1/2)^2}{18} + \dfrac{(-1/2)^2}{20}\right]}$

$= (-17.10) - (3.70)$

lower limit $= -20.80$

upper limit $= (-17.10) + (3.70)$

$= -13.40$

Thus, $(-20.80 < \theta_2 < -13.40)$ with 99 percent confidence.

A FINAL NOTE

At the conclusion of this long and important chapter, we would like to emphasize a point that we have made intermittently throughout the text. This point is that the statistical techniques that are appropriate for categorical data and rank-order data may also be employed with interval data. We have already noted that when we have a two-group comparison involving numerical scores, but do not feel that our data meet the assumptions for a t-test for two independent samples, we can use instead the Mann-Whitney U-test. In the corresponding group comparison for more than two groups, the Kruskal-Wallis analysis of variance by ranks may be substituted for the traditional one-way ANOVA when we do not believe it is appropriate to use the ANOVA. When we shift from techniques designed for interval data to techniques for rank-order data, we must translate numerical scores into ranks.

It is also possible to employ techniques for categorical data with rank-order or interval data. For example, we might have two independent groups that have been compared on a variable in which there are numerical scores, but we may not believe that these scores really constitute an interval scale. In such a case, we sometimes split the observations on the variable of interest at the median and rearrange the data into a 2 × 2 contingency table, as follows:

	Group		
	A	B	
Above median	16	4	20
Below median	4	16	20
	20	20	40

We can then employ the Fisher exact probability test to test the hypothesis that the two groups do not differ with respect to the variable of interest. Such a procedure is typically referred to as a *median test*. Rank-order data or data in the form of numerical scores may be similarly treated in comparisons involving more than two groups.

THE CLASSIFICATION MATRIX

In this chapter, we continued our discussion of inferential techniques by looking at those techniques applicable to experimental and pseudo-experimental problems involving independent groups. We added to our matrix techniques for problems with categorical data, rank-order data, and interval data, for both problems with just two groups and those with more than two groups.

TABLE 10.8
Classification Matrix: Inferential Statistics

Type of Problem		Level of Measurement of Variable(s) of Interest		
		Categorical: Nominal or Ordered Categories	Rank-Order	Interval Scale (or Ratio-Scale)
Univariate Problems		Binomial Test Normal Approximation to Binomial χ^2 One-Variable Test *With Ordered Categories:* Kolmogorov-Smirnov One-Variable Test		One-Sample z-Test One-Sample t-Test Confidence Interval for Population Mean (z- or t-Method)
Bivariate Problems	Correlational Problems	χ^2 Test of Association	Test for Significance of Spearman r_s	t-Test for Significance of Pearson r Fisher z_r Transformation for Testing Significance of Pearson r Confidence Interval for ρ
Bivariate Problems	Experimental and Pseudo-Experimental Problems: Independent Groups	Fisher Exact Probability Test χ^2 Test of Homogeneity	Mann-Whitney U-Test Kruskal-Wallis Analysis of Variance by Ranks	t-Test for Independent Samples One-Way Analysis of Variance Scheffé Contrast
Bivariate Problems	Dependent Measures: Pretest and Posttest or Matched Groups			

Key Terms

Be sure you can define each of these terms.

Fisher exact probability test
hypergeometric model
chi-square test of homogeneity
Mann-Whitney U-test
Kruskal-Wallis analysis of variance by ranks
H-statistic
t-test for two independent samples
pooled sample variance
confidence interval for the difference
  between two population means
one-way analysis of variance (ANOVA)
F-distribution
F-statistic
mean square within groups
mean square between groups
Scheffé contrast

Summary

In Chapter 10, we examined inferential techniques for bivariate experimental and pseudo-experimental problems in which we compare two or more independent groups on some variable of interest. We pointed out that, in such problems, one variable is always categorical and is used to form groups, which are then compared on another variable, the dependent variable. We looked at techniques for comparing groups on categorical, rank-order, and interval variables. In this chapter, our concern was with groups in which all observations were selected independently of each other. We either tested hypotheses that our groups were not from different populations or we set up confidence intervals to estimate the differences between our groups.

We examined two tests that can be used if our dependent variable is categorical. With small sample sizes, we use the Fisher exact probability test to test the hypothesis that our groups are not from different populations.

With larger samples, we use the chi-square test of homogeneity. We also use the chi-square test if we have more than two groups or if our dependent variable has more than two values.

We also examined two tests that can be used with rank-order dependent variables. When we have two groups, we use the Mann-Whitney U-test to test the hypothesis that our groups are not from different populations. With large samples, we can use the normal curve to approximate the distribution of U. With more than two groups, we use the Kruskal-Wallis analysis of variance by ranks to test our null hypothesis. In this case, when sample sizes are large, we can use the chi-square distribution to approximate the distribution of the Kruskal-Wallis H-statistic.

If we have an interval scale dependent variable and our two groups can be assumed to have been selected from normal populations with equal variances, we also have two statistical tests we can use. When we have two groups, we use the t-test for independent samples. We can also use a critical value of t to set up a confidence interval for the difference between the means of our two populations. With both of these methods, we use our sample standard deviations to compute a pooled estimate of the standard deviation of the difference between our means. With more than two groups, we use the one-way analysis of variance (ANOVA) to test the hypothesis that all our population means are the same. In this case, we compute the F-ratio and compare it to a critical value of F. We pointed out that, in fact, the t-test for two independent samples is the two-group special case of the one-way analysis of variance.

When we are comparing more than two groups in an analysis of variance and our observed value of F is significant, we saw that we use Scheffé contrasts to determine which specific differences between our means are significant. We saw that with this method, we can look at a variety of contrasts and still be sure that our risk of a Type I error remains at our chosen level of significance for all the contrasts we make. We pointed out that if our analysis of variance yielded a significant value of F, then at least one Scheffé confidence interval would not contain the value zero.

Review Questions

To review the concepts presented in Chapter 10, choose the best answer to each of the following questions.

1 If we wish to compare two independent binomial populations, the probability distribution to use is
    _____a a pooled binomial distribution.
    _____b a hypergeometric distribution.

2 If we wish to compare the means of two independent normal distributions, the probability distribution to use is
    _____a a t-distribution.
    _____b a hypergeometric distribution.

3   If our null hypothesis states that two independent proportions are the same, then we have
    _____a   two independent binomial populations.
    _____b   two independent normal distributions.

4   If we use ranks from two independent samples to test our null hypothesis, we use
    _____a   the t-statistic.
    _____b   the Mann-Whitney U-statistic.

5   If we cross classify a random sample of people on two binomial variables, we have
    _____a   a correlation problem.
    _____b   a group comparison problem.

6   If we wish to compare two independent binomial populations and both samples are large, we can do an approximate test using
    _____a   the t-distribution.
    _____b   the χ^2 distribution.

7   The research situation in which it makes sense to speak of the direction of causation is
    _____a   an experimental problem.
    _____b   a correlation problem.

8   If we have several independent populations that we wish to compare on some categorical variable, we should do
    _____a   an analysis of variance.
    _____b   a χ^2 test for homogeneity.

9   If we have R independent populations measured on a categorical variable with C values, the statistic to use to test the hypothesis of homogeneity is
    _____a χ^2 with $R - 1, C - 1$ df.
    _____b F with $R - 1, C - 1$ df.

10 If we cross classify a large sample of subjects on two nominal scales, we will do a
    _____a   chi-square test of association.
    _____b   chi-square test of homogeneity.

11 In bivariate problems with categorical data, one difference between tests of homogeneity and tests of association is
    _____a   the statement of the null hypothesis.
    _____b   the degrees of freedom.

12 In an experimental problem with four treatment groups, if we do not believe that the dependent variable has a normal distribution under all treatment conditions, instead of analysis of variance we could
    _____a   do six t-tests.
    _____b   use the Kruskal-Wallis method.

448

13 If $H_o: \mu_1 = \mu_2 = \mu_3 = \mu_4$ is rejected at $\alpha = .05$, we can say

_____a $\mu_1 \neq \mu_2 \neq \mu_3 \neq \mu_4$.

_____b the Scheffé method will find at least one contrast that is significant.

14 The analysis of variance will discriminate among several population means even when the population variances are unequal. The name given to this concept is

_____a robustness.

_____b power.

15 The meaning of the assumption of equal variances is that

_____a all sample variances will be the same.

_____b the populations sampled all have the same amount of variation about their means.

16 The F-statistic in the analysis of variance rejects H_o for large values of F. This is because

_____a the alternate hypothesis is nondirectional.

_____b only large values of F support the alternate hypothesis.

17 A 95 percent confidence interval can test $H_o: \mu_1 - \mu_2 = 0$ at a 5 percent level of significance. If 0 is not a value between the confidence limits, then we can say

_____a H_o is rejected in favor of $\mu_1 - \mu_2 > 0$.

_____b H_o is rejected in favor of $\mu_1 - \mu_2 \neq 0$.

18 One assumption that is common to all procedures in Chapter 10 is that

_____a the observations must be independent of each other.

_____b we must have equal population variances.

19 If we are interested in differentiating between two independent population means and we want to estimate how far apart the means are, we should

_____a test $H_o: \mu_1 - \mu_2 = 0$ vs. $H_a: \mu_1 - \mu_2 \neq 0$.

_____b calculate a confidence interval for $\mu_1 - \mu_2$.

20 In the analysis of variance model, the statistic we use is a ratio of two variances. This statistic tests the null hypothesis that

_____a several independent means are equal.

_____b the population variances are the same.

21 When we are interested in finding out whether several independent population means are the same, we use the analysis of variance model rather than doing several t-tests for two means. The reason for this is

_____a to guard against inflating the probability of a Type I error.

_____b that all populations have the same variance.

449

1   To practice using Table C.11, find out which ones of the following 2 × 2 tables are significant at $\alpha = .05$.

a)

	I	II	
Pass	3	7	10
Fail	6	1	7
	9	8	17

b)

	I	II	
Pass	7	4	11
Fail	3	8	11
	10	12	22

c)

	I	II	
Pass	0	10	10
Fail	4	7	11
	4	17	21

d)

	I	II	
Pass	10	2	12
Fail	1	7	8
	11	9	20

2   Suppose we have data in a 2 × 2 contingency table as follows:

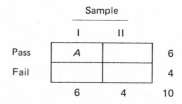

	Sample I	II	
Pass	A		6
Fail			4
	6	4	10

a   Lay out all possible outcomes in a table like Table 10.2 in the text.
b   Calculate the probability for each outcome in a based on the observed value in cell A.
c   Make a probability distribution for the calculations in b (see Table 10.3).
d   Make a probability graph for c (see Figure 10.1).
e   Which outcomes favor the proportion of passes in Population I being greater than the proportion of passes in Population II? Which do not?
f   Which outcomes would cause you to reject H_o: The proportion of passes is the same in both populations, at a 5 percent level of significance?
g   Compare your answer to f to the information in Table C.11. Do you have the same answer?

3   An experimenter has randomly assigned 30 randomly selected rats to an experimental group and a control group, making two groups of 15 rats each. The experimental rats are fed a diet containing a suspected carcenogenic substance and the control rats are fed the same diet without the suspect substance. At the end of the experiment, it is observed that 9 of the experimental rats have developed cancer and only 2 of the control rats have developed cancer. Use these results to test the hypothesis that the suspected substance is "innocent." Use $\alpha = .05$.

4   a   For the 2 × 2 contingency table in Problem 3, calculate the significance probability of the result.

b Next use the chi-square approximation test to analyze the data, that is, calculate χ^2_{ob} for the table.

c What is the approximate significance of χ^2_{ob}?

d Is your answer in c comparable to a?

e Why can the chi-square statistic be used in Problem 3?

5 An investigator wants to see if his strength training program is effective. He takes a random sample of 50 children and gives them a strength test. They are scored as passing or failing. He then trains them for two months and tests them again on the strength test.

The results show:

	Pre	Post	
Pass	15	35	50
Fail	35	15	50
	50	50	100

Would a test based on the hypergeometric model be appropriate to analyze the results? Explain.

6 The Department of Education in a large state is concerned with placing special students in regular classrooms. Before they propose any legislation, they decide to assess certain attitudes that currently exist within different interest groups. They randomly select 100 classroom teachers, 200 parents, 50 school principals, and 50 special education teachers. One part of the assessment consists of a description of a child with special problems. Each subject reads the description and then assigns the child to one of the following categories: the child belongs in a regular classroom; the child could be in a regular class with some time each day in a special class; the child should be in a special class but in a school with regular classes; the child belongs in a special school. Based on the results given in the following table, can the State Department of Education conclude that there is no difference in how the four populations view the school placement of the child? Use $\alpha = .05$.

	Interest Group				
	Classroom teachers	Parents	Principals	Special teachers	Total
Classroom	5	50	15	30	100
Classroom with special time	45	125	20	10	200
Special class	40	15	5	10	70
Special school	10	10	10	0	30
Totals	100	200	50	50	400

Placement

Experimental and Pseudo-Experimental Problems

1 Suppose the Mann-Whitney U is to be used to analyze a data set. For each of the following situations, n_1, n_2, and the level of significance are given. Use Table C.12 to establish a rejection rule.
   a $n_1 = 11$, $n_2 = 14$, $\alpha = .10$, and the test will be two-tailed.
   b $n_1 = 30$, $n_2 = 15$, $\alpha = .05$, and the test will be one-tailed.
   c $n_1 = 30$, $n_2 = 15$, $\alpha = .05$, and the test will be two-tailed.
   d $n_1 = 20$, $n_2 = 40$, $\alpha = .025$, and the test will be one-tailed.
   e $n_1 = 18$, $n_2 = 33$, $\alpha = .01$, and the test will be two-tailed.

2 To practice calculating the Mann-Whitney U, three data sets follow. In each case, calculate the value of U.
   a Group I   12, 14, 20, 18, 13, 21, 7, 5, 29
      Group II   15, 18, 12, 22, 6, 8, 10, 17
   b Group I   90, 35, 40, 30, 31, 25, 29, 21, 42, 45
      Group II   20, 18, 26, 36, 38, 44, 41, 19, 50, 51, 55, 60
   c Group I   1, 4, 9, 16, 18, 20, 21, 23, 30, 31, 42, 48, 49, 50
      Group II   20, 21, 22, 24, 28, 24, 32, 33, 35, 36, 37, 41, 47

3 An experimenter has 10 randomly selected subjects from a population of test-anxious subjects. She randomly assigns 5 to an experimental group and 5 to a control group. She wants to know if her training program reduces test anxiety. The 5 experimental subjects take part in the training program. At the end of the experiment, all 10 subjects are measured on test anxiety, with low scores indicating low anxiety. Using $\alpha = .05$, is the experimenter justified, based on the results that follow, in saying that her program is effective? Use the Mann-Whitney U-test.

   Experimental   3, 5, 7, 8, 15
   Control   10, 11, 19, 20, 21

4 An experimenter has a training program that he believes will improve leadership ability. He wants to try his program and see if the evidence suggests that it is an effective training program. From a population of people who lack leadership ability, he randomly selects 12 and randomly assigns them to experimental and control conditions. The experimental group is trained and the control group is not. At the end of the training, both groups take a leadership inventory measure. High scores mean high leadership ability. Test the hypothesis that the program is ineffective using the Mann-Whitney U-test. Use $\alpha = .025$ based on the following results:

   Experimental   50, 35, 20, 15, 10, 34
   Control   45, 30, 17, 13, 3, 25

5 Use the data in Problem 4 as follows. Rank the scores and find the median for the combined group. Next count the number in the experimental group

who are above the combined median and the number in the control group
who are above the combined median.

a Display the results in a 2 × 2 contingency table.

b Use the hypergeometric model to answer the research question.

c How do the results compare with your answer in 4?

6 To practice reading Table C.13, find a rejection rule for the Kruskal-Wallis
   H-statistic based on the following information.

a $n_1 = 5, n_2 = 4, n_3 = 4, \alpha = .05$

b $n_1 = 4, n_2 = 4, n_3 = 4, \alpha = .05$

c $n_1 = 5, n_2 = 5, n_3 = 5, \alpha = .10$

d $n_1 = 4, n_2 = 3, n_3 = 1, \alpha = .05$

e $n_1 = 5, n_2 = 3, n_3 = 2, \alpha = .01$

f $n_1 = 2, n_2 = 1, n_3 = 1, \alpha = .05$

7 To practice calculating the H-statistic consider the following data set.

	Group	
I	II	III
51	89	16
12	31	19
18	23	25
21	42	38

a Calculate H.

b Using Table C.13, what can we say about the significance of H?

8 A school psychologist is interested in whether different methods of reinforcement will differentially effect the ability to recall information at some
later time. She designs 3 different methods of reinforcement and randomly
assigns 5 subjects to each group. Each subject will be given a list of 15 words
and asked to memorize it. The subjects within each group will be reinforced
according to their treatment group. One hour later, each subject will be
asked to recall as many of the words as possible. A subject's score will be
the number of words the subject recalled.

a Using the H-statistic, set up statistical procedures to enable the psychologist to decide whether the methods of reinforcement differentially effect
recall. Use $\alpha = .05$.

b Based on the following data, what conclusion can be drawn?

	Reinforcement Method	
I	II	III
4	1	11
5	2	10
3	6	9
2	7	12
8	5	2

453

Experimental and Pseudo-Experimental Problems

1 Assume that we have two populations of measures and we have reason to believe that $\sigma_1^2 = \sigma_2^2$. We have one sample, randomly selected, from each population. The results of the sampling are:

Sample 1   21, 37, 24, 60, 61, 5, 9, 39
Sample 2   8, 16, 66, 53, 15, 30, 78, 50, 51

a Calculate $\bar{X}$ and s^2 for each sample.
b Does the assumption that $\sigma_1^2 = \sigma_2^2$ appear reasonable?
c Calculate s_p^2.

2 An experimenter randomly assigns rats to an experimental group and a control group. The experimental group (E) gets a diet with a special supplement and the control group (C) gets only the diet. At the end of the experimental period, all rats will be measured on the variable "weight gain." The experimenter hopes to show that the supplement will increase the average weight gain over the experimental time period.
a What statistical hypothesis should she test?
b What is the alternative hypothesis?
c If she wants to use a t-test for independent samples, what assumptions must she make about the variable "weight gain"?

3 For the experiment in Problem 2, set up a rejection rule to test the hypothesis $H_o: \mu_E = \mu_C$ vs. $H_a: \mu_E > \mu_C$ for each of the following conditions.
a $\alpha = .05$, $n_E = 8$, and $n_C = 10$
b $\alpha = .01$, $n_E = 11$, and $n_C = 10$
c $\alpha = .005$, $n_E = 15$, and $n_C = 15$
d $\alpha = .05$, $n_E = 15$, and $n_C = 15$
e $\alpha = .01$, $n_E = 20$, and $n_C = 18$
f $\alpha = .025$, $n_E = 25$, and $n_C = 25$

4 A school psychologist works in a system that uses individualized instruction. She wonders whether children will achieve more if they set their own goals or whether it is better to have the teacher set goals for the children. A group of 26 sixth-graders is randomly selected from a large group of sixth-graders, and the 26 are randomly assigned to either a "set own goals" group or "teacher sets goals" group. Each group studies arithmetic for 6 weeks. At the end of 6 weeks, the children are tested and the psychologist decides which of the two methods produced more learning in arithmetic.
a Use the method of hypothesis testing and design statistical procedures to enable the psychologist to make a decision. Use $\alpha = .05$.
b If the results are the following, what conclusion can the psychologist make?

	Set own goals	Teacher sets goals
$\bar{X}$	26.3	32.1
s	5.2	6.7

Inferential Statistics

c Use the method of confidence intervals to estimate the difference between μ_1 and μ_2 with 95 percent confidence.
d Interpret your confidence interval.

5 If the assumptions for the t-test for two independent groups do not seem
reasonable, what are two alternate methods that might be used?

6 To practice using the table of critical values of F (Table C.14), set up a
rejection rule for each of the following situations, assuming that the analysis of variance is appropriate.
a $\alpha = .05, n = 30, K = 3$
b $\alpha = .01, n = 50, K = 6$
c $\alpha = .05, n = 130, K = 5$
d $\alpha = .01, n = 64, K = 4$
e $\alpha = .05, n = 30, K = 2$

7 To practice the methods used in calculating the sums of squares in a one-
way ANOVA, use the following data on four samples to do all the calculations shown in Figure 10.10. Follow the steps outlined in the text.

| | Group | | |
I	II	III	IV
30	5	10	21
22	7	25	20
18	18	10	10
35	10	15	18
20		17	11

8 Assume that the data in Problem 7 come from normal distributions with
$\sigma_I^2 = \sigma_{II}^2 = \sigma_{III}^2 = \sigma_{IV}^2$.
a Test the hypothesis that $\mu_I = \mu_{II} = \mu_{III} = \mu_{IV}$.
b Summarize the results as shown in Figure 10.12 in the text.

9 For the four group means mentioned in Problem 7, look at the following
comparisons.
$\theta_1 = \mu_1 - \mu_2$
$\theta_2 = \mu_1 - \mu_3$
$\theta_3 = \mu_2 - \left(\dfrac{\mu_3 + \mu_4}{2}\right)$
$\theta_4 = (\mu_1 + \mu_2) - (\mu_3 + \mu_4)$
$\theta_5 = \mu_1 - 2\mu_2 - \mu_3$
a Which of the above comparisons is a contrast? Why?
b Construct 95 percent Scheffé confidence limits for $\theta_1, \theta_2, \theta_3$, and θ_4.
c What conclusion can you make based on the calculations in b?

10 Suppose an investigator wonders whether average audiovisual integration
ability among schizophrenic children increases with age. She has samples of
schizophrenic children of age 7, age 8, age 9, and age 10. The results are:

455

		Age		
	7	8	9	10
$\bar{X}$	7.3	7.9	9.5	14.2
s	3.2	4.1	2.9	3.6
n	15	15	15	15

a Use $\alpha = .05$ and set up procedures to help the investigator answer her question.

b Carry out the test set up in a.

c Interpret the results.

d Should the investigator perform Scheffé contrasts? Explain.

e Use the Scheffé method and do 95 percent confidence intervals for comparisons of all pairs of means.

f Do these results suggest that audiovisual integration is age related in schizophrenic children?

g Would there be any other significant Scheffé contrasts?

Set 4

For each of the following situations answer the following questions.

a What are the groups to be compared?

b What is the variable on which they will be compared?

c What is the level of measurement of the variable named in b?

d What statistical procedure might be used to analyze the data?

e If you feel other procedures might also be used, list them.

1 A certain IQ test has a mean of 100 and a standard deviation of 15, based on a nationwide norming sample. Suppose an investigator wishes to determine whether average IQ is different for a population of East Coast eighth-graders than for a population of West Coast eighth-graders. The investigator has a random sample of East Coast eighth-graders and a random sample of West Coast eighth-graders. Each sample consists of 200 pupils and both samples will be given the IQ test.

2 A psychology professor believes that he can increase average IQ by training people using a series of exercises that are similar to the kinds of tasks involved in the IQ test. To test his theory, he randomly selects a group of eighth-graders and randomly assigns them to an experimental group and a control group. At the end of the experimental period, all subjects will be given the IQ test.

3 Three different methods of treating a certain disease are known to exist. A doctor wants to know which of the three methods is most effective. A random sample of 18 patients with the disease are randomly assigned to the three treatment groups. The success of each method will be measured by the number of days it takes to cure a patient.

4 To determine whether men or women are better dart throwers, an experimenter selects 40 men and 40 women to take part in a study. Each subject will throw 30 darts and a score will be average distance from the bull's-eye.

5 A sample of 100 subjects is randomly split into an experimental group and a control group. The experimental group will learn problem-solving strategies and the control group will play games with no apparent strategies involved. At the end of the experiment, each subject will be given the same problem to solve. The experimenter will record the number in each group who correctly solve the problem. He will use this evidence to help him decide whether teaching problem-solving strategies improves the ability to solve problems.

457

11

Inferential Statistics for Problems Involving Two Sets of Dependent Measures

In this final chapter, we consider the techniques of inferential statistics that are appropriate to research problems involving two sets of dependent measures. You will recall from Chapter 1 that there are two common research situations in which we deal with two sets of dependent measures. One of these is the situation in which each member of a group of independently selected subjects is measured twice, with some event or treatment intervening. This is known as the pretest and posttest experimental design. In this research situation, each subject has both a pretest and a posttest measure. Because subjects are selected randomly and independently, each pair of observations pertaining to a given subject is independent of any other pair. However, the two observations that make up a pair are not independent. Thus, the pretest measures, taken as a group, are not independent of the posttest measures, taken as a group.

The second research situation yielding two sets of dependent measures is that situation in which two groups of subjects are each measured once, but these two groups of subjects are not independent of each other. This is common in studies where subjects have been "matched" with respect to a particular variable in order to ensure the comparability of the two groups on this variable. In this case, a group of subjects is randomly selected and measured on the matching or control variable. Then, the two subjects with the highest scores on the control variable are considered a matched pair, the subjects with the two next highest scores on the variable are considered another matched pair, and so on until all the subjects are paired with a subject having a similar score on the control variable. Then, one subject from each pair is selected at random to be placed in one experimental group, and the other subject from that pair is automatically placed in another experimental group. The subjects in the two groups receive different treatments, after which they are measured on the variable of interest. In this research situation, there is a dependency between the scores on the variable of interest in the two matched groups. This dependency may be understood as follows. Assume there is a relationship between scores on the matching or control variable and the variable of interest. Assume that this is a positive relationship, i.e., that subjects scoring high on the control variable tend to score high on the variable of interest as well. But then, subjects included in matched pairs will tend to have similar scores on the variable of interest because they have similar scores on the control variable. Thus, in the matched groups research situation, we also note that each pair of observations is independent of any other pair, but that the two observations that make up a pair are not independent of each other. That is, the posttreatment measures of subjects receiving one treatment, taken as a group, are dependent upon the posttreatment measures of subjects receiving the other treatment because one member of each pair of matched subjects is in each group.

Although the pretest and posttest and the matched groups designs are conceptually quite different, they are statistically equivalent. Both designs result in the generation of two sets of measurements that are not independent of each other. The dependency that exists between the two sets of measures must be taken into account mathematically. The techniques that we employ to accomplish this goal are identical, regardless of which of the two designs we have in mind. The techniques we use depend upon the level of measurement of the variable of interest. Following our familiar pattern, we first consider the techniques of inferential statistics appropriate for problems involving two sets of dependent measures when the measures are on a categorical variable.

INFERENTIAL STATISTICS FOR PROBLEMS WITH TWO SETS OF DEPENDENT MEASURES: CATEGORICAL DATA

The McNemar Test

The *McNemar test* is applicable to problems involving two sets of dependent measures on a dichotomous variable of interest. The variable of interest may be inherently dichotomous, or it may be a categorical, ordinal, or interval scale variable that has been collapsed to form two categories. In order to use the McNemar test, it is required that each pair of measures be independent of any other pair of measures.

A Research Example.   Suppose that Political Scientist Andrews is interested in the possible effectiveness of a speech that he has been hired to write for a candidate for the state senate. The speech is on the issue of a tax reform proposal strongly favored by his candidate. The object of Professor Andrews' research will be to determine if hearing this speech on radio or television will cause voters to

459

shift toward support for the tax reform proposal. Professor Andrews has his candidate prepare a five-minute videotape in which he delivers the speech.

Professor Andrews selects a random sample of 20 voters from the state. He asks each of these subjects whether they favor or oppose the tax reform proposal. Then, he has the voters view the videotape. (Subjects are instructed not to discuss the tape among themselves.) Following the viewing, he again asks the voters their opinion on the issue.

Because subjects are asked their opinion both before and after the film, we have a pretest and posttest design. Therefore, the two sets of measures, opinion before and opinion after, are dependent. Because subjects were selected randomly and independently, any pair of before and after measures is independent of any other pair. The variable of interest in the study is a nominal scale variable with two categories: favor and oppose. For these reasons, the research problem is appropriate for analysis using the McNemar test.

The Rationale of the McNemar Test. The results of Professor Andrews' study are shown in Table 11.1. We note in Table 11.1 that of the 20 subjects, 8 favored the proposal and 12 opposed the proposal before viewing the tape. After viewing the tape, 17 favored the proposal and 3 opposed it. We further note that 9 of the subjects did not experience a change in viewpoint as a result of viewing the tape: 7 subjects favored the proposal both before and after the tape, and 2 subjects opposed the proposal both before and after the tape.

The McNemar test does not consider those subjects who show no change. It considers only those subjects who show a switch in their opinion from the pretest to the posttest. The idea behind the test is that if the film is not effective, then we would expect as many subjects to switch in the direction from favor to oppose as in the direction from oppose to favor. On the other hand, if the treat-

TABLE 11.1
Opinion of 20 Voters on Tax Reform Proposal Before and After Viewing the Videotape of the Candidate's Speech

		Opinion After		
		Favor	Oppose	Total
Opinion Before	Favor	7	1	8
	Oppose	10	2	12
	Total	17	3	20

ment is effective, then we would expect the majority of subjects who do switch to switch in a particular direction. In the case of Professor Andrews' research, he is attempting to establish that subjects viewing the tape will tend to switch toward favoring the tax reform proposal. Thus, the null hypothesis for the test would be that the proportion of switching subjects who switch from oppose to favor $(O \rightarrow F)$ after viewing the film is the same as the proportion of switching subjects who switch from favor to oppose $(F \rightarrow O)$. That is, each of these proportions would be equal to .50. In brief notation, $H_o: P(O \rightarrow F) = .50$. The alternative hypothesis for this test would be directional because Professor Andrews wishes to show that more than 50 percent of voters who switch opinions after viewing the film will switch from oppose to favor. Thus, $H_a: P(O \rightarrow F) > .50$.

You may have already guessed that in considering only those cases in which a switch did occur, we have really transformed the research problem into a binomial problem. Each of the subjects who switch must switch in one of the two ways, either from favor to oppose or from oppose to favor. If each subject is selected at random, independently from the other subjects, and if subjects are kept from influencing each other during the course of the experiment, we may regard each of the switching subjects as an independent binomial trial.

Determining the Significance of the McNemar Test. As long as the number of switching subjects is less than or equal to 25, the significance of the test is most easily evaluated by reference to the table of binomial distributions (Table C.1). In the case of Professor Andrews' study, we would consult the table for $n = 11$ (because 11 subjects switched) and $P = .5$. There, we find that the chance of observing 10 or more successes (in this case, 10 out of 11 switches were from oppose to favor) out of 11 trials is only .005 if H_o is true. Thus, the test is significant beyond the $\alpha = .01$ level. Professor Andrews would reject the null hypothesis and conclude that this speech was indeed more likely to cause voters to switch from oppose to favor than to cause them to switch in the opposite direction.

When the number of subjects exceeds the size of available binomial tables, we can use the normal approximation to the binomial (see Chapter 8) or we can use the statistic:

$$\chi_{ob}^2 = \frac{(|A - D| - 1)^2}{A + D}$$

where A is the number of subjects who switch in one direction and D the number of subjects who switch in the opposite direction. This statistic is distributed approximately as chi-square with one degree of freedom. It should be used only when the expected frequency in each of the cells is at least 5. The significance of the observed statistic is evaluated by reference to the table of chi-square distributions (Table C.3). The McNemar test corresponding to Professor Andrews' data has been summarized, according to our seven-step outline, in Figure 11.1.

FIGURE 11.1
McNemar Test for Professor Andrews' Study

Step 1

Because we have pretest and posttest data on a dichotomous variable, we consider the McNemar test.

Step 2

H_o: $P(O \rightarrow F) = .50$, $P(F \rightarrow O) = .50$. That is, among the population of voters who might hear the speech, as many would change their opinion from oppose to favor as would change in the opposite direction.

Step 3

H_a: $P(O \rightarrow F) > .50$

Step 4

The test is the McNemar test. Because n here is $\leqslant 25$, the significance of the test may be evaluated by reference to the table of binomial distributions (Table C.1).

Step 5

The binomial table for $n = 11$, $P = .5$, and $r = 10$ tells us that the chance of finding at least 10 subjects out of a sample of 11 switching from oppose to favor, if the population proportion switching in this direction were .5, would be only .005.

Step 6

In our study, $r = 10$.

Step 7

We can reject H_o beyond the .01 level of significance. We conclude that voters hearing this speech are more likely to switch from oppose to favor than from favor to oppose.

462

INFERENTIAL STATISTICS FOR PROBLEMS WITH TWO SETS OF DEPENDENT MEASURES: RANK-ORDER DATA

The Wilcoxon Signed-Ranks Test

In the McNemar test just described, we compared the measures in each dependent pair. We looked at the pretreatment measure and the posttreatment measure to determine if there had been any change. If there had been a change, we noted whether it was from favor to oppose or from oppose to favor. Because we were dealing with the crosstabulation of dichotomous categorical scale measures, there was no question as to the *size* of the changes that occurred. Either a subject changed from favor to oppose or from oppose to favor.

In some research situations, it is not only meaningful to consider the direction of the differences between paired measures, but it is also possible to consider the relative size of these differences. In these situations, it is often possible to employ techniques of inferential statistics that are more powerful than the McNemar test just described. There are several such powerful tests available, including the *Wilcoxon signed-ranks test* and the *t-test for two dependent samples*. Which test we would employ depends on the research situation.

The most common research situation that allows us to consider the relative size of the differences between paired measures is that in which the variable of interest is an interval scale variable. Then, we can compute the difference between each pair of scores. This will yield a set of difference scores that will also be interval scale measures. This research situation typically calls for the use of a *t*-test for two dependent samples, to be considered later in this chapter. However, it is also possible to envision a research situation in which the differences between paired observations do not constitute an interval scale, but do constitute an ordinal scale. That is, we can sometimes rank order the differences between paired observations, even though these differences do not constitute an interval scale. In this situation, we employ the Wilcoxon signed-ranks test.

A Research Example. Coach Meredith of the Central Valley High School debating team has developed a training program that she believes can improve the ability of debaters to think quickly and to organize strong logical arguments. Coach Meredith would like to determine the efficacy of this program. That is, she would like to test the null hypothesis that the debating skill of student debaters trained under this program is the same as the skill of similar student debaters trained using the traditional method. She would like to test this null hypothesis against the alternative hypothesis that student debaters trained using her method are superior.

With this purpose in mind, Coach Meredith performs the following experimental procedure. At the start of the academic year, she randomly selects a

463

group of 18 students from among those who have expressed an interest in debating. On the basis of the amount of prior experience in debating, Coach Meredith divides these 18 students into two matched groups. That is, she takes the two subjects with the greatest amount of prior experience and makes them a pair; then, she takes the two subjects with the next greatest amount of experience and makes them a pair. She continues until all the subjects have been paired in this manner. Then, she randomly selects one member of each pair to be assigned to Group A, which will receive the new training program. The nonselected member of each pair will be placed in Group B, which receives an equal amount of training using the old method. In this manner, Coach Meredith has sought to ensure the initial comparability of the two groups of subjects on the control variable "prior experience." She has done this because she feels certain that prior experience will be an important factor in determining debating skill. In forming matched groups in this manner, however, Coach Meredith has also created a dependency between the two samples. In whatever manner she chooses to measure posttraining debating skill, there may be a tendency for paired subjects to be closer in skill to each other than to other subjects of differing levels of experience, regardless of group membership. Thus, Coach Meredith must employ a technique of inferential statistics that is appropriate for two sets of dependent measures.

At the conclusion of four weeks of training, Coach Meredith has each of the 18 subjects perform in a mock debate. This situation is structured in such a way that each subject is presented with precisely the same arguments, to which the subjects must respond as effectively as possible within a limited time period. Coach Meredith asks a judge to rate each of the 18 debaters on a 10-point scale ranging from 0, which signifies a totally ineffective reply, to 10, which signifies an extremely effective set of replies. The ratings achieved by each of the 18 subjects are presented in Figure 11.2. Because the measurement of debating skill used in this study is a rating scale created on an *ad hoc* basis with no careful analysis of the components of debating skill, and because ratings are made by a single judge, Coach Meredith does not feel that the variable of interest constitutes an interval scale measure. However, she does believe that a subject with a higher rating is indeed a better debater than one with a lower rating. In addition, she believes that the numerical differences between the ratings of paired debaters will constitute an ordinal scale. That is, if the difference between the ratings assigned to one pair of debaters is greater than the difference between the ratings assigned to another pair of debaters, the difference in debating skill is truly greater among the pair with the greater difference in ratings. This last assumption is an essential prerequisite for the use of the Wilcoxon signed-ranks test, which Coach Meredith determines to use here. In using the signed-ranks test, we compute the *Wilcoxon T-statistic*.

Computation of the *T*-Statistic.   The calculation of the *T*-statistic is illustrated in Figure 11.2. The procedure is as follows.

464

Step 1   Organize the data so that paired observations are tabled adjacent to each other, as illustrated in Figure 11.2.

Step 2   Keeping track of signs, find the difference, d, between the scores of each pair of subjects by subtracting the score of the pair member in one group from the score of the pair member in the other group. When ties occur ($d = 0$), do the following: If the number of zero differences is even, assign half of these a ($+$) and half a ($-$). If the number of zero differences is odd, drop one from the analysis, reduce n by 1, and assign the rest half ($+$) and half ($-$) as before.

Step 3   Ignoring for a moment the signs of these differences, rank the absolute value of the differences smallest to largest. That is, assign the rank 1 to the difference that has the smallest absolute value, assign the rank 2 to the difference that has the next smallest absolute value, and so on. In the event that the absolute value of the differences between pairs are the same, assign these differences the average of the ranks they would have occupied, had there been no tie. Thus in the example in Figure 11.2, the differences $+1$ and -1, having identical absolute values, are each assigned the rank 1.5, the average of ranks 1 and 2. Similarly, the differences $+3$ and $+3$, which also have the same absolute value, are each assigned the rank 4.5, the average of ranks 4 and 5.

Step 4   Having ranked the differences without respect to sign, place the sign of the original difference between each pair of scores next to the rank assigned to that difference. Determine which sign occurs least frequently among these "signed ranks." In this example, the minus sign occurs only once and is, therefore, the least frequent sign.

Step 5   Add up all the ranks having the less frequent sign. Here there is only a single score having the less frequent (minus) sign. Our statistic, T_{ob}, is the absolute value of the sum of the ranks with the less frequent sign. Here, $T_{ob} = |-1.5| = 1.5$.

Having followed these procedures to obtain the value of T_{ob}, we determine the significance of T_{ob} in one of two manners, depending upon the sample size.

Significance of T for Small Samples.   When the number of *pairs* of observations employed in the test is less than or equal to 25, the significance of T_{ob} may be evaluated by reference to the table of critical values of T in the Wilcoxon signed-ranks test (Table C.15). This table provides the critical values of T for one- and two-tailed hypothesis tests at various levels of significance. In the case of Coach Meredith's study, we are dealing with a one-tailed test of a directional hypothesis. Coach Meredith sought to test the null hypothesis that debaters trained according to her new method were no different from debaters trained according

465

FIGURE 11.2
Calculation of Wilcoxon T-Statistic in the Signed-Ranks Test
Comparing Two Matched Groups of Debaters in Coach Meredith's Study

	Step 1		Step 2	Step 3	Step 4
Subject pair	Rating given debater from Group A	Rating given debater from Group B	$\text{Rank}_A - \text{Rank}_B = d$	Rank of d	Ranks with signs
1	10	7	+3	4.5	+4.5
2	10	8	+2	3	+3
3	9	8	+1	1.5	+1.5
4	8	4	+4	6	+6
5	8	5	+3	4.5	+4.5
6	8	8	0	--	--
7	7	8	−1	1.5	−1.5
8	7	1	+6	8	+8
9	7	2	+5	7	+7

Step 5

$$T = \begin{array}{c} \text{Absolute value of} \\ \text{sum of those} \\ \text{ranked differences} \\ \text{having less} \\ \text{frequent sign} \end{array} = |-1.5| = 1.5$$

to the old method, against the alternative that debaters trained according to the new method are superior.

In terms of the signed ranks just described, this is the same as testing the hypothesis that the sum of the positive ranks is equal to the sum of the negative ranks in the population of differences between the ratings of matched subjects from the two different training techniques. Keep in mind that ranks with positive signs represent matched pairs in which the new method subject had a *higher* rating than the paired old method subject, and that ranks with negative signs represent matched pairs where the new method subject had a *lower* rating than the paired old method subject. Keep in mind, also, that the size of the rank is an index of the magnitude of the difference between the matched pair. The greater the difference, the larger the rank. Thus, again in terms of the signed ranks, the alternative hypothesis would be that the sum of the negative ranks is smaller than the sum of the positive ranks in the population of differences.

In consulting the table of critical values for T in the signed-ranks test to determine the significance of T_{ob} in Coach Meredith's study, we will look under

466

the headings for one-tailed tests. Furthermore, we will look down the left-hand column of the table to locate the number of pairs actually employed in calculating T. In this case, even though we began with 9 pairs of subjects, we actually used the data for only 8 pairs. This is because there was one pair of observations in which the Group A and the Group B subjects had the same rating. Looking in the table under $n = 8$ pairs, we find that the critical values of T for one-tailed tests are as follows. For $\alpha = .005$, $T_{cr} = 0$; for $\alpha = .01$, $T_{cr} = 2$; and for $\alpha = .025$, $T_{cr} = 4$. In this test, T_{ob} must be *equal to or smaller than* T_{cr} in order for the test to be significant at a given level of significance. Thus, in the case of Coach Meredith's study, the signed-ranks test is significant beyond the $\alpha = .01$ level, because T_{ob} is 1.5, which is less than 2 but greater than 0. Coach Meredith would reject H_o at the .01 level of significance. She would conclude that debaters trained under her new training program are superior to matched debaters trained under the old program. The signed-ranks test for Coach Meredith's data has been laid out, according to our hypothesis testing outline, in Figure 11.3.

Significance of T in Large Samples. When the number of pairs of observations exceeds 25, the significance of T_{ob} is evaluated by a statistic that is a normal approximation. When $n > 25$, and the null hypothesis of no difference is true, it can be shown that the distribution of T is approximately normal with

$$\mu_T = \frac{n(n + 1)}{4}$$

and

$$\sigma_T = \sqrt{\frac{n(n + 1)(2n + 1)}{24}}.$$

It is therefore possible to apply the regular z-score formula,

$$z = \frac{X - \mu}{\sigma}$$

to the distribution of sample values of T to arrive at the statistic

$$z_{ob} = \frac{T_{ob} - \mu_T}{\sigma} = \frac{T_{ob} - \frac{n(n + 1)}{4}}{\sqrt{\frac{n(n + 1)(2n + 1)}{24}}},$$

which has approximately the standard normal distribution if H_o is true. The significance of this test statistic, as any z-statistic, is evaluated by reference to the table of the standard normal distribution (Table C.2).

467

FIGURE 11.3
Example of Wilcoxon Signed-Ranks Test for Coach Meredith's Study

Step 1

Because we have matched groups, we are dealing with two sets of dependent measures. Because subjects were initially selected randomly and because paired subjects were randomly assigned to Group A or Group B, each pair of observations is independent of any other pair. We have at least ordinal data, and we believe we can accurately rank order the differences between paired subjects. Therefore, we use the Wilcoxon signed-ranks test.

Step 2

H_o: The sum of the positive ranks equals the sum of the negative ranks for the population sampled.

Step 3

H_a: The sum of the positive ranks is greater than the sum of the negative ranks for the population sampled.

Step 4

The test statistic is the Wilcoxon T-statistic.

Step 5

$\alpha = .01$. This is a one-tailed test of a directional hypothesis. Because $n < 25$ here, we can obtain a value of T_{cr} from the table of critical values for T in the signed-ranks test (Table C.15). We find $T_{cr} = 2$. We will reject H_o if T_{ob} is *less than* $T_{cr} = 2$.

Step 6

$T_{ob} = 1.5$.

Step 7

Because $T_{ob} < T_{cr}$, we reject H_o. We conclude that the sum of the positive ranks exceeds the sum of the negative ranks in the population of interest. That is, subjects in Group A tend to be rated higher than their matched pairs in Group B.

INFERENTIAL STATISTICS FOR PROBLEMS
WITH TWO SETS OF DEPENDENT MEASURES:
INTERVAL DATA

When we have two sets of dependent measures on a variable that we believe to be an interval scale variable, we must make a decision whether to use the Wilcoxon signed-ranks test just described or to use the *t*-test for two dependent samples. You have already read several times in this text that tests that are appropriate for a given research problem involving ordinal data would also be appropriate if the same research problem involved interval scale data. Thus, if Coach Meredith had measured debating ability on an interval scale variable, she could have used the signed-ranks test nonetheless. However, if Coach Meredith had interval data that she believed fit certain assumptions, she might have chosen to use the *t*-test for two dependent samples rather than the signed-ranks test. The reason for choosing the *t*-test for two dependent samples is that if the assumptions involved in using the *t*-test are met, the test will tend to be somewhat more powerful than the comparable Wilcoxon signed-ranks test.

The *t*-Test for Two Dependent Samples

The *t*-test for two dependent samples is applicable to pretest and posttest designs and to matched group designs in which each pair of observations is independent of any other pair, the variable of interest is an interval scale variable, *and* we believe that the distribution of differences in the population of paired observations is normal or at least approximately normal. You will note that the assumptions regarding the design are identical to the assumptions underlying the McNemar test or the Wilcoxon signed-ranks test, i.e., observations are paired, creating a dependency between the two sets of measures; and each pair of observations is independent of any other pair. The additional requirements for the use of the *t*-test for two dependent samples are that the variable of interest be measured on an interval scale, and that the distribution of differences between paired observations for the population from which our paired observations are taken is normal or nearly normal. We will consider this last assumption in somewhat greater detail in the following sections.

A Research Example.   Educational Psychologist Fredericks has developed a technique for encouraging poor readers of high-school age to do more reading. This technique involves the use of various action-oriented comic books that contain pictures as well as textual material. Dr. Fredericks believes that the use of this method over the course of several months will result in an improvement in the reading achievement level of remedial reading students. To verify this theory, Dr. Fredericks selects a random sample of 25 students from among the population of remedial reading students in Phoenix high schools. He administers the standardized Blackman Reading Achievement Test to each of these 25 subjects to obtain a measure of their initial reading achievement level. He then uses the new materials with this group of subjects for a two-month period. At the conclusion of the treatment period, Dr. Fredericks administers an equivalent alternate form of the Blackman Reading Achievement Test to the 25 subjects as a measure of their posttreatment reading achievement level. The pretest and posttest data for the 25 subjects are presented in Figure 11.4.

In this research situation, it is appropriate to employ the *t*-test for two dependent samples to test the null hypothesis that Dr. Fredericks' intervention has no effect on the reading achievement level of remedial students vs. the alternative hypothesis that the intervention improves the reading achievement level of remedial readers. The *t*-test for two dependent samples is appropriate for the following reasons: (1) Dr. Fredericks has designed his study in such a way that subjects are given a pretest and a posttest, which means that the pair of scores pertaining to any one subject are not independent of each other. (2) However, because the 25 subjects were each selected randomly and independently, the pair of scores pertaining to any one subject is independent of the pair of scores pertaining to any other subject. (3) Because the Blackman Reading Achievement Test is a nationally administered commercial test instrument that has been carefully prepared by experts in the field of measurement, it is certainly reasonable to believe that scores on this test form an interval scale. (4) Also, because scores on the Blackman Reading Achievement Test have been shown to be normally distributed among the norming group as well as among other specialized populations considered in research using the test, it is not unreasonable to assume that the distribution of differences between pretest and posttest measures for the population of remedial readers would be normally distributed as well. But what is this population of difference scores, and why do we care how it is distributed?

Rationale of the *t*-Test for Two Dependent Samples.   The reason that we are concerned with difference scores may be understood from the purpose of Dr. Fredericks' study. He wishes to show that his new materials produce positive gains in reading achievement level. That is, he wishes to show that subjects using these materials will tend to have higher reading achievement scores after the treatment than before it. Thus, it is the difference between a subject's pretest score and the subject's posttest score that he is interested in.

The rationale of the *t*-test for two dependent samples is that if the treatment has no tendency to produce gains in reading achievement, then the average differences between pretest and posttest scores for any sample of subjects should tend to be about zero. We know that not all the subjects would get the same score each time. However, if the treatment does not help, we expect that about as many subjects would have higher scores on the posttest as lower scores on the posttest. In the *t*-test for two dependent samples, we test the null hypothesis that the treatment is not effective by testing the null hypothesis that the average difference score, μ_D, is zero. As in all inferential statistics, the hypothesis we test

FIGURE 11.4

Calculation of t_{ob} in t-Test for Two Dependent Samples for Dr. Fredericks' Pretest and Posttest Remedial Reading Study

Subject	Pretest score	Posttest score	Difference (post-pre)
1	62	80	+18
2	41	42	+ 1
3	56	63	+ 7
4	57	55	− 2
5	69	74	+ 5
6	43	70	+27
7	32	38	+ 6
8	45	57	+12
9	51	53	+ 2
10	52	50	− 2
11	57	65	+ 8
12	52	64	+12
13	63	69	+ 6
14	61	72	+11
15	62	79	+17
16	56	71	+15
17	55	68	+13
18	60	69	+ 9
19	47	61	+14
20	46	53	+ 7
21	55	57	+ 2
22	43	55	+12
23	63	68	+ 5
24	69	68	− 1
25	56	72	+16

$$n = 25$$

$$\Sigma X_D = 220$$

$$\Sigma X_D{}^2 = 3{,}124$$

$$s_D = \sqrt{\frac{\Sigma X_D{}^2 - \frac{(\Sigma X_D)^2}{n}}{n-1}}$$

$$s_D = \sqrt{\frac{3{,}124 - \frac{(220)^2}{25}}{25-1}}$$

$$s_D = 7.04$$

$$\bar{X}_{ob} = \frac{220}{25} = 8.8$$

$$t_{ob} = \frac{\bar{X}_{ob_D} - \mu_{o_D}}{\frac{s_D}{\sqrt{n}}} = \frac{8.8 - 0}{\frac{7.04}{\sqrt{25}}} = +6.25$$

refers to a population parameter. That is, we are testing the hypothesis that the average difference between pretest and posttest score is zero among the entire population of remedial reading students who might use the new materials. This is why we are concerned with the "population of difference scores." Furthermore, we must be willing to assume that this population is normal or nearly normal because the statistic we use to test the hypothesis is a t-statistic, which we have seen requires the assumption of normality.

471

In fact, the t-test for two dependent samples is really just a one-sample t-test that we apply to the special distribution that is the distribution of difference scores. You will recall from Chapter 8 that the formula for the one-sample t-statistic is

$$t_{ob} = \frac{\bar{X}_{ob} - \mu_o}{\frac{s}{\sqrt{n}}}.$$

When we apply this formula to the differences between pretest and posttest scores, the formula becomes:

$$t_{ob} = \frac{\bar{X}_{ob_D} - \mu_{o_D}}{\frac{s_D}{\sqrt{n}}},$$

where $\bar{X}_{ob_D}$ is the mean of the set of difference scores for the sample, μ_{o_D} is the expected value for this sample mean under the null hypothesis, s_D is the sample standard deviation of the set of difference scores from the sample data, and n is the number of differences or the number of *pairs* of observations. As in the general case of the one-sample t-test, the t_{ob} from this formula will have a t-distribution with $n - 1$ df if H_o is true.

The Calculation of t_{ob}.   In the example of Dr. Fredericks' data, we have worked through the procedures for obtaining t_{ob} in Figure 11.4. In determining pretest to posttest differences, we have chosen to subtract the pretest score for each subject from the posttest score rather than the other way around. This is because we see that the posttest scores are in fact generally higher, which means we will have fewer negative differences this way. This is just for convenience. Once we have obtained the set of 25 difference scores, we find the observed sample mean of these difference scores, $\bar{X}_{ob_D}$, and the sample standard deviation of these difference scores, s_D. Then, we plug these values into the formula for the t-statistic to obtain a value of t_{ob}.

Evaluating the Significance of the t-Statistic.   Because this is a one-sample t-test, we evaluate the significance of the observed result in exactly the manner described in Chapter 8, by reference to the table of the t-distributions (Table C.5). In the case of Dr. Fredericks' research problem, we have a one-tailed test of a directional hypothesis. That is, Dr. Fredericks wishes to show that the average gain score among the population of remedial readers using his program is greater than zero. Therefore, the test is of the null hypothesis $H_o: \mu_D = 0$ against $H_a: \mu_D > 0$. This means that we will be looking for critical values of t in the table of the t-distributions under the headings for level of significance for a one-tailed test. Let us assume Dr. Fredericks had chosen to employ the .01 level of significance. We would then look in the table under $\alpha = .01$ for a one-tailed test. Because there are 25 pairs of observations, the df for the test would be $n - 1 = 24$. We find that t_{cr} for this test is $+2.49$. Because t_{ob}, $+6.25$, exceeds t_{cr}, we reject H_o at the .01 level of significance. Dr. Fredericks would conclude that the average gain in reading achievement level among remedial readers using his materials is greater than zero; that is, his program is effective. The t-test for two dependent samples for these data has been laid out according to our hypothesis testing outline in Figure 11.5.

THE CLASSIFICATION MATRIX

In this chapter, we completed our discussion of inferential techniques by looking at those techniques applicable to problems with two sets of dependent measures. We added the McNemar test, the Wilcoxon signed-ranks test, and the t-test for two dependent samples to our matrix. The complete classification matrix for inferential statistics is presented in Table 11.2.

We hope that you will be able to refer to this matrix in the future as an aid in the selection of appropriate statistical tests for various situations. You should keep in mind that there are a great many statistical tests that we did not cover in this introductory text. Those represented here are a sample of the tests most commonly employed with elementary statistical problems. You should also keep in mind that each test has its own unique assumptions that must be checked prior to using the test. For example, just because you have an experimental design with two groups and interval data, you may not be able to use a t-test for two independent samples. Other requirements must be satisfied, such as the requirement of homogeneity of variances and normal distributions. Therefore, you should always use the matrix to find out what statistical procedures may be applicable to the data, and then review the description of these procedures contained in the text to determine if the data fit all of the assumptions of the procedure. It may well be that you will need to employ a technique appropriate for ordinal or nominal scale measures with interval scale data because other requirements for the procedures contemplated are not met.

FIGURE 11.5
Example of *t*-Test for Two Dependent Samples
for Dr. Fredericks' Remedial Reading Study

Step 1

Because we have a pretest and posttest design, we are dealing with two sets of dependent measures. Because subjects were selected randomly, each pair of observations is independent of any other pair. We have interval scale data. We are willing to assume that the population of difference scores is normally distributed. Therefore, we use the *t*-test for two dependent samples.

Step 2

$$H_o : \mu_{o_D} = 0$$

Step 3

$$H_a : \mu_D > 0$$

Step 4

The test is the one-sample *t*-test, applied to the distribution of difference scores in the sample. The statistic will have a *t*-distribution with $n - 1$ df if H_o is true.

Step 5

$\alpha = .01$. The alternative hypothesis is directional, so the test is one-tailed. Looking in the table of the *t*-distributions (Table C.5), we find $t_{cr} = +2.49$. We will reject H_o at the .01 level if t_{ob} exceeds 2.49.

Step 6

$$t_{ob} = \frac{\bar{X}_{ob_D} - \mu_{o_D}}{\dfrac{s_D}{\sqrt{n}}} = +6.25$$

Step 7

Because $t_{ob} > t_{cr}$, reject H_o. Conclude that the average gain in the population is greater than zero.

Inferential Statistics

TABLE 11.2
Classification Matrix: Inferential Statistics

	Level of Measurement of Variable(s) of Interest		
Type of Problem	Categorical: Nominal or Ordered Categories	Rank-Order	Interval Scale (or Ratio-Scale)
Univariate Problems	Binomial Test Normal Approximation to Binomial χ^2 One-Variable Test *With Ordered Categories:* Kolmogorov-Smirnov One-Variable Test		One-Sample z-Test One-Sample t-Test Confidence Interval for Population Mean (z- or t-Method)
Bivariate Problems — Correlational Problems	χ^2 Test of Association	Test for Significance of Spearman r_s	t-Test for Significance of Pearson r Fisher z_r Transformation for Testing Significance of Pearson r Confidence Interval for ρ
Bivariate Problems — Experimental and Pseudo-Experimental Problems: Independent Groups	Fisher Exact Probability Test χ^2 Test of Homogeneity	Mann-Whitney U-Test Kruskal-Wallis Analysis of Variance by Ranks	t-Test for Independent Samples One-Way Analysis of Variance Scheffé Contrast
Bivariate Problems — Dependent Measures: Pretest and Posttest or Matched Groups	McNemar Test	Wilcoxon Signed-Ranks Test	t-Test for Dependent Samples

Key Terms

Be sure you can define each of these terms.

McNemar test
Wilcoxon signed-ranks test
Wilcoxon T-statistic
t-test for two dependent samples

Summary

In Chapter 11, we completed our discussion of inferential statistics by looking at techniques for problems involving two sets of dependent measures. We looked at two different research situations: the pretest and posttest design and the matched groups design. Though different in a practical sense, these designs are statistically equivalent. Both involve independent pairs of observations, the scores within each pair being dependent. Thus, we have two dependent samples. Our null hypothesis in each case is that these two samples come from identical populations.

When our observations are made on a dichotomous categorical variable, we use the McNemar test to test whether our proportions of "successes" are the same for our two groups. With rank-order data, we use the Wilcoxon signed-ranks test to test the hypothesis that our two samples come from identical populations. With small samples, we use the Wilcoxon T-statistic and compare it to a critical value obtained from a table. With large samples, we can use a normal approximation for T.

With an interval scale variable from an approximately normal distribution, we use the t-test for two dependent samples. In this case, we test hypotheses regarding the mean of the differences between pairs of scores. We obtain our critical value for t by looking at the t-distribution with the number of degrees of freedom equal to one less than the number of pairs of scores.

11
Student Guide

Review Questions

To review the concepts presented in Chapter 11, choose the best answer to each of the following questions.

1 If 15 pairs of identical twins are randomly split into two groups, with one twin of each pair in one group and one twin of each pair in the other group, and if both groups are measured on the same variable, we will have
_____a two independent sets of measures.
_____b two dependent sets of measures.

2 The name given to the design mentioned in Problem 1 is
_____a the pre-post design.
_____b the matched pairs design.

3 If 50 students are given a standardized achievement test at the beginning of the school year and are retested at the end of the year on an equivalent test, then we have
_____a the pre-post design.
_____b the matched pairs design.

4 A teacher wishes to know which of two items on a test is more difficult. Students' answers are marked either right or wrong for each item. A random sample of 50 students is asked to answer both questions. To answer the teacher's question, we would use
_____a the McNemar test.
_____b the t-test for dependent samples.

5 One assumption that is common to the three tests in Chapter 11 is that
_____a the n pairs of observations are independent of each other.
_____b the variances in the populations are the same.

6 The difference between the t-statistic for dependent means and the t-statistic for independent means is that
_____a the numerators of the two look the same but are different.
_____b the denominators are different.

7 One thing that is alike about the 2×2 χ^2 test for homogeneity and the McNemar test is that
_____a both test the null hypothesis that two proportions are equal.
_____b both assume that there are two independent groups.

8 Both the McNemar test and the 2×2 χ^2 test for homogeneity involve two binomial populations. The thing that is different about the two tests is
_____a in the McNemar test, the two binomial populations are independent.
_____b in the McNemar test, the two binomial populations are dependent.

477

Problems Involving Dependent Measures

9 All the methods presented in Chapter 11 are for comparing two sets of dependent measures. These statistical tests could all be called
_____a one-sample tests.
_____b two-sample tests.

10 Suppose we have 24 SAT scores. Half of the scores are measures taken on students in the eleventh grade and the other half are measures on the same students in the twelfth grade. In this case, we have
_____a 24 independent observations.
_____b 12 pairs of dependent observations.

11 In the example in Question 10, if we want to determine if the results suggest that average SAT score is the same or different for eleventh- and twelfth-graders, we could do a t-test. The appropriate degrees of freedom for the t-distribution would be
_____a 11.
_____b 22.

12 One reason we might choose the Wilcoxon signed-ranks test rather than the t-test for dependent samples is
_____a if we feel the samples are not random.
_____b if we don't believe the differences have a normal distribution.

Problems for Chapter 11: Set 1

1 Suppose 15 pairs of identical twins are randomly assigned to a control group and an experimental group with one twin of each pair in each group. The experimental group will be trained in problem-solving strategies and the control group will not. At the end of the experiment, both groups will be given a problem to solve. The subjects will be scored as either passing or failing. The experimenter wishes to know if teaching problem-solving strategies facilitates problem solving. The observed data follow. Use the hypothesis testing method to help the experimenter decide whether the method is successful. Set $\alpha = .05$.

		Control Twin		
		Solved	Did not solve	
Experimental Twin	Solved	3	6	9
	Did not solve	4	2	6
		7	8	15

2 To see which of two test items is more difficult, an investigator asks a random sample of 50 subjects to answer both questions. Subjects are scored as either correct or incorrect on each item. The results follow. Using $\alpha = .05$, what conclusion can be drawn concerning the difficulty of the two items?

First Item

	Correct	Incorrect	
Correct	11	29	40
Incorrect	9	1	10
	20	30	50

(Second Item — rows)

3 A market researcher wants to know if a piece of advertising will change people's attitude about buying a certain new product. He selects 100 people at random and asks them to choose either the new product or a current product on the market. The people are shown the advertising piece and a week later are asked to choose between the two products again. Based on the results that follow, what conclusion can the market researcher make about the effectiveness of the advertising? Use $\alpha = .05$.

Before

	New product	Old product	
New product	40	40	80
Old product	10	10	20
	50	50	100

(After — rows)

Set 2

1 To practice reading Table C.15, set up rejection rules for the following tests.
   a There are 10 matched pairs. We want a two-tailed test with $\alpha = .05$.
   b There are 10 matched pairs. We want a one-tailed test with $\alpha = .005$.
   c There are 15 matched pairs. We want a two-tailed test with $\alpha = .02$.
   d There are 7 matched pairs. We want a one-tailed test with $\alpha = .01$.
   e There are 7 matched pairs. We want a two-tailed test with $\alpha = .01$.

2 An experimenter is interested in the effect of a certain drug on intellectual functioning. She has 9 subjects who are willing to take part in an experiment. Each subject will be his or her own control. In the experimental situation, the subjects take a pill containing the drug. In the control situation, the subjects take a placebo. In either case, they are unaware of which pill they have taken. After 30 minutes, all subjects take the same test of intellectual functioning. Low scores on the test indicate low intellectual functioning. The results of the experiment follow. Use the Wilcoxon signed-ranks test to test the null hypothesis that the drug has no effect on intellectual functioning. Set $\alpha = .025$.

Problems Involving Dependent Measures

Subject	1	2	3	4	5	6	7	8	9
Control	5	27	3	11	8	4	14	13	8
Experimental	3	1	4	7	5	9	7	30	2

3 A psychologist believes that he has a method of reducing the anxiety that people have towards statistics courses. To test his method, he selects a random sample of statistics students and measures them on a statistics anxiety inventory. He then treats them according to his method and remeasures them on the inventory at the end of the treatment. His results follow (low scores mean low anxiety). Use the Wilcoxon signed-ranks test to test the hypothesis that the psychologist's method does not lower anxiety. Use $\alpha = .01$.

Subject	1	2	3	4	5	6	7	8	9	10	11	12	13
Pretest	5	10	12	12	19	20	30	36	18	16	17	19	25
Posttest	7	9	9	13	15	15	17	25	9	10	25	10	8

4 A football coach at a large university has developed a training program for his defense squad. He feels that as a result of his program, his players will be smarter and more aggressive in their play. To test his program, he plans an experiment that will take place in spring training. Using a point system developed over many years, he rates all his defensive players and assigns them to Training Group A or Training Group B, using a matching scheme. Group A will get the new program and Group B will get the traditional program. The players are unaware of the differences in their training schedules. At the end of the training season, all players participate in an intrasquad game. The coach will rate the players according to his system and compare the results of those players trained by Method A with those trained by Method B. Using $\alpha = .05$ and the results that follow, test the null hypothesis that the new method is no better than the old. Use the Wilcoxon signed-ranks test.

Pair	1	2	3	4	5	6	7	8	9	10	11	12	13	14	15	16
A	50	55	60	40	75	32	50	54	60	62	70	71	35	37	40	55
B	45	56	40	30	50	46	27	90	47	56	45	52	39	25	37	53

Set 3

1 Refer to Problem 2 in Set 2. Assume that the scores on intellectual functioning constitute two sets of measures from normal distributions. Then test $H_o: \mu_D = 0$ vs. $H_a: \mu_D > 0$, where D = Control score − Experimental score. Use $\alpha = .025$.

2 Refer to Problem 3 in Set 2. Assume that scores on anxiety towards statistics are from normal distributions. Test the null hypothesis that the psychologist's method is ineffective.

3 A medical researcher wants to know if it makes any difference which arm is used in taking blood pressure measurements. She randomly selects 12 people and records their blood pressure using the left arm and using the right arm. The measurements are taken under identical conditions. In 6 cases, the left arm is measured first, and in 6 cases, the right arm first. Mean arterial blood pressure is recorded for each arm of each subject. The results follow. Do the results suggest that there is a difference in blood pressure measurements in the left and right arms? Use $\alpha = .05$.

Subject	1	2	3	4	5	6	7	8	9	10	11	12
Left arm	118	90	103.7	86.3	114	95.7	100.3	98.3	84.3	99.7	86.0	98.3
Right arm	120	92	100	88	113.3	91.5	96.7	100.7	88.3	94.3	84.0	94.7

4 After years of work, a chronic sufferer of poison ivy comes up with an ointment that he believes will dry up poison ivy in less time than the best product on the market. To test his product, he selects 10 people with poison ivy. The subjects use his ointment on half of the rash and the current best product on the other half. He records the time, in days, that it takes the rash to dry up under both conditions. The results follow. Does the evidence suggest that his new ointment is better than the current best product? Use $\alpha = .01$.

Subject	1	2	3	4	5	6	7	8	9	10
New	1.9	3.9	6.4	2.7	2.4	3.5	1.4	2.7	8.2	2.1
Old	2.8	5.6	5.3	6.3	6.4	3.9	3.8	2.4	2.4	1.8

Set 4

In Set 2 of Chapter 1, we considered a set of problem situations and classified them according to our classification scheme. We also suggested summary descriptive techniques that would be appropriate. We now refer to those situations. For those problems that are inferential, describe
a the null hypothesis to be tested.
b the alternate hypothesis.
c the test statistic.
d the probability distribution of the statistic if H_o is true.
e If no hypothesis seems to be appropriate, tell how the method of confidence intervals could be used.
(The inferential problems are 1, 3, 5, 6, 7, 8, 9, 10.)

Appendices

Appendix A
Review of
Basic Math, Symbols, and Formulas

Signed Numbers:

1 Addition and subtraction:
   a The sum of two numbers having the same sign may be found by combining their absolute values and assigning to the result the common sign of the two numbers:

$$4 + 5 = 9$$
$$(-4) + (-5) = -9$$

   b The sum of two numbers having opposite signs may be found by *subtracting* the smaller absolute value from the larger absolute value and assigning the result the sign of the larger:

$$(4) + (-5) = -1$$
$$7 + (-2) = 5$$

   c The difference between two numbers may be found by changing the sign of the number to be subtracted, and then proceeding as in addition:

$$5 - 3 = 5 + (-3) = 2$$
$$7 - 9 = 7 + (-9) = -2$$

2 Multiplication and division:
   If the two numbers to be multiplied or divided have the same sign, the answer is positive. If they have opposite signs, the answer is negative:

$$(5) \cdot (3) = 15$$
$$(-5) \cdot (3) = -15$$
$$(5) \cdot (-3) = -15$$
$$(-5) \cdot (-3) = 15$$

$$(6)/(2) = 3$$
$$(-6)/(2) = -3$$
$$(6)/(-2) = -3$$
$$(-6)/(-2) = 3$$

Algebraic Terms

1 In the expression $3X$:
   a 3 is a *constant;*

484

b X is a *variable;*
c 3 is known as the *coefficient* of X;
d 3 and X are the two *factors* of the expression.

2 $\sqrt{X}$ is referred to as the square root of X. It is the number which, when multiplied by itself, yields X:
$$\sqrt{9} = 3, \text{ since } 3 \times 3 = 9.$$

3 In the expression X^2:
a X is referred to as the *base;*
b 2 is referred to as the *exponent;*
c $X^2 = (X) \cdot (X)$;
d $X^3 = (X) \cdot (X) \cdot (X)$;
e $X^1 = X$;
f $X^0 = 1$.

For example, $3^2 = 3 \cdot 3 = 9$
$$3^3 = 3 \cdot 3 \cdot 3 = 27$$
$$3^1 = 3$$
$$3^0 = 1$$

Multiplication and Division of Numbers Having Exponents:

1 When two numbers having the same base are multiplied, their product may be found by adding the exponents:
$$(3^3) \cdot (3^2) = (3^{3+2}) = 3^5$$

2 When two numbers having the same base are divided, their quotient may be found by subtracting the exponents:
$$(3^3)/(3^2) = (3^{3-2}) = 3^1 = 3$$

Algebraic Operations

1 We may add (or subtract) the same term to each side of an equation, and still preserve the equality. This operation is sometimes used in *transposing* terms from one side of an equation to the other. When a term is transposed from one side to the other, its sign is changed. For example:
$$X - Y = 2$$
$$X - Y + Y = 2 + Y$$
$$X = 2 + Y$$

or

$$a - b = c$$
$$a - b + b = c + b$$
$$a = c + b$$

2 We may multiply (or divide) both sides of an equation by the same term, and still preserve the equality. For example, we may wish to solve the following equation for a:

485

Basic Math

$$y = \frac{a}{b}$$

$$y(b) = \frac{a}{b}(b)$$

$$a = yb$$

Summation Notation

In statistics, we use the Greek letter capital sigma (Σ) to represent a sum. We write the name of the variable to be summed to the right of Σ. Thus,

ΣX_i means sum up the values of the variable X_i.

We sometimes use subscripts to indicate particular observations on the variable X. If we had measured to subjects on X, the subscript i would run from 1 to 6, where X_1 indicates the first observation, X_2 indicates the second observation, and so on until we come to X_6, indicating the sixth observation. Let's suppose that we have six observations on X, as follows:

$$X_1 = 2$$
$$X_2 = 4$$
$$X_3 = 5$$
$$X_4 = 3$$
$$X_5 = 2$$
$$X_6 = 6$$

We can also use subscripts and superscripts with the summation sign Σ to direct us to add certain observations on X. In this case, we use a subscript beneath Σ to instruct us as to the point where we should begin adding the observations on X and a superscript over Σ to tell us where we should stop adding. Thus,

$$\sum_{i=1}^{4} X_i$$

tells us to add all the observations on the variable X from i = 1 to i = 4, that is, the first four observations. Therefore,

$$\sum_{L=1}^{4} X_i = X_1 + X_2 + X_3 + X_4$$

$$\sum_{L=1}^{4} X_i = 2 + 4 + 5 + 3$$

$$\sum_{L=1}^{4} X_i = 14.$$

It is possible to employ this notation to instruct us to begin and end the process of summation at any one point. For example,

$$\sum_{L=3}^{5} X_i = X_3 + X_4 + X_5$$

$$\sum_{L=3}^{5} X_i = 5 + 3 + 2$$

$$\sum_{L=3}^{5} X_i = 10.$$

Most often in statistics, we sum up *all* of the scores in a set. In this case, we often omit the subscript and superscript notation. Thus, ΣX is assumed to say, "add up all the observations on X." In this text we sometimes emphasize this point using the superscript "all scores." For example,

$$\sum^{\text{all scores}} X$$

is another way of instructing us to add up *all* the observations on X.

Symbols: Greek Letters

α	level of significance
β	probability of a Type II error
θ	contrast among group means
θ'	estimated value of a contrast based on sample data
μ	mean of a distribution in a descriptive problem
ρ	Pearson correlation coefficient
ρ_s	Spearman rank-order correlation coefficient
σ	standard deviation in a descriptive problem
σ^2	population variance
$\sigma_{Y \cdot X}$	standard error of estimate
Σ	summation sign
ϕ'	index of contingency

Symbols: Latin Letters

a	*Y*-intercept of the regression line
b	slope of the regression line
D	Kolmogorov-Smirnov statistic
Ex	expected cell frequency
F	*F*-statistic
H	Kruskal-Wallis statistic
H_a	alternative hypothesis
H_o	null hypothesis
MSB	mean square between groups
MSW	mean square within groups
n	sample size
N	population size
Ob	observed cell frequency

487

P	percentile rank
P	population proportion
P	probability of success in a binomial problem
pr	probability that a given event will occur
Q	probability of failure in a binomial problem
r	sample Pearson correlation coefficient
r_s	sample Spearman correlation coefficient
s	sample standard deviation
$s_p{}^2$	pooled estimate of population variance
t	t-statistic
T	Wilcoxen signed-ranks statistic
U	Mann-Whitney U-statistic
X	variable

Formulas: Descriptive Statistics

The numbers in bold type following each formula refer to the chapter in which the formula is discussed.

$$P(\text{percentile rank}) = \frac{[N_{below} + 1/2N_{at}]}{N_{total}} \times 100 \quad \textbf{(4)}$$

$$\mu = \frac{\overset{\text{all scores}}{\sum} X}{N} \quad \textbf{(4)}$$

$$\sigma = \sqrt{\frac{\Sigma(X-\mu)^2}{N}} = \sqrt{\frac{\Sigma X^2}{N} - \mu^2} \quad \textbf{(4)}$$

$$z = \frac{X-\mu}{\sigma} \quad \textbf{(4)}$$

$$\phi' = \sqrt{\frac{\overset{\text{all rows}}{\sum} \overset{\text{all columns}}{\sum} \frac{(Ob-Ex)^2}{Ex}}{N(L-1)}} \quad \textbf{(5)}$$

$$\rho_s = 1 - \frac{6\Sigma D^2}{N(N^2-1)} \quad \textbf{(5)}$$

$$\rho = \frac{N(\Sigma XY)-(\Sigma X)(\Sigma Y)}{\sqrt{[N\Sigma X^2-(\Sigma X)^2][N\Sigma Y^2-(\Sigma Y)^2]}} = \frac{\Sigma z_X z_Y}{N} \quad \textbf{(5)}$$

$$Y_{PRE} = a + bx \quad \textbf{(5)}$$

$$b = \rho \frac{\sigma_Y}{\sigma_X} \quad \textbf{(5)}$$

$$a = \mu_Y - b\mu_X \quad \textbf{(5)}$$

$$\sigma_{Y \cdot X} = \sqrt{\frac{\Sigma(Y - Y_{PRE})}{N}} \quad \textbf{(5)}$$

Formulas: Inferential Statistics

$$\left(\frac{n}{r}\right) = \frac{n!}{r!(n-r)!} \quad \textbf{(6)}$$

$$z = \frac{r - nP}{\sqrt{nPQ}} \quad \textbf{(6)}$$

$$z = \frac{p - P}{\sqrt{PQ/n}} \quad \textbf{(6)}$$

$$z = \frac{\bar{X}_{ob} - \mu_o}{\sigma/\sqrt{n}} \quad \textbf{(7)}$$

$$\mu_{\bar{X}} = \mu_X \quad \textbf{(7)}$$

$$\sigma_{\bar{X}} = \frac{\sigma_X}{\sqrt{n}} \quad \textbf{(7)}$$

$$\chi^2 = \overset{\underset{\text{all}}{\text{cells}}}{\sum} \frac{(Ob - Ex)^2}{Ex} \quad \textbf{(8)}$$

$$s = \sqrt{\frac{\Sigma(X - \bar{X})^2}{n-1}} = \sqrt{\frac{n\Sigma X^2 - (\Sigma X)^2}{n(n-1)}} \quad \textbf{(8)}$$

$$t = \frac{\bar{X}_{ob} - \mu_o}{s/\sqrt{n}} \quad \textbf{(8)}$$

$$r_s = 1 - \frac{6\Sigma d^2}{n(n^2 - 1)} \quad \textbf{(9)}$$

$$r = \frac{n\Sigma XY - \Sigma X\Sigma Y}{\sqrt{[n\Sigma X^2 - (\Sigma X)^2][n\Sigma Y^2 - (\Sigma Y)^2]}} \quad \textbf{(9)}$$

$$t = \frac{r\sqrt{n-2}}{\sqrt{1-r^2}} \quad \textbf{(9)}$$

489

$$z = \frac{z_r - z_{r_0}}{1/\sqrt{n-3}} \quad \textbf{(9)}$$

$$\chi^2 = \frac{n(|AD - BC| - n/2)^2}{(A+B)(C+D)(A+C)(B+D)} \quad \textbf{(10)}$$

$$U = n_1 n_2 + \frac{n_1(n_1 + 1)}{z} - R_1 \quad \textbf{(10)}$$

$$z = \frac{U - n_1 n_2/2}{\sqrt{\dfrac{n_1 n_2(n_1 + n_2 - 1)}{12}}} \quad \textbf{(10)}$$

$$H = \left[\frac{12}{n(n+1)}\right]\left[\sum \frac{R_j^2}{n_j}\right] - 3(n+1) \quad \textbf{(10)}$$

$$s_p^2 = \frac{(n_1 - 1)s_1^2 + (n_2 - 1)s_2^2}{n_1 + n_2 - 2} \quad \textbf{(10)}$$

$$t = \frac{(\bar{X}_1 - \bar{X}_2) - (\mu_1 - \mu_2)}{s_p\sqrt{1/n_1 + 1/n_2}} \quad \textbf{(10)}$$

$$MSB = \frac{\sum n_j(\bar{X}_j - \bar{X}..)^2}{K - 1} \quad \textbf{(10)}$$

$$MSW = \frac{\sum\sum(X - \bar{X}_j)^2}{n - K} \quad \textbf{(10)}$$

$$F = \frac{MSB}{MSW} \quad \textbf{(10)}$$

$$\chi^2 = \frac{[|A - D| - 1]^2}{A + D} \quad \textbf{(11)}$$

$$z = \frac{T - \dfrac{n(n+1)}{4}}{\sqrt{\dfrac{n(n+1)(2n+1)}{24}}} \quad \textbf{(11)}$$

$$t = \frac{\bar{X}_d - \mu_d}{s_d/\sqrt{n}} \quad \textbf{(11)}$$

490

Appendix A

Appendix B
Glossary of
Terms _____

The numbers in bold type following each definition refer to the chapter in which the item is discussed.

alternative hypothesis   A hypothesis which will be retained if the null hypothesis is rejected. **(6)**

apparent class limits   The values recorded on a continuous variable. **(2)**

bar graph   A graph used to represent the frequency distribution or relative frequency distribution of a categorical variable. In such a graph, a rectangular bar is drawn over each value. The height of this bar represents the frequency or relative frequency of the value. **(2)**

bimodal distribution   A distribution with two modes. **(4)**

binomial experiment   A procedure which produces one of two possible outcomes. Each outcome has a probability that remains constant over repeated trials of the experiment. The trials themselves are independent. **(6)**

binomial table   A table of binomial distributions used to find the probabilities of obtaining each of the various possible numbers of successful trials in a binomial experiment. **(6)**

binomial variable   A variable with only two values. **(6)**

bivariate frequency distribution   A technique used to summarize the scores on both of the variables when subjects have been measured on two variables. **(3)**

bivariate problem   A research problem in which there are two variables of interest. **(1)**

categorical variable   A variable whose values are names of categories. **(1)**

Central Limit Theorem   The theorem that holds that the distribution of sample means approaches a normal distribution as the sample size gets larger. **(7)**

chi-square test of association   A statistical test used to test the significance of the relationship between two categorical variables in the population of interest, based on a comparison of observed and expected frequencies. **(9)**

chi-square test of homogeneity   A statistical test employed to test hypotheses about population parameters, using two-independently selected samples with respect to a categorical variable. This test is used when the sample size is large. **(10)**

491

circle graph A graph used to represent the frequency distribution or relative frequency distribution of a nominal scale variable. The circle is divided into segments, each of which represents the proportion of total observations recorded for every value in the data set. **(2)**

class mark The midpoint of a score interval. **(2)**

computational formula A formula that simplifies the computation of a statistic. **(4)**

conditional distribution A frequency distribution showing how the observations of one value on a variable distribute with respect to the values on a second variable. **(3)**

confidence interval for *P* A statistical test used for a binomial experiment when an initial hypothesis regarding the population proportion is not available. **(8)**

confidence interval for Pearson correlation An interval estimate of a Pearson correlation, reported with a specified amount of confidence that the interval actually contains the population correlation. **(9)**

confidence interval for the difference between two population means An interval estimate for the difference between two population means, reported with a specified amount of confidence that the interval actually contains this difference. **(10)**

confidence interval (*t*-method) An interval estimate for a population parameter using the *t*-distributions and an estimated value for the population standard deviation, reported with a specified amount of confidence that the interval actually contains the parameter. **(8)**

confidence interval (*z*-method) An interval estimate of a population parameter using the normal curve model and the population standard deviation reported with a specified amount of confidence that the interval actually contains the parameter. **(7)**

continuity correction factor The process by which a result is made more precise, when a continuous distribution is used to estimate probabilities on a discrete distribution. **(8)**

continuous interval variable An interval variable with values that can be measured as precisely (between whole units) as our measuring device will permit. **(2)**

coordinate system A graphic technique in which points are used to represent pairs of numbers. The first number in the pair indicates a position on the horizontal axis, and the second indicates a position on the vertical axis. **(3)**

correlational problem A bivariate problem in which one group of subjects is measured on two variables. **(1)**

Cramér's index of contingency A measure of relationship for two categorical variables. **(5)**

criterion variable In a linear regression problem, the variable whose values are to be predicted or estimated. **(5)**

crosstabulation table A bivariate frequency distribution for two categorical variables showing the frequency with which each value of one variable occurs with each value of the other variable. **(3)**

cumulative frequency The number of observations at or below a specified value. **(2)**

492

cumulative frequency distribution A chart showing a list of the values on an interval variable and the cumulative frequency of each value. **(2)**

cumulative frequency graph A graph of a cumulative frequency distribution in which a dot is placed above each upper real limit at a height representing the cumulative frequency of that interval. The dots thus formed are then connected by straight lines. This graph may also be constructed using cumulative relative frequencies. **(2)**

cumulative percentage The cumulative frequency of a value divided by the total number of observations. **(2)**

cumulative percentage distribution A chart showing a list of the values on an interval variable and the cumulative percentage for each value. **(2)**

definitional formula A formula that clearly indicates the true mathematical meaning of the statistic being computed. **(4)**

degrees of freedom A parameter of certain probability distributions. The choice of which distribution to use is based on knowledge of the degrees of freedom. **(8)**

dependent measures Measures taken in such a way that the outcome of one observation influences the outcome of the other. **(1)**

dependent variable The variable in an experimental problem that is affected by manipulation of the independent variable. **(1)**

descriptive problem A research problem in which the entire population is measured. **(1)**

descriptive statistic A numerical index used to describe a distribution of scores. **(4)**

deviation from regression The difference between the actual value of the criterion variable and the value of the variable predicted by the regression equation. **(5)**

directional hypothesis An alternative hypothesis that predicts an increase or decrease in the probability of a successful outcome. **(6)**

direction of causation In a bivariate correlational problem, the case in which one variable can possibly influence the outcome of the other variable, but not the reverse. **(3)**

discrete graph A graph used to indicate the frequency distribution of a discrete interval variable. The height of each vertical line on this graph represents the frequency with which a value occurs. **(2)**

discrete interval variable An interval variable with values that fall at whole unit steps. **(2)**

distribution of sample means A distribution of all values of the sample mean, across all samples of the same size. **(7)**

exact significance probability The probability of observing a result as extreme as the observed value when the null hypothesis is true. **(7)**

expected cell frequency The number of observations expected in a cell if there is no relationship between the variables. **(5)**

experimental problem A bivariate problem in which one variable is manipulated by the experimenter so that the effect on the other variable can be measured. **(1)**

external validity The extent to which experimental results can be generalized. **(1)**

***F*-distribution** A theoretical distribution employed as the sampling distribution in the analysis of variance. **(10)**

***F*-statistic** The statistic employed in the analysis of variance, consisting of a ratio of the mean square between groups to the mean square within groups. **(10)**

Fisher exact probability test A statistical test employed to test hypotheses about population parameters, using two independently selected samples with respect to a dichotomous variable. This test is used when the sample size is small. **(10)**

Fisher *z*-transformation A technique used to test hypotheses about nonzero Pearson correlations. **(9)**

frequency The number of times a value occurs in a set of measures. **(2)**

frequency distribution A chart showing the list of possible values on a variable and the frequency with which each value occurs in the data set. **(2)**

frequency polygon A graph used to picturize the frequency distribution or relative frequency distribution of a continuous interval variable, or the grouped frequency distribution of a discrete interval variable. A dot is placed above each class mark at a height representing the frequency of observations in the interval. The dots thus formed are then connected with straight lines. **(2)**

grouped frequency distribution A frequency distribution that lists the values of a variable in equally sized groups or intervals, rather than as separate values. Each frequency represents the number of observations recorded in each interval. **(2)**

***H*-statistic** The statistic used in the Kruskal-Wallis analysis of variance. **(10)**

histogram A bar graph used to represent the frequency distribution or relative frequency distribution of a continuous interval variable, or the grouped frequency distribution of a discrete interval variable. The bars sit on the real class limits, and the height of each bar represents the frequency or relative frequency of observations in the interval. **(2)**

hypergeometric model The mathematical model underlying the Fisher exact probability test. **(10)**

hypothesis testing A procedure used to test the null hypothesis, consisting of an initial assumption, development of an expectation based on the null hypothesis, comparison of expectation to factual observation, and a conclusion. **(6)**

independent measures Measures taken in such a way that the outcome of one of the observations does not influence the outcome of the other. **(1)**

independent variable The variable in an experimental problem that is manipulated by the experimenter. **(1)**

inferential problem A research problem in which the observations on a random sample are used to draw an inference about the entire population. **(1)**

internal validity The extent to which differences between two experimental groups can be considered real. **(1)**

interquartile range The distance between the twenty-fifth and seventy-fifth percentiles in a distribution. **(4)**

interval estimation A procedure, used to test the null hypothesis, in which the observed value of a sample statistic is employed to estimate a population parameter, allowing for a margin of error. **(7)**

494

interval scale variable   A variable whose values are numbers that can be used to indicate the amount by which two values differ. The unit of measure on an interval scale is the same throughout the scale. **(1)**

intervening variables   In a pseudo-experimental study, variables other than the "group" variable that are not part of the study but that possibly affect the dependent variable, and thus the outcome of the experiment. **(3)**

Kolmogorov-Smirnov one-variable test   A statistical test used in a research situation in which independent observations have been obtained on a single categorical variable, where the categories have a clear order. **(8)**

Kruskal-Wallis analysis of variance by ranks   A statistical test employed to test hypotheses about central tendencies in populations, using three or more independently selected samples with respect to a rank-order variable. **(10)**

law of large numbers   The law which states that the larger a random sample taken, the higher is the probability that the sample will look like the population sampled. **(7)**

level of significance   The probability of a Type I error. **(6)**

linear relationship   On a scatterdiagram, the case in which the points tend to gather lengthwise around an imaginary straight line. **(3)**

Mann-Whitney *U*-test   A statistical test employed to test hypotheses about the central tendencies of two populations, using two independently selected samples with respect to a rank-order variable. **(10)**

marginal distribution   The frequency of each value on one variable across all values on the other, recorded in the margins of a bivariate frequency distribution. The marginal frequency distribution itself is a univariate frequency distribution. **(3)**

marginal frequency   On a crosstabulation table, the total of all cell frequencies in a given column or row. **(3)**

McNemar test   A statistical test used to test hypotheses regarding two sets of dependent measures with respect to a dichotomous variable. **(11)**

mean   The sum of all the observations divided by the total number of observations. **(4)**

mean square between groups   In the analysis of variance, the measure of the extent to which sample means differ from each other. **(10)**

mean square within groups   In the analysis of variance, the measure of the extent to which scores within several samples spread out around the means of those samples. **(10)**

measurement process   The process by which a variable is observed in the population of interest. **(1)**

measures of central tendency   Numerical indices that represent the typical score in a distribution. **(4)**

measures of location   Numerical indices that relate a score to the other scores in a group. **(4)**

measures of relationship   Numerical indices that reflect the extent to which two variables are related in a bivariate correlation problem. **(5)**

measures of variability   Numerical indices that indicate how the scores in a distribution differ from one another. **(4)**

median   The value which falls exactly in the middle of a distribution, that is, the middle-ranked value. **(4)**

495

mode   The value which occurs most frequently in a distribution. **(4)**

multiplication rule for probabilities   The rule which applies when we are determining the product of two probabilities. If we know the probability of a particular outcome of some experimental procedure, and we also know the probability of the outcome of a second, independent experimental procedure, then the probability that both outcomes will occur on two successive procedures is the product of the two probabilities. **(6)**

negatively skewed distribution   A distribution in which the few extreme scores fall *below* the bulk of scores. **(4)**

negative relationship   In a bivariate correlation problem, the case in which the bivariate frequency distribution indicates a tendency for low values on one variable to occur with high values on the other. **(3)**

nominal scale variable   A variable whose values are names of categories. **(1)**

nondirectional hypothesis   An alternative hypothesis that predicts a change in the probability of a successful outcome without indicating whether the change is in the direction of increasing the probability or decreasing the probability. **(6)**

normal approximation to test for significance of Spearman correlation   A statistical test that uses the normal curve model to test the significance of the relationship between two rank-order variables. **(9)**

normal approximation to the binomial   A statistical test used in binomial experiments where the sample size is greater than 25. **(8)**

normal curve model   A theoretically derived family of bell-shaped curves often used as probability distributions. **(7)**

normally distributed variable   A variable that tends to have a bell-shaped distribution. **(7)**

null hypothesis   A hypothesis that is assumed to be true and is under test in the hypothesis testing model. **(6)**

observed cell frequency   The number of actual observations recorded in a cell. **(5)**

one-sample *t*-test   A statistical test used to test hypotheses about population means when the population standard deviation is unknown. **(8)**

one-sample *z*-test   A test using the normal curve model that is employed to test a hypothesis regarding a population mean on the basis of the mean of a random sample drawn from that population. **(7)**

one-tailed hypothesis test   A test based on a rejection rule in one tail of the sampling distribution. **(6)**

one-way analysis of variance (ANOVA)   A technique used to test hypotheses regarding the means of two or more independently selected samples with respect to an interval scale variable. **(10)**

operational definition   The definition of each variable in terms of the method by which the variable is measured. **(1)**

ordered categorical variable   A variable whose values are names of categories that have been ranked. **(1)**

ordinal scale variable   A variable whose values are ranks. **(1)**

parameter   The attribute of a population for which an estimate is sought. **(1)**

Pearson product-moment correlation coefficient   A measure of the extent to which two interval variables are linearly related. **(5)**

496

percentile rank   The percentage of the total number of observations falling at or below a given score. **(4)**

pictograph   A type of bar graph in which pictures are used to represent the frequencies of each value. **(2)**

pooled sample variance   The estimate of the population variance used in the calculation of the t-statistic in the t-test for two independent samples. **(10)**

population of interest   The group upon which observations of a variable are taken. **(1)**

population parameter   The numerical descriptive measure of a population that we seek to estimate by means of the sample statistic. **(7)**

positively skewed distribution   A distribution in which the few extreme scores fall *above* the bulk of scores. **(4)**

positive relationship   In a bivariate correlation problem, the case in which the bivariate frequency distribution indicates a tendency for low values on one variable to occur with low values on the other, and for high values on one variable to occur with high values on the other. **(3)**

power of a statistical test   The probability of correctly rejecting the null hypothesis. **(6)**

prediction line   Also called the "least squares" line, a straight line from which an estimate or predicted value is made for a subject's score on the criterion variable, from that subject's score on the predictor variable. **(5)**

predictor variable   In a linear regression problem, the variable whose values are used to predict, or estimate, the values on the criterion variable. **(5)**

probability   The chance that a given event will occur. **(6)**

probability distribution   A list of the values of a variable along with the probability with which each value might occur. **(6)**

pseudo-experimental problem   A bivariate problem in which two or more independent groups are compared on some variable of interest. **(1)**

random selection   The selection of subjects such that each subject in the population of interest is equally likely to be chosen. **(1)**

range   The distance between the two most extreme scores in a distribution. **(4)**

rank-order variable   A variable whose values are ranks. **(1)**

ratio scale variable   A variable whose values are numbers that indicate the amount of the variable present. There is a true zero point on the ratio scale that represents the absence of the variable. **(1)**

real class limits   For a continuous variable, the points on the scale that divide one score interval from another. **(2)**

reciprocal causation   In a bivariate correlation problem, the case in which either variable can influence the outcome of the other variable. **(3)**

regression equation   A linear formula used in predicting values on one variable given a value on the other. **(5)**

rejection rule   A set of outcomes that is considered unlikely under the null hypothesis. **(6)**

relative frequency   The frequency of a value divided by the total number of observations. **(2)**

relative frequency distribution   A chart showing the list of possible values and the relative frequency with which each value occurs in the data set. **(2)**

497

reliability   The consistency, or repeatability, of the results of a particular measurement process. **(1)**

sample standard deviation   An unbiased estimate of the population standard deviation, based on the sample data. **(8)**

sample statistic   An index computed on a sample, used to draw an inference about a population parameter. **(7)**

sampling error   The standard deviation of a sampling distribution of a sample statistic. **(7)**

scatterdiagram   A graph used to represent a bivariate frequency distribution in which both variables are interval scale variables. **(3)**

Scheffé contrast   A linear comparison between two or more means in which the coefficients of the means add up to zero. **(10)**

scientific method   A three-step inference-making procedure in which inferences (i.e., decisions or predictions) are made about a population on the basis of sample data. **(1)**

skewed distribution   A distribution in which the bulk of the scores fall in one area, with a few scores lying far away in one direction. **(4)**

smooth curve   A graph that can be used in place of a histogram or frequency polygon to emphasize the continuous nature of the variable under observation. **(2)**

Spearman rank-order correlation coefficient   A measure of relationship for two rank-ordered variables. **(5)**

standard deviation   An approximation of the average distance by which scores in a distribution differ in either direction from the mean. **(4)**

standard error of estimate   The estimated amount by which the actual Y-values differ on the average from the predicted Y-values. **(5)**

standard normal distribution   A normal distribution in which all of the scores have been standardized, i.e., converted to z-scores. **(7)**

standard score   The distance by which a score differs from the mean, measured in standard deviations. **(4)**

statistic   The sample value that is used to estimate a parameter. **(1)**

step function   A graph used to indicate the relative frequency distribution of a discrete interval variable. **(2)**

symmetric distribution   A distribution with a shape that can be divided into two identical parts. **(4)**

test for significance of Spearman correlation   A statistical test used to test the significance of the relationship between two rank-order variables. **(9)**

test of significance for Pearson correlation   A statistical test used to test the significance of the relationship between two interval variables. **(9)**

t-**distribution**   Any one of a family of probability distributions employed in place of the normal curve model when the population standard deviation is unknown. **(8)**

tree diagram   A method of depicting all possible outcomes on n trials of a binomial experiment. **(6)**

t-**statistic**   A statistic used in place of the z-statistic when the population standard deviation is unknown. **(8)**

498

t-test for two dependent samples   A statistical test used to test hypotheses regarding the mean of the differences between two sets of dependent measures with respect to an interval scale variable. **(11)**

t-test for two independent samples   A statistical test used to test hypotheses about population means, employing the means of two independently selected samples with respect to an interval scale variable. **(10)**

two-tailed hypothesis test   A test based on a rejection rule with outcomes in both tails of the sampling distribution. **(6)**

Type I error   The rejection of a true null hypothesis. **(6)**

Type II error   The retention of a false null hypothesis. **(6)**

unbiased estimator   A statistic that equals the population parameter in the long run, i.e., on the average. **(7)**

univariate problem   A research problem in which there is one variable of interest. **(1)**

unrelated variables   On a scatterdiagram, the case in which the points representing values on the variables form a circular pattern rather than a straight line. **(3)**

validity   The extent to which a measurement process measures what we intended it to. **(1)**

variable   The name of a characteristic with a set of associated values. **(1)**

Wilcoxon signed-ranks test   A statistical test used to test hypotheses regarding two sets of independent measures with respect to a rank-order variable. **(11)**

Wilcoxon *t*-statistic   The statistic employed in the Wilcoxon signed-ranks test. **(11)**

χ^2 **distribution**   A family of distributions to which we refer when determining critical values in the various chi-square tests. **(8)**

χ^2 **one-variable test**   A statistical test for population proportions involving a single categorical variable with more than two values. **(8)**

z-**score**   A standard score expressing the number of standard deviations by which a score lies above or below the mean. **(4)**

Appendix C
Tables _____

TABLE C.1

Three-Place Tables of the Binomial Distribution

n	r	.01	.05	.10	.20	.30	.40	p .50	.60	.70	.80	.90	.95	.99	x
2	0	980	902	810	640	490	360	250	160	090	040	010	002	0+	0
	1	020	095	180	320	420	480	500	480	420	320	180	095	020	1
	2	0+	002	010	040	090	160	250	360	490	640	810	902	980	2
3	0	970	857	729	512	343	216	125	064	027	008	001	0+	0+	0
	1	029	135	243	384	441	432	375	288	189	096	027	007	0+	1
	2	0+	007	027	096	189	288	375	432	441	384	243	135	029	2
	3	0+	0+	001	008	027	064	125	216	343	512	729	857	970	3
4	0	961	815	656	410	240	130	062	026	008	002	0+	0+	0+	0
	1	039	171	292	410	412	346	250	154	076	026	004	0+	0+	1
	2	001	014	049	154	265	346	375	346	265	154	049	014	001	2
	3	0+	0+	004	026	076	154	250	346	412	410	292	171	039	3
	4	0+	0+	0+	002	008	026	062	130	240	410	656	815	961	4
5	0	951	774	590	328	168	078	031	010	002	0+	0+	0+	0+	0
	1	048	204	328	410	360	259	156	077	028	006	0+	0+	0+	1
	2	001	021	073	205	309	346	312	230	132	051	008	001	0+	2
	3	0+	001	008	051	132	230	312	346	309	205	073	021	001	3
	4	0+	0+	0+	006	028	077	156	259	360	410	328	204	048	4
	5	0+	0+	0+	0+	002	010	031	078	168	328	590	774	951	5
6	0	941	735	531	262	118	047	016	004	001	0+	0+	0+	0+	0
	1	057	232	354	393	303	187	094	037	010	002	0+	0+	0+	1
	2	001	031	098	246	324	311	234	138	060	015	001	0+	0+	2
	3	0+	002	015	082	185	276	312	276	185	082	015	002	0+	3
	4	0+	0+	001	015	060	138	234	311	324	246	098	031	001	4
	5	0+	0+	0+	002	010	037	094	187	303	393	354	232	057	5
	6	0+	0+	0+	0+	001	004	016	047	118	262	531	735	941	6
7	0	932	698	478	210	082	028	008	002	0+	0+	0+	0+	0+	0
	1	066	257	372	367	247	131	055	017	004	0+	0+	0+	0+	1
	2	002	041	124	275	318	261	164	077	025	004	0+	0+	0+	2
	3	0+	004	023	115	227	290	273	194	097	029	003	0+	0+	3
	4	0+	0+	003	029	097	194	273	290	227	115	023	004	0+	4
	5	0+	0+	0+	004	025	077	164	261	318	275	124	041	002	5
	6	0+	0+	0+	0+	004	017	055	131	247	367	372	257	066	6
	7	0+	0+	0+	0+	0+	002	008	028	082	210	478	698	932	7
8	0	923	663	430	168	058	017	004	001	0+	0+	0+	0+	0+	0
	1	075	279	383	336	198	090	031	008	001	0+	0+	0+	0+	1
	2	003	051	149	294	296	209	109	041	010	001	0+	0+	0+	2
	3	0+	005	033	147	254	279	219	124	047	009	0+	0+	0+	3
	4	0+	0+	005	046	136	232	273	232	136	046	005	0+	0+	4
	5	0+	0+	0+	009	047	124	219	279	254	147	033	005	0+	5
	6	0+	0+	0+	001	010	041	109	209	296	294	149	051	003	6
	7	0+	0+	0+	0+	001	008	031	090	198	336	383	279	075	7
	8	0+	0+	0+	0+	0+	001	004	017	058	168	430	663	923	8

501

TABLE C.1 (continued)

n	r	.01	.05	.10	.20	.30	.40	p .50	.60	.70	.80	.90	.95	.99	x
9	0	914	630	387	134	040	010	002	0+	0+	0+	0+	0+	0+	0
	1	083	299	387	302	156	060	018	004	0+	0+	0+	0+	0+	1
	2	003	063	172	302	267	161	070	021	004	0+	0+	0+	0+	2
	3	0+	008	045	176	267	251	164	074	021	003	0+	0+	0+	3
	4	0+	001	007	066	172	251	246	167	074	017	001	0+	0+	4
	5	0+	0+	001	017	074	167	246	251	172	066	007	001	0+	5
	6	0+	0+	0+	003	021	074	164	251	267	176	045	008	0+	6
	7	0+	0+	0+	0+	004	021	070	161	267	302	172	063	003	7
	8	0+	0+	0+	0+	0+	004	018	060	156	302	387	299	083	8
	9	0+	0+	0+	0+	0+	0+	002	010	040	134	387	630	914	9
10	0	904	599	349	107	028	006	001	0+	0+	0+	0+	0+	0+	0
	1	091	315	387	268	121	040	010	002	0+	0+	0+	0+	0+	1
	2	004	075	194	302	233	121	044	011	001	0+	0+	0+	0+	2
	3	0+	010	057	201	267	215	117	042	009	001	0+	0+	0+	3
	4	0+	001	011	088	200	251	205	111	037	006	0+	0+	0+	4
	5	0+	0+	001	026	103	201	246	201	103	026	001	0+	0+	5
	6	0+	0+	0+	006	037	111	205	251	200	088	011	001	0+	6
	7	0+	0+	0+	001	009	042	117	215	267	201	057	010	0+	7
	8	0+	0+	0+	0+	001	011	044	121	233	302	194	075	004	8
	9	0+	0+	0+	0+	0+	002	010	040	121	268	387	315	091	9
	10	0+	0+	0+	0+	0+	0+	001	006	028	107	349	599	904	10
11	0	895	569	314	086	020	004	0+	0+	0+	0+	0+	0+	0+	0
	1	099	329	384	236	093	027	005	001	0+	0+	0+	0+	0+	1
	2	005	087	213	295	200	089	027	005	001	0+	0+	0+	0+	2
	3	0+	014	071	221	257	177	081	023	004	0+	0+	0+	0+	3
	4	0+	001	016	111	220	236	161	070	017	002	0+	0+	0+	4
	5	0+	0+	002	039	132	221	226	147	057	010	0+	0+	0+	5
	6	0+	0+	0+	010	057	147	226	221	132	039	002	0+	0+	6
	7	0+	0+	0+	002	017	070	161	236	220	111	016	001	0+	7
	8	0+	0+	0+	0+	004	023	081	177	257	221	071	014	0+	8
	9	0+	0+	0+	0+	001	005	027	089	200	295	213	087	005	9
	10	0+	0+	0+	0+	0+	001	005	027	093	236	384	329	099	10
	11	0+	0+	0+	0+	0+	0+	0+	004	020	086	314	569	895	11
12	0	886	540	282	069	014	002	0+	0+	0+	0+	0+	0+	0+	0
	1	107	341	377	206	071	017	003	0+	0+	0+	0+	0+	0+	1
	2	006	099	230	283	168	064	016	002	0+	0+	0+	0+	0+	2
	3	0+	017	085	236	240	142	054	012	001	0+	0+	0+	0+	3
	4	0+	002	021	133	231	213	121	042	008	001	0+	0+	0+	4
	5	0+	0+	004	053	158	227	193	101	029	003	0+	0+	0+	5
	6	0+	0+	0+	016	079	177	226	177	079	016	0+	0+	0+	6
	7	0+	0+	0+	003	029	101	193	227	158	053	004	0+	0+	7
	8	0+	0+	0+	001	008	042	121	213	231	133	021	002	0+	8
	9	0+	0+	0+	0+	001	012	054	142	240	236	085	017	0+	9

TABLE C.1 (continued)

n	r	.01	.05	.10	.20	.30	.40	p .50	.60	.70	.80	.90	.95	.99	x
12	10	0+	0+	0+	0+	0+	002	016	064	168	283	230	099	006	10
	11	0+	0+	0+	0+	0+	0+	003	017	071	206	377	341	107	11
	12	0+	0+	0+	0+	0+	0+	0+	002	014	069	282	540	886	12
13	0	878	513	254	055	010	001	0+	0+	0+	0+	0+	0+	0+	0
	1	115	351	367	179	054	011	002	0+	0+	0+	0+	0+	0+	1
	2	007	111	245	268	139	045	010	001	0+	0+	0+	0+	0+	2
	3	0+	021	100	246	218	111	035	006	001	0+	0+	0+	0+	3
	4	0+	003	028	154	234	184	087	024	003	0+	0+	0+	0+	4
	5	0+	0+	006	069	180	221	157	066	014	001	0+	0+	0+	5
	6	0+	0+	001	023	103	197	209	131	044	006	0+	0+	0+	6
	7	0+	0+	0+	006	044	131	209	197	103	023	001	0+	0+	7
	8	0+	0+	0+	001	014	066	157	221	180	069	006	0+	0+	8
	9	0+	0+	0+	0+	003	024	087	184	234	154	028	003	0+	9
	10	0+	0+	0+	0+	001	006	035	111	218	246	100	021	0+	10
	11	0+	0+	0+	0+	0+	001	010	045	139	268	245	111	007	11
	12	0+	0+	0+	0+	0+	0+	002	011	054	179	367	351	115	12
	13	0+	0+	0+	0+	0+	0+	0+	001	010	055	254	513	878	13
14	0	869	488	229	044	007	001	0+	0+	0+	0+	0+	0+	0+	0
	1	123	359	356	154	041	007	001	0+	0+	0+	0+	0+	0+	1
	2	008	123	257	250	113	032	006	001	0+	0+	0+	0+	0+	2
	3	0+	026	114	250	194	085	022	003	0+	0+	0+	0+	0+	3
	4	0+	004	035	172	229	155	061	014	001	0+	0+	0+	0+	4
	5	0+	0+	008	086	196	207	122	041	007	0+	0+	0+	0+	5
	6	0+	0+	001	032	126	207	183	092	023	002	0+	0+	0+	6
	7	0+	0+	0+	009	062	157	209	157	062	009	0+	0+	0+	7
	8	0+	0+	0+	002	023	092	183	207	126	032	001	0+	0+	8
	9	0+	0+	0+	0+	007	041	122	207	196	086	008	0+	0+	9
	10	0+	0+	0+	0+	001	014	061	155	229	172	035	004	0+	10
	11	0+	0+	0+	0+	0+	003	022	085	194	250	114	026	0+	11
	12	0+	0+	0+	0+	0+	001	006	032	113	250	257	123	008	12
	13	0+	0+	0+	0+	0+	0+	001	007	041	154	356	359	123	13
	14	0+	0+	0+	0+	0+	0+	0+	001	007	044	229	488	869	14
15	0	860	463	206	035	005	0+	0+	0+	0+	0+	0+	0+	0+	0
	1	130	366	343	132	031	005	0+	0+	0+	0+	0+	0+	0+	1
	2	009	135	267	231	092	022	003	0+	0+	0+	0+	0+	0+	2
	3	0+	031	129	250	170	063	014	002	0+	0+	0+	0+	0+	3
	4	0+	005	043	188	219	127	042	007	001	0+	0+	0+	0+	4
	5	0+	001	010	103	206	186	092	024	003	0+	0+	0+	0+	5
	6	0+	0+	002	043	147	207	153	061	012	001	0+	0+	0+	6
	7	0+	0+	0+	014	081	177	196	118	035	003	0+	0+	0+	7
	8	0+	0+	0+	003	035	118	196	177	081	014	0+	0+	0+	8
	9	0+	0+	0+	001	012	061	153	207	147	043	002	0+	0+	9

TABLE C.1 (continued)

n	r	.01	.05	.10	.20	.30	.40	p .50	.60	.70	.80	.90	.95	.99	x
15	10	0+	0+	0+	0+	003	024	092	186	206	103	010	001	0+	10
	11	0+	0+	0+	0+	001	007	042	127	219	188	043	005	0+	11
	12	0+	0+	0+	0+	0+	002	014	063	170	250	129	031	0+	12
	13	0+	0+	0+	0+	0+	0+	003	022	092	231	267	135	009	13
	14	0+	0+	0+	0+	0+	0+	0+	005	031	132	343	366	130	14
	15	0+	0+	0+	0+	0+	0+	0+	0+	005	035	206	463	860	15
16	0	851	440	185	028	003	0+	0+	0+	0+	0+	0+	0+	0+	0
	1	138	371	329	113	023	003	0+	0+	0+	0+	0+	0+	0+	1
	2	010	146	275	211	073	015	002	0+	0+	0+	0+	0+	0+	2
	3	0+	036	142	246	146	047	009	001	0+	0+	0+	0+	0+	3
	4	0+	006	051	200	204	101	028	004	0+	0+	0+	0+	0+	4
	5	0+	001	014	120	210	162	067	014	001	0+	0+	0+	0+	5
	6	0+	0+	003	055	165	198	122	039	006	0+	0+	0+	0+	6
	7	0+	0+	0+	020	101	189	175	084	019	001	0+	0+	0+	7
	8	0+	0+	0+	006	049	142	196	142	049	006	0+	0+	0+	8
	9	0+	0+	0+	001	019	084	175	189	101	020	0+	0+	0+	9
	10	0+	0+	0+	0+	006	039	122	198	165	055	003	0+	0+	10
	11	0+	0+	0+	0+	001	014	067	162	210	120	014	001	0+	11
	12	0+	0+	0+	0+	0+	004	028	101	204	200	051	006	0+	12
	13	0+	0+	0+	0+	0+	001	009	047	146	246	142	036	0+	13
	14	0+	0+	0+	0+	0+	0+	002	015	073	211	275	146	010	14
	15	0+	0+	0+	0+	0+	0+	0+	003	023	113	329	371	138	15
	16	0+	0+	0+	0+	0+	0+	0+	0+	003	028	185	440	851	16
17	0	843	418	167	023	002	0+	0+	0+	0+	0+	0+	0+	0+	0
	1	145	374	315	096	017	002	0+	0+	0+	0+	0+	0+	0+	1
	2	012	158	280	191	058	010	001	0+	0+	0+	0+	0+	0+	2
	3	001	041	156	239	125	034	005	0+	0+	0+	0+	0+	0+	3
	4	0+	008	060	209	187	080	018	002	0+	0+	0+	0+	0+	4
	5	0+	001	017	136	208	138	047	008	001	0+	0+	0+	0+	5
	6	0+	0+	004	068	178	184	094	024	003	0+	0+	0+	0+	6
	7	0+	0+	001	027	120	193	148	057	009	0+	0+	0+	0+	7
	8	0+	0+	0+	008	064	161	185	107	028	002	0+	0+	0+	8
	9	0+	0+	0+	002	028	107	185	161	064	008	0+	0+	0+	9
	10	0+	0+	0+	0+	009	057	148	193	120	027	001	0+	0+	10
	11	0+	0+	0+	0+	003	024	094	184	178	068	004	0+	0+	11
	12	0+	0+	0+	0+	001	008	047	138	208	136	017	001	0+	12
	13	0+	0+	0+	0+	0+	002	018	080	187	209	060	008	0+	13
	14	0+	0+	0+	0+	0+	0+	005	034	125	239	156	041	001	14
	15	0+	0+	0+	0+	0+	0+	001	010	058	191	280	158	012	15
	16	0+	0+	0+	0+	0+	0+	0+	002	017	096	315	374	145	16
	17	0+	0+	0+	0+	0+	0+	0+	0+	002	023	167	418	843	17

TABLE C.1 (continued)

n	r	.01	.05	.10	.20	.30	.40	p .50	.60	.70	.80	.90	.95	.99	x
18	0	835	397	150	018	002	0+	0+	0+	0+	0+	0+	0+	0+	0
	1	152	376	300	081	013	001	0+	0+	0+	0+	0+	0+	0+	1
	2	013	168	284	172	046	007	001	0+	0+	0+	0+	0+	0+	2
	3	001	047	168	230	105	025	003	0+	0+	0+	0+	0+	0+	3
	4	0+	009	070	215	168	061	012	001	0+	0+	0+	0+	0+	4
	5	0+	001	022	151	202	115	033	004	0+	0+	0+	0+	0+	5
	6	0+	0+	005	082	187	166	071	015	001	0+	0+	0+	0+	6
	7	0+	0+	001	035	138	189	121	037	005	0+	0+	0+	0+	7
	8	0+	0+	0+	012	081	173	167	077	015	001	0+	0+	0+	8
	9	0+	0+	0+	003	039	128	185	128	039	003	0+	0+	0+	9
	10	0+	0+	0+	001	015	077	167	173	081	012	0+	0+	0+	10
	11	0+	0+	0+	0+	005	037	121	189	138	035	001	0+	0+	11
	12	0+	0+	0+	0+	001	015	071	166	187	082	005	0+	0+	12
	13	0+	0+	0+	0+	0+	004	033	115	202	151	022	001	0+	13
	14	0+	0+	0+	0+	0+	001	012	061	168	215	070	009	0+	14
	15	0+	0+	0+	0+	0+	0+	003	025	105	230	168	047	001	15
	16	0+	0+	0+	0+	0+	0+	001	007	046	172	284	168	013	16
	17	0+	0+	0+	0+	0+	0+	0+	001	013	081	300	376	152	17
	18	0+	0+	0+	0+	0+	0+	0+	0+	002	018	150	397	835	18
19	0	826	377	135	014	001	0+	0+	0+	0+	0+	0+	0+	0+	0
	1	159	377	285	068	009	001	0+	0+	0+	0+	0+	0+	0+	1
	2	014	179	285	154	036	005	0+	0+	0+	0+	0+	0+	0+	2
	3	001	053	180	218	087	017	002	0+	0+	0+	0+	0+	0+	3
	4	0+	011	080	218	149	047	007	001	0+	0+	0+	0+	0+	4
	5	0+	002	027	164	192	093	022	002	0+	0+	0+	0+	0+	5
	6	0+	0+	007	095	192	145	052	008	001	0+	0+	0+	0+	6
	7	0+	0+	001	044	153	180	096	024	002	0+	0+	0+	0+	7
	8	0+	0+	0+	017	098	180	144	053	008	0+	0+	0+	0+	8
	9	0+	0+	0+	005	051	146	176	098	022	001	0+	0+	0+	9
	10	0+	0+	0+	001	022	098	176	146	051	005	0+	0+	0+	10
	11	0+	0+	0+	0+	008	053	144	180	098	017	0+	0+	0+	11
	12	0+	0+	0+	0+	002	024	096	180	153	044	001	0+	0+	12
	13	0+	0+	0+	0+	001	008	052	145	192	095	007	0+	0+	13
	14	0+	0+	0+	0+	0+	002	022	093	192	164	027	002	0+	14
	15	0+	0+	0+	0+	0+	001	007	047	149	218	080	011	0+	15
	16	0+	0+	0+	0+	0+	0+	002	017	087	218	180	053	001	16
	17	0+	0+	0+	0+	0+	0+	0+	005	036	154	285	179	014	17
	18	0+	0+	0+	0+	0+	0+	0+	001	009	068	285	377	159	18
	19	0+	0+	0+	0+	0+	0+	0+	0+	001	014	135	377	826	19
20	0	818	358	122	012	001	0+	0+	0+	0+	0+	0+	0+	0+	0
	1	165	377	270	058	007	0+	0+	0+	0+	0+	0+	0+	0+	1
	2	016	189	285	137	028	003	0+	0+	0+	0+	0+	0+	0+	2
	3	001	060	190	205	072	012	001	0+	0+	0+	0+	0+	0+	3
	4	0+	013	090	218	130	035	005	0+	0+	0+	0+	0+	0+	4

TABLE C.1 (continued)

n	r	.01	.05	.10	.20	.30	.40	p .50	.60	.70	.80	.90	.95	.99	x
20	5	0+	002	032	175	179	075	015	001	0+	0+	0+	0+	0+	5
	6	0+	0+	009	109	192	124	037	005	0+	0+	0+	0+	0+	6
	7	0+	0+	002	055	164	166	074	015	001	0+	0+	0+	0+	7
	8	0+	0+	0+	022	114	180	120	035	004	0+	0+	0+	0+	8
	9	0+	0+	0+	007	065	160	160	071	012	0+	0+	0+	0+	9
	10	0+	0+	0+	002	031	117	176	117	031	002	0+	0+	0+	10
	11	0+	0+	0+	0+	012	071	160	160	065	007	0+	0+	0+	11
	12	0+	0+	0+	0+	004	035	120	180	114	022	0+	0+	0+	12
	13	0+	0+	0+	0+	001	015	074	166	164	055	002	0+	0+	13
	14	0+	0+	0+	0+	0+	005	037	124	192	109	009	0+	0+	14
	15	0+	0+	0+	0+	0+	001	015	075	179	175	032	002	0+	15
	16	0+	0+	0+	0+	0+	0+	005	035	130	218	090	013	0+	16
	17	0+	0+	0+	0+	0+	0+	001	012	072	205	190	060	001	17
	18	0+	0+	0+	0+	0+	0+	0+	003	028	137	285	189	016	18
	19	0+	0+	0+	0+	0+	0+	0+	0+	007	058	270	377	165	19
	20	0+	0+	0+	0+	0+	0+	0+	0+	001	012	122	358	818	20
21	0	810	341	109	009	001	0+	0+	0+	0+	0+	0+	0+	0+	0
	1	172	376	255	048	005	0+	0+	0+	0+	0+	0+	0+	0+	1
	2	017	198	284	121	022	002	0+	0+	0+	0+	0+	0+	0+	2
	3	001	066	200	192	058	009	001	0+	0+	0+	0+	0+	0+	3
	4	0+	016	100	216	113	026	003	0+	0+	0+	0+	0+	0+	4
	5	0+	003	038	183	164	059	010	001	0+	0+	0+	0+	0+	5
	6	0+	0+	011	122	188	105	026	003	0+	0+	0+	0+	0+	6
	7	0+	0+	003	065	172	149	055	009	0+	0+	0+	0+	0+	7
	8	0+	0+	001	029	129	174	097	023	002	0+	0+	0+	0+	8
	9	0+	0+	0+	010	080	168	140	050	006	0+	0+	0+	0+	9
	10	0+	0+	0+	003	041	134	168	089	018	001	0+	0+	0+	10
	11	0+	0+	0+	001	018	089	168	134	041	003	0+	0+	0+	11
	12	0+	0+	0+	0+	006	050	140	168	080	010	0+	0+	0+	12
	13	0+	0+	0+	0+	002	023	097	174	129	029	001	0+	0+	13
	14	0+	0+	0+	0+	0+	009	055	149	172	065	003	0+	0+	14
	15	0+	0+	0+	0+	0+	003	026	105	188	122	011	0+	0+	15
	16	0+	0+	0+	0+	0+	001	010	059	164	183	038	003	0+	16
	17	0+	0+	0+	0+	0+	0+	003	026	113	216	100	016	0+	17
	18	0+	0+	0+	0+	0+	0+	001	009	058	192	200	066	001	18
	19	0+	0+	0+	0+	0+	0+	0+	002	022	121	284	198	017	19
	20	0+	0+	0+	0+	0+	0+	0+	0+	005	048	255	376	172	20
	21	0+	0+	0+	0+	0+	0+	0+	0+	001	009	109	341	810	21
22	0	802	324	098	007	0+	0+	0+	0+	0+	0+	0+	0+	0+	0
	1	178	375	241	041	004	0+	0+	0+	0+	0+	0+	0+	0+	1
	2	019	207	281	107	017	001	0+	0+	0+	0+	0+	0+	0+	2
	3	001	073	208	178	047	006	0+	0+	0+	0+	0+	0+	0+	3
	4	0+	018	110	211	096	019	002	0+	0+	0+	0+	0+	0+	4

TABLE C.1 (continued)

n	r	.01	.05	.10	.20	.30	.40	p .50	.60	.70	.80	.90	.95	.99	x
22	5	0+	003	044	190	149	046	006	0+	0+	0+	0+	0+	0+	5
	6	0+	001	014	134	181	086	018	001	0+	0+	0+	0+	0+	6
	7	0+	0+	004	077	177	131	041	005	0+	0+	0+	0+	0+	7
	8	0+	0+	001	036	142	164	076	014	001	0+	0+	0+	0+	8
	9	0+	0+	0+	014	095	170	119	034	003	0+	0+	0+	0+	9
	10	0+	0+	0+	005	053	148	154	066	010	0+	0+	0+	0+	10
	11	0+	0+	0+	001	025	107	168	107	025	001	0+	0+	0+	11
	12	0+	0+	0+	0+	010	066	154	148	053	005	0+	0+	0+	12
	13	0+	0+	0+	0+	003	034	119	170	095	014	0+	0+	0+	13
	14	0+	0+	0+	0+	001	014	076	164	142	036	001	0+	0+	14
	15	0+	0+	0+	0+	0+	005	041	131	177	077	004	0+	0+	15
	16	0+	0+	0+	0+	0+	001	018	086	181	134	014	001	0+	16
	17	0+	0+	0+	0+	0+	0+	006	046	149	190	044	003	0+	17
	18	0+	0+	0+	0+	0+	0+	002	019	096	211	110	018	0+	18
	19	0+	0+	0+	0+	0+	0+	0+	006	047	178	208	073	001	19
	20	0+	0+	0+	0+	0+	0+	0+	001	017	107	281	207	019	20
	21	0+	0+	0+	0+	0+	0+	0+	0+	004	041	241	375	178	21
	22	0+	0+	0+	0+	0+	0+	0+	0+	0+	007	098	324	802	22
23	0	794	307	089	006	0+	0+	0+	0+	0+	0+	0+	0+	0+	0
	1	184	372	226	034	003	0+	0+	0+	0+	0+	0+	0+	0+	1
	2	020	215	277	093	013	001	0+	0+	0+	0+	0+	0+	0+	2
	3	001	079	215	163	038	004	0+	0+	0+	0+	0+	0+	0+	3
	4	0+	021	120	204	082	014	001	0+	0+	0+	0+	0+	0+	4
	5	0+	004	051	194	133	035	004	0+	0+	0+	0+	0+	0+	5
	6	0+	001	017	145	171	070	012	001	0+	0+	0+	0+	0+	6
	7	0+	0+	005	088	178	113	029	003	0+	0+	0+	0+	0+	7
	8	0+	0+	001	044	153	151	058	009	0+	0+	0+	0+	0+	8
	9	0+	0+	0+	018	109	168	097	022	002	0+	0+	0+	0+	9
	10	0+	0+	0+	006	065	157	136	046	005	0+	0+	0+	0+	10
	11	0+	0+	0+	002	033	123	161	082	014	0+	0+	0+	0+	11
	12	0+	0+	0+	0+	014	082	161	123	033	002	0+	0+	0+	12
	13	0+	0+	0+	0+	005	046	136	157	065	006	0+	0+	0+	13
	14	0+	0+	0+	0+	002	022	097	168	109	018	0+	0+	0+	14
	15	0+	0+	0+	0+	0+	009	058	151	153	044	001	0+	0+	15
	16	0+	0+	0+	0+	0+	003	029	113	178	088	005	0+	0+	16
	17	0+	0+	0+	0+	0+	001	012	070	171	145	017	001	0+	17
	18	0+	0+	0+	0+	0+	0+	004	035	133	194	051	004	0+	18
	19	0+	0+	0+	0+	0+	0+	001	014	082	204	120	021	0+	19
	20	0+	0+	0+	0+	0+	0+	0+	004	038	163	215	079	001	20
	21	0+	0+	0+	0+	0+	0+	0+	001	013	093	277	215	020	21
	22	0+	0+	0+	0+	0+	0+	0+	0+	003	034	226	372	184	22
	23	0+	0+	0+	0+	0+	0+	0+	0+	0+	006	089	307	794	23

TABLE C.1 (continued)

n	r	.01	.05	.10	.20	.30	.40	p .50	.60	.70	.80	.90	.95	.99	x
24	0	786	292	080	005	0+	0+	0+	0+	0+	0+	0+	0+	0+	0
	1	190	369	213	028	002	0+	0+	0+	0+	0+	0+	0+	0+	1
	2	022	223	272	081	010	001	0+	0+	0+	0+	0+	0+	0+	2
	3	002	086	221	149	031	003	0+	0+	0+	0+	0+	0+	0+	3
	4	0+	024	129	196	069	010	001	0+	0+	0+	0+	0+	0+	4
	5	0+	005	057	196	118	027	003	0+	0+	0+	0+	0+	0+	5
	6	0+	001	020	155	160	056	008	0+	0+	0+	0+	0+	0+	6
	7	0+	0+	006	100	176	096	021	002	0+	0+	0+	0+	0+	7
	8	0+	0+	001	053	160	136	044	005	0+	0+	0+	0+	0+	8
	9	0+	0+	0+	024	122	161	078	014	001	0+	0+	0+	0+	9
	10	0+	0+	0+	009	079	161	117	032	003	0+	0+	0+	0+	10
	11	0+	0+	0+	003	043	137	149	061	008	0+	0+	0+	0+	11
	12	0+	0+	0+	001	020	099	161	099	020	001	0+	0+	0+	12
	13	0+	0+	0+	0+	008	061	149	137	043	003	0+	0+	0+	13
	14	0+	0+	0+	0+	003	032	117	161	079	009	0+	0+	0+	14
	15	0+	0+	0+	0+	001	014	078	161	122	024	0+	0+	0+	15
	16	0+	0+	0+	0+	0+	005	044	136	160	053	001	0+	0+	16
	17	0+	0+	0+	0+	0+	002	021	096	176	100	006	0+	0+	17
	18	0+	0+	0+	0+	0+	0+	008	056	160	155	020	001	0+	18
	19	0+	0+	0+	0+	0+	0+	003	027	118	196	057	005	0+	19
	20	0+	0+	0+	0+	0+	0+	001	010	069	196	129	024	0+	20
	21	0+	0+	0+	0+	0+	0+	0+	003	031	149	221	086	002	21
	22	0+	0+	0+	0+	0+	0+	0+	001	010	081	272	223	022	22
	23	0+	0+	0+	0+	0+	0+	0+	0+	002	028	213	369	190	23
	24	0+	0+	0+	0+	0+	0+	0+	0+	0+	005	080	292	786	24
25	0	778	277	072	004	0+	0+	0+	0+	0+	0+	0+	0+	0+	0
	1	196	365	199	024	001	0+	0+	0+	0+	0+	0+	0+	0+	1
	2	024	231	266	071	007	0+	0+	0+	0+	0+	0+	0+	0+	2
	3	002	093	226	136	024	002	0+	0+	0+	0+	0+	0+	0+	3
	4	0+	027	138	187	057	007	0+	0+	0+	0+	0+	0+	0+	4
	5	0+	006	065	196	103	020	002	0+	0+	0+	0+	0+	0+	5
	6	0+	001	024	163	147	044	005	0+	0+	0+	0+	0+	0+	6
	7	0+	0+	007	111	171	080	014	001	0+	0+	0+	0+	0+	7
	8	0+	0+	002	062	165	120	032	003	0+	0+	0+	0+	0+	8
	9	0+	0+	0+	029	134	151	061	009	0+	0+	0+	0+	0+	9
	10	0+	0+	0+	012	092	161	097	021	001	0+	0+	0+	0+	10
	11	0+	0+	0+	004	054	147	133	043	004	0+	0+	0+	0+	11
	12	0+	0+	0+	001	027	114	155	076	011	0+	0+	0+	0+	12
	13	0+	0+	0+	0+	011	076	155	114	027	001	0+	0+	0+	13
	14	0+	0+	0+	0+	004	043	133	147	054	004	0+	0+	0+	14
	15	0+	0+	0+	0+	001	021	097	161	092	012	0+	0+	0+	15
	16	0+	0+	0+	0+	0+	009	061	151	134	029	0+	0+	0+	16
	17	0+	0+	0+	0+	0+	003	032	120	165	062	002	0+	0+	17
	18	0+	0+	0+	0+	0+	001	014	080	171	111	007	0+	0+	18
	19	0+	0+	0+	0+	0+	0+	005	044	147	163	024	001	0+	19

TABLE C.1 (continued)

n	r	.01	.05	.10	.20	.30	.40	p .50	.60	.70	.80	.90	.95	.99	x
25	20	0+	0+	0+	0+	0+	0+	002	020	103	196	065	006	0+	20
	21	0+	0+	0+	0+	0+	0+	0+	007	057	187	138	027	0+	21
	22	0+	0+	0+	0+	0+	0+	0+	002	024	136	226	093	002	22
	23	0+	0+	0+	0+	0+	0+	0+	0+	007	071	266	231	024	23
	24	0+	0+	0+	0+	0+	0+	0+	0+	001	024	199	365	196	24
	25	0+	0+	0+	0+	0+	0+	0+	0+	0+	004	072	277	778	25

TABLE C.2
Areas and Ordinates of the Standard Normal Curve

(1) z	(2) Area in larger portion	(3) Area in smaller portion	(4) Area from μ to z
0.00	.5000	.5000	.0000
0.01	.5040	.4960	.0040
0.02	.5080	.4920	.0080
0.03	.5120	.4880	.0120
0.04	.5160	.4840	.0160
0.05	.5199	.4801	.0199
0.06	.5239	.4761	.0239
0.07	.5279	.4721	.0279
0.08	.5319	.4681	.0319
0.09	.5359	.4641	.0359
0.10	.5398	.4602	.0398
0.11	.5438	.4562	.0438
0.12	.5478	.4522	.0478
0.13	.5517	.4483	.0517
0.14	.5557	.4443	.0557
0.15	.5596	.4404	.0596
0.16	.5636	.4364	.0636
0.17	.5675	.4325	.0675
0.18	.5714	.4286	.0714
0.19	.5753	.4247	.0753
0.20	.5793	.4207	.0793
0.21	.5832	.4168	.0832
0.22	.5871	.4129	.0871
0.23	.5910	.4090	.0910
0.24	.5948	.4052	.0948
0.25	.5987	.4013	.0987
0.26	.6026	.3974	.1026
0.27	.6064	.3936	.1064
0.28	.6103	.3897	.1103
0.29	.6141	.3859	.1141
0.30	.6179	.3821	.1179
0.31	.6217	.3783	.1217
0.32	.6255	.3745	.1255
0.33	.6293	.3707	.1293
0.34	.6331	.3669	.1331

510

TABLE C.2 (continued)

(1) z	(2) Area in larger portion	(3) Area in smaller portion	(4) Area from μ to z
0.35	.6368	.3632	.1368
0.36	.6406	.3594	.1406
0.37	.6443	.3557	.1443
0.38	.6480	.3520	.1480
0.39	.6517	.3483	.1517
0.40	.6554	.3446	.1554
0.41	.6591	.3409	.1591
0.42	.6628	.3372	.1628
0.43	.6664	.3336	.1664
0.44	.6700	.3300	.1700
0.45	.6736	.3264	.1736
0.46	.6772	.3228	.1772
0.47	.6808	.3192	.1808
0.48	.6844	.3156	.1844
0.49	.6879	.3121	.1879
0.50	.6915	.3085	.1915
0.51	.6950	.3050	.1950
0.52	.6985	.3015	.1985
0.53	.7019	.2981	.2019
0.54	.7054	.2946	.2054
0.55	.7088	.2912	.2088
0.56	.7123	.2877	.2123
0.57	.7157	.2843	.2157
0.58	.7190	.2810	.2190
0.59	.7224	.2776	.2224
0.60	.7257	.2743	.2257
0.61	.7291	.2709	.2291
0.62	.7324	.2676	.2324
0.63	.7357	.2643	.2357
0.64	.7389	.2611	.2389
0.65	.7422	.2578	.2422
0.66	.7454	.2546	.2454
0.67	.7486	.2514	.2486
0.68	.7517	.2483	.2517
0.69	.7549	.2451	.2549

TABLE C.2 (continued)

(1) z	(2) Area in larger portion	(3) Area in smaller portion	(4) Area from μ to z
0.70	.7580	.2420	.2580
0.71	.7611	.2389	.2611
0.72	.7642	.2358	.2642
0.73	.7673	.2327	.2673
0.74	.7704	.2296	.2704
0.75	.7734	.2266	.2734
0.76	.7764	.2236	.2764
0.77	.7794	.2206	.2794
0.78	.7823	.2177	.2823
0.79	.7852	.2148	.2852
0.80	.7881	.2119	.2881
0.81	.7910	.2090	.2910
0.82	.7939	.2061	.2939
0.83	.7967	.2033	.2967
0.84	.7995	.2005	.2995
0.85	.8023	.1977	.3023
0.86	.8051	.1949	.3051
0.87	.8078	.1922	.3078
0.88	.8106	.1894	.3106
0.89	.8133	.1867	.3133
0.90	.8159	.1841	.3159
0.91	.8186	.1814	.3186
0.92	.8212	.1788	.3212
0.93	.8238	.1762	.3238
0.94	.8264	.1736	.3264
0.95	.8289	.1711	.3289
0.96	.8315	.1685	.3315
0.97	.8340	.1660	.3340
0.98	.8365	.1635	.3365
0.99	.8389	.1611	.3389
1.00	.8413	.1587	.3413
1.01	.8438	.1562	.3438
1.02	.8461	.1539	.3461
1.03	.8485	.1515	.3485
1.04	.8508	.1492	.3508

Appendix C

TABLE C.2 (continued)

(1) z	(2) Area in larger portion	(3) Area in smaller portion	(4) Area from μ to z
1.05	.8531	.1469	.3531
1.06	.8554	.1446	.3554
1.07	.8577	.1423	.3577
1.08	.8599	.1401	.3599
1.09	.8621	.1379	.3621
1.10	.8643	.1357	.3643
1.11	.8665	.1335	.3665
1.12	.8686	.1314	.3686
1.13	.8708	.1292	.3708
1.14	.8729	.1271	.3729
1.15	.8749	.1251	.3749
1.16	.8770	.1230	.3770
1.17	.8790	.1210	.3790
1.18	.8810	.1190	.3810
1.19	.8830	.1170	.3830
1.20	.8849	.1151	.3849
1.21	.8869	.1131	.3869
1.22	.8888	.1112	.3888
1.23	.8907	.1093	.3907
1.24	.8925	.1075	.3925
1.25	.8944	.1056	.3944
1.26	.8962	.1038	.3962
1.27	.8980	.1020	.3980
1.28	.8997	.1003	.3997
1.29	.9015	.0985	.4015
1.30	.9032	.0968	.4032
1.31	.9049	.0951	.4049
1.32	.9066	.0934	.4066
1.33	.9082	.0918	.4082
1.34	.9099	.0901	.4099
1.35	.9115	.0885	.4115
1.36	.9131	.0869	.4131
1.37	.9147	.0853	.4147
1.38	.9162	.0838	.4162
1.39	.9177	.0823	.4177

TABLE C.2 (continued)

(1) z	(2) Area in larger portion	(3) Area in smaller portion	(4) Area from μ to z
1.40	.9192	.0808	.4192
1.41	.9207	.0793	.4207
1.42	.9222	.0778	.4222
1.43	.9236	.0764	.4236
1.44	.9251	.0749	.4251
1.45	.9265	.0735	.4265
1.46	.9279	.0721	.4279
1.47	.9292	.0708	.4292
1.48	.9306	.0694	.4306
1.49	.9319	.0681	.4319
1.50	.9332	.0668	.4332
1.51	.9345	.0655	.4345
1.52	.9357	.0643	.4357
1.53	.9370	.0630	.4370
1.54	.9382	.0618	.4382
1.55	.9394	.0606	.4394
1.56	.9406	.0594	.4406
1.57	.9418	.0582	.4418
1.58	.9429	.0571	.4429
1.59	.9441	.0559	.4441
1.60	.9452	.0548	.4452
1.61	.9463	.0537	.4463
1.62	.9474	.0526	.4474
1.63	.9484	.0516	.4484
1.64	.9495	.0505	.4495
1.65	.9505	.0495	.4505
1.66	.9515	.0485	.4515
1.67	.9525	.0475	.4525
1.68	.9535	.0465	.4535
1.69	.9545	.0455	.4545
1.70	.9554	.0446	.4554
1.71	.9564	.0436	.4564
1.72	.9573	.0427	.4573
1.73	.9582	.0418	.4582
1.74	.9591	.0409	.4591

514

TABLE C.2 (continued)

(1) z	(2) Area in larger portion	(3) Area in smaller portion	(4) Area from μ to z
1.75	.9599	.0401	.4599
1.76	.9608	.0392	.4608
1.77	.9616	.0384	.4616
1.78	.9625	.0375	.4625
1.79	.9633	.0367	.4633
1.80	.9641	.0359	.4641
1.81	.9649	.0351	.4649
1.82	.9656	.0344	.4656
1.83	.9664	.0336	.4664
1.84	.9671	.0329	.4671
1.85	.9678	.0322	.4678
1.86	.9686	.0314	.4686
1.87	.9693	.0307	.4693
1.88	.9699	.0301	.4699
1.89	.9706	.0294	.4706
1.90	.9713	.0287	.4713
1.91	.9719	.0281	.4719
1.92	.9726	.0274	.4726
1.93	.9732	.0268	.4732
1.94	.9738	.0262	.4738
1.95	.9744	.0256	.4744
1.96	.9750	.0250	.4750
1.97	.9756	.0244	.4756
1.98	.9761	.0239	.4761
1.99	.9767	.0233	.4767
2.00	.9772	.0228	.4772
2.01	.9778	.0222	.4778
2.02	.9783	.0217	.4783
2.03	.9788	.0212	.4788
2.04	.9793	.0207	.4793
2.05	.9798	.0202	.4798
2.06	.9803	.0197	.4803
2.07	.9808	.0192	.4808
2.08	.9812	.0188	.4812
2.09	.9817	.0183	.4817

TABLE C.2 (continued)

(1) z	(2) Area in larger portion	(3) Area in smaller portion	(4) Area from μ to z
2.10	.9821	.0179	.4821
2.11	.9826	.0174	.4826
2.12	.9830	.0170	.4830
2.13	.9834	.0166	.4834
2.14	.9838	.0162	.4838
2.15	.9842	.0158	.4842
2.16	.9846	.0154	.4846
2.17	.9850	.0150	.4850
2.18	.9854	.0146	.4854
2.19	.9857	.0143	.4857
2.20	.9861	.0139	.4861
2.21	.9864	.0136	.4864
2.22	.9868	.0132	.4868
2.23	.9871	.0129	.4871
2.24	.9875	.0125	.4875
2.25	.9878	.0122	.4878
2.26	.9881	.0119	.4881
2.27	.9884	.0116	.4884
2.28	.9887	.0113	.4887
2.29	.9890	.0110	.4890
2.30	.9893	.0107	.4893
2.31	.9896	.0104	.4896
2.32	.9898	.0102	.4898
2.33	.9901	.0099	.4901
2.34	.9904	.0096	.4904
2.35	.9906	.0094	.4906
2.36	.9909	.0091	.4909
2.37	.9911	.0089	.4911
2.38	.9913	.0087	.4913
2.39	.9916	.0084	.4916
2.40	.9918	.0082	.4918
2.41	.9920	.0080	.4920
2.42	.9922	.0078	.4922
2.43	.9925	.0075	.4925
2.44	.9927	.0073	.4927

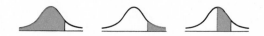

TABLE C.2 (continued)

(1) z	(2) Area in larger portion	(3) Area in smaller portion	(4) Area from μ to z
2.45	.9929	.0071	.4929
2.46	.9931	.0069	.4931
2.47	.9932	.0068	.4932
2.48	.9934	.0066	.4934
2.49	.9936	.0064	.4936
2.50	.9938	.0062	.4938
2.51	.9940	.0060	.4940
2.52	.9941	.0059	.4941
2.53	.9943	.0057	.4943
2.54	.9945	.0055	.4945
2.55	.9946	.0054	.4946
2.56	.9948	.0052	.4948
2.57	.9949	.0051	.4949
2.58	.9951	.0049	.4951
2.59	.9952	.0048	.4952
2.60	.9953	.0047	.4953
2.61	.9955	.0045	.4955
2.62	.9956	.0044	.4956
2.63	.9957	.0043	.4957
2.64	.9959	.0041	.4959
2.65	.9960	.0040	.4960
2.66	.9961	.0039	.4961
2.67	.9962	.0038	.4962
2.68	.9963	.0037	.4963
2.69	.9964	.0036	.4964
2.70	.9965	.0035	.4965
2.71	.9966	.0034	.4966
2.72	.9967	.0033	.4967
2.73	.9968	.0032	.4968
2.74	.9969	.0031	.4969
2.75	.9970	.0030	.4970
2.76	.9971	.0029	.4971
2.77	.9972	.0028	.4972
2.78	.9973	.0027	.4973
2.79	.9974	.0026	.4974

TABLE C.2 (continued)

(1) z	(2) Area in larger portion	(3) Area in smaller portion	(4) Area from μ to z
2.80	.9974	.0026	.4974
2.81	.9975	.0025	.4975
2.82	.9976	.0024	.4976
2.83	.9977	.0023	.4977
2.84	.9977	.0023	.4977
2.85	.9978	.0022	.4978
2.86	.9979	.0021	.4979
2.87	.9979	.0021	.4979
2.88	.9980	.0020	.4980
2.89	.9981	.0019	.4981
2.90	.9981	.0019	.4981
2.91	.9982	.0018	.4982
2.92	.9982	.0018	.4982
2.93	.9983	.0017	.4983
2.94	.9984	.0016	.4984
2.95	.9984	.0016	.4984
2.96	.9985	.0015	.4985
2.97	.9985	.0015	.4985
2.98	.9986	.0014	.4986
2.99	.9986	.0014	.4986
3.00	.9987	.0013	.4987
3.01	.9987	.0013	.4987
3.02	.9987	.0013	.4987
3.03	.9988	.0012	.4988
3.04	.9988	.0012	.4988
3.05	.9989	.0011	.4989
3.06	.9989	.0011	.4989
3.07	.9989	.0011	.4989
3.08	.9990	.0010	.4990
3.09	.9990	.0010	.4990
3.10	.9990	.0010	.4990
3.11	.9991	.0009	.4991
3.12	.9991	.0009	.4991
3.13	.9991	.0009	.4991
3.14	.9992	.0008	.4992

518

TABLE C.2 (continued)

(1) z	(2) Area in larger portion	(3) Area in smaller portion	(4) Area from μ to z
3.15	.9992	.0008	.4992
3.16	.9992	.0008	.4992
3.17	.9992	.0008	.4992
3.18	.9993	.0007	.4993
3.19	.9993	.0007	.4993
3.20	.9993	.0007	.4993
3.21	.9993	.0007	.4993
3.22	.9994	.0006	.4994
3.23	.9994	.0006	.4994
3.24	.9994	.0006	.4994
3.30	.9995	.0005	.4995
3.40	.9997	.0003	.4997
3.50	.9998	.0002	.4998
3.60	.9998	.0002	.4998
3.70	.9999	.0001	.4999

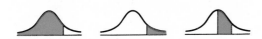

TABLE C.3
Critical Values of the Chi-Square Distribution

.10	.05	.025	.02	.01	.005	.001	α/df
2.7	3.8	5.0	5.4	6.6	7.9	10.8	1
4.6	6.0	7.4	7.8	9.2	10.6	13.8	2
6.3	7.8	9.4	9.8	11.3	12.8	16.3	3
7.8	9.5	11.1	11.7	13.3	14.9	18.5	4
9.2	11.1	12.8	13.4	15.1	16.7	20.5	5
10.6	12.6	14.4	15.0	16.8	18.5	22.5	6
12.0	14.1	16.0	16.6	18.5	20.3	24.3	7
13.4	15.5	17.5	18.2	20.1	22.0	26.1	8
14.7	16.9	19.0	19.7	21.7	23.6	27.9	9
16.0	18.3	20.5	21.2	23.2	25.2	29.6	10
17.3	19.7	21.9	22.6	24.7	26.8	31.3	11
18.5	21.0	23.3	24.1	26.2	28.3	32.9	12
19.8	22.4	24.7	25.5	27.7	29.8	34.5	13
21.1	23.7	26.1	26.9	29.1	31.3	36.1	14
22.3	25.0	27.5	28.3	30.6	32.8	37.7	15
23.5	26.3	28.8	29.6	32.0	34.3	39.3	16
24.8	27.6	30.2	31.0	33.4	35.7	40.8	17
26.0	28.9	31.5	32.3	34.8	37.2	42.3	18
27.2	30.1	32.9	33.7	36.2	38.6	43.8	19
28.4	31.4	34.2	35.0	37.6	40.0	45.3	20
29.6	32.7	35.5	36.3	38.9	41.4	46.8	21
30.8	33.9	36.8	37.7	40.3	42.8	48.3	22
32.0	35.2	38.1	39.0	41.6	44.2	49.7	23
33.2	36.4	39.4	40.3	43.0	45.6	51.2	24
34.4	37.7	40.6	41.6	44.3	46.9	52.6	25
35.6	38.9	41.9	42.9	45.6	48.3	54.0	26
36.7	40.1	43.2	44.1	47.0	49.6	55.5	27
37.9	41.3	44.5	45.4	48.3	51.0	56.9	28
39.1	42.6	45.7	46.7	49.6	52.3	58.3	29
40.3	43.8	47.0	48.0	50.9	53.7	59.7	30
51.8	55.8	59.3	60.4	63.7	66.8	73.5	40
74.4	79.1	83.3	84.6	88.4	92.0	99.7	60
118.5	124.3	129.6	131.1	135.8	140.2	149.5	100

TABLE C.4
Critical Values of D in the Kolmogorov-Smirnov One-Sample Test

Sample size (n)	Level of significance for D				
	.20	.15	.10	.05	.01
1	.900	.925	.950	.975	.995
2	.684	.726	.776	.842	.929
3	.565	.597	.642	.708	.828
4	.494	.525	.564	.624	.733
5	.446	.474	.510	.565	.669
6	.410	.436	.470	.521	.618
7	.381	.405	.438	.486	.577
8	.358	.381	.411	.457	.543
9	.339	.360	.388	.432	.514
10	.322	.342	.368	.410	.490
11	.307	.326	.352	.391	.468
12	.295	.313	.338	.375	.450
13	.284	.302	.325	.361	.433
14	.274	.292	.314	.349	.418
15	.266	.283	.304	.338	.404
16	.258	.274	.295	.328	.392
17	.250	.266	.286	.318	.381
18	.244	.259	.278	.309	.371
19	.237	.252	.272	.301	.363
20	.231	.246	.264	.294	.356
25	.21	.22	.24	.27	.32
30	.19	.20	.22	.24	.29
35	.18	.19	.21	.23	.27
Over 35	$\dfrac{1.07}{\sqrt{n}}$	$\dfrac{1.14}{\sqrt{n}}$	$\dfrac{1.22}{\sqrt{n}}$	$\dfrac{1.36}{\sqrt{n}}$	$\dfrac{1.63}{\sqrt{n}}$

TABLE C.5
Critical Values of *t*

For any given df, the table shows the values of *t* corresponding to various levels of probability. Obtained *t* is significant at a given level if it is equal to or greater than the value shown in the table.

df	\.10	\.05	\.025	\.01	\.005	\.0005
	.20	.10	.05	.02	.01	.001
1	3.078	6.314	12.706	31.821	63.657	636.619
2	1.886	2.920	4.303	6.965	9.925	31.598
3	1.638	2.353	3.182	4.541	5.841	12.941
4	1.533	2.132	2.776	3.747	4.604	8.610
5	1.476	2.015	2.571	3.365	4.032	6.859
6	1.440	1.943	2.447	3.143	3.707	5.959
7	1.415	1.895	2.365	2.998	3.499	5.405
8	1.397	1.860	2.306	2.896	3.355	5.041
9	1.383	1.833	2.262	2.821	3.250	4.781
10	1.372	1.812	2.228	2.764	3.169	4.587
11	1.363	1.796	2.201	2.718	3.106	4.437
12	1.356	1.782	2.179	2.681	3.055	4.318
13	1.350	1.771	2.160	2.650	3.012	4.221
14	1.345	1.761	2.145	2.624	2.977	4.140
15	1.341	1.753	2.131	·2.602	2.947	4.073
16	1.337	1.746	2.120	2.583	2.921	4.015
17	1.333	1.740	2.110	2.567	2.898	3.965
18	1.330	1.734	2.101	2.552	2.878	3.922
19	1.328	1.729	2.093	2.539	2.861	3.883
20	1.325	1.725	2.086	2.528	2.845	3.850
21	1.323	1.721	2.080	2.518	2.831	3.819
22	1.321	1.717	2.074	2.508	2.819	3.792
23	1.319	1.714	2.069	2.500	2.807	3.767
24	1.318	1.711	2.064	2.492	2.797	3.745
25	1.316	1.708	2.060	2.485	2.787	3.725
26	1.315	1.706	2.056	2.479	2.779	3.707
27	1.314	1.703	2.052	2.473	2.771	3.690
28	1.313	1.701	2.048	2.467	2.763	3.674
29	1.311	1.699	2.045	2.462	2.756	3.659
30	1.310	1.697	2.042	2.457	2.750	3.646
40	1.303	1.684	2.021	2.423	2.704	3.551
60	1.296	1.671	2.000	2.390	2.660	3.460
120	1.289	1.658	1.980	2.358	2.617	3.373
∞	1.282	1.645	1.960	2.326	2.576	3.291

Column headings above: first row under "Level of significance for one-tailed test": .10, .05, .025, .01, .005, .0005; second set under "Level of significance for two-tailed test": .20, .10, .05, .02, .01, .001

TABLE C.6
Values of Spearman Correlation Coefficient
Significant at $\alpha = .05$ and $\alpha = .01$

n	.05	.01	n	.05	.01
5	.900	1.000	16	.425	.601
6	.829	.943	18	.399	.564
7	.714	.893	20	.377	.534
8	.643	.833	22	.359	.508
9	.600	.783	24	.343	.485
10	.564	.746	26	.329	.465
12	.506	.712	28	.317	.448
14	.456	.645	30	.306	.432

TABLE C.7
Table for Transforming r to z_r

r	.00	.01	.02	.03	.04	.05	.06	.07	.08	.09
.0	.00000	.01000	.02000	.03001	.04002	.05004	.06007	.07012	.08017	.09024
.1	.10034	.11045	.12058	.13074	.14093	.15114	.16139	.17167	.18198	.19234
.2	.20273	.21317	.22366	.23419	.24477	.25541	.26611	.27686	.28768	.29857
.3	.30952	.32055	.33165	.34283	.35409	.36544	.37689	.38842	.40006	.41180
.4	.42365	.43561	.44769	.45990	.47223	.48470	.49731	.51007	.52298	.53606
.5	.54931	.56273	.57634	.59014	.60415	.61838	.63283	.64752	.66246	.67767
.6	.69315	.70892	.72500	.74142	.75817	.77530	.79281	.81074	.82911	.84795
.7	.86730	.88718	.90764	.92873	.95048	.97295	.99621	1.02033	1.04537	1.07143
.8	1.09861	1.12703	1.15682	1.18813	1.22117	1.25615	1.29334	1.33308	1.37577	1.42192
.9	1.47222	1.52752	1.58902	1.65839	1.73805	1.83178	1.94591	2.09229	2.29756	2.64665

For negative values of r put a minus sign in front of the tabled numbers.

TABLE C.8
Table for Transforming ζ (Zeta) to ρ (Rho)

ζ	.00	.01	.02	.03	.04	.05	.06	.07	.08	.09
.0	.0000	.0100	.0200	.0300	.0400	.0500	.0599	.0699	.0798	.0898
.1	.0997	.1096	.1194	.1293	.1391	.1489	.1587	.1684	.1781	.1878
.2	.1974	.2070	.2165	.2260	.2355	.2449	.2543	.2636	.2729	.2821
.3	.2913	.3004	.3095	.3185	.3275	.3364	.3452	.3540	.3627	.3714
.4	.3800	.3885	.3969	.4053	.4136	.4219	.4301	.4382	.4462	.4542
.5	.4621	.4700	.4777	.4854	.4930	.5005	.5080	.5154	.5227	.5299
.6	.5370	.5441	.5511	.5581	.5649	.5717	.5784	.5850	.5915	.5980
.7	.6044	.6107	.6169	.6231	.6291	.6352	.6411	.6469	.6527	.6584
.8	.6640	.6696	.6751	.6805	.6858	.6911	.6963	.7014	.7064	.7114
.9	.7163	.7211	.7259	.7306	.7352	.7398	.7443	.7487	.7531	.7574
1.0	.7616	.7658	.7699	.7739	.7779	.7818	.7857	.7895	.7932	.7969
1.1	.8005	.8041	.8076	.8110	.8144	.8178	.8210	.8243	.8275	.8306
1.2	.8337	.8367	.8397	.8426	.8455	.8483	.8511	.8538	.8565	.8591
1.3	.8617	.8643	.8668	.8693	.8717	.8741	.8764	.8787	.8810	.8832
1.4	.8854	.8875	.8896	.8917	.8937	.8957	.8977	.8996	.9015	.9033
1.5	.9052	.9069	.9087	.9104	.9121	.9138	.9154	.9170	.9186	.9202
1.6	.9217	.9232	.9246	.9261	.9275	.9289	.9302	.9316	.9329	.9342
1.7	.9354	.9367	.9379	.9391	.9402	.9414	.9425	.9436	.9447	.9458
1.8	.9468	.9478	.9488	.9498	.9508	.9518	.9527	.9536	.9545	.9554
1.9	.9562	.9571	.9579	.9587	.9595	.9603	.9611	.9619	.9626	.9633
2.0	.9640	.9647	.9654	.9661	.9668	.9674	.9680	.9687	.9693	.9699
2.1	.9705	.9710	.9716	.9722	.9727	.9732	.9738	.9743	.9748	.9753
2.2	.9757	.9762	.9767	.9771	.9776	.9780	.9785	.9789	.9793	.9797
2.3	.9801	.9805	.9809	.9812	.9816	.9820	.9823	.9827	.9830	.9834
2.4	.9837	.9840	.9843	.9846	.9849	.9852	.9855	.9858	.9861	.9863
2.5	.9866	.9869	.9871	.9874	.9876	.9879	.9881	.9884	.9886	.9888
2.6	.9890	.9892	.9895	.9897	.9899	.9901	.9903	.9905	.9906	.9908
2.7	.9910	.9912	.9914	.9915	.9917	.9919	.9920	.9922	.9923	.9925
2.8	.9926	.9928	.9929	.9931	.9932	.9933	.9935	.9936	.9937	.9938
2.9	.9940	.9941	.9942	.9943	.9944	.9945	.9946	.9947	.9949	.9950
3.0	.9951									
4.0	.9993									
5.0	.9999									

For negative values of ζ put a minus sign in front of the tabled value of ρ.

TABLE C.9
Chart of 90 Percent Confidence Intervals for the
Population Correlation Coefficient, ρ

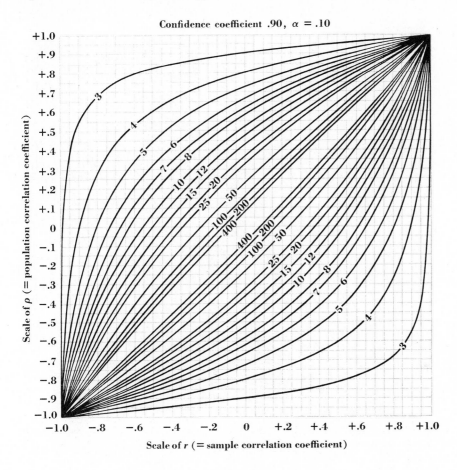

Confidence coefficient .90, $\alpha = .10$

Scale of ρ (= population correlation coefficient)

Scale of r (= sample correlation coefficient)

TABLE C.10
Chart of 95 Percent Confidence Intervals for the
Population Correlation Coefficient, ρ

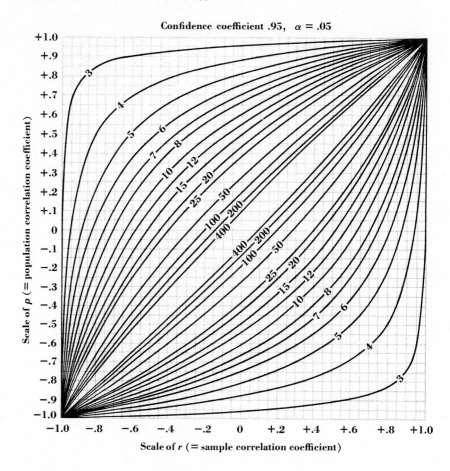

TABLE C.11
Critical Values of D (or C) in the Fisher Test

Totals at bottom		B (or A)†	Level of significance			
			.05	.025	.01	.005
$A + B = 3$	$C + D = 3$	3	0	—	—	—
$A + B = 4$	$C + D = 4$	4	0	0	—	—
	$C + D = 3$	4	0	—	—	—
$A + B = 5$	$C + D = 5$	5	1	1	0	0
		4	0	0	—	—
	$C + D = 4$	5	1	0	0	—
		4	0	—	—	—
	$C + D = 3$	5	0	0	—	—
	$C + D = 2$	5	0	—	—	—
$A + B = 6$	$C + D = 6$	6	2	1	1	0
		5	1	0	0	—
		4	0	—	—	—
	$C + D = 5$	6	1	0	0	0
		5	0	0	—	—
		4	0	—	—	—
	$C + D = 4$	6	1	0	0	0
		5	0	0	—	—
	$C + D = 3$	6	0	0	—	—
		5	0	—	—	—
	$C + D = 2$	6	0	—	—	—
$A + B = 7$	$C + D = 7$	7	3	2	1	1
		6	1	1	0	0
		5	0	0	—	—
		4	0	—	—	—
	$C + D = 6$	7	2	2	1	1
		6	1	0	0	0
		5	0	0	—	—
		4	0	—	—	—
	$C + D = 5$	7	2	1	0	0
		6	1	0	0	—
		5	0	—	—	—
	$C + D = 4$	7	1	1	0	0
		6	0	0	—	—
		5	0	—	—	—
	$C + D = 3$	7	0	0	0	—
		6	0	—	—	—
	$C + D = 2$	7	0	—	—	—

†When B is entered in the middle column, the significance levels are for D. When A is used in place of B, the significance levels are for C.

TABLE C.11 (continued)

Totals at bottom		B (or A)†	Level of significance			
			.05	.025	.01	.005
$A + B = 8$	$C + D = 8$	8	4	3	2	2
		7	2	2	1	0
		6	1	1	0	0
		5	0	0	—	—
		4	0	—	—	—
	$C + D = 7$	8	3	2	2	1
		7	2	1	1	0
		6	1	0	0	—
		5	0	0	—	—
	$C + D = 6$	8	2	2	1	1
		7	1	1	0	0
		6	0	0	0	—
		5	0	—	—	—
	$C + D = 5$	8	2	1	1	0
		7	1	0	0	0
		6	0	0	—	—
		5	0	—	—	—
	$C + D = 4$	8	1	1	0	0
		7	0	0	—	—
		6	0	—	—	—
	$C + D = 3$	8	0	0	0	—
		7	0	0	—	—
	$C + D = 2$	8	0	0	—	—
$A + B = 9$	$C + D = 9$	9	5	4	3	3
		8	3	3	2	1
		7	2	1	1	0
		6	1	1	0	0
		5	0	0	—	—
		4	0	—	—	—
	$C + D = 8$	9	4	3	3	2
		8	3	2	1	1
		7	2	1	0	0
		6	1	0	0	—
		5	0	0	—	—
	$C + D = 7$	9	3	3	2	2
		8	2	2	1	0
		7	1	1	0	0
		6	0	0	—	—
		5	0	—	—	—

TABLE C.11 (continued)

Totals at bottom		B (or A)†	Level of significance			
			.05	.025	.01	.005
$A + B = 9$	$C + D = 6$	9	3	2	1	1
		8	2	1	0	0
		7	1	0	0	—
		6	0	0	—	—
		5	0	—	—	—
	$C + D = 5$	9	2	1	1	1
		8	1	1	0	0
		7	0	0	—	—
		6	0	—	—	—
	$C + D = 4$	9	1	1	0	0
		8	0	0	0	—
		7	0	0	—	—
		6	0	—	—	—
	$C + D = 3$	9	1	0	0	0
		8	0	0	—	—
		7	0	—	—	—
	$C + D = 2$	9	0	0	—	—
$A + B = 10$	$C + D = 10$	10	6	5	4	3
		9	4	3	3	2
		8	3	2	1	1
		7	2	1	1	0
		6	1	0	0	—
		5	0	0	—	—
		4	0	—	—	—
	$C + D = 9$	10	5	4	3	3
		9	4	3	2	2
		8	2	2	1	1
		7	1	1	0	0
		6	1	0	0	—
		5	0	0	—	—
	$C + D = 8$	10	4	4	3	2
		9	3	2	2	1
		8	2	1	1	0
		7	1	1	0	0
		6	0	0	—	—
		5	0	—	—	—
	$C + D = 7$	10	3	3	2	2
		9	2	2	1	1
		8	1	1	0	0
		7	1	0	0	—
		6	0	0	—	—
		5	0	—	—	—

†When B is entered in the middle column, the significance levels are for D. When A is used in place of B, the significance levels are for C.

TABLE C.11 (continued)

Totals at bottom		B (or A)†	Level of significance			
			.05	.025	.01	.005
$A + B = 10$	$C + D = 6$	10	3	2	2	1
		9	2	1	1	0
		8	1	1	0	0
		7	0	0	—	—
		6	0	—	—	—
	$C + D = 5$	10	2	2	1	1
		9	1	1	0	0
		8	1	0	0	—
		7	0	0	—	—
		6	0	—	—	—
	$C + D = 4$	10	1	1	0	0
		9	1	0	0	0
		8	0	0	—	—
		7	0	—	—	—
	$C + D = 3$	10	1	0	0	0
		9	0	0	—	—
		8	0	—	—	—
	$C + D = 2$	10	0	0	—	—
		9	0	—	—	—
$A + B = 11$	$C + D = 11$	11	7	6	5	4
		10	5	4	3	3
		9	4	3	2	2
		8	3	2	1	1
		7	2	1	0	0
		6	1	0	0	—
		5	0	0	—	—
		4	0	—	—	—
	$C + D = 10$	11	6	5	4	4
		10	4	4	3	2
		9	3	3	2	1
		8	2	2	1	0
		7	1	1	0	0
		6	1	0	0	—
		5	0	—	—	—
	$C + D = 9$	11	5	4	4	3
		10	4	3	2	2
		9	3	2	1	1
		8	2	1	1	0
		7	1	1	0	0
		6	0	0	—	—
		5	0	—	—	—

TABLE C.11 (continued)

Totals at bottom		B (or A)†	Level of significance			
			.05	.025	.01	.005
$A + B = 11$	$C + D = 8$	11	4	4	3	3
		10	3	3	2	1
		9	2	2	1	1
		8	1	1	0	0
		7	1	0	0	—
		6	0	0	—	—
		5	0	—	—	—
	$C + D = 7$	11	4	3	2	2
		10	3	2	1	1
		9	2	1	1	0
		8	1	1	0	0
		7	0	0	—	—
		6	0	0	—	—
	$C + D = 6$	11	3	2	2	1
		10	2	1	1	0
		9	1	1	0	0
		8	1	0	0	—
		7	0	0	—	—
		6	0	—	—	—
	$C + D = 5$	11	2	2	1	1
		10	1	1	0	0
		9	1	0	0	0
		8	0	0	—	—
		7	0	—	—	—
	$C + D = 4$	11	1	1	1	0
		10	1	0	0	0
		9	0	0	—	—
		8	0	—	—	—
	$C + D = 3$	11	1	0	0	0
		10	0	0	—	—
		9	0	—	—	—
	$C + D = 2$	11	0	0	—	—
		10	0	—	—	—
$A + B = 12$	$C + D = 12$	12	8	7	6	5
		11	6	5	4	4
		10	5	4	3	2
		9	4	3	2	1
		8	3	2	1	1
		7	2	1	0	0
		6	1	0	0	—
		5	0	0	—	—
		4	0	—	—	—

†When B is entered in the middle column, the significance levels are for D. When A is used in place of B, the significance levels are for C.

TABLE C.11 (continued)

Totals at bottom		B (or A)†	Level of significance			
			.05	.025	.01	.005
$A + B = 12$	$C + D = 11$	12	7	6	5	5
		11	5	5	4	3
		10	4	3	2	2
		9	3	2	2	1
		8	2	1	1	0
		7	1	1	0	0
		6	1	0	0	—
		5	0	0	—	—
	$C + D = 10$	12	6	5	5	4
		11	5	4	3	3
		10	4	3	2	2
		9	3	2	1	1
		8	2	1	0	0
		7	1	0	0	0
		6	0	0	—	—
		5	0	—	—	—
	$C + D = 9$	12	5	5	4	3
		11	4	3	3	2
		10	3	2	2	1
		9	2	2	1	0
		8	1	1	0	0
		7	1	0	0	—
		6	0	0	—	—
		5	0	—	—	—
	$C + D = 8$	12	5	4	3	3
		11	3	3	2	2
		10	2	2	1	1
		9	2	1	1	0
		8	1	1	0	0
		7	0	0	—	—
		6	0	0	—	—
	$C + D = 7$	12	4	3	3	2
		11	3	2	2	1
		10	2	1	1	0
		9	1	1	0	0
		8	1	0	0	—
		7	0	0	—	—
		6	0	—	—	—

TABLE C.11 (continued)

Totals at bottom		B (or A)†	Level of significance			
			.05	.025	.01	.005
$A + B = 12$	$C + D = 6$	12	3	3	2	2
		11	2	2	1	1
		10	1	1	0	0
		9	1	0	0	0
		8	0	0	—	—
		7	0	0	—	—
		6	0	—	—	—
	$C + D = 5$	12	2	2	1	1
		11	1	1	1	0
		10	1	0	0	0
		9	0	0	0	—
		8	0	0	—	—
		7	0	—	—	—
	$C + D = 4$	12	2	1	1	0
		11	1	0	0	0
		10	0	0	0	—
		9	0	0	—	—
		8	0	—	—	—
	$C + D = 3$	12	1	0	0	0
		11	0	0	0	—
		10	0	0	—	—
		9	0	—	—	—
	$C + D = 2$	12	0	0	—	—
		11	0	—	—	—
$A + B = 13$	$C + D = 13$	13	9	8	7	6
		12	7	6	5	4
		11	6	5	4	3
		10	4	4	3	2
		9	3	3	2	1
		8	2	2	1	0
		7	2	1	0	0
		6	1	0	0	—
		5	0	0	—	—
		4	0	—	—	—
	$C + D = 12$	13	8	7	6	5
		12	6	5	5	4
		11	5	4	3	3
		10	4	3	2	2
		9	3	2	1	1
		8	2	1	1	0
		7	1	1	0	0
		6	1	0	0	—
		5	0	0	—	—

†When B is entered in the middle column, the significance levels are for D. When A is used in place of B, the significance levels are for C.

TABLE C.11 (continued)

Totals at bottom		B (or A)†	Level of significance			
			.05	.025	.01	.005
$A + B = 13$	$C + D = 11$	13	7	6	5	5
		12	6	5	4	3
		11	4	4	3	2
		10	3	3	2	1
		9	3	2	1	1
		8	2	1	0	0
		7	1	0	0	0
		6	0	0	—	—
		5	0	—	—	—
	$C + D = 10$	13	6	6	5	4
		12	5	4	3	3
		11	4	3	2	2
		10	3	2	1	1
		9	2	1	1	0
		8	1	1	0	0
		7	1	0	0	—
		6	0	0	—	—
		5	0	—	—	—
	$C + D = 9$	13	5	5	4	4
		12	4	4	3	2
		11	3	3	2	1
		10	2	2	1	1
		9	2	1	0	0
		8	1	1	0	0
		7	0	0	—	—
		6	0	0	—	—
		5	0	—	—	—
	$C + D = 8$	13	5	4	3	3
		12	4	3	2	2
		11	3	2	1	1
		10	2	1	1	0
		9	1	1	0	0
		8	1	0	0	—
		7	0	0	—	—
		6	0	—	—	—
	$C + D = 7$	13	4	3	3	2
		12	3	2	2	1
		11	2	2	1	1
		10	1	1	0	0
		9	1	0	0	0
		8	0	0	—	—
		7	0	0	—	—
		6	0	—	—	—

TABLE C.11 (continued)

Totals at bottom		B (or A)†	Level of significance			
			.05	.025	.01	.005
$A + B = 13$	$C + D = 6$	13	3	3	**2**	2
		12	2	2	1	1
		11	2	1	1	0
		10	1	1	0	0
		9	1	0	0	—
		8	0	0	—	—
		7	0	—	—	—
	$C + D = 5$	13	2	2	1	1
		12	2	1	1	0
		11	1	1	0	0
		10	1	0	0	—
		9	0	0	—	—
		8	0	—	—	—
	$C + D = 4$	13	2	1	1	0
		12	1	1	0	0
		11	0	0	0	—
		10	0	0	—	—
		9	0	—	—	—
	$C + D = 3$	13	1	1	0	0
		12	0	0	0	—
		11	0	0	—	—
		10	0	—	—	—
	$C + D = 2$	13	0	0	0	—
		12	0	—	—	—
$A + B = 14$	$C + D = 14$	14	10	9	8	7
		13	8	7	6	5
		12	6	6	5	4
		11	5	4	3	3
		10	4	3	2	2
		9	3	2	2	1
		8	2	2	1	0
		7	1	1	0	0
		6	1	0	0	—
		5	0	0	—	—
		4	0	—	—	—

†When B is entered in the middle column, the significance levels are for D. When A is used in place of B, the significance levels are for C.

TABLE C.11 (continued)

Totals at bottom		B (or A)†	Level of significance			
			.05	.025	.01	.005
$A + B = 14$	$C + D = 13$	14	9	8	7	6
		13	7	6	5	5
		12	6	5	4	3
		11	5	4	3	2
		10	4	3	2	2
		9	3	2	1	1
		8	2	1	1	0
		7	1	1	0	0
		6	1	0	—	—
		5	0	0	—	—
	$C + D = 12$	14	8	7	6	6
		13	6	6	5	4
		12	5	4	4	3
		11	4	3	3	2
		10	3	3	2	1
		9	2	2	1	1
		8	2	1	0	0
		7	1	0	0	—
		6	0	0	—	—
		5	0	—	—	—
	$C + D = 11$	14	7	6	6	5
		13	6	5	4	4
		12	5	4	3	3
		11	4	3	2	2
		10	3	2	1	1
		9	2	1	1	0
		8	1	1	0	0
		7	1	0	0	—
		6	0	0	—	—
		5	0	—	—	—
	$C + D = 10$	14	6	6	5	4
		13	5	4	4	3
		12	4	3	3	2
		11	3	3	2	1
		10	2	2	1	1
		9	2	1	0	0
		8	1	1	0	0
		7	0	0	0	—
		6	0	0	—	—
		5	0	—	—	—

TABLE C.11 (continued)

Totals at bottom		B (or A)†	Level of significance			
			.05	.025	.01	.005
$A + B = 14$	$C + D = 9$	14	6	5	4	4
		13	4	4	3	3
		12	3	3	2	2
		11	3	2	1	1
		10	2	1	1	0
		9	1	1	0	0
		8	1	0	0	—
		7	0	0	—	—
		6	0	—	—	—
	$C + D = 8$	14	5	4	4	3
		13	4	3	2	2
		12	3	2	2	1
		11	2	2	1	1
		10	2	1	0	0
		9	1	0	0	0
		8	0	0	0	—
		7	0	0	—	—
		6	0	—	—	—
	$C + D = 7$	14	4	3	3	2
		13	3	2	2	1
		12	2	2	1	1
		11	2	1	1	0
		10	1	1	0	0
		9	1	0	0	—
		8	0	0	—	—
		7	0	—	—	—
	$C + D = 6$	14	3	3	2	2
		13	2	2	1	1
		12	2	1	1	0
		11	1	1	0	0
		10	1	0	0	—
		9	0	0	—	—
		8	0	0	—	—
		7	0	—	—	—
	$C + D = 5$	14	2	2	1	1
		13	2	1	1	0
		12	1	1	0	0
		11	1	0	0	0
		10	0	0	—	—
		9	0	0	—	—
		8	0	—	—	—

†When B is entered in the middle column, the significance levels are for D. When A is used in place of B, the significance levels are for C.

TABLE C.11 (continued)

Totals at bottom		B (or A)†	Level of significance			
			.05	.025	.01	.005
$A + B = 14$	$C + D = 4$	14	2	1	1	1
		13	1	1	0	0
		12	1	0	0	0
		11	0	0	—	—
		10	0	0	—	—
		9	0	—	—	—
	$C + D = 3$	14	1	1	0	0
		13	0	0	0	—
		12	0	0	—	—
		11	0	—	—	—
	$C + D = 2$	14	0	0	0	—
		13	0	0	—	—
		12	0	—	—	—
$A + B = 15$	$C + D = 15$	15	11	10	9	8
		14	9	8	7	6
		13	7	6	5	5
		12	6	5	4	4
		11	5	4	3	3
		10	4	3	2	2
		9	3	2	1	1
		8	2	1	1	0
		7	1	1	0	0
		6	1	0	0	—
		5	0	0	—	—
		4	0	—	—	—
	$C + D = 14$	15	10	9	8	7
		14	8	7	6	6
		13	7	6	5	4
		12	6	5	4	3
		11	5	4	3	2
		10	4	3	2	1
		9	3	2	1	1
		8	2	1	1	0
		7	1	1	0	0
		6	1	0	—	—
		5	0	—	—	—

TABLE C.11 (continued)

Totals at bottom		B (or A)†	Level of significance			
			.05	.025	.01	.005
$A + B = 15$	$C + D = 13$	15	9	8	7	7
		14	7	7	6	5
		13	6	5	4	4
		12	5	4	3	3
		11	4	3	2	2
		10	3	2	2	1
		9	2	2	1	0
		8	2	1	0	0
		7	1	0	0	—
		6	0	0	—	—
		5	0	—	—	—
	$C + D = 12$	15	8	7	7	6
		14	7	6	5	4
		13	6	5	4	3
		12	5	4	3	2
		11	4	3	2	2
		10	3	2	1	1
		9	2	1	1	0
		8	1	1	0	0
		7	1	0	0	—
		6	0	0	—	—
		5	0	—	—	—
	$C + D = 11$	15	7	7	6	5
		14	6	5	4	4
		13	5	4	3	3
		12	4	3	2	2
		11	3	2	2	1
		10	2	2	1	1
		9	2	1	0	0
		8	1	1	0	0
		7	1	0	0	—
		6	0	0	—	—
		5	0	—	—	—
	$C + D = 10$	15	6	6	5	5
		14	5	5	4	3
		13	4	4	3	2
		12	3	3	2	2
		11	3	2	1	1
		10	2	1	1	0
		9	1	1	0	0
		8	1	0	0	—
		7	0	0	—	—
		6	0	—	—	—

† When B is entered in the middle column, the significance levels are for D.
When A is used in place of B, the significance levels are for C.

TABLE C.11 (continued)

Totals at bottom		B (or A)†	Level of significance			
			.05	.025	.01	.005
$A + B = 15$	$C + D = 9$	15	6	5	4	4
		14	5	4	3	3
		13	4	3	2	2
		12	3	2	2	1
		11	2	2	1	1
		10	2	1	0	0
		9	1	1	0	0
		8	1	0	0	—
		7	0	0	—	—
		6	0	—	—	—
	$C + D = 8$	15	5	4	4	3
		14	4	3	3	2
		13	3	2	2	1
		12	2	2	1	1
		11	2	1	1	0
		10	1	1	0	0
		9	1	0	0	—
		8	0	0	—	—
		7	0	—	—	—
		6	0	—	—	—
	$C + D = 7$	15	4	4	3	3
		14	3	3	2	2
		13	2	2	1	1
		12	2	1	1	0
		11	1	1	0	0
		10	1	0	0	0
		9	0	0	—	—
		8	0	0	—	—
		7	0	—	—	—
	$C + D = 6$	15	3	3	2	2
		14	2	2	1	1
		13	2	1	1	0
		12	1	1	0	0
		11	1	0	0	0
		10	0	0	0	—
		9	0	0	—	—
		8	0	—	—	—
	$C + D = 5$	15	2	2	2	1
		14	2	1	1	1
		13	1	1	0	0
		12	1	0	0	0
		11	0	0	0	—
		10	0	0	—	—
		9	0	—	—	—

†When B is entered in the middle column, the significance levels are for D. When A is used in place of B, the significance levels are for C.

TABLE C.12
Mann-Whitney U One-Tailed Test at .05 Level; Two-Tailed Test at .10 Level

n_2	\multicolumn{20}{c}{n_1}

n_2	1	2	3	4	5	6	7	8	9	10	11	12	13	14	15	16	17	18	19	20
1	—																			
2	—	—																		
3	—	—	0																	
4	—	—	0	1																
5	—	0	1	2	4															
6	—	0	2	3	5	7														
7	—	0	2	4	6	8	11													
8	—	1	3	5	8	10	13	15												
9	—	1	4	6	9	12	15	18	21											
10	—	1	4	7	11	14	17	20	24	27										
11	—	1	5	8	12	16	19	23	27	31	34									
12	—	2	5	9	13	17	21	26	30	34	38	42								
13	—	2	6	10	15	19	24	28	33	37	42	47	51							
14	—	3	7	11	16	21	26	31	36	41	46	51	56	61						
15	—	3	7	12	18	23	28	33	39	44	50	55	61	66	72					
16	—	3	8	14	19	25	30	36	42	48	54	60	65	71	77	83				
17	—	3	9	15	20	26	33	39	45	51	57	64	70	77	83	89	96			
18	—	4	9	16	22	28	35	41	48	55	61	68	75	82	88	95	102	109		
19	0	4	10	17	23	30	37	44	51	58	65	72	80	87	94	101	109	116	123	
20	0	4	11	18	25	32	39	47	54	62	69	77	84	92	100	107	115	123	130	138
21	0	5	11	19	26	34	41	49	57	65	73	81	89	97	105	113	121	130	138	146
22	0	5	12	20	28	36	44	52	60	68	77	85	94	102	111	119	128	136	145	154
23	0	5	13	21	29	37	46	54	63	72	81	90	98	107	116	125	134	143	152	161
24	0	6	13	22	30	39	48	57	66	75	85	94	103	113	122	131	141	150	160	162
25	0	6	14	23	32	41	50	60	69	79	89	98	108	118	128	137	147	157	167	177
26	0	6	15	24	33	43	53	62	72	82	92	103	113	123	133	143	154	164	174	185
27	0	7	15	25	35	45	55	65	75	86	96	107	117	128	139	149	160	171	182	192
28	0	7	16	26	36	46	57	68	78	89	100	111	122	133	144	156	167	178	189	200
29	0	7	17	27	38	48	59	70	82	93	104	116	127	138	150	162	173	185	196	208
30	0	7	17	28	39	50	61	73	85	96	108	120	132	144	156	168	180	192	204	216
31	0	8	18	29	40	52	64	76	88	100	112	124	136	149	161	174	186	199	211	224
32	0	8	19	30	42	54	66	78	91	103	116	128	141	154	167	180	193	206	218	231
33	0	8	19	31	43	56	68	81	94	107	120	133	146	159	172	186	199	212	226	239
34	0	9	20	32	45	57	70	84	97	110	124	137	151	164	178	192	206	219	233	247
35	0	9	21	33	46	59	73	86	100	114	128	141	156	170	184	198	212	226	241	255
36	0	9	21	34	48	61	75	89	103	117	131	146	160	175	189	204	219	233	248	263
37	0	10	22	35	49	63	77	91	106	121	135	150	165	180	195	210	225	240	255	271
38	0	10	23	36	50	65	79	94	109	124	139	154	170	185	201	216	232	247	263	278
39	1	10	23	38	52	67	82	97	112	128	143	159	175	190	206	222	238	254	270	286
40	1	11	24	39	53	68	84	99	115	131	147	163	179	196	212	228	245	261	278	294

TABLE C.12 (continued)

Mann-Whitney U One-Tailed Test at .025 Level; Two-Tailed Test at .05 Level

n_2	\ n_1 1	2	3	4	5	6	7	8	9	10	11	12	13	14	15	16	17	18	19	20
1	—																			
2	—	—																		
3	—	—	—																	
4	—	—	—	0																
5	—	—	0	1	2															
6	—	—	1	2	3	5														
7	—	—	1	3	5	6	8													
8	—	0	2	4	6	8	10	13												
9	—	0	2	4	7	10	12	15	17											
10	—	0	3	5	8	11	14	17	20	23										
11	—	0	3	6	9	13	16	19	23	26	30									
12	—	1	4	7	11	14	18	22	26	29	33	37								
13	—	1	4	8	12	16	20	24	28	33	37	41	45							
14	—	1	5	9	13	17	22	26	31	36	40	45	50	55						
15	—	1	5	10	14	19	24	29	34	39	44	49	54	59	64					
16	—	1	6	11	15	21	26	31	37	42	47	53	59	64	70	75				
17	—	2	6	11	17	22	28	34	39	45	51	57	63	69	75	81	87			
18	—	2	7	12	18	24	30	36	42	48	55	61	67	74	80	86	93	99		
19	—	2	7	13	19	25	32	38	45	52	58	65	72	78	85	92	99	106	113	
20	—	2	8	14	20	27	34	41	48	55	62	69	76	83	90	98	105	112	119	127
21	—	3	8	15	22	29	36	43	50	58	65	73	80	88	96	103	111	119	126	134
22	—	3	9	16	23	30	38	45	53	61	69	77	85	93	101	109	117	125	133	141
23	—	3	9	17	24	32	40	48	56	64	73	81	89	98	106	115	123	132	140	149
24	—	3	10	17	25	33	42	50	59	67	76	85	94	102	111	120	129	138	147	156
25	—	3	10	18	27	35	44	53	62	71	80	89	98	107	117	126	135	145	154	163
26	—	4	11	19	28	37	46	55	64	74	83	93	102	112	122	132	141	151	161	171
27	—	4	11	20	29	38	48	57	67	77	87	97	107	117	127	137	147	158	168	178
28	—	4	12	21	30	40	50	60	70	80	90	101	111	122	132	143	154	164	175	186
29	—	4	13	22	32	42	52	62	73	83	94	105	116	127	138	149	160	171	182	193
30	—	5	13	23	33	43	54	65	76	87	98	109	120	131	143	154	166	177	189	200
31	—	5	14	24	34	45	56	67	78	90	101	113	125	136	148	160	172	184	196	208
32	—	5	14	24	35	46	58	69	81	93	105	117	129	141	153	166	178	190	203	215
33	—	5	15	25	37	48	60	72	84	96	108	121	133	146	159	171	184	197	210	222
34	—	5	15	26	38	50	62	74	87	99	112	125	138	151	164	177	190	203	217	230
35	—	6	16	27	39	51	64	77	89	103	116	129	142	156	169	183	196	210	224	237
36	—	6	16	28	40	53	66	79	92	106	119	133	147	161	174	188	202	216	231	245
37	—	6	17	29	41	55	68	81	95	109	123	137	151	165	180	194	209	223	238	252
38	—	6	17	30	43	56	70	84	98	112	127	141	156	170	185	200	215	230	245	259
39	0	7	18	31	44	58	72	86	101	115	130	145	160	175	190	206	321	236	252	267
40	0	7	18	31	45	59	74	89	103	119	134	149	165	180	196	211	227	243	258	274

Appendix C

TABLE C.12 (continued)
Mann-Whitney U One-Tailed Test at .01 Level; Two-Tailed Test at .02 Level

n_2	n_1																			
	1	2	3	4	5	6	7	8	9	10	11	12	13	14	15	16	17	18	19	20
1	—																			
2	—	—																		
3	—	—	—																	
4	—	—	—	—																
5	—	—	—	0	1															
6	—	—	—	1	2	3														
7	—	—	0	1	3	4	6													
8	—	—	0	2	4	6	7	9												
9	—	—	1	3	5	7	9	11	14											
10	—	—	1	3	6	8	11	13	16	19										
11	—	—	1	4	7	9	12	15	18	22	25									
12	—	—	2	5	8	11	14	17	21	24	28	31								
13	—	0	2	5	9	12	16	20	23	27	31	35	39							
14	—	0	2	6	10	13	17	22	26	30	34	38	43	47						
15	—	0	3	7	11	15	19	24	28	33	37	42	47	51	56					
16	—	0	3	7	12	16	21	26	31	36	41	46	51	56	61	66				
17	—	0	4	8	13	18	23	28	33	38	44	49	55	60	66	71	77			
18	—	0	4	9	14	19	24	30	36	41	47	53	59	65	70	76	82	88		
19	—	1	4	9	15	20	26	32	38	44	50	56	63	69	75	82	88	94	101	
20	—	1	5	10	16	22	28	34	40	47	53	60	67	73	80	87	93	100	107	114
21	—	1	5	11	17	23	30	36	43	50	57	64	71	78	85	92	99	106	113	121
22	—	1	6	11	18	24	31	38	45	53	60	67	75	82	90	97	105	112	120	127
23	—	1	6	12	19	26	33	40	48	55	63	71	79	87	94	102	110	118	126	134
24	—	1	6	13	20	27	35	42	50	58	66	75	83	91	99	108	116	124	133	141
25	—	1	7	13	21	29	36	45	53	61	70	78	87	95	104	113	122	130	139	148
26	—	1	7	14	22	30	38	47	55	64	73	82	91	100	109	118	127	136	146	155
27	—	2	7	15	23	31	40	49	58	67	76	85	95	104	114	123	133	142	152	162
28	—	2	8	16	24	33	42	51	60	70	79	89	99	109	119	129	139	149	159	169
29	—	2	8	16	25	34	43	53	63	73	83	93	103	113	123	134	144	155	165	176
30	—	2	9	17	26	35	45	55	65	76	86	96	107	118	128	139	150	161	172	182
31	—	2	9	18	27	37	47	57	68	78	89	100	111	122	133	144	156	167	178	189
32	—	2	9	18	28	38	49	59	70	81	92	104	115	127	138	150	161	173	185	196
33	—	2	10	19	29	40	50	61	73	84	96	107	119	131	143	155	167	179	191	203
34	—	3	10	20	30	41	52	64	75	87	99	111	123	135	148	160	173	185	198	210
35	—	3	11	20	31	42	54	66	78	90	102	115	127	140	153	165	178	191	204	217
36	—	3	11	21	32	44	56	68	80	93	106	118	131	144	158	171	184	197	211	224
37	—	3	11	22	33	45	57	70	83	96	109	122	135	149	162	176	190	203	217	231
38	—	3	12	22	34	46	59	72	85	99	112	126	139	153	167	181	195	209	224	238
39	—	3	12	23	35	48	61	74	88	101	115	129	144	158	172	187	201	216	230	245
40	—	3	13	24	36	49	63	76	90	104	119	133	148	162	177	192	207	222	237	252

TABLE C.12 (continued)
Mann-Whitney U One-Tailed Test at .005 Level; Two-Tailed Test at .01 Level

n_2	n_1																			
	1	2	3	4	5	6	7	8	9	10	11	12	13	14	15	16	17	18	19	20
1	—																			
2	—	—																		
3	—	—	—																	
4	—	—	—	—																
5	—	—	—	—	0															
6	—	—	—	0	1	2														
7	—	—	—	0	1	3	4													
8	—	—	—	1	2	4	6	7												
9	—	—	0	1	3	5	7	9	11											
10	—	—	0	2	4	6	9	11	13	16										
11	—	—	0	2	5	7	10	13	16	18	21									
12	—	—	1	3	6	9	12	15	18	21	24	27								
13	—	—	1	3	7	10	13	17	20	24	27	31	34							
14	—	—	1	4	7	11	15	18	22	26	30	34	38	42						
15	—	—	2	5	8	12	16	20	24	29	33	37	42	46	51					
16	—	—	2	5	9	13	18	22	27	31	36	41	45	50	55	60				
17	—	—	2	6	10	15	19	24	29	34	39	44	49	54	60	65	70			
18	—	—	2	6	11	16	21	26	31	37	42	47	53	58	64	70	75	81		
19	—	0	3	7	12	17	22	28	33	39	45	51	57	63	69	74	81	87	93	
20	—	0	3	8	13	18	24	30	36	42	48	54	60	67	73	79	86	92	99	105
21	—	0	3	8	14	19	25	32	38	44	51	58	64	71	78	84	91	98	105	112
22	—	0	4	9	14	21	27	34	40	47	54	61	68	75	82	89	96	104	111	118
23	—	0	4	9	15	22	29	35	43	50	57	64	72	79	87	94	102	109	117	125
24	—	0	4	10	16	23	30	37	45	52	60	68	75	83	91	99	107	115	123	131
25	—	0	5	10	17	24	32	39	47	55	63	71	79	87	96	104	112	121	129	138
26	—	0	5	11	18	25	33	41	49	58	66	74	83	92	100	109	118	127	135	144
27	—	1	5	12	19	27	35	43	52	60	69	78	87	96	105	114	123	132	142	151
28	—	1	5	12	20	28	36	45	54	63	72	81	91	100	109	119	128	138	148	157
29	—	1	6	13	21	29	38	47	56	66	75	85	94	104	114	124	134	144	154	164
30	—	1	6	13	22	30	40	49	58	68	78	88	98	108	119	129	150	160	170	
31	—	1	6	14	22	32	41	51	61	71	81	92	102	113	123	134	145	155	166	177
32	—	1	7	14	23	33	43	53	63	74	84	95	106	117	128	139	150	161	172	184
33	—	1	7	15	24	34	44	55	65	76	87	98	110	121	132	144	155	167	179	190
34	—	1	7	16	25	35	46	57	68	79	90	102	113	125	137	149	161	173	185	197
35	—	1	8	16	26	37	47	59	70	82	93	105	117	129	142	154	166	179	191	203
36	—	1	8	17	27	38	49	60	72	84	96	109	121	134	146	159	172	184	197	210
37	—	1	8	17	28	39	51	62	75	87	99	112	125	138	151	164	177	190	203	217
38	—	1	9	18	29	40	52	64	77	90	102	116	129	142	155	169	182	196	210	223
39	—	2	9	19	30	41	54	66	79	92	106	119	133	146	160	174	188	202	216	230
40	—	2	9	19	31	43	55	68	81	95	109	122	136	150	165	179	193	208	222	237

544

TABLE C.13
Probabilities Associated with Values as Large as Observed Values of H
in the Kruskal-Wallis One-Way Analysis of Variance by Ranks

Sample sizes			H	p	Sample sizes			H	p
n_1	n_2	n_3			n_1	n_2	n_3		
2	1	1	2.7000	.500	4	3	2	6.4444	.008
								6.3000	.011
2	2	1	3.6000	.200				5.4444	.046
								5.4000	.051
2	2	2	4.5714	.067				4.5111	.098
			3.7143	.200				4.4444	.102
3	1	1	3.2000	.300	4	3	3	6.7455	.010
3	2	1	4.2857	.100				6.7091	.013
			3.8571	.133				5.7909	.046
								5.7273	.050
3	2	2	5.3572	.029				4.7091	.092
			4.7143	.048				4.7000	.101
			4.5000	.067	4	4	1	6.6667	.010
			4.4643	.105				6.1667	.022
3	3	1	5.1429	.043				4.9667	.048
			4.5714	.100				4.8667	.054
			4.0000	.129				4.1667	.082
3	3	2	6.2500	.011				4.0667	.102
			5.3611	.032	4	4	2	7.0364	.006
			5.1389	.061				6.8727	.011
			4.5556	.100				5.4545	.046
			4.2500	.121				5.2364	.052
3	3	3	7.2000	.004				4.5545	.098
			6.4889	.011				4.4455	.103
			5.6889	.029	4	4	3	7.1439	.010
			5.6000	.050				7.1364	.011
			5.0667	.086				5.5985	.049
			4.6222	.100				5.5758	.051
4	1	1	3.5714	.200				4.5455	.099
4	2	1	4.8214	.057				4.4773	.102
			4.5000	.076	4	4	4	7.6538	.008
			4.0179	.114				7.5385	.011
4	2	2	6.0000	.014				5.6923	.049
			5.3333	.033				5.6538	.054
			5.1250	.052				4.6539	.097
			4.4583	.100				4.5001	.104
			4.1667	.105	5	1	1	3.8571	.143
4	3	1	5.8333	.021	5	2	1	5.2500	.036
			5.2083	.050				5.0000	.048
			5.0000	.057				4.4500	.071
			4.0556	.093				4.2000	.095
			3.8889	.129				4.0500	.119

TABLE C.13 (continued)

n_1	n_2	n_3	H	p	n_1	n_2	n_3	H	p
5	2	2	6.5333	.008				5.6308	.050
			6.1333	.013				4.5487	.099
			5.1600	.034				4.5231	.103
			5.0400	.056	5	4	4	7.7604	.009
			4.3733	.090				7.7440	.011
			4.2933	.122				5.6571	.049
5	3	1	6.4000	.012				5.6176	.050
			4.9600	.048				4.6187	.100
			4.8711	.052				4.5527	.102
			4.0178	.095	5	5	1	7.3091	.009
			3.8400	.123				6.8364	.011
5	3	2	6.9091	.009				5.1273	.046
			6.8218	.010				4.9091	.053
			5.2509	.049				4.1091	.086
			5.1055	.052				4.0364	.105
			4.6509	.091	5	5	2	7.3385	.010
			4.4945	.101				7.2692	.010
5	3	3	7.0788	.009				5.3385	.047
			6.9818	.011				5.2462	.051
			5.6485	.049				4.6231	.097
			5.5152	.051				4.5077	.100
			4.5333	.097	5	5	3	7.5780	.010
			4.4121	.109				7.5429	.010
5	4	1	6.9545	.008				5.7055	.046
			6.8400	.011				5.6264	.051
			4.9855	.044				4.5451	.100
			4.8600	.056				4.5363	.102
			3.9873	.098	5	5	4	7.8229	.010
			3.9600	.102				7.7914	.010
5	4	2	7.2045	.009				5.6657	.049
			7.1182	.010				5.6429	.050
			5.2727	.049				4.5229	.099
			5.2682	.050				4.5200	.101
			4.5409	.098	5	5	5	8.0000	.009
			4.5182	.101				7.9800	.010
5	4	3	7.4449	.010				5.7800	.049
			7.3949	.011				5.6600	.051
			5.6564	.049				4.5600	.100
								4.5000	.102

TABLE C.14
Critical Values of F

The obtained F is significant at a given level if it is equal to or greater than the value shown in the table.
Upper entry is critical value for $\alpha = .05$ and lower entry is critical value for $\alpha = .01$.

Degrees of freedom for numerator

df (denom)	1	2	3	4	5	6	7	8	9	10	11	12	14	16	20	24	30	40	50	75	100	200	500	∞
1	161 / 4052	200 / 4999	216 / 5403	225 / 5625	230 / 5764	234 / 5859	237 / 5928	239 / 5981	241 / 6022	242 / 6056	243 / 6082	244 / 6106	245 / 6142	246 / 6169	248 / 6208	249 / 6234	250 / 6258	251 / 6286	252 / 6302	253 / 6323	253 / 6334	254 / 6352	254 / 6361	254 / 6366
2	18.51 / 98.49	19.00 / 99.01	19.16 / 99.17	19.25 / 99.25	19.30 / 99.30	19.33 / 99.33	19.36 / 99.34	19.37 / 99.36	19.38 / 99.38	19.39 / 99.40	19.40 / 99.41	19.41 / 99.42	19.42 / 99.43	19.43 / 99.44	19.44 / 99.45	19.45 / 99.46	19.46 / 99.47	19.47 / 99.48	19.47 / 99.48	19.48 / 99.49	19.49 / 99.49	19.49 / 99.49	19.50 / 99.50	19.50 / 99.50
3	10.13 / 34.12	9.55 / 30.81	9.28 / 29.46	9.12 / 28.71	9.01 / 28.24	8.94 / 27.91	8.88 / 27.67	8.84 / 27.49	8.81 / 27.34	8.78 / 27.23	8.76 / 27.13	8.74 / 27.05	8.71 / 26.92	8.69 / 26.83	8.66 / 26.69	8.64 / 26.60	8.62 / 26.50	8.60 / 26.41	8.58 / 26.30	8.57 / 26.27	8.56 / 26.23	8.54 / 26.18	8.54 / 26.14	8.53 / 26.12
4	7.71 / 21.20	6.94 / 18.00	6.59 / 16.69	6.39 / 15.98	6.26 / 15.52	6.16 / 15.21	6.09 / 14.98	6.04 / 14.80	6.00 / 14.66	5.96 / 14.54	5.93 / 14.45	5.91 / 14.37	5.87 / 14.24	5.84 / 14.15	5.80 / 14.02	5.77 / 13.93	5.74 / 13.83	5.71 / 13.74	5.70 / 13.69	5.68 / 13.61	5.66 / 13.57	5.65 / 13.52	5.64 / 13.48	5.63 / 13.46
5	6.61 / 16.26	5.79 / 13.27	5.41 / 12.06	5.19 / 11.39	5.05 / 10.97	4.95 / 10.67	4.88 / 10.45	4.82 / 10.27	4.78 / 10.15	4.74 / 10.05	4.70 / 9.96	4.68 / 9.89	4.64 / 9.77	4.60 / 9.68	4.56 / 9.55	4.53 / 9.47	4.50 / 9.38	4.46 / 9.29	4.44 / 9.24	4.42 / 9.17	4.40 / 9.13	4.38 / 9.07	4.37 / 9.04	4.36 / 9.02
6	5.99 / 13.74	5.14 / 10.92	4.76 / 9.78	4.53 / 9.15	4.39 / 8.75	4.28 / 8.47	4.21 / 8.26	4.15 / 8.10	4.10 / 7.98	4.06 / 7.87	4.03 / 7.79	4.00 / 7.72	3.96 / 7.60	3.92 / 7.52	3.87 / 7.39	3.84 / 7.31	3.81 / 7.23	3.77 / 7.14	3.75 / 7.09	3.72 / 7.02	3.71 / 6.99	3.69 / 6.94	3.68 / 6.90	3.67 / 6.88
7	5.59 / 12.25	4.74 / 9.55	4.35 / 8.45	4.12 / 7.85	3.97 / 7.46	3.87 / 7.19	3.79 / 7.00	3.73 / 6.84	3.68 / 6.71	3.63 / 6.62	3.60 / 6.54	3.57 / 6.47	3.52 / 6.35	3.49 / 6.27	3.44 / 6.15	3.41 / 6.07	3.38 / 5.98	3.34 / 5.90	3.32 / 5.85	3.29 / 5.78	3.28 / 5.75	3.25 / 5.70	3.24 / 5.67	3.23 / 5.65
8	5.32 / 11.26	4.46 / 8.65	4.07 / 7.59	3.84 / 7.01	3.69 / 6.63	3.58 / 6.37	3.50 / 6.19	3.44 / 6.03	3.39 / 5.91	3.34 / 5.82	3.31 / 5.74	3.28 / 5.67	3.23 / 5.56	3.20 / 5.48	3.15 / 5.36	3.12 / 5.28	3.08 / 5.20	3.05 / 5.11	3.03 / 5.06	3.00 / 5.00	2.98 / 4.96	2.96 / 4.91	2.94 / 4.88	2.93 / 4.86
9	5.12 / 10.56	4.26 / 8.02	3.86 / 6.99	3.63 / 6.42	3.48 / 6.06	3.37 / 5.80	3.29 / 5.62	3.23 / 5.47	3.18 / 5.35	3.13 / 5.26	3.10 / 5.18	3.07 / 5.11	3.02 / 5.00	2.98 / 4.92	2.93 / 4.80	2.90 / 4.73	2.86 / 4.64	2.82 / 4.56	2.80 / 4.51	2.77 / 4.45	2.76 / 4.41	2.73 / 4.36	2.72 / 4.33	2.71 / 4.31
10	4.96 / 10.04	4.10 / 7.56	3.71 / 6.55	3.48 / 5.99	3.33 / 5.64	3.22 / 5.39	3.14 / 5.21	3.07 / 5.06	3.02 / 4.95	2.97 / 4.85	2.94 / 4.78	2.91 / 4.71	2.86 / 4.60	2.82 / 4.52	2.77 / 4.41	2.74 / 4.33	2.70 / 4.25	2.67 / 4.17	2.64 / 4.12	2.61 / 4.05	2.59 / 4.01	2.56 / 3.96	2.55 / 3.93	2.54 / 3.91
11	4.84 / 9.65	3.98 / 7.20	3.59 / 6.22	3.36 / 5.67	3.20 / 5.32	3.09 / 5.07	3.01 / 4.88	2.95 / 4.74	2.90 / 4.63	2.86 / 4.54	2.82 / 4.46	2.79 / 4.40	2.74 / 4.29	2.70 / 4.21	2.65 / 4.10	2.61 / 4.02	2.57 / 3.94	2.53 / 3.86	2.50 / 3.80	2.47 / 3.74	2.45 / 3.70	2.42 / 3.66	2.41 / 3.62	2.40 / 3.60
12	4.75 / 9.33	3.88 / 6.93	3.49 / 5.95	3.26 / 5.41	3.11 / 5.06	3.00 / 4.82	2.92 / 4.65	2.85 / 4.50	2.80 / 4.39	2.76 / 4.30	2.72 / 4.22	2.69 / 4.16	2.64 / 4.05	2.60 / 3.98	2.54 / 3.86	2.50 / 3.78	2.46 / 3.70	2.42 / 3.61	2.40 / 3.56	2.36 / 3.49	2.35 / 3.46	2.32 / 3.41	2.31 / 3.38	2.30 / 3.36
13	4.67 / 9.07	3.80 / 6.70	3.41 / 5.74	3.18 / 5.20	3.02 / 4.86	2.92 / 4.62	2.84 / 4.44	2.77 / 4.30	2.72 / 4.19	2.67 / 4.10	2.63 / 4.02	2.60 / 3.96	2.55 / 3.85	2.51 / 3.78	2.46 / 3.67	2.42 / 3.59	2.38 / 3.51	2.34 / 3.42	2.32 / 3.37	2.28 / 3.30	2.26 / 3.27	2.24 / 3.21	2.22 / 3.18	2.21 / 3.16
14	4.60 / 8.86	3.74 / 6.51	3.34 / 5.56	3.11 / 5.03	2.96 / 4.69	2.85 / 4.46	2.77 / 4.28	2.70 / 4.14	2.65 / 4.03	2.60 / 3.94	2.56 / 3.86	2.53 / 3.80	2.48 / 3.70	2.44 / 3.62	2.39 / 3.51	2.35 / 3.43	2.31 / 3.34	2.27 / 3.26	2.24 / 3.21	2.21 / 3.14	2.19 / 3.11	2.16 / 3.06	2.14 / 3.02	2.13 / 3.00

Degrees of freedom for denominator

TABLE C.14 (continued)

Degrees of freedom for numerator

Degrees of freedom for denominator	∞	500	200	100	75	50	40	30	24	20	16	14	12	11	10	9	8	7	6	5	4	3	2	1
15	2.07 / 2.87	2.08 / 2.89	2.10 / 2.92	2.12 / 2.97	2.15 / 3.00	2.18 / 3.07	2.21 / 3.12	2.25 / 3.20	2.29 / 3.29	2.33 / 3.36	2.39 / 3.48	2.43 / 3.56	2.48 / 3.67	2.51 / 3.73	2.55 / 3.80	2.59 / 3.89	2.64 / 4.00	2.70 / 4.14	2.79 / 4.32	2.90 / 4.56	3.06 / 4.89	3.29 / 5.42	3.68 / 6.36	4.54 / 8.68
16	2.01 / 2.75	2.02 / 2.77	2.04 / 2.80	2.07 / 2.86	2.09 / 2.89	2.13 / 2.96	2.16 / 3.01	2.20 / 3.10	2.24 / 3.18	2.28 / 3.25	2.33 / 3.37	2.37 / 3.45	2.42 / 3.55	2.45 / 3.61	2.49 / 3.69	2.54 / 3.78	2.59 / 3.89	2.66 / 4.03	2.74 / 4.20	2.85 / 4.44	3.01 / 4.77	3.24 / 5.29	3.63 / 6.23	4.49 / 8.53
17	1.96 / 2.65	1.97 / 2.67	1.99 / 2.70	2.02 / 2.76	2.04 / 2.79	2.08 / 2.86	2.11 / 2.92	2.15 / 3.00	2.19 / 3.08	2.23 / 3.16	2.29 / 3.27	2.33 / 3.35	2.38 / 3.45	2.41 / 3.52	2.45 / 3.59	2.50 / 3.68	2.55 / 3.79	2.62 / 3.93	2.70 / 4.10	2.81 / 4.34	2.96 / 4.67	3.20 / 5.18	3.59 / 6.11	4.45 / 8.40
18	1.92 / 2.57	1.93 / 2.59	1.95 / 2.62	1.98 / 2.68	2.00 / 2.71	2.04 / 2.78	2.07 / 2.83	2.11 / 2.91	2.15 / 3.00	2.19 / 3.07	2.25 / 3.19	2.29 / 3.27	2.34 / 3.37	2.37 / 3.44	2.41 / 3.51	2.46 / 3.60	2.51 / 3.71	2.58 / 3.85	2.66 / 4.01	2.77 / 4.25	2.93 / 4.58	3.16 / 5.09	3.55 / 6.01	4.41 / 8.28
19	1.88 / 2.49	1.90 / 2.51	1.91 / 2.54	1.94 / 2.60	1.96 / 2.63	2.00 / 2.70	2.02 / 2.76	2.07 / 2.84	2.11 / 2.92	2.15 / 3.00	2.21 / 3.12	2.26 / 3.19	2.31 / 3.30	2.34 / 3.36	2.38 / 3.43	2.43 / 3.52	2.48 / 3.63	2.55 / 3.77	2.63 / 3.94	2.74 / 4.17	2.90 / 4.50	3.13 / 5.01	3.52 / 5.93	4.38 / 8.18
20	1.84 / 2.42	1.85 / 2.44	1.87 / 2.47	1.90 / 2.53	1.92 / 2.56	1.96 / 2.63	1.99 / 2.69	2.04 / 2.77	2.08 / 2.86	2.12 / 2.94	2.18 / 3.05	2.23 / 3.13	2.28 / 3.23	2.31 / 3.30	2.35 / 3.37	2.40 / 3.45	2.45 / 3.56	2.52 / 3.71	2.60 / 3.87	2.71 / 4.10	2.87 / 4.43	3.10 / 4.94	3.49 / 5.85	4.35 / 8.10
21	1.81 / 2.36	1.82 / 2.38	1.84 / 2.42	1.87 / 2.47	1.89 / 2.51	1.93 / 2.58	1.96 / 2.63	2.00 / 2.72	2.05 / 2.80	2.09 / 2.88	2.15 / 2.99	2.20 / 3.07	2.25 / 3.17	2.28 / 3.24	2.32 / 3.31	2.37 / 3.40	2.42 / 3.51	2.49 / 3.65	2.57 / 3.81	2.68 / 4.04	2.84 / 4.37	3.07 / 4.87	3.47 / 5.78	4.32 / 8.02
22	1.78 / 2.31	1.80 / 2.33	1.81 / 2.37	1.84 / 2.42	1.87 / 2.46	1.91 / 2.53	1.93 / 2.58	1.98 / 2.67	2.03 / 2.75	2.07 / 2.83	2.13 / 2.94	2.18 / 3.02	2.23 / 3.12	2.26 / 3.18	2.30 / 3.26	2.35 / 3.35	2.40 / 3.45	2.47 / 3.59	2.55 / 3.76	2.66 / 3.99	2.82 / 4.31	3.05 / 4.82	3.44 / 5.72	4.30 / 7.94
23	1.76 / 2.26	1.77 / 2.28	1.79 / 2.32	1.82 / 2.37	1.84 / 2.41	1.88 / 2.48	1.91 / 2.53	1.96 / 2.62	2.00 / 2.70	2.04 / 2.78	2.10 / 2.89	2.14 / 2.97	2.20 / 3.07	2.24 / 3.14	2.28 / 3.21	2.32 / 3.30	2.38 / 3.41	2.45 / 3.54	2.53 / 3.71	2.64 / 3.94	2.80 / 4.26	3.03 / 4.76	3.42 / 5.66	4.28 / 7.88
24	1.73 / 2.21	1.74 / 2.23	1.76 / 2.27	1.80 / 2.33	1.82 / 2.36	1.86 / 2.44	1.89 / 2.49	1.94 / 2.58	1.98 / 2.66	2.02 / 2.74	2.09 / 2.85	2.13 / 2.93	2.18 / 3.03	2.22 / 3.09	2.26 / 3.17	2.30 / 3.25	2.36 / 3.36	2.43 / 3.50	2.51 / 3.67	2.62 / 3.90	2.78 / 4.22	3.01 / 4.72	3.40 / 5.61	4.26 / 7.82
25	1.71 / 2.17	1.72 / 2.19	1.74 / 2.23	1.77 / 2.29	1.80 / 2.32	1.84 / 2.40	1.87 / 2.45	1.92 / 2.54	1.96 / 2.62	2.00 / 2.70	2.06 / 2.81	2.11 / 2.89	2.16 / 2.99	2.20 / 3.05	2.24 / 3.13	2.27 / 3.21	2.34 / 3.32	2.41 / 3.46	2.49 / 3.63	2.60 / 3.86	2.76 / 4.18	2.99 / 4.68	3.38 / 5.57	4.24 / 7.77
26	1.69 / 2.13	1.70 / 2.15	1.72 / 2.19	1.76 / 2.25	1.78 / 2.28	1.82 / 2.36	1.85 / 2.41	1.90 / 2.50	1.95 / 2.58	1.99 / 2.66	2.05 / 2.77	2.10 / 2.86	2.15 / 2.96	2.18 / 3.02	2.22 / 3.09	2.25 / 3.17	2.32 / 3.29	2.39 / 3.42	2.47 / 3.59	2.59 / 3.82	2.74 / 4.14	2.96 / 4.64	3.37 / 5.53	4.22 / 7.72
27	1.67 / 2.10	1.68 / 2.12	1.71 / 2.16	1.74 / 2.21	1.76 / 2.25	1.80 / 2.33	1.84 / 2.38	1.88 / 2.47	1.93 / 2.55	1.97 / 2.63	2.03 / 2.74	2.08 / 2.83	2.13 / 2.93	2.16 / 2.98	2.20 / 3.06	2.25 / 3.14	2.30 / 3.26	2.37 / 3.39	2.46 / 3.56	2.57 / 3.79	2.73 / 4.11	2.96 / 4.60	3.35 / 5.49	4.21 / 7.68
28	1.65 / 2.06	1.67 / 2.09	1.69 / 2.13	1.72 / 2.18	1.75 / 2.22	1.78 / 2.30	1.81 / 2.35	1.87 / 2.44	1.91 / 2.52	1.96 / 2.60	2.02 / 2.71	2.06 / 2.80	2.12 / 2.90	2.15 / 2.95	2.19 / 3.03	2.24 / 3.11	2.29 / 3.23	2.36 / 3.36	2.44 / 3.53	2.56 / 3.76	2.71 / 4.07	2.95 / 4.57	3.34 / 5.45	4.20 / 7.64
29	1.64 / 2.03	1.65 / 2.06	1.68 / 2.10	1.71 / 2.15	1.73 / 2.19	1.77 / 2.27	1.80 / 2.32	1.85 / 2.41	1.90 / 2.49	1.94 / 2.57	2.00 / 2.68	2.05 / 2.77	2.10 / 2.87	2.14 / 2.92	2.18 / 3.00	2.22 / 3.08	2.28 / 3.20	2.35 / 3.33	2.43 / 3.50	2.54 / 3.73	2.70 / 4.04	2.93 / 4.54	3.33 / 5.52	4.18 / 7.60
30	1.62 / 2.01	1.64 / 2.03	1.66 / 2.07	1.69 / 2.13	1.72 / 2.16	1.76 / 2.24	1.79 / 2.29	1.84 / 2.38	1.89 / 2.47	1.93 / 2.55	1.99 / 2.66	2.04 / 2.74	2.09 / 2.84	2.12 / 2.90	2.16 / 2.98	2.21 / 3.06	2.27 / 3.17	2.34 / 3.30	2.42 / 3.47	2.53 / 3.70	2.69 / 4.02	2.92 / 4.51	3.32 / 5.39	4.17 / 7.56

TABLE C.14 (continued)

Degrees of freedom for numerator

Degrees of freedom for denominator	1	2	3	4	5	6	7	8	9	10	11	12	14	16	20	24	30	40	50	75	100	200	500	∞
32	4.15 / 7.50	3.30 / 5.34	2.90 / 4.46	2.67 / 3.97	2.51 / 3.66	2.40 / 3.42	2.32 / 3.25	2.25 / 3.12	2.19 / 3.01	2.14 / 2.94	2.10 / 2.86	2.07 / 2.80	2.02 / 2.70	1.97 / 2.62	1.91 / 2.51	1.86 / 2.42	1.82 / 2.34	1.76 / 2.25	1.74 / 2.20	1.69 / 2.12	1.67 / 2.08	1.64 / 2.02	1.61 / 1.98	1.59 / 1.96
34	4.13 / 7.44	3.28 / 5.29	2.88 / 4.42	2.65 / 3.93	2.49 / 3.61	2.38 / 3.38	2.30 / 3.21	2.23 / 3.08	2.17 / 2.97	2.12 / 2.89	2.08 / 2.82	2.05 / 2.76	2.00 / 2.66	1.95 / 2.58	1.89 / 2.47	1.84 / 2.38	1.80 / 2.30	1.74 / 2.21	1.71 / 2.15	1.67 / 2.08	1.64 / 2.04	1.61 / 1.98	1.59 / 1.94	1.57 / 1.91
36	4.11 / 7.39	3.26 / 5.25	2.86 / 4.38	2.63 / 3.89	2.48 / 3.58	2.36 / 3.35	2.28 / 3.18	2.21 / 3.04	2.15 / 2.94	2.10 / 2.86	2.06 / 2.78	2.03 / 2.72	1.98 / 2.62	1.93 / 2.54	1.87 / 2.43	1.82 / 2.35	1.78 / 2.26	1.72 / 2.17	1.69 / 2.12	1.65 / 2.04	1.62 / 2.00	1.59 / 1.94	1.56 / 1.90	1.55 / 1.87
38	4.10 / 7.35	3.25 / 5.21	2.85 / 4.34	2.62 / 3.86	2.46 / 3.54	2.35 / 3.32	2.26 / 3.15	2.19 / 3.02	2.14 / 2.91	2.09 / 2.82	2.05 / 2.75	2.02 / 2.69	1.96 / 2.59	1.92 / 2.51	1.85 / 2.40	1.80 / 2.32	1.76 / 2.22	1.71 / 2.14	1.67 / 2.08	1.63 / 2.00	1.60 / 1.97	1.57 / 1.90	1.54 / 1.86	1.53 / 1.84
40	4.08 / 7.31	3.23 / 5.18	2.84 / 4.31	2.61 / 3.83	2.45 / 3.51	2.34 / 3.29	2.25 / 3.12	2.18 / 2.99	2.12 / 2.88	2.07 / 2.80	2.04 / 2.73	2.00 / 2.66	1.95 / 2.56	1.90 / 2.49	1.84 / 2.37	1.79 / 2.29	1.74 / 2.20	1.69 / 2.11	1.66 / 2.05	1.61 / 1.97	1.59 / 1.94	1.55 / 1.88	1.53 / 1.84	1.51 / 1.81
42	4.07 / 7.27	3.22 / 5.15	2.83 / 4.29	2.59 / 3.80	2.44 / 3.49	2.32 / 3.26	2.24 / 3.10	2.17 / 2.96	2.11 / 2.86	2.06 / 2.77	2.02 / 2.70	1.99 / 2.64	1.94 / 2.54	1.89 / 2.46	1.82 / 2.35	1.78 / 2.26	1.73 / 2.17	1.68 / 2.08	1.64 / 2.02	1.60 / 1.94	1.57 / 1.91	1.54 / 1.85	1.51 / 1.80	1.49 / 1.78
44	4.06 / 7.24	3.21 / 5.12	2.82 / 4.26	2.58 / 3.78	2.43 / 3.46	2.31 / 3.24	2.23 / 3.07	2.16 / 2.94	2.10 / 2.84	2.05 / 2.75	2.01 / 2.68	1.98 / 2.62	1.92 / 2.52	1.88 / 2.44	1.81 / 2.32	1.76 / 2.24	1.72 / 2.15	1.66 / 2.06	1.63 / 2.00	1.58 / 1.92	1.56 / 1.88	1.52 / 1.82	1.50 / 1.78	1.48 / 1.75
46	4.05 / 7.21	3.20 / 5.10	2.81 / 4.24	2.57 / 3.76	2.42 / 3.44	2.30 / 3.22	2.22 / 3.05	2.14 / 2.92	2.09 / 2.82	2.04 / 2.73	2.00 / 2.66	1.97 / 2.60	1.91 / 2.50	1.87 / 2.42	1.80 / 2.30	1.75 / 2.22	1.71 / 2.13	1.65 / 2.04	1.62 / 1.98	1.57 / 1.90	1.54 / 1.86	1.51 / 1.80	1.48 / 1.76	1.46 / 1.72
48	4.04 / 7.19	3.19 / 5.08	2.80 / 4.22	2.56 / 3.74	2.41 / 3.42	2.30 / 3.20	2.21 / 3.04	2.14 / 2.90	2.08 / 2.80	2.03 / 2.71	1.99 / 2.64	1.96 / 2.58	1.90 / 2.48	1.86 / 2.40	1.79 / 2.28	1.74 / 2.20	1.70 / 2.11	1.64 / 2.02	1.61 / 1.96	1.56 / 1.88	1.53 / 1.84	1.50 / 1.78	1.47 / 1.73	1.45 / 1.70
50	4.03 / 7.17	3.18 / 5.06	2.79 / 4.20	2.56 / 3.72	2.40 / 3.41	2.29 / 3.18	2.20 / 3.02	2.13 / 2.88	2.07 / 2.78	2.02 / 2.70	1.98 / 2.62	1.95 / 2.56	1.90 / 2.46	1.85 / 2.39	1.78 / 2.26	1.74 / 2.18	1.69 / 2.10	1.63 / 2.00	1.60 / 1.94	1.55 / 1.86	1.52 / 1.82	1.48 / 1.76	1.46 / 1.71	1.44 / 1.68
55	4.02 / 7.12	3.17 / 5.01	2.78 / 4.16	2.54 / 3.68	2.38 / 3.37	2.27 / 3.15	2.18 / 2.98	2.11 / 2.85	2.05 / 2.75	2.00 / 2.66	1.97 / 2.59	1.93 / 2.53	1.88 / 2.43	1.83 / 2.35	1.76 / 2.23	1.72 / 2.15	1.67 / 2.06	1.61 / 1.96	1.58 / 1.90	1.52 / 1.82	1.50 / 1.78	1.46 / 1.71	1.43 / 1.66	1.41 / 1.64
60	4.00 / 7.08	3.15 / 4.98	2.76 / 4.13	2.52 / 3.65	2.37 / 3.34	2.25 / 3.12	2.17 / 2.95	2.10 / 2.82	2.04 / 2.72	1.99 / 2.63	1.95 / 2.56	1.92 / 2.50	1.86 / 2.40	1.81 / 2.32	1.75 / 2.20	1.70 / 2.12	1.65 / 2.03	1.59 / 1.93	1.56 / 1.87	1.50 / 1.79	1.48 / 1.74	1.44 / 1.68	1.41 / 1.63	1.39 / 1.60
65	3.99 / 7.04	3.14 / 4.95	2.75 / 4.10	2.51 / 3.62	2.36 / 3.31	2.24 / 3.09	2.15 / 2.93	2.08 / 2.79	2.02 / 2.70	1.98 / 2.61	1.94 / 2.54	1.90 / 2.47	1.85 / 2.37	1.80 / 2.30	1.73 / 2.18	1.68 / 2.09	1.63 / 2.00	1.57 / 1.90	1.54 / 1.84	1.49 / 1.76	1.46 / 1.71	1.42 / 1.64	1.39 / 1.60	1.37 / 1.56
70	3.98 / 7.01	3.13 / 4.92	2.74 / 4.08	2.50 / 3.60	2.35 / 3.29	2.23 / 3.07	2.14 / 2.91	2.07 / 2.77	2.01 / 2.67	1.97 / 2.59	1.93 / 2.51	1.89 / 2.45	1.84 / 2.35	1.79 / 2.28	1.72 / 2.15	1.67 / 2.07	1.62 / 1.98	1.56 / 1.88	1.53 / 1.82	1.47 / 1.74	1.45 / 1.69	1.40 / 1.62	1.37 / 1.56	1.35 / 1.53
80	3.96 / 6.96	3.11 / 4.88	2.72 / 4.04	2.48 / 3.56	2.33 / 3.25	2.21 / 3.04	2.12 / 2.87	2.05 / 2.74	1.99 / 2.64	1.95 / 2.55	1.91 / 2.48	1.88 / 2.41	1.82 / 2.32	1.77 / 2.24	1.70 / 2.11	1.65 / 2.03	1.60 / 1.94	1.54 / 1.84	1.51 / 1.78	1.45 / 1.70	1.42 / 1.65	1.38 / 1.57	1.35 / 1.52	1.32 / 1.49

TABLE C.14 (continued)

Degrees of freedom for numerator

df (denom.)	1	2	3	4	5	6	7	8	9	10	11	12	14	16	20	24	30	40	50	75	100	200	500	∞
100	3.94 / 6.90	3.09 / 4.82	2.70 / 3.98	2.46 / 3.51	2.30 / 3.20	2.19 / 2.99	2.10 / 2.82	2.03 / 2.69	1.97 / 2.59	1.92 / 2.51	1.88 / 2.43	1.35 / 2.36	1.79 / 2.26	1.75 / 2.19	1.68 / 2.06	1.63 / 1.98	1.57 / 1.89	1.51 / 1.79	1.48 / 1.73	1.42 / 1.64	1.39 / 1.59	1.34 / 1.51	1.30 / 1.46	1.28 / 1.43
125	3.92 / 6.84	3.07 / 4.78	2.68 / 3.94	2.44 / 3.47	2.29 / 3.17	2.17 / 2.95	2.08 / 2.79	2.01 / 2.65	1.95 / 2.56	1.90 / 2.47	1.86 / 2.40	1.83 / 2.33	1.77 / 2.23	1.72 / 2.15	1.65 / 2.03	1.60 / 1.94	1.55 / 1.85	1.49 / 1.75	1.45 / 1.68	1.39 / 1.59	1.36 / 1.54	1.31 / 1.46	1.27 / 1.40	1.25 / 1.37
150	3.91 / 6.81	3.06 / 4.75	2.67 / 3.91	2.43 / 3.44	2.27 / 3.13	2.16 / 2.92	2.07 / 2.76	2.00 / 2.62	1.94 / 2.53	1.89 / 2.44	1.85 / 2.37	1.82 / 2.30	1.76 / 2.20	1.71 / 2.12	1.64 / 2.00	1.59 / 1.91	1.54 / 1.83	1.47 / 1.72	1.44 / 1.66	1.37 / 1.56	1.34 / 1.51	1.29 / 1.43	1.25 / 1.37	1.22 / 1.33
200	3.89 / 6.76	3.04 / 4.71	2.65 / 3.88	2.41 / 3.41	2.26 / 3.11	2.14 / 2.90	2.05 / 2.73	1.98 / 2.60	1.92 / 2.50	1.87 / 2.41	1.83 / 2.34	1.80 / 2.28	1.74 / 1.17	1.69 / 2.09	1.62 / 1.97	1.57 / 1.88	1.52 / 1.79	1.45 / 1.69	1.42 / 1.62	1.35 / 1.53	1.32 / 1.48	1.26 / 1.39	1.22 / 1.33	1.19 / 1.28
400	3.86 / 6.70	3.02 / 4.66	2.62 / 3.83	2.39 / 3.36	2.23 / 3.06	2.12 / 2.85	2.03 / 2.69	1.96 / 2.55	1.90 / 2.46	1.85 / 2.37	1.81 / 2.29	1.78 / 2.23	1.72 / 2.12	1.67 / 2.04	1.60 / 1.92	1.54 / 1.84	1.49 / 1.74	1.42 / 1.64	1.38 / 1.57	1.32 / 1.47	1.28 / 1.42	1.22 / 1.32	1.16 / 1.24	1.13 / 1.19
1000	3.85 / 6.66	3.00 / 4.62	2.61 / 3.80	2.38 / 3.34	2.22 / 3.04	2.10 / 2.82	2.02 / 2.66	1.95 / 2.53	1.89 / 2.43	1.84 / 2.34	1.80 / 2.26	1.76 / 2.20	1.70 / 2.09	1.65 / 2.01	1.58 / 1.89	1.53 / 1.81	1.47 / 1.71	1.41 / 1.61	1.36 / 1.54	1.30 / 1.44	1.26 / 1.38	1.19 / 1.28	1.13 / 1.19	1.08 / 1.11
∞	3.84 / 6.64	2.99 / 4.60	2.60 / 3.78	2.37 / 3.32	2.21 / 3.02	2.09 / 2.80	2.01 / 2.64	1.94 / 2.51	1.88 / 2.41	1.83 / 2.32	1.79 / 2.24	1.75 / 2.18	1.69 / 2.07	1.64 / 1.99	1.57 / 1.87	1.52 / 1.79	1.46 / 1.69	1.40 / 1.59	1.35 / 1.52	1.28 / 1.41	1.24 / 1.36	1.17 / 1.25	1.11 / 1.15	1.00 / 1.00

Degrees of freedom for denominator

550

TABLE C.15
Critical Values of T in the Wilcoxon Matched-Pairs Signed-Ranks Test

	Level of significance for one-tailed test		
	.025	.01	.005
n	Level of significance for two-tailed test		
	.05	.02	.01
6	0	—	—
7	2	0	—
8	4	2	0
9	6	3	2
10	8	5	3
11	11	7	5
12	14	10	7
13	17	13	10
14	21	16	13
15	25	20	16
16	30	24	20
17	35	28	23
18	40	33	28
19	46	38	32
20	52	43	38
21	59	49	43
22	66	56	49
23	73	62	55
24	81	69	61
25	89	77	68

Appendix D
Answers to
Review Questions and
Selected Problems

Answers to Review Questions in Chapter 1

1 b   2 b   3 c   4 b   5 a   6 b   7 a   8 a   9 b   10 b
11 b   12 b   13 b

Answers to Selected Problems in Chapter 1: Set 1

1 Population of interest	Variable name	Value set	Type of scale
a adult humans	blood type	A, B, O, AB	nominal
b students in class	statistics achievement	high, low	ordered categories
c graduate students in psychology	statistics aptitude	scores: 0, 1, 2, . . . , 50	interval
d twenty girls	"blondness"	1, 2, . . . , 20	rank-order
e elementary schools	type of school	public, private	nominal
f individuals	short-term memory	unit = 1 sec: 0, 1, 2, ...	interval
g small children	handwriting ability	has it, doesn't have it	nominal
h small children	handwriting ability	unit = 1 word: 0, 1, 2, . . . , 10	interval
i families	employment of fathers	have jobs, do not have jobs	nominal
j families	family income	unit = 1 dollar	interval
k families	economic status	list of jobs and professions ranked by status	rank-order
l high schools	educational effectiveness	unit = 1 student: 0, 1, 2, . . .	interval
m American adults	daily diet	unit = 1 calorie: 0, 1, 2, . . .	interval
n American adults	car color preference	color names	nominal

3 Because each question does not measure equal amounts of achievement, two subjects might get scores of 5, yet one may have more achievement than the other. A score of 7 might mean more or less achievement than 6, depending on *which questions* were answered correctly. As a variable, the scale would not be very accurate at measuring achievement because the values wouldn't really differentiate among subjects.

Set 2

1 Inferential. Univariate. The number of cigarettes smoked per day is an interval scale variable.

2 Descriptive. Univariate. Car color is a nominal variable.

5 Inferential. Univariate. The rating scale is a set of ordered categories.

6 Inferential. Bivariate. Experimental. The independent variable is training program, with the values control and experimental. The dependent variable is the leadership scale, which is probably an interval scale. Because the subjects are matched, there will be two sets of dependent measures.

10 Inferential. Bivariate. Pseudo-experimental. The independent variable is grade level with the values 9 and 12. The dependent variable is career choice with the values "same as choice 15 years earlier" or "different from choice 15 years earlier."

Answers to Review Questions in Chapter 2

1 b   2 b   3 a   4 b   5 b   6 a   7 a   8 b   9 a   10 b
11 a   12 b   13 a   14 b   15 b   16 a   17 b   18 b

Answers to Selected Problems in Chapter 2: Set 1

Favorite color	Frequency
Red	95
Blue	59
Yellow	31
Green	17
Other	35
Total	237

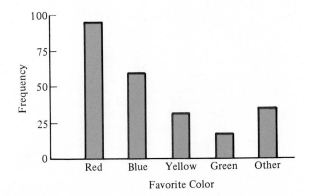

4 Category of expenditure	Amount in thousand dollars
Professorial Salaries	92.0
Overhead	92.0
Assistants' Salaries	15.1
Secretarial Salaries	8.4
Supplies	2.0
Total	209.5

Set 2

1 The following are discrete interval variables: B, C, D, E, G, H, I. The only question here is that family income might be considered continuous if we consider that theoretically the unit can be subdivided because money works on the decimal system.

These are continuous interval variables: A, F, J.

The discrete interval variables that represent underlying continuous variables are as follows: D, H. I.

2 a

Frequency Distribution for Income 125 Families, Tranquility, New York

Annual income in thousands of dollars	Frequency	Relative frequency
4,000 up to 6,000	2	.016
6,000 up to 8,000	5	.040
8,000 up to 10,000	8	.064
10,000 up to 12,000	12	.096
12,000 up to 14,000	10	.080
14,000 up to 16,000	11	.088
16,000 up to 18,000	13	.104
18,000 up to 20,000	11	.088
20,000 up to 22,000	16	.128
22,000 up to 24,000	10	.080
24,000 up to 26,000	12	.096
26,000 up to 28,000	9	.072
28,000 up to 30,000	4	.032
30,000 up to 32,000	2	.016
	125	1.000

b i   The real limits are $4,000, $6,000, $8,000, and so on up to $32,000.
ii   12 percent   iii   12 percent   iv   32 percent   v   The answers will be the same. See the following graph.

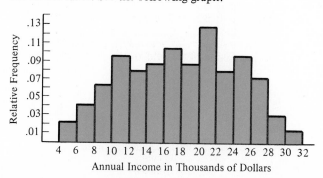

vi The real limits would be changed. For example, a recording of 4,000 could be anywhere from 3,000 up to 5,000. Thus, the even numbers 4,000, 6,000, 8,000 would become the midpoints of the bars, rather than the endpoints. This would change the shape of the distribution. To see how the shape would change, you might draw the histogram.

4 a

Interval	Apparent Limits	Real Limits*
1	0, 49	−.5, 49.5
2	50, 99	49.5, 99.5
3	100, 149	99.5, 149.5
4	150, 199	149.5, 199.5

*Note that in this case the real limits have no meaning because the subjects can't be continuously divided. Also, it is hard to imagine any underlying continuum that this variable is measuring.

b

Interval	Apparent Limits	Real Limits
1	55, 64	54.5, 64.5
2	65, 74	64.5, 74.5
3	75, 84	74.5, 84.5
4	85, 94	84.5, 94.5
5	95, 104	94.5, 104.5
6	105, 114	104.5, 114.5
7	115, 124	114.5, 124.5
8	125, 134	124.5, 134.5

The real limits show the cut-off points on the underlying continuous scale of weight, thus enabling us to classify each person weighed into one and only one of our score intervals.

c The endpoints are the real limits. The information is a picture of what is stated in the previous answer. It shows that any point between two real limits is a possible weight value, even though the measures are to the nearest kilo.

6 a 6   b 34   c $50 - 15 = 35$ d $43 - 9 = 34$ e 73.5   f 87

7 Relabel the vertical axis in Fig. 2.8 by doubling each frequency value. Because there are 50 scores, this converts from frequency to percent.
a 62   b 85   c 80   d 11   e 35

8 a By converting the answer back to frequencies and dividing each percentage by 2.
b By multiplying frequencies by 2 to convert to percentages.

10 a Most family incomes fall within a reasonable range but a few families have very large incomes. We would therefore expect the distribution to look like the following:

555

b Because heads and tails are equally likely to occur the frequency distribution for the results might look like the following:

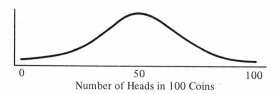

Number of Heads in 100 Coins

c The frequency distribution for SAT scores would look something like the following:

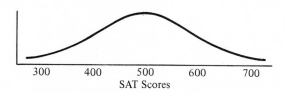

SAT Scores

d If everyone is guessing on all the questions, then not very many people are going to get high scores. The distribution would thus look something like this:

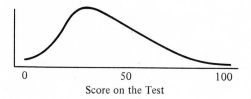

Score on the Test

Answers to Review Questions in Chapter 3

1 a   2 a   3 b   4 b   5 a   6 a   7 a   8 b   9 b   10 b
11 b   12 a   13 a   14 b   15 a   16 b

Answers to Selected Problems in Chapter 3: Set 1

1 a $\frac{33}{163}$ = .202 or 20.2 percent   b $\frac{101}{163}$ = .6196 or 62.0 percent

   c $\frac{64}{163}$ = .3926 or 39.3 percent   d $\frac{33}{101}$ = .3267 or 32.7 percent

   e They are not the same. The proportion of those in favor within the population is not the same as the proportion of those in favor within the male population.

   f

Sex	Proportion in favor
Male	.516
Female	.484
Total	1.000

Sex	Proportion opposed	Proportion with no opinion
Male	.697	.652
Female	.303	.348
Total	1.000	1.000

h We notice that the three conditional distributions show three different distributions on the variable "sex." This illustrates what we mean when we say the two variables are related. To establish the fact that the two variables are related, we could also have looked at the two conditional distributions for opinion.

2 a In this population, 4 percent are male agriculture graduates.
   b In this population, 65 percent of the population are male. We wish to know what percentage are in the agriculture school. We make an equation:

$$(X)(65 \text{ percent}) = 4 \text{ percent};$$

then, $X = \dfrac{4 \text{ percent}}{65 \text{ percent}} = .062 \text{ or } 6.2 \text{ percent.}$

   c In this population, 9 percent of the graduates are male psychologists.
   d We wish to know what percentage of the 65 percent who are male graduates are also psychologists. We make an equation:

$$(X)(65 \text{ percent}) = 9 \text{ percent};$$

then, $X = \dfrac{9 \text{ percent}}{65 \text{ percent}} = 13.8 \text{ percent.}$

   e In this population, 15 percent of all graduates are in psychology. We wish to know what percentage of the 15 percent are males. We make an equation:

$$(X)(15 \text{ percent}) = 9 \text{ percent};$$

then, $X = \dfrac{9 \text{ percent}}{15 \text{ percent}} = 60 \text{ percent.}$

f and g

	Males	Females	Marginal
Agriculture	6.2	2.9	5
Psychology	13.8	17.1	15
English	38.5	31.4	36
Science	35.4	22.9	31
Nursing	6.2	25.7	13
Total	100.0*	100.0	100

*Sum is not exact because of rounding.

h and i The marginal distribution tells us what percentage of the total number of graduates is in each school. The conditional distribution for males tells us what percentage of all males are in each school and the conditional distribution for females tells us what percentage of all females are in each school. The results show these distributions are not the same. Thus, the distribution of school in university depends on which group you are talking about, males or females. For this reason we say the two variables are related.

Set 2

1 a Because high-school rank occurs in time before success in college, we would consider high-school rank as the independent variable.

   b Because the number of passess occurs in time before the outcome of game, we might consider outcome of the game as the dependent variable. As we noted earlier, however, the plausible direction of causation might go in the other direction as well, so it is not possible to establish one direction of causation.

   c It is not possible to establish a direction of causation.

   d Pretest would be considered the independent variable.

   e If we believe that intelligence is innate and leadership is trainable, then intelligence is the independent variable.

   f Level of education would be the independent variable.

   g Mental alertness could be a function of sleep deprivation, so sleep deprivation would be the independent variable.

   h Causation is not really applicable here, but for data analysis we would consider sex as the independent variable because it occurs first in time.

   i Causation could go either way in this example.

   j Causation could go either way in this example.

   k If the question is "does smoking cigarettes cause lung cancer?" we would choose smoking as the independent variable.

   l Age is the independent variable.

   m Age is the independent variable.

   n Causation is not applicable in this example.

Set 3

1 a A relationship will be linear if, as the values of one variable increase, the values of the other variable increase or if, as the values of one variable increase, the values of the other variable decrease, and if these changes are at a constant rate; that is, if $Y = a + bX$. Otherwise the relationship is nonlinear.

   b If as one variable increases, so do values on the other variable.

   c If as one variable increases, the values on the other decrease.

   d If as one variable increases, there is no consistent change in the other variable.

3 a Nonlinear   b Nonlinear   c No relationship   d No relationship

   e Possibly nonlinear. People in the low intelligence group will probably show little improvement. People in the high intelligence group would probably show little improvement because they presumably scored well on the pretest. People in the average intelligence group will probably show the largest amounts of improvement. These reasons could make the scatterdiagram look nonlinear. This example illustrates one of the disadvantages of improvement scores as a measure of performance.

   f possibly a negative relationship   g Nonlinear   h A positive linear relationship

Set 4

1 a The posttest scores tend to be higher than the pretest scores, thus suggesting that the training program improves visual motor integration.

558

Answers

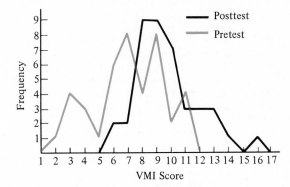

The graphs show how the distribution of VMI scores on the posttest is higher on the VMI scale. Notice that the major concentration of scores on the pretest is from 5 to 10, and on the posttest from 7 to 11.

4 a Use relative frequencies because the groups are unequal in size.

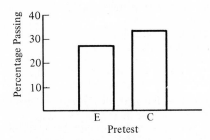

The bar graph shows little difference in the initial performance of the two groups.

b

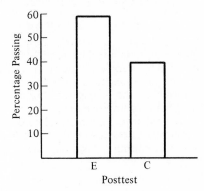

This shows that the experimental group has scored 20 percent higher than the control group on the posttest.

c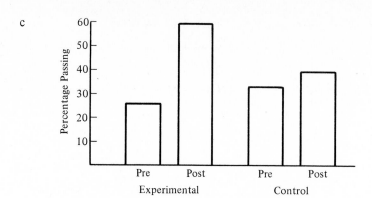

This shows the experimental group improving from 26.2 percent to 59.5 percent from pretest to posttest, whereas the control group improved from 33.3 percent to 40 percent from pretest to posttest.

d Control   e Experimental   f Experimental

g Part of the improvement could be attributed to development because the control group also improved. But for the most part, the improvement can be attributed to the training if all other variables are controlled.

Set 5

1 Make a frequency distribution. Because the sample is large, a grouped frequency distribution will be effective. Because the variable is interval, a histogram or frequency polygon would be useful, if the data is grouped.

2 Make a frequency distribution showing the list of colors and how frequently each color is chosen. Because the numbers are large, percentages or relative frequencies will be useful. A bar graph would adequately picture this.

5 Make a relative frequency distribution showing the values 1, 2, 3, 4 and the percentage of the total sample favoring each option. Because this is a set of ordered categories, a bar graph could be used to picture the result where the bars are in order (1, 2, 3, 4) corresponding to the choices.

6 List the score of each trained politician next to the score of the untrained counterpart. Use two frequency polygons.

10 Cross-classify the subjects on grade level and career choice agreement. Use bar graphs to compare the conditional distributions of career choice for each grade level.

Answers to Review Questions in Chapter 4

1 a	2 b	3 a	4 b	5 b	6 a	7 b	8 b	9 a	10 b
11 b	12 a	13 a	14 a	15 a	16 b	17 b	18 a		
19 a	20 a	21 b	22 a	23 b					

Answers to Selected Problems in Chapter 4: Set 1

1 The mode is male because it is the value that occurs most frequently in the distribution.

2 The typical subject would be right-handed because right-handed is the mode of the distribution.

Set 2

1 $P = \dfrac{33 + (1/2 \times 1)}{40} \times 100 = 83.75$

2 $P = \dfrac{27 + (1/2 \times 3)}{40} \times 100 = 71.25$

3 $P = \dfrac{0 + .5}{40} \times 100 = 1.25$

4 Joe has percentile rank

$$P = \frac{53 + (1/2 \times 6)}{70} \times 100 = 80.$$

Sam has percentile rank

$$P = \frac{26 + (1/2 \times 4)}{70} \times 100 = 40.$$

Joe's percentile rank is twice as large as Sam's, but these ranks tell us only the proportion of students falling below their scores. The ranks say nothing about the amount of fitness that the students have.

Set 3

1 Mean = 40/10 = 4; median = 2; mode = 0. The median is probably more accurate in describing a typical score because the distribution is skewed.

2 Squares are 0, 0, 0, 0, 1, 9, 25, 49, 49, 289. Average of square scores = 422/10 = 42.2. Note: We will make use of this average in Set 4 of these problems.

3 Deviations are −4, −4, −4, −4, −3, −1, 1, 3, 3, 13. Average deviation is 0 since sum of deviation scores is zero. This illustrates a fundamental property of the mean of a distribution.

5 a Mean score = 399/200 = 2.00.
   b The median is the score ranked 100 or 101, that is, 2.

6 a Because the median is the fiftieth percentile, we can use the method of arithmetic estimation to estimate the median. The median class interval is 47–45. So we estimate the median to be at least 44.5 but not more

than 47.5. To sharpen this estimate we use the formula for estimating percentiles:

$$\text{Median} = 44.5 + \frac{3(33 - 31)}{20} = 44.5 + .3 = 44.8$$

b Because the distribution has extreme scores on the lower end of the scale, we could say the distribution is negatively skewed.

Set 4

1 a $\sigma = \sqrt{\dfrac{\Sigma(X - 4)^2}{10}}$

$$\sigma = \sqrt{\frac{[(-4)^2 + (-4)^2 + (-4)^2 + (-4)^2 + (-3)^2 + (-1)^2 + 1^2 + 3^2 + 3^2 + 13^2)]}{10}}$$

$$= \sqrt{\frac{16 + 16 + 16 + 16 + 9 + 1 + 1 + 9 + 9 + 169}{10}}$$

$$= \sqrt{\frac{262}{10}}$$

$$= \sqrt{26.2}$$

$\sigma = 5.12$

b $\sigma = \sqrt{\dfrac{\Sigma X^2}{10} - 4^2} = \sqrt{\dfrac{422}{10} - 16} = \sqrt{26.2} = 5.12$

2 a The range goes from 5 to 21 years. If we consider number of years as a discrete variable, we report $(21 - 5)$ as the range. If we consider the variable as a continuous variable, we report $(21.5 - 4.5)$ as the range.

b The interquartile range is the distance between the twenty-fifth and seventy-fifth percentiles. We will consider the variable to be continuous. We see that 25 percent of $125 = 31.25$. This is the number of scores at or below the twenty-fifth percentile. From the frequency distribution we see 19 scores at or below 9.5 and 32 scores at or below 10.5. This tells us that the twenty-fifth percentile is between 9.5 and 10.5 years of education. To sharpen our estimate we use the method of arithmetic estimation.

$$P_{25} = 9.5 + 1 \frac{(31.25 - 19)}{32 - 19} = 9.5 + .94 = 10.44.$$

We use the same logic to get P_{75}, so

$$P_{75} = 15.5 + 1 \frac{(93.75 - 90)}{105 - 90} = 15.5 + .25 = 15.75.$$

$$P_{75} - P_{25} = 15.75 - 10.44 = 5.31.$$

The middle 50 percent of the observations are in this range. The middle half of the population have between 10.44 and 15.75 years of education. Note that the median is almost in the middle of this range, thus reflecting the lack of skewness of the distribution.

c Use the computational formula:

$$\Sigma X^2 = 22{,}594 \qquad\qquad \mu^2 = 168.3765$$

$$\frac{\Sigma X^2}{125} = 180.75 \qquad\qquad \sigma = \sqrt{(180.75 - 168.3765)} = \sqrt{12.3737}$$

$$\sigma = 3.52$$

Although the head of the household has an average of 12.89 years of education, we estimate that any head whose years of education is within 3.5 years of 12.89 could be considered within the average in the group. That is, those heads of households whose years of education are greater than 9.37 but less than 16.41 years would be considered in the average group on this variable.

5 a The mean score under the two conditions are very similar, thus suggesting that on the average the two groups perform about the same. However, the scores in the experimental group vary much more than those in the control group, thus suggesting that the treatment possibly helped some students and possibly hindered others.

b In this case, the results suggest that the treatment improved scores on the average and that, because the scores varied in the same way in both distributions, the scores in the experimental group were, in general, higher than those in the control group. We would thus expect to find that treating test anxiety is helpful in improving performances in statistics.

Set 5

1 a 74.5 is the twenty-ninth percentile.
b 94.5 is the 74.5th percentile.
c 79 has a percentile rank of approximately 41.
d The percentile rank for 86 kilos is approximately 57.5.
e The median is approximately 82.5 kilograms.
f $P_{25} = 73.5$ while $P_{75} = 95$, so the interquartile range is $(95 - 73.5)$. (Again, these answers depend on how carefully you draw the graph.)

4 a i $z = \dfrac{44 - 36}{6} = \dfrac{8}{6} = 1.33$ iv $z = \dfrac{39 - 36}{6} = .5$

ii $z = \dfrac{27 - 36}{6} = \dfrac{-9}{6} = -1.5$ v $z = \dfrac{31 - 36}{6} = \dfrac{-5}{6} = -.83$

iii $z = \dfrac{36 - 36}{6} = \dfrac{0}{6} = 0$ vi $z = \dfrac{57 - 36}{6} = \dfrac{21}{6} = 3.5$

b i $X = (1)(6) + 36 = 42$ iv $X = (3.2)(6) + 36 = 55.2$
ii $X = (-1.5)(6) + 36 = 27$ v $X = (-.67)(6) + 36 = 31.98$
iii $X = (0)(6) + 36 = 36$ vi $X = (-2.1)(6) + 36 = 23.4$

5 a μ_X corresponds to $z = 0$, so $\mu_X = 36$.
b A score that is 1 standard deviation above the mean has a z-score of 1. Thus, 42 has a z-score of 1. Because $42 - 36 = 6$, $\sigma_X = 6$.
c $a = -2, b = -1.5, c = -1.0, d = -.5, e = .5, f = 1.5, g = 2$.

d $-.5 - (-2) = 1.5; 33 - 24 = 9$. Notice that $9 = (1.5)(6.)$
e 2 units, on the z-scale, 12 units on the X-scale. (Notice that $12 = 2 \times 6$.)
f Length on X-scale = (length on z-scale) $\times \sigma$.

Answers to Review Questions in Chapter 5

1 b	2 b	3 b	4 a	5 b	6 a	7 b	8 b	9 b	10 b
11 b	12 a	13 a	14 b	15 b	16 a	17 b	18 a		
19 a	20 a								

Answers to Selected Problems in Chapter 5: Set 1

1 a Because $N = 2000$, the first step is to get the *observed frequency* in each cell. For example, 4 percent $\times$ 2000 = 80.00 in the upper right cell.

	Male	Female	Total
Agriculture	80	20	100
Psychology	180	120	300
Engineering	500	220	720
Science	460	160	620
Nursing	80	180	260
Total	1300	700	2000

b Calculate the *expected* frequency for each cell based on the marginal totals. For example, in the upper right cell, we expect $1300 \times \dfrac{100}{2000} = 65$ if there is no relationship.

	Male	Female	Total
Agriculture	65	35	100
Psychology	195	105	300
Engineering	468	252	720
Science	403	217	620
Nursing	169	91	260
Total	1300	700	2000

c Next, we compare the table of observed frequencies with the expected frequencies. For example, in the upper right cell we have $(80 - 65)^2/65 = 3.4615$. We do this for the 8 cells of the cross tabulation table. Then, we add the results:

$$3.4615 + 6.4286 + 1.1538 + 2.1429 + 2.1880 + 4.0635 + 8.0620$$
$$+ 14.9724 + 46.8698 + 87.0439 = 176.3865.$$

Then $\phi' = \sqrt{\dfrac{176.3865}{1 \times 2000}} = .297$.

d Because $\phi' = .297$, there is some relationship between the variables "sex of degree recipient" and "university school." But because .297 is on the low end of the scale from 0 to 1, we conclude that the relationship is not strong. In other words, the percentages graduating from each school are not the same as the percentages of males graduating from each school and they are not the same as the percentages of females graduating from each school. The value .297 is a measure of these disagreements. The fact that it is small indicates that the degree of disagreement is not large, based on 2000 observations.

3
$$\phi' = \sqrt{\frac{27.92}{1 \times 1371}} = .14.$$

This result suggests a very small relationship between performance on the exam and the sex of the applicant. The overall pass rate for all applicants is .62, whereas the pass rate for males is .65 and the pass rate for females is .48. If there were no relationship, both pass rates would be .62. The value .14 indicates that the departure from the expected pass rate of .62 is not all that large.

Set 2

1 $\Sigma D^2 = 46; N = 9, N^2 - 1 = 80.$

Substituting in the formula $\rho_s = 1 - \dfrac{6\Sigma d^2}{N(N^2 - 1)}$, we have

$\rho_s = 1 - \dfrac{6 \times 46}{9 \times 80} = 1 - \dfrac{276}{720} = 1 - .38 = .62$; thus, $\rho_s = .62$, which suggests that those who attend longer tend to lose the most weight.

Set 3

1 Let X = pretest and let Y = points of gain.
  a $\rho = -.95$
  b $\mu_X = 3, \sigma_X = 2, \mu_Y = 4, \sigma_Y = 2$

Standard Scores	
Pretest	Gain
−1.5	1.5
−1.5	1.0
− .5	.5
− .5	1.0
− .5	.5
.5	− .5
.5	− .5
1.0	−1.5
1.0	− .5
1.5	−1.5

  c $\rho = -.95$. This shows that ρ is the same whether you correlate raw scores or standard scores.

3 Let X = arithmetic score and let Y = midterm score.
  a $\Sigma X = 934$ $\Sigma X^2 = 40,266$ $\Sigma XY = 25,218$
     $\Sigma Y = 589$ $\Sigma Y^2 = 16,183$ $N = 22$
     $\rho = .42$

b Slope $= \rho \dfrac{\sigma_Y}{\sigma_X} = .42 \times \dfrac{4.34}{5.28} = .35$, and

$a = \mu_Y - .35\,\mu_X = 26.77 - (.35)\,42.45 = 12.08$;
so $Y_{PRE} = 12.08 + (.35)\,X$.

c Standard error of estimate $= \sqrt{\sigma_Y^2(1 - \rho^2)} = \sqrt{18.81(1 - .18)}$
$$= \sqrt{(18.81)(.82)} = \sqrt{15.43}$$
$$= 3.93$$

d The standard deviation of midterm scores is 4.34. This indicates the approximate average amount by which the scores differ from their mean. Using arithmetic as a predictor, we see that the approximate average amount by which the true midterm scores differ from the predicted values is 3.93. There is not much predictive value in the arithmetic scores.

e $\rho^2 = .18$. This means that 18 percent of the variation of midterm scores is explained by the linear relationship between the two variables. Thus, 82 percent of the variation is unexplained by the linear relationship. This tells us that the relationship is not strong.

Set 4

1 The mean and standard deviation would be useful. However, if the distribution is skewed, the median and range would be good descriptors. It would also be important to note whether the distribution is bimodal.

2 Because the scale is nominal, the mode is the only measure to use.

5 You could find the median response because we are dealing here with an ordinal scale. But because the scale is small (only four values) it would probably be more useful just to present the mode.

6 For each matched pair of politicians, calculate the difference between their final leadership scores by subtracting the score achieved by the untrained politician from the score achieved by the trained politician. Average these differences over all 10 pairs. If the average is greater than 0, trained politicians have scored higher on the average. Report the standard deviation of the differences.

10 You could look at the difference in the proportions of agreement and at the difference in the proportions of nonagreement to compare the two groups.

12 Bivariate. Correlational. The variables are arithmetic ability with values that form an interval scale, and statistics achievement with values that form an interval scale. Make a scatter diagram and calculate ρ.

14 Correlation. One variable is judge, a nominal scale variable, and the other is rank of pie, an ordinal scale variable. Use a list showing the rank given to each pie by each judge. To measure the relationship, ρ_s can be used.

Answers to Review Questions in Chapter 6

1 a	2 a	3 a	4 a	5 b	6 a	7 a	8 a	9 b	10 b
11 a	12 b	13 a	14 b	15 a	16 b	17 a	18 a		
19 b	20 b	21 b							

Answers to Selected Problems in Chapter 6: Set 1

1 a .95   b .05

2 a .10   b .90

4 a .84   b .10   c .29   d .03   3 .28
   f We need to know what percentage of the total group are either full professors or females or possibly both. In this case, 16 percent are female
and 29 percent are full professors, totaling 45 percent; however, the 2
percent who fall into both categories are counted twice so we subtract
2 percent, leaving 43 percent. The answer is .43.

5 and 6
   a Problem 1 is binomial if we say the seed either will germinate or will not
germinate. If we define success as the seed germinates, $pr(S) = .95$ and
$pr(F) = .05$.
   b Problem 2 is binomial if we say a person either has the trait or does not
have the trait. If we define success as having the trait, $pr(S) = .1$ and
$pr(F) = .9$.
   c If we define success in Problem 3 as getting a question correct, if $pr(S) =$
.5 and $pr(F) = .5$ for each question, and if the questions are independent,
the score on the test is the variable used to summarize a binomial experiment with $n = 10$. But "score" itself is not binomial because score has 11
values.
   d Problem 4 is a bivariate problem. One variable is binomial, the other is not.

8 a S = has trait; F = does not have trait.
   b Selecting one person at random and determining S or F.
   c $n = 5$. There are 5 trials.
   d $r = 0, 1, 2, 3, 4, 5.$
   e $P = .1; Q = .9$
   f

Value of r	Number of ordered outcomes
0	1
1	5
2	10
3	10
4	5
5	1

   g $pr(r = 0) = .590; pr(r = 1) = .328, pr(r = 2) = .073, pr(r = 3) = .008,$
$pr(r = 4) = .000, pr(r = 5) = .000.$
   h

Value of r	$pr(r)$
0	.590
1	.328
2	.073
3	.008
4	.000+
5	.000+

i

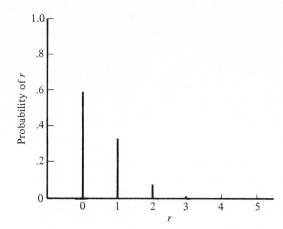

Set 2

1 a $H_0: P = .95$ (The proportion of seeds that germinates is .95.)
   b $H_a: P < .95$ (The proportion of seeds that germinates is less than .95.)
   c Success is "the seed germinates."
   d r = number of seeds that germinate in a random sample of 100.
   e If $P = .95$, the probability distribution of r is the binomial distribution with $P = .95, n = 100$.
   f If we say that less than 95 percent of the seeds germinate when in reality the company's claim is correct, we make a Type I error.
   g If we say that the company's claim is correct when in fact less than 95 percent of corn seeds germinate, we make a Type II error.
   h The probability of rejecting the hypothesis that the company's claim is correct when in fact fewer than 95 percent of corn seeds germinate is the meaning of power.

Set 3

1 a $pr(r = 0) = .130, pr(r = 1) = .346, pr(r = 2) = .346$
   b $pr(r = 0) = .000, pr(r = 1) = .007, pr(r = 2) = .135$
   c $pr(r = 0) = .168, pr(r = 1) = .360, pr(r = 2) = .309$
   d $pr(r = 2) = .044, pr(r = 8) = .044, pr(r = 5) = .246$
   e They are identical. The methods we used in section 1 were used to construct the tables.
   f $pr(r \leqslant 3) = pr(r = 0 \text{ or } r = 1 \text{ or } r = 2 \text{ or } r = 3) =$
   $0 + 0 + 0 + .002 = .002$
   $pr(r \geqslant 12) = pr(r = 12 \text{ or } r = 13 \text{ or } r = 14 \text{ or } r = 15) =$
   $.063 + .022 + .005 + 0 = .090$
   $pr(4 \leqslant r \leqslant 10) = pr(r = 4 \text{ or } 5 \text{ or } 6 \text{ or } 7 \text{ or } 8 \text{ or } 9 \text{ or } 10) = .780$
   g $pr(r = 10) = .161, pr(r \leqslant 5) = .029, pr(r \geqslant 13) = .153$

4 Remember $r \leqslant 2$ or $r \geqslant 8$ means $r = \{0, 1, 2, 8, 9, 10\}$. So $pr(r \leqslant 2 \text{ or } r \geqslant 8)$
   $= pr(r = 0) + pr(r = 1) + pr(r = 2) + pr(r = 8) + pr(r = 9) + pr(r = 10)$.

Because for $P = .5$, $pr(r = 0) + pr(r = 1) + pr(r = 2) = .055$ and $pr(r = 8) +$ $pr(r = 9) + pr(r = 10) = .055$, the answer is $.055 + .055 = .110$ for $P = .5$.
$pr(r \leqslant 2$ or $r \geqslant 8$ if $P = .4) = .167 + .013 = .18$.
$pr(r \leqslant 2$ or $r \geqslant 8$ if $P = .6) = .013 + .167 = .18$.
$pr(r \leqslant 2$ or $r \geqslant 8$ if $P = .3) = .382 + .001 = .383$.
$pr(r \leqslant 2$ or $r \geqslant 8$ if $P = .2) = .677 + .000 = .677$.
$pr(r \leqslant 2$ or $r \geqslant 8$ if $P = .8) = .000 + .677 = .677$.

Set 4

1 a Use the binomial model with 25 trials, corresponding to the 25 parts selected; success will be the part is bad, so $P = .01$ and $Q = .99$.
 b $P = .01$, that is, 1 percent of the machine parts produced are faulty.
 c $P > .01$, that is, more than 1 percent of machine parts produced are faulty.
 d Because stopping production means reject H_o, set $\alpha \leqslant .01$.
 e This will be a one-tailed test. Only a significantly large proportion of faulty parts will cause rejection of H_o.
 f $r =$ the number of faulty parts in a random sample of 25.
 g If $P = .01$, r has a probability distribution that is the binomial distribution where $n = 25$, $P = .01$.
 h $pr(r \geqslant 3$ if $P = .01) = .002$.
 $pr(r \geqslant 2$ if $P = .01) = .026$.
 We want $\alpha \leqslant .01$, so the best rule will be to reject H_o if $r \geqslant 3$ in a sample of 25.
 i This rule gives an exact probability of .002.

4 H_o will be rejected if $r \geqslant 3$; thus, if 5 percent of the parts produced are faulty, $pr(r \geqslant 3$ if $P = .05) = .127$. This means that the experiment set up in number 2 has just .127 chances of correctly telling us to repair machines if 5 percent of all parts produced are faulty. However, if $pr(r \geqslant 3$ if $P = .10)$ $= .463$, the chances are roughly 46 in 100 that the experiment will result in an outcome that will correctly force machine repair when the proportion of faulty parts is 10 percent. And if 20 percent of all parts produced are faulty, $pr(r \geqslant 3$ if $P = .2) = .901$; that is, the chances are roughly 90 in 100 when the proportion of faulty parts is 20 percent. Is the experiment in Problem 1 a good experiment? That depends on the manufacturer's feeling about the seriousness of a Type II error when P is close to .01.

7 If the sampling yields 15 out of 25 faulty parts, then the observed r is 15 and 15 is an outcome that leads to rejection of H_o. Thus, the manufacturer should conclude that if the machines are producing more than 1 percent faulty parts, he should stop production and repair the machines.

Answers to Review Questions in Chapter 7

1 a	2 a	3 b	4 a	5 b	6 b	7 a	8 b	9 b	10 a
11 a	12 a	13 b	14 b	15 b	16 a	17 a	18 b		
19 b	20 b	21 a	22 a	23 b	24 b	25 a			

Answers to Selected Problems in Chapter 7: Set 1

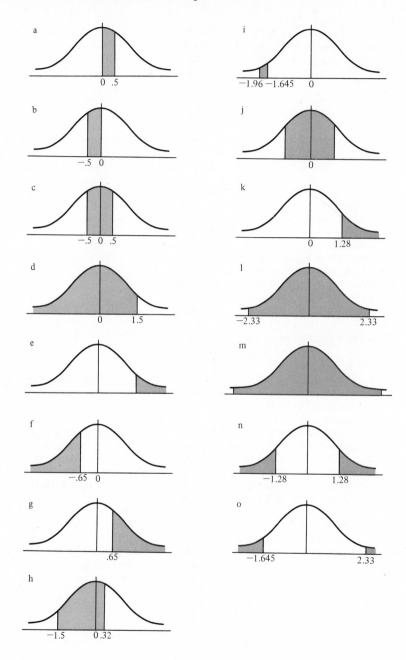

a   0   .5

b   −.5   0

c   −.5   0   .5

d   0   1.5

e

f   −.65   0

g   .65

h   −1.5   0.32

i   −1.96 −1.645   0

j   0

k   0   1.28

l   −2.33   2.33

m

n   −1.28   1.28

o   −1.645   2.33

3   a   88   b   50   c   82   d   99+   e   2   f   23   g   25

Set 2

1 For Problem 2 of Set 1:
  a Look in Column 4 of Table C.2 for .0500. Notice that .05 is exactly half-way between .0505 and .0495, so our z-score is halfway between 1.64 and 1.65. The z-score is 1.645, so the answer is -1.645.
  b Look in Column 4 for .0100. The closest z-score is 2.33, so the answer is -2.33.
  c Look in Column 3 for .9750. The z-score is 1.96 and that is the answer.
  d Look in Column 3 for .7700. The z-score is .74.
  e Look in Column 3 for .5000. The z-score is 0.
  f Look in Column 4 for .3900. The closest z-score is .28, so the answer is $-.28$.
  Notice that the estimates from our graph are quite good!

  For Problem 3 of Set 1:
  a Read answer in Column 3: .8944, so percentile rank is 89.44.
  b Read answer in Column 3: .5000, so percentile rank is 50.
  c Read answer in Column 3: .8159, so percentile rank is 81.59.
  d Read answer in Column 3: .9987, so percentile rank is 99.87.
  e Read answer in Column 4: .0228, so percentile rank is 2.28.
  f Read answer in Column 4: .2266, so percentile rank is 22.66.
  g Read answer in Column 4: .2514, so percentile rank is 25.14.
  Again, notice that the estimates from our graph are very good.

  For Problem 4 of Set 1:
  In these problems, we locate z in the first column and get our answer from Column 2, 3, or 4—whichever our picture indicates.

a .1915	e .0668	i .025	m .9974
b .1915	f .2578	j .6826	n .2006
c .3830	g .2578	k .1003	o .0599
d .9332	h .5587	l .9802	

2 To solve these problems, draw the normal curve and shade the area to help you decide whether your answer is reasonable.
  a This is a proportion to score problem.
    i First, we find the z-score that has percentile rank 10. Look for .10 in Column 3. The closest z-score is 1.28. The score we are looking for is 1.28 standard deviations below the mean.
    ii Score $= 100 - (1.28)\,15 = 80.8$. The score that has a percentile rank of 10 is 80.8.
  b This is a proportion to score problem.
    i First, we find the z-score that has percentile rank 80. Then, look for .80 in Column 2. We see that .84 is the closest z-score. The score we are looking for is .84 standard deviations above the mean.
    ii Score $= 100 + (.84)(15) = 112.60$.
  c This is a score to proportion problem.
    i First, we find z for the score of 120.

$$z = \frac{120 - 100}{15} = \frac{20}{15} = 1.33.$$

    Thus, our score is 1.33 standard deviations above the mean.
    ii Use column 2 to find the area over $z \leqslant 1.33$. This area is .9082.

571

iii We want the area over $z > 1.33$ so we subtract from 1:

$$1 - .9082 = .0918.$$

9.18 percent of the scores are above 120.

d This is a score to proportion problem.

i $z = \dfrac{135 - 100}{15} = \dfrac{35}{15} = 2.33.$

ii In Column 2, the area over the range $z \leqslant 2.33$ is .9901. This means that 135 has a percentile rank of 99.01.

e This is a score to proportion problem.

i $z = \dfrac{90 - 100}{15} = \dfrac{-10}{15} = -.67.$

ii Use $z = .67$ and Column 3 and find that the area over $z \leqslant -.67$ is .2514. This means that the score 90 has a percentile rank of 25.14.

f This is a proportion to score problem.

i First, we find .25 in Column 4 and read $z = .67$. This means that we are looking for the scores that fall within .67 standard deviations of the mean.

ii $100 \pm (.67)(15)$. The lower score is 90.0 and the upper score is 110.0. The middle 50 percent of scores fall between 90 and 110.

g This is a score to proportion problem.

i First, we find the z-score limits of the range:

Lower $z = \dfrac{90 - 100}{15} = \dfrac{-10}{15} = -.67.$

Upper $z = \dfrac{120 - 100}{15} = \dfrac{20}{15} = 1.33.$

ii Next, we find the area between $z = -.67$ and 0 using Column 4. This area is .2486. Then, we find the area between $z = 1.33$ and 0 using Column 4. This area is .4082.

iii The answer is $.2486 + .4082 = .6568$, so 65.68 percent of the scores fall between 90 and 120.

h This is a proportion to score problem.

i First, we find .025 in Column 3 and read $z = 1.96$. The scores we are looking for have z-scores of 1.96 and -1.96.

ii The lower score is $100 - (1.96)(15) = 70.6$. The upper score is $100 + (1.96)(15) = 129.4$. So the scores we want are above 129.4 or below 70.6.

5 a These are score to proportion problems. First get z for each score and then get the percentile rank.

Test	Percentile rank
intelligence	99.87
math aptitude	98.93
verbal aptitude	53.98
artistic aptitude	4.75
vocational preference	
math-science	50.00
fine arts	96.41
social service	28.77
business	92.36

b Not totally. Her aptitudes indicate strength in math and weakness in the arts. Her preferences are reversed. She also specifies a preference for business. This interest, coupled with her high aptitude in math, suggests that she might enjoy and be successful in some quantitative business field.

Set 3

1 a $\sigma_{\bar{X}} = \dfrac{2000}{2} = \$1000, \mu_{\bar{X}} = \mu_X = \$12,000$

b $\sigma_{\bar{X}} = \dfrac{2000}{5} = \$400, \quad \mu_{\bar{X}} = \mu_X = \$12,000$

c $\sigma_{\bar{X}} = \dfrac{2000}{10} = \$200, \quad \mu_{\bar{X}} = \mu_X = \$12,000$

d $\sigma_{\bar{X}} = \dfrac{2000}{20} = \$100, \quad \mu_{\bar{X}} = \mu_X = \$12,000$

e $\sigma_{\bar{X}} = \dfrac{2000}{40} = \$50, \quad \mu_{\bar{X}} = \mu_X = \$12,000$

4 a The probability of a sample mean being 105 or more is .047, based on 25 scores. This is because 105 is 1.67 deviations above the mean. Thus, we would be surprised: 105 constitutes an extreme score.

b However, if $n = 400$, a sample mean of 105 or more constitutes an event of very small probability, i.e., .0000+. This means that a sample mean of 105 or more based on 400 observations is an extremely rare event.

Set 4

1 a i Because IQ is an interval scale variable, we will use a one-sample z-test assuming the standard deviation is 15.

ii $H_o: \mu = 100$

iii $H_a: \mu \neq 100$

iv The test statistic is $z_{ob} = \dfrac{\bar{X}_{ob} - 100}{15/\sqrt{400}}$, which will have the standard normal distribution if H_o is true.

v $\alpha = .05$, so reject H_o if $z_{ob} < -1.96$ or $z_{ob} > 1.96$.

b If the mean is still 100, but our sample causes us to reject H_o, a Type I error is committed.

c If the mean is no longer 100, but our sample causes us to accept H_o, we have made a Type II error.

d If $\bar{X}_{ob} = 103.7$, then $z_{ob} = 4.93$. Therefore, H_o is rejected. Conclude that the average IQ in the normed population is no longer 100.

e The significance probability of the observed result is .0000.

Set 5

1 a $105 \pm 1.96 \times \dfrac{15}{10} = 105 \pm 2.94$

$102.06 < \mu < 107.94$ with 95 percent confidence.

b The investigator has 95 percent confidence that the population mean IQ is at least 102.06 and no greater than 107.94.

c $105 \pm 2.58 \times \frac{15}{10} = 105 \pm 3.87$

$101.13 < \mu < 108.87$ with 99 percent confidence.
The increased confidence lengthens the interval, thus decreasing the precision of the estimate.

2 $14{,}200 \pm 1.645 \times \frac{2000}{8} = 14{,}200 \pm 411.25$

The 90 percent confidence interval places the population average income between $13,788.75 and $14,611.25.

4 a As the level of confidence approaches 100 percent, the width of the interval increases.

b As the sample size increases, the standard error of the mean decreases. Thus, the estimate based on any one sample becomes more reliable and the width of the confidence interval decreases.

Set 6

1 The test statistic is $z = \dfrac{\bar{X} - 32}{8/\sqrt{16}} = \dfrac{\bar{X} - 32}{2}$.

a Reject H_o if $z_{ob} > 1.645$.

Because $z = \dfrac{\bar{X} - 32}{2}$, the rejection rule could also be written as follows:

reject H_o if $\dfrac{\bar{X}_{ob} - 32}{2} > 1.645$.

Solve for $\bar{X}_{ob}$: $\bar{X}_{ob} - 32 > 2 \times 1.645$ or $\bar{X}_{ob} > 32 + (2 \times 1.645)$. Then, reject H_o if $\bar{X}_{ob} > 35.29$.

b i If H_o is false and $\mu = 36$, we want the probability that $\bar{X}_{ob} > 35.29$. This is a score to proportion problem. We find the z-score for 35.29.

$z = \dfrac{35.29 - 36}{2} = -.355$ or $-.36$.

We want to find the probability that $z > -.36$. We use Table C.2, Column 2 to find the probability that $z < .36$ is .6406. So the probability that $z > -.36$ is .6406 (using the symmetry of the normal curve). The power of the test to reject H_o when $\mu = 36$ is .64.

ii If H_o is false and $\mu = 40$, we want $pr(\bar{X}_{ob} > 35.29)$.

$z = \dfrac{35.29 - 40}{2} = -2.36$. $pr(z > -2.36) = pr(z < 2.36)$.

So power is .9909.

iii If $\mu = 38$, then we want $pr(\bar{X}_{ob} > 35.29)$.

$z = \dfrac{35.29 - 38}{2} = -1.36$.

574

$pr(z > -1.36) = pr(z < 1.36) = .9131$. In this case, power is .9131. We see that the farther the true value of μ is from our null hypothesis, the more probable it is that a sample of 16 will lead to rejection of H_o.

Answers to Review Questions in Chapter 8

1 a   2 a   3 a   4 a   5 b   6 b   7 a   8 b   9 b   10 a
11 b   12 a   13 b   14 a   15 a   16 a   17 b   18 b

Answers to Selected Problems in Chapter 8: Set 1

1   a

b

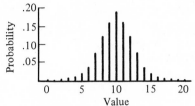

Binomial Distribution ($n = 20, P = .5$)   Binomial Distribution ($n = 20, P = .8$)

c

d

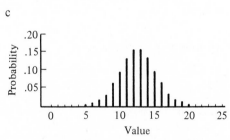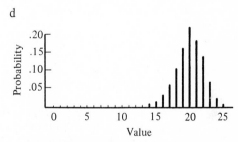

Binomial Distribution ($n = 25, P = .5$)   Binomial Distribution ($n = 25, P = .8$)

2   a   1a is symmetric. 1b is skewed negatively. 1c is symmetric. 1d is skewed negatively, but less so than 1b.
      b   The graph looks more like a normal curve as n increases.
      c   The graph looks more like a normal curve when $P = .5$ than when $P = .8$.

5   a   i   .070   ii .930   iii .058

	Without correction	With correction
i	.0465	.0808
ii	.8686	.9192
iii	.000	.0558

      c   No, not to two decimal places.
      d $nPQ = 3.2$, which is not bigger than 9. P is too large and n is not large enough to make the normal curve with $\mu = nP$ and $\sigma = \sqrt{nPQ}$ fit the binomial distribution where $n = 20$ and $P = .8$.

6 $H_0 : P = .90$ vs. $H_a : P < .90$.
$\alpha = .05$
Use the normal approximation to the binomial with $n = 100$ and $P = .9$.
$$z = \frac{P - .9 + .005}{.03}.$$

Reject H_0 if $z_{ob} < -1.645$.
$$z = \frac{.85 - .9 + .005}{.03} = -1.5.$$ Therefore, H_0 should not be rejected:

The farmer cannot deny the company's claim.

Set 2

2

Category	Number in the sample	Expected number if H_0 is true
Have the trait	16	10
Do not have the trait	9	15
Total	25	25

H_0: the proportion having the trait $(P) = .4$ vs. $H_a : P \neq .4$.

Use $\displaystyle\sum_{}^{\substack{\text{all} \\ \text{cells}}} \frac{(Ob - Ex)^2}{Ex}$. This statistic has the chi-square distribution with 1 degree of freedom if H_0 is true. Reject H_0 if $\chi^2_{ob} > 3.8$ for $\alpha = .05$.

$$\chi^2_{ob} = \frac{(16 - 10)^2}{10} + \frac{(9 - 15)^2}{15} = 3.6 + 2.4 = 6.0$$

Because $6.0 > 3.8$, reject H_0 at a 5 percent level of significance. The hypothesized proportions are not correct.

4 a H_0: The proportion in each response category is .20.
H_a: The proportions in the response categories are not all .20.
With $\alpha = .05$ and 4 degrees of freedom, reject H_0 if $\chi^2_{ob} > 9.5$.

b $\chi^2_{ob} = \dfrac{(230 - 180)^2}{180} + \dfrac{(410 - 180)^2}{180} + \dfrac{(105 - 180)^2}{180} + \dfrac{(130 - 180)^2}{180}$

$\qquad + \dfrac{(25 - 180)^2}{180}$

$\qquad = 13.89 + 293.89 + 31.25 + 13.89 + 133.47$

$\qquad = 486.39.$

$489.39 > 9.5$, so H_0 is rejected.

c The student responses are not proportionately the same in each category. The data suggest the following explanation of the large χ^2 value: More than half the sample either agree or strongly agree with the statement. This suggests that students do not like to hear their teachers talk all the time.

576

Answers

Set 3

1 a $H_o: P_{SA} = .20, P_{MA} = .20, P_N = .20, P_{MD} = .20, P_{SD} = .20$
H_a: The distribution is not as specified in H_o.
For $\alpha = .01$, reject H_o if $D_{ob} > .29$.

b The results:

Category	Cumulative proportion under H_o	Cumulative sample proportion
strongly agree	1.00	1.00
mildly agree	.80	.83
neutral	.60	.33
mildly disagree	.40	.10
strongly disagree	.20	.03

The largest difference is in the mildly disagree category, so
$D_{ob} = |.40 - .10| = .30$.
Because $.30 > .29$, we reject H_o at $\alpha = .01$.

c The evidence suggests that proportionately more students feel anxious during statistics exams than do not.

Set 4

3 Looking at Table C.5, we find that the t-percentiles with 120 degrees of freedom are very close to the z-percentiles, but even when the degrees of freedom are as few as 30, the two correspond quite adequately.

4 $\bar{X} = 137.93$
$s = 29.96$

Set 5

1 a $H_o: \mu = 100$ vs. $H_a: \mu \neq 100$.
$\alpha = .05$

$z = \dfrac{\bar{X} - 100}{15/\sqrt{20}}$ has standard normal distribution if H_o is true.

Reject H_o if $z_{ob} > 1.96$ or $z_{ob} < -1.96$.

$z_{ob} = \dfrac{109.1 - 100}{15/\sqrt{20}} = 2.71$.

We reject H_o and conclude that the mean IQ of the sampled population is not 100.

b Test the same hypothesis as in Part a, but use $t = \dfrac{\bar{X} - 100}{s/\sqrt{20}}$, which has a t distribution with 19 df.
Reject H_o if $t_{ob} > 2.093$ or $t_{ob} < -2.093$.

$t_{ob} = \dfrac{109.1 - 100}{14.39/\sqrt{20}} = 2.83$.

We reject H_o and conclude that the mean IQ of the sampled population is not 100.

2 a Using the z-method:

$$109.1 \pm (1.96)\frac{15}{\sqrt{20}} \text{ or } 109.1 \pm 6.574.$$

With 95 percent confidence, we can say that $102.5 < \mu_{IQ} < 115.7$.

b Using the t-method:

$$109.1 \pm (2.093)\frac{(14.39)}{\sqrt{20}} \text{ or } 109.1 \pm 6.735$$

With 95 percent confidence we can say that $102.4 < \mu_{IQ} < 115.8$.

c Notice that with either method, 100 is not in the confidence interval and, therefore, would not be a reasonable value for μ. This conclusion is consistent with the conclusion of our hypothesis tests in Problem 1.

Answers to Review Questions in Chapter 9

1 a   2 b   3 a   4 b   5 b   6 a   7 b   8 a   9 a   10 b
11 a   12 b   13 b   14 b   15 a   16 b   17 a   18 b
19 b   20 a

Answers to Selected Problems in Chapter 9: Set 1

1 H_o: Dreams are independent of REM vs. H_a: There is a relationship between having or not having dreams and the presence or absence of REM.
$\alpha = .01$
Reject H_o if $\chi^2_{ob} > 6.6$.

	Report having	Report not having
REM	15	35
No REM	15	35

This is the expected frequency table under H_o.

To compute χ^2_{ob}, calculate $\dfrac{(Ob - Ex)^2}{Ex}$ for each cell and add:

$$\frac{(25-15)^2}{15} + \frac{(25-35)^2}{35} + \frac{(5-15)^2}{15} + \frac{(45-35)^2}{35}$$

$$= \frac{100}{15} + \frac{100}{35} + \frac{100}{15} + \frac{100}{35} = \frac{200}{15} + \frac{200}{35} = 19.05.$$

$\chi^2_{ob} = 19.05$, so reject H_o.

There is a relationship between the two variables. The data show that people who report having dreams have a higher incidence of REM than people who report that they do not have dreams. The value of ϕ' is .436, indicating that the relationship is a moderate one.

(Remember from Chapter 5, $\phi' = \sqrt{\dfrac{\chi^2_{ob}}{(L-1)n}}$.

So $\phi' = \sqrt{\dfrac{19.05}{100}} = \sqrt{.1905} = .436$.)

4 Because a family has possibly a mother and a father and possibly one or more children, the observations are not independent given the dependency of choice within families. The independent unit is the family.

5 a The population sampled consists of people who go to the race track.
  b H_0: Betting preference is independent of birth order.
    Reject H_0 if $\chi^2_{ob} > 9.5$.
    $\chi^2_{ob} = 33.3$, so reject H_0.
  c There is a relationship between betting preference and birth order in the population of people who go to the race track. The data suggests that proportionately more first-borns bet on 2:1 odds than do second-borns or those of lower birth order, whereas proportionately more of those of lower birth order bet on 15:1 odds than do either first- or second-borns and proportionately more second-borns wouldn't bet than either first-borns or those of lower birth order.

2 For normal approximation use $z = r_s \sqrt{n-1}$.
  a Reject H_0 if $z_{ob} > 2.58$ or $z_{ob} < -2.58$.
    $z_{ob} = -.6 \sqrt{30-1} = -3.23$.
    Reject H_0 and conclude $\rho_s \neq 0$.
  b Reject H_0 if $z_{ob} > 2.58$ or $z_{ob} < -2.58$.
    $z_{ob} = 1.56$, so do not reject H_0.
  c Reject H_0 if $z_{ob} > 2.58$ or $z_{ob} < -2.58$.
    $z_{ob} = 1.05$, so do not reject H_0.

3 $H_0: \rho_s = 0$ vs. $H_a: \rho_s \neq 0$.
  For $\alpha = .02$, reject H_0 if $r_s > .746$ or $r_s < -.746$.
  $r_{s_{ob}} = .62$, so do not reject H_0.

Set 3

1 For $\alpha = .05$, reject H_0 if $z_r > 1.96$ or $z_r < -1.96$. We use the statistic
  $z_{ob} = (z_{r_{ob}} - z_{\rho_0})(\sqrt{n-3})$.
  a $r_{ob} = .3$ and $\rho_0 = .7$, so $z_{r_{ob}} = .3095$ and $z_{\rho_0} = .8673$.
    Then $z_{ob} = (.3095 - .8673)\sqrt{17} = -2.30$.
    Therefore, reject H_0 and conclude $\rho \neq .7$.
  b $z_{ob} = (.5493 - .8673)\sqrt{47} = -2.18$.
    Therefore, reject H_0 and conclude $\rho \neq .7$.
  c $z_{ob} = (.69315 - .86730)\sqrt{97} = -1.72$.
    Therefore, do not reject H_0.
  d $z_{ob} = (1.0986 - .86730)\sqrt{97} = 2.28$.
    Therefore, reject H_0 and conclude that $\rho \neq .7$.

3 Notice that we cannot be as precise by using the chart.
  a $r = .3$, $n = 20$. Looking at the chart, we find that the lower limit is approximately $-.15$ and the upper limit is approximately $.65$. Therefore, our 95 percent confidence interval is $-.15 < \rho < .65$.
  b $r = -.6$, $n = 50$, so we read approximately $-.7$ and $-.3$. Our 95 percent confidence interval is $-.7 < \rho < -.3$.

579

c $r = .54, n = 100$. We read approximately .4 and .65. Our 95 percent confidence interval is $.4 < \rho < .65$.

5 A 99 percent confidence interval for ρ places it between .46 and .62. Thus, we have 99 percent confidence that ρ is at least .46 and no more than .62.

Answers to Review Questions in Chapter 10

1 b	2 a	3 a	4 b	5 a	6 b	7 a	8 b	9 a	10 a
11 a	12 b	13 b	14 a	15 b	16 b	17 b	18 a		
19 b	20 a	21 a							

Answers to Selected Problems in Chapter 10: Set 1

1 a $A + B = 10, C + D = 7$; then if $B = 7, D = 1$ is a significant result at $\alpha = .05$.
   b $A + B = 11, C + D = 11$; then if $A = 7, C = 1$ or 0 is significant. This is not the case in our table, so b is not a significant result.
   c To use Table C.11, we must define $A + B = 11, C + D = 10$. The table is not significant if we are using $B = 7$ or $D = 10$.
   d $A + B = 12$ and $C + D = 8$; then if $A = 10$ and $C = 1$, the result is significant at $\alpha = .01$.

3

	I	II	
Cancer	A 9	C 2	11
No Cancer	B 6	D 13	19
	15	15	30

This 2 × 2 table shows the results.

In Table C.11, use $A + B = 15$ and $C + D = 15$. Then if $A = 9$, the table indicates that $C = 2$ is significant at .025. The experimenter is justified in saying that the incidence of cancer is greater when the suspected carcinogenic substance is present in the diet than when it is not present.

5 No. One of the assumptions underlying the hypergeometric model is that we have two independent samples. However, because this is a pretest/posttest design, we have two sets of dependent measures. Thus, the assumption of independence is violated and we should not use the model.

6 We test H_o that the distribution of placement classification is the same in all 4 populations against the alternate hypothesis that not all 4 distributions are alike. Because we have a group comparison problem with 4 independent groups, because the observations within each group are independent, and because the expected cell frequencies are at least 5 in more than 80 percent of the cells, we use the χ^2 statistic to test H_o. The degrees of freedom are $3 \times 3 = 9$. Thus, reject H_o if $\chi^2_{ob} > 16.9$ for $\alpha = .05$. $\chi^2_{ob} = 116.57$, so we

reject H_o and conclude that not all 4 distributions are alike. We can say that the 4 interest groups do not agree on the placement of children with special problems.

Set 2

1 a Let $n_1 = 11, n_2 = 14$. We locate 11 in the column heading and 14 in the row heading and read the number 46 from the table. Then if $U_{ob} \leqslant 46$, we reject H_o for $\alpha = .10$ in a two-tailed test.
  b $n_1 = 30$ and $n_2 = 15$. We locate 15 in the column heading and 30 in the row heading and read 156 from the table. Then if $U_{ob} \leqslant 156$, we reject H_o for $\alpha = .05$ in a one-tailed test.
  c $n_1 = 30$ and $n_2 = 15$. We locate 15 in the column heading and 30 in the row heading and read 143 from the table. If $U_{ob} \leqslant 143$, we reject H_o for a two-tailed test with $\alpha = .05$.
  d $n_1 = 20, n_2 = 40, \alpha = .025$, and we want a one-tailed test. From the table we read 274. Reject H_o if $U_{ob} \leqslant 274$.
  e $n_1 = 18, n_2 = 33, \alpha = .01$, and we want a two-tailed test. From the table we read 167. Reject H_o if $U_{ob} \leqslant 167$.

3 We will test the null hypothesis that there is no difference in the anxiety level of the experimental group and the control group against the alternative that the experimental group is less anxious than the control group. We will use the Mann-Whitney U-statistic. The test is one-tailed and $\alpha = .05$, $n_1 = n_2 = 5$. We will reject H_o if $U_{ob} \leqslant 4$.

$$U_{ob} = 5 \times 5 + \frac{5 \times 6}{2} - R_1 \text{ and}$$

$$R_1 = 1 + 2 + 3 + 4 + 7 = 17.$$

So $U_{ob} = 25 + 15 - 17 = 23$.

Because $23 > \frac{25}{2}$, we have U'. Then, $25 - 23 = 2$. So $U_{ob} = 2$. Therefore,

we reject H_o and conclude that the experimental group is less anxious than the control group. If all other factors are controlled, we can say that the training program reduces test anxiety.

6 The null hypothesis in each case will be that the 3 population distributions are identical, against the alternate hypothesis that not all 3 are the same.
  a Look at Table C.13 and locate $n_1 = 5, n_2 = 4, n_3 = 4$.
     Then for $\alpha = .05$, we reject H_o if $H_{ob} \geqslant 5.6176$.
  b To keep $\alpha \leqslant .05$, we reject H_o if $H_{ob} \geqslant 5.6923$.
     This makes the exact significance level .049.
  c Reject H_o if $H_{ob} \geqslant 4.56$.
  d Reject H_o if $H_{ob} \geqslant 5.2083$.
  e For $\alpha = .01$, reject H_o if $H_{ob} \geqslant 6.8218$.
  f For $\alpha \leqslant .05$, it is impossible to establish a rejection rule. When $n_1 = 2$, $n_2 = 1, n_3 = 1, H \geqslant 2.7$ has a probability of .5. This tells us that there is insufficient data to test the null hypothesis and control the risk of a Type I error.

581

8 a We will test H_o that the 3 methods of reinforcement affect recall ability in the same way, against the alternate hypothesis that at least one method has an effect that is different from the others. We use H with $\alpha \leqslant .05$. Thus, reject H_o if $H_{ob} \geqslant 5.78$. This rule has a probability of .049 if H_o is true.

   b $R_1 = 13 + 11 + 10 + 8.5 + 5 = 47.5$
   $R_2 = 15 + 13 + 8.5 + 7 + 6 = 49.5$
   $R_3 = 13 + 4 + 3 + 2 + 1 = 23$

$$H = \frac{12}{15 \times 16}\left[\frac{(47.5)^2}{5} + \frac{(49.5)^2}{5} + \frac{(23)^2}{5}\right] - (3 \times 16) = 4.355.$$

There were 3 tied at 2 words recalled and there were 2 tied at 5 words recalled. To adjust H for ties,

$$t_1 = 3 \text{ and } t_1{}^3 - t_1 = 27 - 3 = 24 = T_{k_1}$$

$$t_2 = 2 \text{ and } t_2{}^3 - t_2 = 8 - 2 = 6 = T_{k_2}$$

$$\Sigma T_k = 24 + 6 = 30$$

$$1 - \frac{30}{3360} = 1 - .008929 = .991071$$

$$\frac{H}{.991071} = H \text{ adjusted for ties} = 4.394.$$

Because $H_{ob} < 5.78$, we accept H_o. Based on the results of the experiment, there is no reason to believe the treatments have differential effects.

Set 3

1 a

	Sample 1	Sample 2
n	8	9
ΣX	256	367
ΣX^2	11,334	19,795
$\overline{X}$	32	40.78
s^2	448.85	603.69

You might want to refer to Chapter 8 to review the computation of s^2, the sample variance.

   b The two sample variances are not exactly the same but the larger s^2 is not even twice as big as the smaller. Such a result might reasonably occur, when $\sigma_1{}^2 = \sigma_2{}^2$, simply because of sampling variability. If the ratio were as large as 4 to 1, we would begin to question the assumption.

   c $$s_p{}^2 = \frac{(n_1 - 1)s_1{}^2 + (n_2 - 1)s_2{}^2}{N_1 + N_2 - 2} = \frac{7 \times 448.85 + 8 \times 603.69}{15}$$

$$= \frac{7,971.56}{15} = 531.44.$$

Now $s_p{}^2$ is the best estimate of σ^2 using the information from the two random samples. Note: Your answer may differ in the decimal places from the one just given because of rounding error.

3 a $\alpha = .05.\, n_1 + n_2 - 2 = 18 - 2 = 16$, so reject H_o if $t_{ob} > 1.734$.

b $\alpha = .01$, df $= 19$, so reject H_o if $t_{ob} > 2.539$.

c $\alpha = .005$, df $= 28$; so reject H_o if $t_{ob} > 2.763$.

d $\alpha = .05$, df $= 28$; so reject H_o if $t_{ob} > 1.701$.

e $\alpha = .01$, df $= 36$. We estimate the critical value using

$t_{40} = 2.423$, $t_{30} = 2.457$

$t_{30} - t_{40} = .0340$, $40 - 30 = 10$, $36 - 30 = 6$,

$\dfrac{6}{10} = \dfrac{X}{.0340}$. Then $X = .0204$.

Next, $2.457 - .0204 = 2.4366$, so t_{36} is approximately 2.437. Reject H_o if $t_{ob} > 2.437$.

f $\alpha = .025$, df $= 48$.

48 is almost halfway between 40 and 60, so t_{cr} is almost halfway between 2.000 and 2.021. Therefore, reject H_o if $t_{ob} > 2.010$.

5 We could use the Mann-Whitney U-test or we could dichotomize the dependent variable and use the hypergeometric model.

6 a F will have 2 and 27 degrees of freedom, so reject H_o if $F_{ob} > 3.35$ for $\alpha = .05$.

b F will have 5 and 44 degrees of freedom, so reject H_o if $F_{ob} > 3.46$ for $\alpha = .01$.

c F will have 4 and 125 degrees of freedom, so reject H_o if $F_{ob} > 2.44$ for $\alpha = .05$.

d F will have 3 and 60 degrees of freedom, so reject H_o if $F_{ob} > 4.13$ for $\alpha = .01$.

e F will have 1 and 28 degrees of freedom, so reject H_o if $F_{ob} > 4.20$ for $\alpha = .05$.

7 For Step 1, lay out the scores.

		Group				Totals
		I	II	III	IV	
Step 2	ΣX_j	125	40	102	80	Step 5 347
Step 3	n_j	5	4	6	5	Step 6 $n = 20$
	$\bar{X}_j$	25	10	17	16	17.35
Step 4	ΣX_j^2	3333	498	1964	1386	Step 7 7181
	s_j	7.21	5.72	6.78	5.15	

Step 8 $\quad$ SST $= 7181 - \dfrac{347^2}{20}) = 7181 - 6020.45 = 1160.55$

Step 9 $SSB = \dfrac{(125)^2}{5} + \dfrac{(40)^2}{4} + \dfrac{(102)^2}{6} + \dfrac{(80)^2}{5} + \dfrac{(347)^2}{20} = 6539 - 6020.45$

$= 518.55$

Step 10 $SSW = SST - SSB = 1160.55 - 518.55 = 642.01$

8 a   To test $H_o: \mu_I = \mu_{II} = \mu_{III} = \mu_{IV}$, we use F with 3 and 16 degrees of freedom, so we reject H_o if $F_{ob} > 3.24$ for $\alpha = .05$. Because $F_{ob} = \dfrac{MSB}{MSW} = \dfrac{172.85}{40.13} = 4.31$, we reject H_o because $4.31 > 3.24$.

   b

Source of Variance	SS	df	Mean square	F
Between groups	518.55	3	172.85	4.31
Within groups	642	16	40.13	
Total	1160.55	19		

9 a θ_1, because the coefficient of $\mu_1 = 1$ and the coefficient of $\mu_2 = -1$ and $1 + (-1) = 0$.
   θ_2, for the same reason.
   θ_3, because the coefficients are $+1, -\dfrac{1}{2}, -\dfrac{1}{2}$.
   θ_4, because the coefficients are $+1, +1, -1, -1$.
   θ_5 is not a contrast because the coefficients are $+1, -2, -1$, and $+1 -2 -1 \neq 0$.

   b   We need $\sqrt{(K-1)F_{cr}} = \sqrt{3 \times 3.24} = 3.12$.

   For θ_1: $(25 - 10) \pm (3.12) \sqrt{40.13\left(\dfrac{1}{5} + \dfrac{1}{4}\right)}$

   $= 15 \pm (3.12)(4.25)$
   $= 15 \pm 13.26$.
   So $1.74 < \mu_1 - \mu_2 < 28.26$.

   For θ_2: $(25 - 17) \pm (3.12) \sqrt{40.13\left(\dfrac{1}{5} + \dfrac{1}{6}\right)}$

   $= 13 \pm (3.12)(3.84)$
   $= 13 \pm 11.97$.
   So $1.03 < \mu_1 - \mu_2 < 24.97$.

   For θ_3: $[10 - (17 + 16)] \pm (3.12) \sqrt{40.13\left(\dfrac{1}{4} + \dfrac{1/4}{6} + \dfrac{1/4}{5}\right)}$

   $= -6.5 \pm (3.12)(3.70)$
   $= -6.5 \pm 11.54$.
   So $-18.04 < \mu_2 - \dfrac{(\mu_3 + \mu_4)}{2} < +5.04$.

   For θ_4: $[(25 + 10) - (17 + 16)] \pm (3.12) \sqrt{40.13\left(\dfrac{1}{5} + \dfrac{1}{4} + \dfrac{1}{6} + \dfrac{1}{5}\right)}$

   $= 2 \pm (3.12)(5.72)$
   $= 2 \pm 17.86$.

So $-15.86 < (\mu_1 + \mu_2) - (\mu_3 + \mu_4) < 19.86$.

We have at least 95 percent confidence that all 4 statements are correct.

c We can conclude that $\mu_1 > \mu_2$ and that $\mu_1 > \mu_3$, but we cannot reject H_o:

$$\mu_2 = \frac{\mu_3 + \mu_4}{2} \text{ and } H_o: \mu_1 + \mu_2 = \mu_3 + \mu_4.$$

Set 4

1 a There are two populations of eighth-graders, one on the east coast and the other on the west coast.
   b IQ
   c Interval scale
   d If we think that the population IQ distributions are normal, then a t-test for independent samples would be appropriate.
   e We could use the U-statistic.

3 a There are three treatment groups.
   b Length of time until cured.
   c Interval scale, at least.
   d If we can make the normal distribution assumptions about number of days until cured, we could use the F-statistic.
   e We might want to use the H-statistic.

5 a This is an experiment with two groups.
   b Success in problem solving.
   c This is a success/failure situation. We have two ordered categories.
   d Because the samples are large, we could use the χ^2 test to see if the proportion of "solvers" is the same under both conditions.

Answers to Review Questions in Chapter 11

1 b   2 b   3 a   4 a   5 a   6 b   7 a   8 b   9 a   10 b
11 a   12 b

Answers to Selected Problems in Chapter 11: Set 1

3 H_o: The advertising is ineffective vs. H_a: The advertising is effective. Because this is a pre-post design, we use McNemar's test. Because $n = 10 + 40 = 50$, we use the χ^2 approximation. We reject H_o if $\chi^2_{ob} > 3.84$ for $\alpha = .05$.

$$\chi^2_{ob} = \frac{(|10 - 40| - 1)^2}{10 + 40} = \frac{(29)^2}{50} = 16.82.$$

Therefore, we reject H_o. The data suggests that proportionately more people selected the product after seeing the advertisement than before.

Set 2

1 a Reject H_o if $T_{ob} \leqslant 8$.
   b Reject H_o if $T_{ob} \leqslant 3$.

c Reject H_o if $T_{ob} \leqslant 20$.

d Reject H_o if $T_{ob} \leqslant 0$.

e The $-$ means that no value of T is significant if $n = 7$ and $\alpha = .01$ in a two-tailed test.

2 H_o: The drug has no effect vs. H_a: The drug has a negative effect on intellectual functioning. If H_o is true, $T_{ob} \leqslant 6$ has a probability of .025. We will do a one-tailed test because we are attempting to show a negative effect of the drug. We will find our difference scores by subtracting the drug scores from the sugar pill scores, i.e., $D = C - E$.

Subjects (S)	1	2	3	4	5	6	7	8	9
Differences (d)	2	26	−1	4	3	−5	7	−17	6
Ranks (R)	2	9	1	4	3	5	7	8	6
Signed ranks (R)	2	9	−1	4	3	−5	7	−8	6

$T = -1 + -5 + -8 = -14$

$T_{ob} = 14$ and 14 is not a significant result; thus, we cannot say that the drug has a negative effect on intellectual functioning.

Set 3

1 H_o: $\mu_D = 0$ vs. H_a: $\mu_D > 0$, where $D =$ control score $-$ experimental score.

Use $\dfrac{\bar{X}_{ob}}{s_D/\sqrt{9}}$, which has the t-distribution with 8 df when H_o is true.

We reject H_o if $t_{ob} > 2.306$ for $\alpha = .025$ in a one-tailed test. To calculate t_{ob}, we first find the D-scores. Then, we calculate the sample mean and s for the D-scores.

$\bar{X}_{ob} = \dfrac{25}{9} = 2.78.$

$s = \sqrt{\dfrac{9 \times 1105 - (25)^2}{9 \times 8}} = 11.38.$

Then, $t_{ob} = \dfrac{2.78}{11.38/3} = .73.$

We cannot reject H_o. The evidence of the experiment does not indicate that the drug lowers intellectual functioning on the average.

2 H_o: $\mu_D = 0$ vs. H_a: $\mu_D > 0$, where $D =$ pre $-$ post.

Use $\dfrac{\bar{X}_D}{s_D/\sqrt{13}}$, which has the t-distribution with 12 df when H_o is true.

Then, reject H_o if $t_{ob} > 2.681$ for $\alpha = .01$ in a one-tailed test. To calculate t_{ob}, first, get the D-scores, and then, calculate the sample mean and the sample standard deviation.

586

$$\bar{X} = \frac{67}{13} = 5.15.$$

$$s = \sqrt{\frac{13 \times 897 - (67)^2}{13 \times 12}} = 6.78.$$

$$\text{Then, } t_{ob} = \frac{5.15}{6.78/\sqrt{13}} = 2.74.$$

Set 4

1 This is a one-sample problem. We could calculate a $(1 - \alpha) \cdot 100$ percent confidence interval for μ, the average number of cigarettes smoked per day.

We use $\bar{X} \pm \left(t_{1-\frac{\alpha}{2};n-1} \right) \frac{s}{\sqrt{n}}$, where n is the number of subjects in the sample.

5 The rating scale constitutes a set of ordered categories. We could use a chi-square one variable test.
H_o: $P_1 = P_2 = P_3 = P_4$ vs. H_a: Not all P's are the same.

Use $\sum \frac{(Ob - Ex)^2}{Ex}$, which is χ^2 with 3 df when H_o is true.

6 This is a matched-pairs design.
H_o: $\mu_D = 0$ vs. H_a: $\mu_D > 0$, where D = experimental − control. Then, if the

D-scores have a normal distribution, we can use $\dfrac{\bar{X}_D}{s_D/\sqrt{n}}$, which has the t-

distribution with $n - 1$ df when H_o is true.

10 Test H_o: The proportion of ninth-graders who correctly choose their career is the same as the proportion of twelfth-graders who correctly choose their career.
That is, H_o: $P_9 = P_{12}$ vs. H_a: $P_9 \neq P_{12}$.
If the sample is small, lay out the 2×2 crosstabulation table and calculate the significance probability using the method covered in the beginning of Chapter 10. If the sample is large enough, use the χ^2 approximation test of homogeneity. Use

$$\sum^{4 \text{ cells}} \frac{(Ob - Ex)^2}{Ex},$$

which has approximately the χ^2 distribution with 1 df.

References

Edwards, A.L., *Experimental Design in Psychological Research*. New York: Holt, Rinehart and Winston, 1968.

Edwards, A.L., *Statistical Analysis*. New York: Holt, Rinehart and Winston, Inc., 1974.

Guenther, W.C., *Concepts of Statistical Inference*. New York: McGraw-Hill, 1965.

Guilford, J.P., & Fruchter, B., *Fundamental Statistics in Psychology and Education*. New York: McGraw-Hill, 1973.

Lieberman, G.I., & Owen, D.B., *Tables of the Hypergeometric Probability Distribution*. Stanford, Calif.: Stanford University Press, 1961.

Lindeman, R., & Merenda, P., *Educational Measurement*. Glenview, Illinois: Scott, Foresman, 1979.

Mosteller, F., Rourke, R.E.K., & Thomas, G.B., *Probability with Statistical Applications*. Reading, Mass.: Addison-Wesley, 1970.

Robbins, H., & Van Ryzin, J., *Introduction to Statistics*. Chicago: Science Research Associates, 1975.

Roscoe, J.T., *Fundamental Research Statistics for the Behavioral Sciences*. New York: Holt, Rinehart and Winston, 1975.

Runyon, R.P., & Haber, A., *Fundamentals of Behavioral Statistics*. Reading, Mass.: Addison-Wesley, 1971.

Scheffé, H., *Analysis of Variance*. New York: John Wiley & Sons., Inc., 1959.

Siegel, S., *Nonparametric Statistics for the Behavioral Sciences*. New York: McGraw-Hill, 1956.

Walker, H.M., & Lev, J., *Elementary Statistical Methods*. New York: Holt, Rinehart and Winston, 1969.

Index

589

Dependent measures, 27–28
Dependent variable, 24
Descriptive problem, 20

Estimation
   arithmetic (for percentile rank),
      159–160
   graphic (for percentile rank), 160–164
   interval, (*see* Confidence intervals)

Fisher exact test, 395–401
Fisher *z*-transformation, 377–378
Frequency
   cell, 88
   marginal, 89
   of a value, 41
Frequency distribution, defined, 40
   bivariate, 86–88
   conditional, 88
   cumulative, 58–59
   discrete interval, 49–50
   grouped, 51–58
   marginal, 89

Graphs of frequency distributions
   bar graph, 44, 70
   circle graph, 45
   cumulative, 65–67
   discrete, 59–60
   frequency polygon, 64–65
   grouped data, 67–70
   histogram, 61–64
   scatterdiagram, 97–102
   smooth curve, 70–71
   step function, 59–61
Group comparisons, graphic
   methods, 110–114

Hypothesis
   alternate, 244
   null, 230, 244
   directional, 253
   nondirectional, 253, 332
Hypothesis testing, introduction to,
   229–230
   one-tailed test, 255
   two-tailed test, 255
   outline of, 255–256
Hypothesis tests
   one categorical variable, 327–332
   one ordered categorical variable,
      332–337

one mean
   z-test, 293–303
   t-test, 339–344
one proportion
   binomial, 242–250
   normal approximation, 318–325
more than two distributions
   categorical variable, 402–408
   rank-order variable, 415–420
more than two means, 427–438
relationship for categorical variables,
   360–368
relationship for rank-order variables,
   368–373
relationship for interval variables,
   374–378
two distributions of dependent ranks,
   463–468
two distributions of independent ranks,
   408–414
two dependent means, 469–473
two independent means, 420–426
two dependent proportions, 461–462
two independent proportions, 395–408

Independent measures, 26
Independent variable, 23
Inferential problem, 21
Inferential statistical tests, (*see* Hypothesis
   tests)
Interval estimation, 229

Kruskal-Wallis *H*-test, 415–420

Law of Large Numbers, 283–284
Level of significance, 247–248
Limits
   apparent class, 55
   real class, 54–55
   confidence, (*see* Confidence intervals)
Linear regression, 206–216
Linear relationship
   positive, 101
   negative, 102
   no relationship, 104

McNemar test, 461–462
Mann Whitney *U*-test, 408–414
Mean
   of interval variable, 146–147
   as sample statistic, (*see* Statistics,
      inferential)
Measurement, 13

590

Index

Classification Matrix: Descriptive Statistics

Level of Measurement of Variable(s) of Interest

	Categorical: Nominal or Ordered Categories	Rank-Order	Interval Scale (or Ratio-Scale)
Univariate Problems	Modal Category Proportion of Observations Falling Outside Modal Category *With Ordered Categories:* Median Category	Percentile Rank	Mode, Median, Mean Range, Interquartile Range, Standard Deviation Percentile Rank, Standard Score
Correlational Problems	Cramer's Index of Contingency	Spearman Rank-Order Correlation	Pearson Product-Moment Correlation
Experimental and Pseudo-Experimental Problems: Independent Groups	Compare Modes in Different Groups	Rank All Scores in All Groups Taken Together, Then Compare the Overall Rankings of the Middle Score in Each Group	Compare Mode, Median, Mean Compare Range, Interquartile Range, Standard Deviation
Dependent Measures: Pretest and Posttest or Matched Groups	Compare Modes (Either Pretest and Posttest Modes or Modes in the Two Groups)	Use Ranking Method Described in the Cell Above This One	Use Methods Described in the Cell Above This One

(Row group labels on the left: Type of Problem — Univariate Problems; Bivariate Problems, comprising Correlational Problems, Experimental and Pseudo-Experimental Problems: Independent Groups, and Dependent Measures: Pretest and Posttest or Matched Groups)